The Renewable Energy Reader

The Renewable Energy Reader

K.K. DuVivier
Professor of Law
University of Denver
Sturm College of Law

Carolina Academic Press
Durham, North Carolina

Library of Congress Cataloging-in-Publication Data

DuVivier, K. K.
  The renewable energy reader / K. K. DuVivier.
      p. cm.
  Includes index.
  ISBN 978-1-59460-873-5 (alk. paper)
  1. Renewable energy sources--Law and legislation--United States. 2. Renewable energy
sources. I. Title.
  KF2120.D88 2011
  346.7304'6794--dc23

                                                                          2011034879

Carolina Academic Press
700 Kent Street
Durham, NC 27701
Telephone (919) 489-7486
Fax (919) 493-5668
www.cap-press.com

Printed in the United States of America

To Marjorie and Ned for making everything possible;
Lance for your love and support; and
Alice and Emmett who make me proud and inspire me to work for a better tomorrow.

Summary of Contents

Contents

Table of Figures

Acknowledgments

This book is truly a collaborative work, and I am deeply grateful to the following people for their input. This book would not be nearly as rich without their help.

Bob Noun, Carol Tombari, John Ashworth, Robin Newmark, James Bosch, Michele Kubik, Donna Heimiller, Sarah Barba, Nancy Prosser-Stovall, etc. from the National Renewable Energy Laboratory for all of their help with content and graphics. Also to the following for their special assistance: Mark Safty and Elizabeth A. Mitchell from Holland & Hart; Connie Rogers from Davis Graham & Stubbs; Rebecca W. Watson from Welborn, Sullivan, Meck & Tooley, P.C.; Randy Stearnes from Tacoma Public Utilities; Matt Futch from IBM (formerly from the Colorado Governor's Energy Office); Catherine M. H. Keske from Colorado State University; and Andrew B. Reid from Springer and Steinberg.

My research assistants Dustin Charapata, Chelsea Huffman, Megan Moriarty, Sarah Stout, and Thomas Scott for your countless hours of work.

To Nicole Lyells, Stacy Bowers, Diane Burkhardt, Joan Policastri, Caryl Shipley, for formatting, copyright permissions, and research assistance.

An additional thank you is due to each of the following as well for their help along the way: Don C. Smith, Jacqueline Weaver, Jim and Jean Buck, Chuck and Kate DuVivier, Joe DuVivier and Ken White, Laurent Meillon, Becky English, Don Tressler, Gerry Todd, Robert Youngberg, Steve Stevens, John A. Herrick, Sarah Quinn, Alan Gilbert, Becky Bye, Susan Osborne, Alan Boles, Theresa I. Corless, Gale Norton, Jack Sinclair, Ron Binz, Suedeen Kelly, Greg Ching, Nancy Laplaca, Ron Lehr, Carol E. Lyons, Mike Zimmer, Roger Feldman, Henry A. Signore, Jerry Sherk, Luke Danielson, Matt Larson, Rod Wetsel, Gordon Draper, Bruce Finley, Matt Baker, Jim Tarpey, Rich Heinemeyer, Mark T. Gran, Ashland City planners, Tim Colton, and Linda Lacy.

The author also gratefully acknowledges the permissions granted by all of the authors, artists, and publishers of the following works reproduced in this book. Unless otherwise indicated, all footnotes from the originals have been excluded for space reasons.

Randall S. Abate, *Public Nuisance Suits for the Climate Justice Movement: The Right Thing and the Right Time*, 85 Washington Law Review 197 (2010).

Hope M. Babcock, *Responsible Environmental Behavior, Energy Conservation, and Compact Fluorescent Bulbs: You Can Lead a Horse to Water but Can You Make it Drink?*, 37 Hofstra Law Review 943 (2009).

Roger Bedard, D.O.E. Hydrokinetic Workshop, slide 10, *4 Primary Types of Wave Energy Conversion* (Oct. 26, 2005). © 2011 Electric Power Research Institute (EPRI), Inc. All rights reserved. Electric Power Research Institute, EPRI, and TOGETHER.... SHAPING THE FUTURE OF ELECTRICITY are registered service marks of the Electric Power Research Institute. (Reprinted with permission from EPRI).

Michael C. Blumm, Erica J. Thorson, & Joshua D. Smith, *Practiced at the Art of Deception: The Failure of Columbia Basin Salmon Recovery under the Endangered Species Act*, 36 Envtl. L. 709 (2006).

Sara C. Bronin, *Solar Rights*, 89 Boston University Law Review 1217 (2009).

N. Carlisle, J. Elling, and T. Penney, *Proposed U.S. offshore wind projects and capacity showing projects with significant progress.* Image provided courtesy of The Alliance for Sustainable Energy, LLC. Table 3-5 of NREL TP-500-40475, Walter Musial & Bonnie Ram, Large Scale Offshore Wind Power in the United States: Assessment of Opportunities and Barriers 29 (2010), *available at* http://www.nrel.gov/wind/pdfs/40745.pdf.

Edward H. Comer, *Transforming the Role of Energy Efficiency*, 23 Natural Resources & Environment 34 (2008-2009).

Kevin L. Doran, *Privacy and Smart Grid: When Progress and Privacy Collide*, 41 University of Toledo Law Review 909 (2010).

Wendell A. Duffield & John H. Sass, *Geothermal Energy: Clean Power from the Earth's Heat*, excerpts reformatted from public domain U.S. GEOLOGICAL SURVEY CIRCULAR 1249.

Rick Eichstaedt, Rebecca Sherman & Adell Amos, *More Dam Process: Relicensing of Dams and the 2005 Energy Policy Act*, 50 Advocate (Idaho) 33 (June/July 2007).

Jody M. Endres, *Clearing the Air: The Meta-Standard Approach to Ensuring Biofuels Environmental and Social Stability*, 28 Virginia Environmental Law Journal 73 (2010).

Steven Ferrey, *Symposium, Smart Brownfield Redevelopment for the 21st Century: Converting Brownfield Environmental Negatives into Energy Positives*, 34 Boston College Environmental Affairs Law Review 417 (2007).

Roger L. Freemen & Ben Kass, *Siting Wind Energy Facilities on Private Land in Colorado: Common Legal Issues*, 39 Colorado Lawyer 43 (May 2010).

Brian Andrew Fuentes, *Impact of setbacks and rooflines on solar access.* Image provided courtesy of Fuentesdesign.com.

L. Leon Geyer, Phillip Chong, & Bill Hxue, *Ethanol, Biomass, Biofuels and Energy: A Profile and Overview*, 12 Drake Journal of Agricultural Law 61 (2007).

Robert Glennon & Andrew M. Reeves, *Solar Energy's Cloudy Future*, 1 Arizona Journal of Environmental Law and Policy 92 (2010).

Lakshman Guruswamy, *Energy Justice and Sustainable Development*, 21 Colorado Journal of International Environmental Law and Policy 231 (2010).

Joseph J. Kalo & Lisa C. Schiavinato, *Wind Over North Carolina Waters: The State's Preparedness to Address Offshore and Coastal Water-Based Wind Energy Projects*, 87 North Carolina Law Review 1819 (2009).

Alice Kaswan, *Greening the Grid and Climate Justice*, 39 Environmental Law 1143 (2009).

Elizabeth Ann Kronk, *Alternative Energy Development in Indian Country: Lighting the Way for the Seventh Generation*, 49 Idaho Law Review 449 (2010).

Colleen McCann Kettles, *A Comprehensive Review of Solar Access Law in the United States: Suggested Standards for a Model Statute Ordinance,* Report for Solar American Board For Codes and Standards (Oct. 2008).

Donald J. Kochan & Tiffany Grant, *In the Heat of the Law, It's Not Just Steam: Geothermal Resources and The Impacts on Thermophile Biodiversity,* 13 Hastings West-Northwest Journal of Environmental Law and Policy 35 (Winter 2007).

David J. Lazerwitz, *Renewable Energy Development on the Federal Public Lands: Catching Up with the New Land Rush,* 55 Rocky Mountain Mineral Law Institute 13-1 (2009).

James W. Moeller, *Electric Demand-Side Management Under Federal Law,* 13 Virginia Environmental Law Journal 57 (Fall 1993).

Andrew P. Morriss, William T. Bogart, Andrew Dorchak, Roger E. Meiners, *Green Jobs Myths,* 16 Missouri Environmental Law and Policy Review 326 (2009).

Bent Ole Gram Mortensen, *International Experiences of Wind Energy,* 2 Environmental and Energy Law and Policy Journal 179 (2008).

Mike New, *Major Components of a Hydro System.* Image provided courtesy of Canyon Hydro.

Martin Nie, *The Use of Co-Management and Protected Land-Use Designations to Protect Tribal Cultural Resources and Reserved Treaty Rights on Federal Lands,* 48 Natural Resources Journal 585 (2008). Reprinted with permission of the *Natural Resources Journal,* University of New Mexico School of Law. Copyright © 2008. (Issue: Volume 48, No. 3, Summer.)

Richard Perez, *A fundamental look at the energy reserves of the planet.* SHC Solar Update Volume 50, pp. 2-3. Image provided courtesy of International Energy Agency.

Eric Plunkett, *Residential State Energy Code Status; The Future of Energy Codes.* Images provided courtesy of the Building Codes Assistance Project.

Elias L. Quinn, *Smart Metering and Privacy: Existing Law and Competing Policies,* Framing Document for Colo. PUC High Profile Dkt. No. 091-593EG (Order C09-0878). Mr. Quinn is currently a Trial Attorney for the United States Department of Justice, Environmental Enforcement Division. The views expressed in the excerpt here are the personal views of Mr. Quinn and do not necessarily reflect the views of the Department of Justice.

Karen Ray, *Are Biofuel Crops the Next Kudzu?,* 17 San Joaquin Agricultural Law Review 247 (2007/2008).

Arnold W. Reitze, Jr., *Biofuels—Snake Oil for the Twenty-First Century,* 87 Oregon Law Review 1183 (2008).

Sarah C. Richardson, Note, *The Changing Political Landscape of Hydropower Project Relicensing,* 25 William and Mary Environmental Law and Policy Review 499 (2000).

Ronald H. Rosenberg, *Diversifying America's Energy Future: The Future of Renewable Wind Power,* 26 Virginia Environmental Law Journal 505 (2008).

Troy A. Rule, *Shadows on the Cathedral: Solar Access Laws in a Different Light,* 2010 University of Illinois Law Review 851 (2010).

Irma S. Russell, *Streamlining NEPA to Combat Global Climate Change: Heresay or Necessity?,* 39 Environmental Law 1049 (2009).

Kurt E. Seel, *Legal Barriers to Geothermal Development,* 2008 American Bar Association Section of Environment, Energy and Resources 16 (2008).

Shari Shapiro, *Who Should Regulate? Federalism and Conflict in Regulation of Green Buildings,* 34 William and Mary Environmental Law and Policy Review 257 (2009).

Sidney A. Shapiro & Joseph P. Tomain, *Rethinking Reform of Electricity Markets*, 40 Wake Forest Law Review 499 (2005).

Christian Steiness, *Horns Rev 1 Wind Farm*. Image provided courtesy of Vattenfall.

Lawrence Susskind, Alejandro E. Camacho & Todd Schenk, *Collaborative Planning and Adaptive Management in Glen Canyon: A Cautionary Tale*, 35 Columbia Journal of Environmental Law 1 (2010).

William A. Tanenbaum, *Practical Steps to Contract for Energy-Efficient Data Centers and IT Operations*, 981 PLI/Pat 247 (PLI Order No. 19120) (2009).

Carol Sue Tombari, POWER OF THE PEOPLE: AMERICA'S NEW ELECTRICITY CHOICES, (Fulcrum Group 2008).

Chris Van Essen, *Subsurface imprint of windfarms.*

Rod Wetsel, Wetsel & Carmichael, LLP Attorneys at Law, (coauthor with K.K. DuVivier) of *Jousting at Windmills: When Wind Power Development Collides with Oil, Gas, and Mineral Development*, 55 Rocky Mountain Mineral Law Institute Chapter 9 (2009).

The Renewable Energy Reader

Chapter 1

Introduction

This Reader is intended to be a concise and accessible sourcebook for U.S. renewable energy law. Each chapter provides historical background as well as illustrations and technology charts to give readers context for better understanding renewable energy sources and related legal issues. Each chapter also includes excerpts from some of the most prominent primary and secondary legal sources highlighting current and potential legal challenges to the advancement of renewable resources.

The goal of this book is not to convert anyone into a renewable energy believer. The numbers speak for themselves as you will see in the pages that follow. Instead, I hope you will find this book a reader-friendly reference, providing an overview of the significant legal implications—positive and negative—of renewable energy development.

A. Catalysts for Change

For millennia, our civilizations were built from the sweat of labor by humans, frequently slaves, and domesticated animals. The sun was our primary source of heat and light. Biomass, in the form of wood, plant wastes, or manure, cooked our food and helped drive away a bit of the night darkness.

The industrial revolution ushered in a dramatic metamorphosis for civilization. Draft animals, harnessed to equipment and straining as they plodded, often in circles, had represented the typical power source for centuries. But in the early 1780s, Scottish inventors James Watt and Matthew Boulton developed breakthroughs for fossil-fuel steam engines. To help purchasers comprehend the enormous potential of their machines, the inventors coined the term "horsepower" to represent the number of draft horses each engine could replace.

As figure 1.1 on the next page illustrates, a human can sustainably generate only 0.1 horsepower, so employing horses had multiplied a human's ability to perform work tenfold. But fossil fuels pushed the envelope of possibilities almost beyond comprehension.

If you have ever tried to generate power mechanically, you may have been shocked at how little we humans are able to create. Recently, environmentally-conscious health clubs have wired their equipment to generate power when members exercise. These efforts are mostly educational as the amount of electricity they generate is small. For example, the

Figure 1.1   Energy Basics

1 watt	International System of Units (SI) unit of power defined as one joule per second. Also equivalent to 3.4 imperial unit British thermal units (btus).
0.1 horsepower	The amount of mechanical power a human can sustainably generate, equivalent to approximately 75 watts. This is in contrast to the average amount of waste body heat a human generates of approximately 100 watts or 350 btus at rest.
100 watts	Electricity consumed by an average incandescent light bulb. However, incandescent light bulbs produce more heat than light. A compact fluorescent bulb that produces the same amount of light as a 100 watt incandescent bulb consumes less than a quarter of the energy to produce that same amount of light, approximately 23 watts.
1 horsepower	Amount of power a typical draft horse could generate—33,000 foot-pounds per minute.
1 kilowatt (kW) = one thousand (10^3) watts	One kilowatt of power is approximately equal to 1.34 horsepower. As a rule of thumb, the mean residential average electricity consumption in the United States is 1,000 kWh/month. As there are 8,760 hours per year, the U.S. average annual household consumption of approximately 8,800 kWh is almost double that of the UK average of 4,700 kWh per year.
1,875 watts	Typical U.S. hairdryer and approximately the amount of supplemental heating capacity needed for a 2,000 square foot home built to Passive House building standards (See Chapter 7).
3-9 kW	Average range for home solar PV array to meet U.S. electricity needs.
1 Megawatt (MW) = one million (10^6) watts	Typical unit for measuring utility scale electricity generation facilities. 10-20 MW—typical industrial-scale solar project or single dish system unit. 50+ MW—typical concentrating solar power plant (see Chapter 8). 150 MW—typical industrial scale terrestrial wind farm in the United States.
500-1000 MW	(Also equivalent to 0.5 – 1.0 GW) Capacity of large coal plants in United States. U.S. nuclear power plants have capacities between 500 and 1,300 MW.
1 Gigawatt (GW) = one billion (10^9) watts	This unit is sometimes used for large power plants or power grids. The installed capacity of wind power in Germany was 25.8 GW in 2010.
1,121 GW	U.S. total electric nameplate capacity in 2009 (Providers = Utilities 62%; Independent producers 35%; Commercial/industrial 3%).
1 Terawatt (TW) = one trillion (10^{12}) watts	The total power used by humans worldwide is commonly measured in this unit: approximately 132,000 TWh in 2008. Lasers and lightning can produce up to 1 terawatt for very short periods of time.
Power v. Energy	The terms "power" and "energy" are often confused. Using the metaphor of water going through a hose: "Power" would indicate the diameter of the hose (e.g., 1 kW), and "energy" would be the volume of water the goes through the hose during a particular time period (e.g., 1 kWh), or in other words, the volume or amount of energy consumed or generated. As another example, when a light bulb with a power rating of 100W is turned on for one hour, the energy used is 100Wh. This same amount of energy would light a 50-watt bulb for two hours or a 25-watt bulb for four hours. Power station capacity is rated in watts, but annual energy sales would be in watt-hours. MWa means megawatts per year (or annually).
1 kWh	A kilowatt-hour is the amount of energy needed for a steady power of 1 kilowatt running for one hour. 1kWh is equivalent to 3,600,000 joules; 3,410 btus; .00008598 tonnes of oil equivalent; .00614 barrel of oil equivalent; .001228 of hard coal; or 860 kilocalories. See, e.g., Unitjuggler.com

Figure 1.2   World primary energy consumption

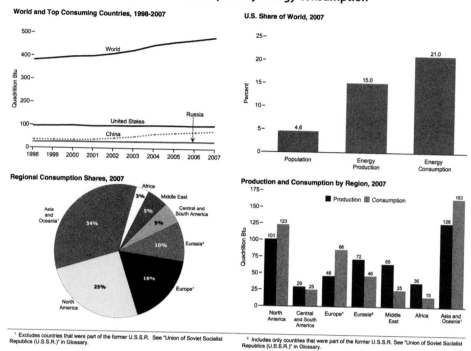

University of Oregon estimates that 3,000 people a day on the twenty machines in its gym facility would generate only about 6,000 kilowatt hours a year. Compare this to the capacities of fossil-fuel power plants listed in Figure 1.1, which can generate millions of times the work of one human. For your convenience, Figure 1.1 also lists the common units of power and conversion factors.

Fossil fuels have allowed humans to exponentially leverage the amount of work they could perform, thus laying the foundation for modern life in developed countries. In a 2005 essay, "How much of a slave master am I?" French consultant Jean-Marc Jancovici, of Manicore.com, noted that the human body consumes about four to five kWh per day and can create only approximately 0.5 kWh of mechanical energy over that same time period. Consequently, "we start to understand that our species has performed a fantastic 'power breakthrough' when domesticating fossil fuels: with 1 euro ... I can buy 1 litre of petrol (or gas), that contains about 10 kWh of energy, which is about the equivalent of two 'slaves' working for a full day.... [And] the result is that our modern energy consumption is the equivalent to giving each French (or European) ... about 100 slaves!"

U.S. per capita energy consumption is higher than any other country in the world except Canada. See Figures 1.2 and 7.3. So our quota of slaves would be enough to make a pharaoh proud.

1. Higher Prices as World Competition Increases Demand for Depleting Reserves

Because energy is so fundamental to modern civilization, it is no wonder we have found fossil fuels so seductive. But our attraction to fossil fuels is problematic. While the discovery of concentrated energy from fossil fuels accelerated our productivity and the growth of civilization, it also unfortunately has created an imbalance of the human footprint on our planet.

Patty Limerick and Howard Geller set it out nicely in their introduction to WHAT EVERY WESTERNER SHOULD KNOW ABOUT ENERGY EFFICIENCY AND CONSERVATION: A GUIDE TO A NEW RELATIONSHIP, CENTER OF AMERICAN WEST, REPORT #8 (2007):

> We may not know your name, but we already know one pretty private thing about you. You have been involved in a tempestuous relationship, pursuing a mad romance with fossil fuel. But now, thanks to a spectrum of big changes, from global climate change to rising energy prices, your love affair with petroleum is winding down. It's time to go in a search of a new relationship, one with better prospects for long-term happiness.
>
> . . .
>
> Let's be clear: fossil fuel is not going to disappear from your life any time soon. The intense combustion of the romance will die down, but you will remain good friends. It may even be useful to think of this transition as an unusually good-natured triangle, in which you, fossil fuel, and conservation and efficiency live congenially together until one roommate finally moves out and renewable energy moves in. You and fossil fuel have had, after all, thousands of good times together, and there is no justification for ingratitude or for the denial of those pleasant memories. There is certainly no reason to waste time in bitterness, condemnation, or recriminations over the ending of a relationship that has delivered so much pleasure.

Love is blind, and we are most comfortable clinging to the familiar, warts and all. But catalysts are shoving us in new directions. We had a glorious run of it, but here are some of the specific reasons this fossil-fuel infatuation needs to change.

First, the days of cheap fossil fuels are over. The obvious and easy finds are behind us. We are drilling more exploration wells for every discovery. We are drilling in deeper and more inhospitable environments. Thus, we are expending more energy for every barrel produced, as discussed in Chapter 5.B, and this raises the costs per barrel. Tar sands, shale gas, and shale oil, now producible because of new technologies such as horizontal drilling and fracing, may increase worldwide reserve figures, but they also involve additional costs in terms of energy and water for their production. In its World Energy Outlook 2010 report (WEO 2010), the International Energy Agency (IEA) predicts that crude oil import prices will more than double under current policies, from approximately $60 per barrel in 2008 to $135 per barrel in 2035.

Second, no matter how great the reserves, a non-renewable reserve is depletable by definition. Although we will never run out of conventional oil completely, at some point, worldwide production will peak and decline. Experts may debate whether we already have reached global "Peak Oil" or whether human ingenuity can push it out decades, but a major factor contributing to the IEA's projected increase in price is the significant increase in competition for dwindling reserves. According to the WEO 2010, worldwide demand for energy will grow by almost 50 percent (from 12,271 MToes (million tonnes of oil equivalent) to 18,048 Mtoes by 2035 under current policies. *See* Figure 1.3.

Most of this worldwide growth in energy demand will come from countries outside of the Organization for Economic Cooperation and Development (OECD). For instance, China, which is a non-OECD country, surpassed the United States in 2009 to become the world's largest energy user. Even though China's per-capita energy consumption is only

Figure 1.3   Growth in world energy demand and consumption

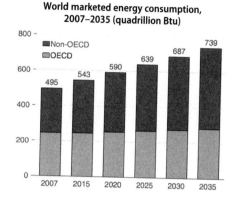

World marketed energy consumption, 2007–2035 (quadrillion Btu)

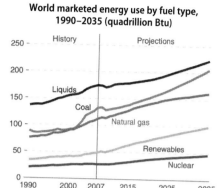

World marketed energy use by fuel type, 1990–2035 (quadrillion Btu)

Current OECD member countries (as of March 10, 2010) are the United States, Canada, Mexico, Austria, Belgium, Czech Republic, Denmark, Finland, France, Germany, Greece, Hungary, Iceland, Ireland, Italy, Luxembourg, the Netherlands, Norway, Poland, Portugal, Slovakia, Spain, Sweden, Switzerland, Turkey, the United Kingdom, Japan, South Korea, Australia, and New Zealand. Chile became a member on May 7, 2010, but its membership is not reflected in *IEO2010.*

Source: U.S. Energy Info. Admin., Int'l Energy Outlook 2010 1 (2010).

about one-third of the OECD average, its population of 1.3 billion, the world's highest, makes any incremental change significant in the aggregate. Consequently, the IEA projects that non-OECD countries, primarily China and India, will account for 93 percent of the estimated increase in world primary energy demand.

The IEA also notes that world demand for electricity will continue to grow more than any other final form of energy, with more than 80 percent of the increase coming from non-OECD countries. Renewable sources will play a critical role in meeting this increased electricity demand. The IEA's "New Policies Scenario" reflects future actions by governments to meet commitments they have made to tackle climate change and growing energy insecurity. Under this scenario, the WEO 2010 projects that renewables-based generation will triple between 2008 and 2035, growing to almost a third of all global electricity generation.

Increased world demand and the scramble for a dwindling pool of energy resources has translated into higher prices for all energy sources. The rising price of fossil fuels makes renewables, which generally involve newer less-established technologies, more cost competitive. Figure 1.4 shows NREL's levelized cost of renewable electricity technologies. At six to twelve cents per kWh, wind is generally considered to be at "grid parity" or competitive with conventional fuels currently feeding the grid.

Measuring grid parity is complicated by the difficulty of making a fair comparison between renewables and conventional fuels. Most renewable technologies are still struggling to establish a toehold in the market without the same financial subsidies and benefits in terms of infrastructure that centuries-old competitors may enjoy. For example, a 2011 study from the New York Academy of Sciences concluded that the true monetizable cost of generating electricity from coal was many times what you might see on your utility bill—between nine and 26.89 cents per kWh. Here is the study's reasoning:

> Each stage in the life cycle of coal—extraction, transport, processing, and combustion—generates a waste stream and carries multiple hazards for health and the environment. These costs are external to the coal industry and are thus often considered "externalities." We estimate that the life cycle effects of

Figure 1.4   Levelized Cost of Energy (LCOE) of renewable electricity by technology

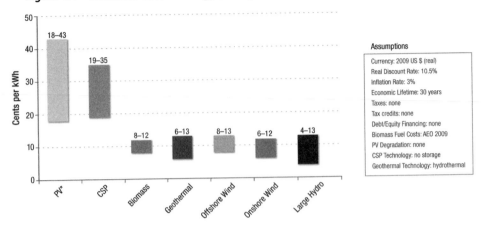

* Current range of utility scale (greater than 5MW) PV in the U.S.

National Renewable Energy Lab., 2009 Renewable Energy Databook 13 (2010)

coal and the waste stream generated are costing the U.S. public a third to over one-half of a trillion dollars annually. Many of these so-called externalities are, moreover, cumulative. Accounting for the damages conservatively doubles to triples the price of electricity from coal per kWh generated, making wind, solar, and other forms of nonfossil fuel power generation, along with investments in efficiency and electricity conservation methods, economically competitive....

Paul R. Epstein et al., *Full Cost Accounting for the Life Cycle of Coal,* 1219 ANNALS N.Y. ACAD. SCI., 73, 73 (2011). *See also* CATHERINE M. H. KESKE ET AL., REPORT TO COLORADO GOVERNOR'S ENERGY OFFICE, DESIGNING A TECHNOLOGY NEUTRAL, BENEFIT-PRICING POLICY FOR THE ELECTRIC POWER SECTOR IN COLORADO (2010) (proposing a method of calculating total social costs of electricity generation technologies).

Another cost issue for renewables is that they are " generally more capital-intensive than fossil fuels, so the investment needed to provide the extra renewables capacity is large...." WEO 2010, *supra*, at 51. The IEA estimates that a cumulative investment of $5.7 trillion (in year 2009 dollars) will be needed between 2010 and 2035, and further states, "Investment needs are greatest in China, which has now emerged as a leader in wind power and photovoltaic production, as well as a major supplier of the equipment." *Id.*

A common myth is that renewable energy development requires inordinate government subsidies. In fact, every single energy resource in the United States has enjoyed, and continues to enjoy, government subsidies. A 1992 study by the Energy Information Administration noted in its Executive Summary:

> The Government has not been content to let energy markets function without interference.... Government has consistently selected from a menu of policy alternatives, various means to tilt the playing field in favor of certain producers or consumers of energy over others.

· · ·

This quick overview indicates that the scope of Government subsidies is vast.

Energy Info. Admin., SR/EMEU/92-02, Federal Energy Subsidies: Direct and Indirect Interventions in Energy Markets, at ix, x (1992).

A more recent comprehensive study reported a total of $644 billion in federal energy subsidies between 1983-2003. Roger H. Bezdek & Robert M. Wendling, *A Half Century of US Federal Government Energy Incentives: Value, Distribution, and Policy Implications*, 27 Int'l J. Global Energy Issues 42 (2007). Petroleum received almost half—$302 billion. Coal, natural gas, and nuclear each received more than double what all renewables combined received over the same fifty year period. *Id.* at 43.

While coal remains one of the least expensive sources of energy on utility bills, coal production is the one of the highest directly-subsidized electricity generation resources. *See* Energy Info. Admin., SR/CNEAF/2008-01 Federal Financial Interventions and Subsidies in Energy Markets 2007 at 16 (2008). Coal-related subsidies in the United States reached $2.7 billion in 2007 alone. *Id.*

Conventional energy sources also enjoy significantly higher subsidies on the world stage. For example, the IEA notes, "[Among] the thousands of numbers in the World Energy Outlook 2010, despite their disparity, [these numbers] are worth putting alongside each other: $312 billion—the cost of consumption subsidies to fossil fuels in 2009; $57 billion—the cost of support given to renewable energy in 2009." WEO 2010, *supra*, at 3. Thus, fossil fuels are currently receiving almost six times the subsidies of yet-to be-developed renewable energy, which the WEO 2010 projects will triple to meet worldwide generation needs. Although the IEA articulates no conclusion from this juxtaposition, at the very least, it suggests change is inevitable.

2. Geopolitics and Security

The U.S.'s historic reliance on petroleum from the Middle East has contributed to serious destabilization in the region and has compromised our national security by putting us at the mercy of unfriendly suppliers. Thomas L. Friedman made the following observation in his book, Hot, Flat, and Crowded 79-80 (2008):

> [The bombing of the World Trade Center in New York City on 9-11-2001—Ed.] illuminated . . . that our oil addiction is not just changing the climate system; it is also changing the international system in four fundamental ways. First, and most important, through our energy purchases, we are helping to strengthen the most intolerant, antimodern, anti-Western, anti-women's rights, and antipluralistic strain of Islam—the strain propagated by Saudi Arabia.

> Second, our oil addiction is helping to finance a reversal of the democratic trends in Russia, Latin America, and elsewhere that were set in motion by the fall of the Berlin Wall and the end of Communism. . . .

> Third, our growing dependence on oil is fueling an ugly global energy scramble that brings out the worst in nations. . . .

> Finally, through our energy purchases we are funding both sides of the war on terror. . . . We are financing the U.S. Army, Navy, Air Force, and Marine Corps with our tax dollars, and we are indirectly financing, with our energy purchases, al-Qaeda, Hamas, Hezbollah, and Islamic Jihad.

The United States shifted its policy after 9/11 to rely on Mexican oil and Canadian tar sands for about half of our imports. This means we are still currently sending hefty sums to the Middle East. Furthermore, the IEA projects that non-OPEC production will remain fairly constant only until about 2025. By 2035, OPEC nations will dominate the global output, funneling additional dollars and power into Saudi Arabia, Iran, and Iraq.

One of the largest consumers of energy in the United States is our military, which spent over $4.1 billion for fuel and electricity in 2008 alone. Recognizing that "reliable access to affordable, stable energy supplies is a significant challenge for the [military] and the nation," the branches have started to implement aggressive strategies to reduce consumption and increase the use of renewable and alternative energy. ARMY SENIOR ENERGY COUNCIL, ARMY ENERGY SECURITY IMPLEMENTATION STRATEGY at i-ii (2009). The Army alone has set goals of increasing electricity generation by renewable sources to 25 percent by 2025 and reducing fossil fuel use in new and renovated buildings by 100 percent by 2030 relative to its 2003 level. *Id.* at 11.

The vulnerability of U.S.'s aging electric grid raises a final issue with security. The centralized power plant model currently dominates our electricity delivery system and is discussed further in Chapter 7. This system is susceptible to cascading blackouts. While the culprits for past blackouts mostly have been power surges and natural causes such as storms or squirrels, we may become a conscious target in the future.

A graduate engineering student in China caught Congress's attention when he published a paper on "Cascade-based Attack Vulnerability of the U.S. Power Grid." John Markoff & David Barboza, *Academic Paper in China Sets Off Alarms in U.S.*, N.Y. TIMES, Mar. 20, 2010. The student said he picked the United States as the focus of his study simply because its vulnerability was well known and the data was readily available to the public. Will renewables make our grid more or less vulnerable to cyberattacks? At least, the proliferation of distributed sources of generation will make it more difficult to debilitate an area by taking out one large target. Furthermore distributed renewables can provide back-up power if a large centralized plant goes down.

3. Environmental Concerns

Limerick and Geller above urge us to terminate our "tempestuous relationship ... with fossil fuel." But doesn't every relationship require some accommodations for quirks and dirty laundry?

We don't have enough space here to fully explore the dirty laundry related to the use of conventional fuels, but here is just a short sampling:

(a) Petroleum

Pipeline spills similar to the 1,000 barrels of crude that spilled into the Yellowstone River in July of 2011 happen every year. This is the type of risk we have accepted to have oil delivered where we need it. But as we drill deeper and in more fragile environments, such as the Arctic, are we not also increasing the potential for unprecedented disasters such as the April 2010 blowout of BP's Macondo well, which killed eleven workers and released an estimated 4.9 million barrels of crude oil into the Gulf of Mexico during the three months it took to bring the spill under control?

(b) Coal

The good news from a national security standpoint is that the United States may have over two hundred years of coal reserves. Unfortunately, relying on coal-fired generation to produce the majority of our electricity also has myriad negative environmental consequences.

(i) Mining problems: First, there is the cost of human lives in mine accidents—forty-eight fatalities in the United States in 2010, the highest number since 2006. International fatality rates are much higher. While surface mining results in fewer lost miners, it wreaks more extensive havoc with the environment. Hundreds of thousands of acres are being strip mined in the Rocky Mountain west, but Appalachia is receiving the most attention recently because of the surface mining technique called Mountaintop Removal Mining (MRM).

With MRM, a coal company removes the entire summit of a mountain in order to more easily access the coal seams. Because not all of the overburden can be replaced to approximate the original mountain contours, waste is dumped into nearby valleys. This "holler fill" within watersheds can degrade or destroy habitat and human water supplies. In a 2005 report, EPA estimated that, without further restrictions, 2,200 square miles of Appalachian forests would be eliminated by the year 2012.

(ii) Air pollution: Coal is an organic substance containing a number of toxins that are emitted when it is burned. Unless these air pollutants can be removed in the combustion process, they are dispersed into the air and can have harmful health impacts for those exposed to them including respiratory diseases, developmental problems, and increased risk for cancers.

Some of the toxins emitted by the burning of coal include sulfur dioxide, radioactive trace elements, and mercury. These air pollutants can travel for miles and create acid rain or contaminate waterways when they come into contact with water. For example, the DOI announced that mercury from airborne coal emissions was found in every fish it tested from 291 streams in the United States, including fish in remote and isolated waterways. *See* Cornelia Dean, *Mercury Found in Every Fish Tested, Scientists Say*, N.Y. TIMES, Aug. 19, 2009. Furthermore, 25 percent of these fish contained mercury above levels EPA has determined are safe for human consumption.

(iii) Particulate matter and fly ash: Another problem related to coal combustion is the emission of particulates, such as sulfates and nitrates that can lead to respiratory problems if released into the air. Modern scrubbing technology now collects many of particulates that do not combust with the coal. However, the storing of the resulting coal fly ash has simply converted a widespread air pollution problem into a more localized and concentrated groundwater and surface pollution problem.

Vast quantities of coal fly ash are stored in waste ponds and dry waste piles at power plant sites. These waste sites can contaminate groundwaters with the arsenic, heavy metals, and other toxins contained in the fly ash. A 2010 study by Environmental Integrity Project, the Sierra Club, and Earthjustice uncovered additional groundwater sites across the United States contaminated by power plant-produced coal ash, which brought the total number of contaminated sites up to 137.

In addition to the potential for polluting groundwater, fly ash ponds have breached their impoundments, resulting in floods of wastes. A stunning example was the spill of 1.1 billion gallons of coal fly ash from the Tennessee Valley Authority's Kingston Fossil Plant in December of 2008. The spill took out three homes and contaminated the Emory River. It covered 300 acres with the fly ash sludge, and cleanup costs were estimated to be over $1.2 billion.

(iv) Climate Change: Last, but not least, coal-generation has raised concerns over the impact of anthropogenic greenhouse gas emissions. The Intergovernmental Panel on Climate Change (IPCC)'s Fourth Assessment Report on climate change (AR4), completed in 2007, brought together research from specialists in over 130 countries.

The more than 450 lead authors, 800 contributing authors, and 2,500 expert reviewers worked over a six year period and processed over 90,000 comments to reach the conclusions of IPCC AR4. The overwhelming consensus of scientists who have conducted research in the area found that it is "unequivocal" that the earth's climate is warming and that the increase is "very likely" (greater than a 90 percent chance) caused by anthropogenic greenhouse gas concentrations. *See* http://www.ipcc.ch/ and http://www.climate.gov.

Dozens of scientific organizations across the world have endorsed the AR4 findings. *See, e.g.*, http://www.ucsusa.org/ssi/climate-change/scientific-consensus-on.html. Furthermore, efforts to discredit the IPCC's findings have been unsuccessful. For example, a group of physicists and statisticians attempting to challenge the AR4 findings had to admit to the House Science & Technology Committee at a hearing in April of 2011, "We see a global warming trend that is very similar to that previously reported by the other groups." Margot Roosevelt, *Critics' Review Unexpectedly Supports Scientific Consensus on Global Warming*, L.A. TIMES, Apr. 4, 2011.

Efforts to reduce greenhouse gas (GHG) emissions have especially targeted coal because it is the greatest single-source contributor to worldwide CO_2 output. But any fossil fuel that emits carbon upon burning, including petroleum and natural gas, are also problematic from a climate-change perspective.

(c) Natural Gas

Although methane or natural gas emits less CO_2 than coal when it is burned, some argue that an uncombusted molecule of methane (CH_4) has more than twenty times the negative climate changing impact of a molecule of CO_2. Consequently, escaping methane during the production of natural gas may be a serious concern.

In addition to climate change concerns, natural gas can explode causing fires and loss of life. According to a September 25, 2010, story by Andrew Lehren in the *New York Times* immediately following a pipeline explosion in San Bruno, California, lack of oversight of gas pipelines has resulted in hundreds of pipeline explosions in the United States that have caused injuries to more than 280 people and the deaths of sixty-seven from 2005 to 2010 alone.

As with oil pipelines mentioned above, spills and explosions seem to be risks we have assumed in return for handy delivery of oil or gas by pipeline. However, new focus is being placed on risks related to the exploration and production of shale gas. Some argue that the encroachment of natural gas development in populated areas and the use of hydraulic fracturing, also called "fracking" or "fracing," may be placing humans at risk and contaminating groundwater.

(d) Nuclear Power

Because nuclear fission creates electricity without emitting carbon dioxide, climate-change concerns have brought it back to the forefront as a "clean" alternative to fossil fuels. However, the Fukushima Daiichi reactor disaster has dampened enthusiasm for nuclear power in many parts of the world. Long after the March 11, 2011 earthquake and tsunami that caused portions of the reactor to explode, Japan has been struggling to deal with treatment of plant workers, concerns about the contamination of food and water supplies, and large-scale evacuations of areas exposed to radiation releases. Although France, with the majority of its electricity generation from nuclear reactors, remains committed, Germany has vowed to close its nuclear reactors and wean itself completely from nuclear power by

2022. In the United States, expansion of nuclear power may be further delayed by public fear of anything "radioactive" and the immutable problems of waste disposal.

As you will see in the following chapters, renewable energy resources also have environmental drawbacks. While none of them include the death and widespread contamination issues listed above, they seem to be receiving as much resistance, if not more in some instances, as conventional fuels have encountered.

In addition, aesthetics appears to be one of the major environmental objections to renewable development. Large conventional power plants can be sited far from where the electricity is delivered, and they have rarely been shelved on the basis of aesthetic criteria, such as looks or noise, alone. The fact that renewable projects are smaller scale and more dispersed may be creating more opportunities for people to raise aesthetic concerns.

But if we want our cellphones, laptops, televisions, appliances, and other electricity-fed gadgets, experts agree that we will need to develop all energy resources. This includes aggressive pursuit of renewable resources to play an important role both in the current energy equation and more so in the future.

B. The Rise of Renewables

Even if you take issue with some of the points in the previous section, enough of these concerns have resonated with enough people in the world to result in a worldwide rise in the development of renewable energy sources.

Figure 1.5 shows U.S. energy production and consumption by percentage of source. The U.S. total energy consumption of 94.9 quadrillion btus in 2009 was down from a

Figure 1.5 U.S. energy production and consumption

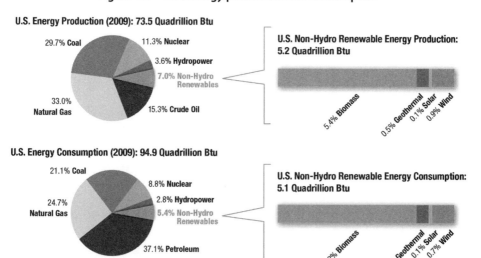

Note: Because hydropower is considered a conventional source of energy, it is accounted for separately from other new renewable sources of energy. Energy consumption is higher than energy production due to oil imports.

National Renewable Energy Lab., 2009 Renewable Energy Databook 7 (2010).

Figure 1.6   U.S. renewable electricity capacity and generation

U.S. Electric Nameplate Capacity (2009): 1,121 GW

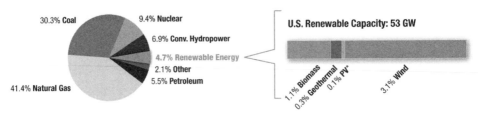

U.S. Electric Net Generation (2009): 3,954 billion kWh

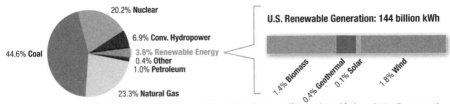

Other includes: pumped storage, batteries, chemicals, hydrogen, pitch, purchased steam, sulfur, tire-derived fuels, and miscellaneous technologies.

* Includes on- and off-grid capacity.   National Renewable Energy Lab., 2009 Renewable Energy Databook 10 (2010).

high of 101.6 in 2007 before the recession hit. This graphic reflects all energy use including petroleum and biofuels for transportation purposes.

Figure 1.6 focuses just on *electricity* capacity and generation — the area in which most renewables other than biofuels have the greatest potential. The percentage share for non-hydro renewable energy in the entire U.S. electricity generation mix is small: 3.6 percent. However, that percentage represents dramatic growth over the last ten years as the share of U.S. electric-generating capacity attributable to non-hydro renewables has more than doubled from 1.9 percent in 2000 to 4.7 percent in 2009. *See* Figure 4.1.

Furthermore this growth has occurred not as a result of a concerted federal effort, but instead primarily because of state-by-state initiatives often in the form of Renewable Portfolio Standards or Renewable Energy Standards as discussed further in Chapters 5 and 7. Figure 1.7 shows the states that are leading the way.

Most dramatic however, is Figure 1.8, which shows the new electric capacity added worldwide by source. At 50 percent of the new electric capacity worldwide, renewable energy now exceeds the new electricity capacity from fossil fuels. Much of this added renewable electricity capacity is a result of the aggressive renewable energy policies of the European Union, Japan, India, and China.

Now let's explore the legal landscape of renewable development in the United States with this book as our guide. We will have to keep the bus moving; there are so many sights to see and topics to explore on this whirlwind tour that we do not have time to savor them all deeply.

The journey begins with the five major renewable energy resources themselves: solar, wind, hydro, biomass, and geothermal. Logically, we start with the sun or solar power in Chapter 2 because virtually all renewable energy sources derive from the sun. Next, Chapter 3 addresses wind power, but we would not have winds if not for changes in surface temperatures on earth caused by the sun's heat. Chapter 4 reviews hydropower. As with wind,

Figure 1.7 Top states for renewable electricity installed nameplate capacity

Solar PV
❶ California
❷ New Jersey
❸ Colorado
❹ Arizona
❺ Florida

CSP
❶ California
❷ Nevada
❸ Hawaii
❹ Arizona

Biomass
❶ California
❷ Florida
❸ Maine
❹ Virginia
❺ Georgia

Wind
❶ Texas
❷ Iowa
❸ California
❹ Washington
❺ Minnesota

Geothermal
❶ California
❷ Nevada
❸ Utah
❹ Hawaii
❺ Idaho

Hydropower
❶ Washington
❷ California
❸ Oregon
❹ New York
❺ Alabama

National Renewable Energy Lab., 2009 Renewable Energy Databook 33 (2010).

the sun is still a key player for hydropower because the hydrologic cycle on earth is powered by evaporation of water by the sun. Even the biomass discussed in Chapter 5 begins with plant and algal materials produced through photosynthesis with the sun. Only geothermal energy, which is addressed in Chapter 6, comes not from the sun, but from heat inside the earth, and astronomers believe the earth itself was once part of the same molecular cloud as our star—the sun.

The last three chapters of the book shift our gaze away from the sun. Energy efficiency, discussed in Chapter 7, immediately follows the renewable fuel sources because it is often called "the fifth fuel" after coal, petroleum, nuclear, and renewables. Then, Chapter 8 focuses on the specific concerns of developing renewable resources on federal lands. Finally,

Figure 1.8 New electricity capacity added worldwide

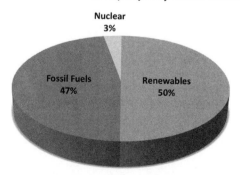

Data from Renewable Energy Policy Network, Renewables 2011 Global Status Report 95 (2011)

Chapter 9 brings us first to the role of Native Nations in renewable development and then back home again with a discussion of the concepts of environmental, climate, and energy justice.

In addition to the context provided by the history and technology information unique to each resource, you also might notice themes that cross the boundaries of various resource types. Some of these include the following:

- The role of government and the level of government—federal, state, or local— in drafting policies, creating financial incentives, or executing and enforcing.

- The balance of environmental concerns so that "green v. green" fighting does not result in promoting options that may be more environmentally damaging.

- The hurdles renewables face before the playing field can be leveled in comparison to conventional energy sources that enjoyed government giveaways and legal support in the eras during which they gained dominance and which linger in the form of exemptions, infrastructure, and subsidies that continue to give conventional sources advantages over fledgling renewable technologies.

- The inability of past paradigms to meet the needs of renewable development and the possibility that instead of using precedents from other regimes, we may need to develop new *sui generis* laws for new technologies.

- The environmental, climate, and energy justice recognition that not only do we have an obligation to share the earth's energy resources, but that we also may now have to bear some of the environmental and aesthetic consequences of our energy choices which in the past we may have foisted off on the less wealthy or less privileged.

Our journey may leave the impression that the legal challenges to renewable energy deployment in the United States are daunting. This is certainly a legitimate take away. The frequent comparison of the launch of renewables to the Apollo program is apropos. The stakes are no less urgent. If we allow other countries, such as China, to outpace us in the growing global demand for renewables, "we will swap dependence on foreign oil for dependence on foreign technology." John Ingold, *Energy Chief Says U.S. Lags*, Denver Post, Feb. 20, 2010, at 2B (quoting Secretary of Energy Steven Chu).

But, seemingly insurmountable goals can sometimes provide the greatest opportunities. U.S. innovation and ingenuity have risen to such challenges in the past. We must not forget that the cosmonauts beat us into orbit, and many of our rockets exploded on the launch pad before we finally won the race to the moon.

Chapter 2

Solar

Our journey through renewable energy sources appropriately starts with our sun. Not only is the sun the source from which virtually all other renewable energy resources derive, but solar power also holds the promise of being humankind's greatest energy resource. Figure 2.1 shows the significant potential for energy from the sun that strikes the earth if we could harness it at 100 percent efficiency.

Figure 2.1: A fundamental look at energy reserves of the planet

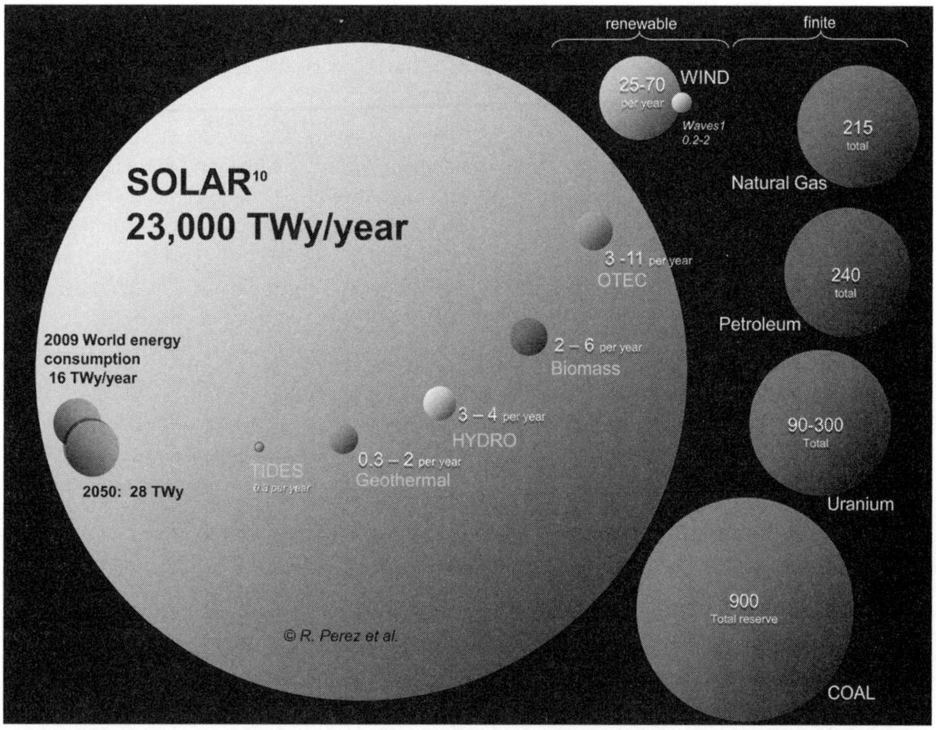

Perez R. & M. Perez, *A fundamental look at the energy reserves of the planet,*
Illustration in International Energy Agency SHC Solar Update 2-3 (Vol. 50 2009).

The sun has long been used for growing crops or to provide natural light for illumination inside and outside structures. Ancients also positioned their buildings to take advantage of the heat provided by the sun. Coming full circle, modern window and building designs have greatly increased our ability to capture sunlight entering a structure to provide passive solar heating and lighting to reduce total energy demand up to 85 percent in comparison to conventional buildings. Chapter 7 will discuss energy efficient building and the role of passive solar heating in more detail.

Figure 2.2 describes some of the common solar technologies. In addition to passive solar, which does not require any machinery to store or utilize the solar heat, there are also active solar systems, which use pumps or fans to move heat in a building. The fastest growing application of solar power is photovoltaics or PV—direct generation of electricity from semiconducting solar panels or thin films. According to Reuters, grid-connected solar PV power has grown worldwide by an average of 60 percent every year since 2000. Figure 2.3 illustrates the recent growth of solar PV installations in the United States alone.

Large, centralized power plants currently supply most of the United States' electricity as will be discussed further in Chapter 7. According to Harold L. Platt in his book The Electric City from University of Chicago Press (1991), the centralized-plant paradigm arose from a battle between George Westinghouse and Thomas Edison in the late 1800s. Perhaps most significantly, Edison's apprentice Samuel Insull devised a metering structure that lured dispersed electricity generators to join into a consolidated power network.

Figure 2.2: Common solar technologies

Photovoltaics or Solar PV
Transforms the sun's energy directly into electricity. Photons in sunlight hit solar panels freeing electrons from the photovoltaic material so they flow out of the cell as electric current.

Type	Description	Attributes & Issues
Building-scale arrays	Currently, the most common types of solar panels are silicon, but alternative thin films and other technologies are fast developing. Thin film materials allow coatings on windows and walls from ground floors to rooftops.	With 2010 technologies, an average home in Colorado needing 1,000 kWh of electricity per month would need approximately 750 square feet of rooftop or ground mounted panels to provide their complete energy load.
Solar Gardens	Colorado's statute allows groups of ten or more who cannot afford their own solar panels to share in solar arrays of less than 2 MW.	Tax credit, securities regulation, ownership, and placement issues are currently impediments.
Utility scale & Concentrating PV (The acronym CSP can include Concentrating PV or CST described below.)	Solar farms or parks use large fields of photovoltaic panels to generate electricity at a commercial scale. Newer concentrated PV systems use lenses to magnify the sun's light into solar cells with multiple layers of sun-to-electricity transforming materials instead of the traditional panels with only one such layer.	Spain is a world leader in solar power. Its Olmedilla PV Park generates 60 MW with 160,000 panels. Cogentrix Energy is proposing to build the biggest concentrating PV plant in the world on 225 acres near Alamosa, Colorado.

Solar Thermal
Transforms the sun's energy directly into thermal energy (heat) instead of electricity.

Type	Description	Attributes & Issues
Passive Solar	Used by humans since ancient times for illumination and for heating space by simply orienting buildings to take advantage of sunlight entering windows.	The heat is used in place. European energy efficient building standards (e.g. Passiv Haus) rely heavily on passive solar heating (using no furnace). To take advantage of passive solar heating, a building would benefit most with access to the full solar envelope. More info on this in Chapter 7.
Active Solar	The distinction between active and passive solar is that active solar systems use pumps or fans to move and circulate heat captured in a collector of some kind (such as water tanks or "thermal mass" like rock or concrete walls).	Solar thermal systems in Colorado regularly can meet 60 to 70 % of a building's space heating needs. Ideally, a building would have access to the full solar envelope.
Solar Domestic Hot Water (DHW)	Open loop systems circulate potable water through the solar collector directly to the hot water tank. Closed loop systems circulate a non-freezing fluid through the collector and then the heat is exchanged into the DHW storage tank.	Nationwide, the range is 60 to 95% of DHW needs. Colorado is one of the best locations: a 4' x 8' flat panel collector can generate above 35 kbtus. Because a typical water heater is only about 65%, 35kbtus collected can translate to avoided gas consumption of approximately 54 kbtus. A backup heat source can provide for cloudy days.

Concentrating (aka Concentrated) Solar Thermal or CST
Large-scale arrays of devices concentrate the sun's heat to generate steam to produce electricity. In addition to allowing higher efficiencies than rooftop solar, CST is often coupled with heat storage mechanisms that allow the continued production of electricity even when the sun is not shining. This technology is used for utility-scale, centralized power generation and will be discussed further in Chapter 8.

CSP stands for Concentrating Solar Power and can apply to both CST and Concentrating PV described above.

Although some energy generated by a large utility-scale plant is lost through transmission and distribution to consumers, this centralized model had several advantages for the owners: (1) larger plants created efficiencies of scale, (2) affluent customers were eager to join because polluting smokestacks could be moved out of their neighborhoods (an issue now for energy justice advocates), and (3) centralization required customer dependence, resulting in rate control and guaranteed profits for the utilities. The majority of energy sources we will discuss in this book fit into the centralized, industrial-scale model, including concentrating solar power or CSP technologies explored in Chapter 8. *See also* Figure 2.2.

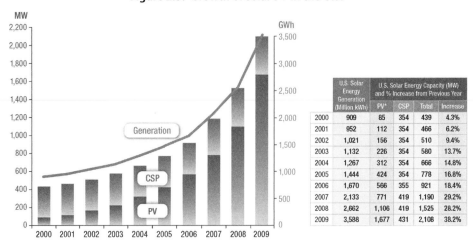

Figure 2.3: Growth of solar PV in the U.S.

	U.S. Solar Energy Generation (Million kWh)	U.S. Solar Energy Capacity (MW) and % Increase from Previous Year			
		PV*	CSP	Total	Increase
2000	909	85	354	439	4.3%
2001	952	112	354	466	6.2%
2002	1,021	156	354	510	9.4%
2003	1,132	226	354	580	13.7%
2004	1,267	312	354	666	14.8%
2005	1,444	424	354	778	16.8%
2006	1,670	566	355	921	18.4%
2007	2,133	771	419	1,190	29.2%
2008	2,662	1,106	419	1,525	28.2%
2009	3,588	1,677	431	2,108	38.2%

National Renewable Energy Lab., 2009 Renewable Energy Databook 68 (2010)

In contrast, this chapter will focus on distributed generation, a system that delivers power locally, directly at the point it is generated. As The Electric City noted, most early electricity generation was distributed, but many consumers were happy to switch to a centralized model so that smoky combustion plants could be moved away from them.

At that time, no one had the option of producing electricity on location from the sun—silently and without pollution. Solar PV is one of the newest electricity-generating technologies, coming more than a century after Michael Faraday's creation of the first electric motor. Bell Laboratories announced its "power photocell" in April of 1954, but most applications were limited to the space program because of the prohibitive costs. In From Space To Earth, The Story Of Solar Electricity from Harvard University Press 2000, author John Perrin notes that a solar array to power the average home would have cost $1,430,000 in 1956.

By the mid-1970s, space-program manufacturing brought the price of solar panels down to $10 to $20 per watt. While more expensive than grid-generated electricity, solar PV became the least expensive alternative for remote applications, such as irrigation pumps and roadside lighting or signage, that otherwise would include the cost of constructing additional transmission or distribution lines. Solar PV also became a manageable expense for some homes that are "off the grid," meaning that all of their heat and electricity needs are met without connecting to a centralized utility.

Solar PV still is undergoing transformative changes. New technologies and world competition in manufacturing continue to bring the price down. In addition, utilities have begun to install net meters that measure both input and output of electricity. With a net meter, distributed power sources, such as the grid-connected solar sources discussed in this chapter, are used in conjunction with a utility's electric grid to cut back on demand needed from traditional centralized fossil-fuel plants. In addition, when these solar devices generate more electricity than needed by the particular building on which they are located, they not only cut demand from a utility but also can contribute additional power to the grid. Chapter 7 discusses the distinction between centralized and distributed generation in more detail.

This chapter concentrates on the legal issues encountered in protecting solar rights at individual building sites under a distributed generation model. After providing some history and context in Section A, we will shift our focus to how those rights are defined and treated under the common law in Section B. Section C will consider methods governments have employed to provide further support and protection for distributed solar energy. Finally, Section D will address some of the peculiar conflicts that can arise between trees and distributed solar power.

A. History & Context

This first excerpt gives us the history of solar rights and an overview of some current issues.

Sara C. Bronin, *Solar Rights*
89 B.U. L. Rev. 1217, 1218-25 (2009)

The rights to access and to harness the rays of the sun—solar rights—have significant economic consequences. Solar rights dictate whether a property owner can grow crops, illuminate her space without electricity, dry wet clothes, reap the health benefits of natural light, and, perhaps most significantly in our modern era, operate solar collectors—devices used to transform solar energy into thermal, chemical, or electrical energy.

For at least two thousand years, people have attempted to assign solar rights in a fair and efficient manner. Ancient Romans protected the right to solar heat and light through prescriptive easements, government allocations, and court decrees. Ancient Greeks protected solar rights through rigid land planning schemes that oriented streets and buildings to take advantage of light and passive solar heat. More recent rules—such as the so-called "ancient lights" rule established in medieval England or the permit system currently used by Japan—have continued to refine the concept of solar rights. Each regime has recognized that sunlight, in reaching any one parcel, may travel across multiple parcels, and its route may vary throughout the day and from day to day. By necessity, then, the creation of solar rights implicates the rights of neighbors, both immediate and further afield.

In the United States, solar rights have fallen short, either because they do not exist or because, where they do exist, they provide inadequate protection to the holders of the rights. In the late 1970s and early 1980s, numerous American legal scholars debated these deficiencies. These commentators agreed that the absence of a coherent legal framework for the treatment of solar rights had negative consequences, chief among which was the dampening effect on the use of solar collectors. In their view, solar collectors produced an environmentally-friendly, inexhaustible, and economically secure alternative to carbon-based fuels. The law, they argued, should encourage the proliferation of clean energy by providing rights to solar collector owners. These scholars advanced several proposals to change the law to meet this goal. Their proposals ranged from revisions to existing statutes, to the use of nuisance suits to bar neighbors from blocking one another's light, to the creation of permit systems or zoning ordinances which administratively allocate solar rights.

... To date, despite scholars' efforts, progress with respect to the clarification and efficient allocation of solar rights has been slow. A few jurisdictions have experimented with their suggestions, but reforms have not been comprehensive, and solar rights are guaranteed in very few jurisdictions. At least in part because of the muddled legal regime, and despite numerous technological advances that have reduced the cost of solar collectors, only one percent of our nation's energy currently comes from the sun....

***

Except in a few limited circumstances, the American legal system has not recognized the solar right—the ability of a property owner to enjoy or utilize a defined amount of sunlight on her parcel and to defend this right as against other property owners. Yet there are at least two strong reasons for this country to do so, especially as such rights might apply to solar collectors.

First, solar access is extremely valuable to the individuals who have it. The quality and amount of sunlight which reaches a structure's interior, for example, affects three economic measures: the resale price of the structure, as buyers will pay premiums for naturally lit space; the productivity of the structure's occupants, who work better with sunlight than artificial light; and the operating costs of heating, cooling, and lighting systems. Similarly, the use of sunlight in outdoor areas can have financial consequences: a property owner can grow garden vegetables, produce commercial crops for resale, or use sunlight instead of electricity to dry laundry—all of which save or generate income. Perhaps most importantly, solar collectors, for which sunlight is the primary and essential ingredient, almost always save owners more in energy costs than the purchase price, and rapid technological developments have rendered them increasingly more valuable and will continue to do so in years to come. The recognition that solar access has value to individuals must serve as the basis for any solar rights regime.

Second, a solar rights regime also has value to the country as a whole.... [O]ur failure to consider solar rights appropriately has dampened investment in domestic solar collectors ... because it is difficult to justify substantial up-front investments in solar collectors without a guarantee of solar access. The reluctance to invest in solar collectors has affirmed our dependence on foreign fossil fuels. The energy conservation and energy security rationales for solar rights go hand in hand and have been discussed for decades.... Although the need for guaranteed property rights in solar access has grown more acute, we have failed to modify the law to provide them.

* * *

While the growing number of large installations may signal that the market has begun to embrace the economies of scale, the need for small installations remains. Individual solar collectors can serve the many end users that are not reachable by large solar installations. In addition, individual solar collectors allow individuals to benefit directly from their investment.... When it comes to the environment, individual solar collectors have a smaller negative impact than do large installations. And finally, individual solar collectors are more efficient than large installations because they are installed near the end user, meaning that little is lost during transmission.

In theory, there is a middle ground between the individual solar collector and the large solar installation: a mid-sized facility, which might, for example, serve a small urban neighborhood with costs divided equally among neighbors within a few blocks.... [The mid-sized facility, like the individual solar collector owner, would somehow have to obtain rights across other parcels to ensure solar access—Ed.] Mid-sized facilities generating power to multiple end users might, for example, have to incorporate as an electric utility, file paperwork with the public utility control commission, submit to the governance of an electric cooperative, or obey other rules. Most states' rules are so onerous that mid-sized solar facilities are rare.

Notes and Questions

1. Many utilities seem to prefer the development of large-scale CSP solar instead of encouraging distributed rooftop solar. Consider some of the reasons for this.

2. Chapter 8 addresses the use of public lands for developing large-scale CSP projects. Some of the major concerns about such development include (1) aesthetics, (2) habitat destruction, (3) the use of water, and (4) dedication of public lands to such an intense, single land use for decades. Consider how distributed solar on existing rooftops in developed areas might avoid some of these concerns. For example, with respect to the fourth item — the inability to use public lands covered with solar panels for other purposes — multiple-use instead of single-use would be the norm when solar panels are placed on existing structures. In fact, the panels can provide additional benefits to the property owner by shading a parking lot or reducing cooling costs for a building by deflecting sun absorption from a traditional roof.

3. Transmission related to the development of renewable energy is a topic worthy of a book of its own. In the context of this chapter, consider whether existing distribution infrastructure could be expanded. If so, do the additional costs of condemning rights-of-way and constructing miles of new transmission lines for CSP suggest that more, smaller distributed solar installations might be a better solution? Also, consider whether the power losses during transmission might be a sufficient incentive to generate more electricity exactly where it is needed — on rooftops within population centers.

4. Bronin addresses "a middle ground between the individual solar collector and the large solar installations: a mid-sized facility...." The City of Ellensburg, Washington established the first U.S. "community renewable park" in 2006. This facility required generous support from local homeowners and businesses including Central Washington University. In exchange for their contributions, subscribers receive from the City of Ellensburg a credit for the value of electricity produced by the renewable systems. What might be some of the advantages and disadvantages of this community solar approach?

5. In 2010, the State of Colorado enacted legislation that recognized just the kind of small neighborhood solar development that Bronin proposes. "Community Solar Gardens" under Colo. Rev. Stat. § 40-2-127 (2010) allow groups of ten or more people who cannot afford their own solar panels to share in solar arrays of less than two MW. Colorado had two existing community solar facilities by the middle of 2011 — Sol Partners Cooperative Solar Farm by United Power and El Jebel by Clean Energy Collective. Neither community created its solar installation under the 2010 statute's model. Why do you think this might be the case?

6. Consider the following impediments currently discouraging the development of community-scale solar gardens. As of 2011, federal tax credits for solar panels only applied to installations on individual properties. Regulators have warned that securities laws may control investment in community solar gardens. Solar gardens may not be as effective as individual rooftop solar to incentivize reductions in energy use because consumers will see a less direct correlation between the electricity they generate and what they consume. Finally, without some protection for solar access as discussed in the following section, the risk that the investment will not pay off is magnified.

B. Solar Access under the Common Law

Erik J.A. Swenson of Fulbright & Jaworski, L.L.P., has stated, "Solar energy easily travels the first 93 million miles. It's the last 93 feet that is the tough part!" Swenson's observation reflects the fact that the common law in the United States has struggled to quantify and protect solar rights. As U.S. law evolved during the birth of the industrial revolution, its courts failed to adopt some of the protections provided in England such as the doctrine of ancient lights, which recognized a claim to the sun to provide passive heat and lighting before the age of light bulbs and centralized gas and electric heating.

Yet, access to the sun is a valuable right, and U.S. common law has used both a contract and a tort model to deal with it. Section 1 describes the main contract approaches; Section 2 deals with tort efforts to allot solar rights; Section 3 includes excerpts from *Prah v. Maretti*, one of the seminal cases addressing modern solar rights issues.

1. Contract Approaches

In an urban environment, one of the only ways to ensure buildings or trees do not obstruct solar panels is to enter into a contract with a neighbor to the south.

Sara C. Bronin, *Solar Rights*
89 B.U. L. Rev. 1217, 1225-37 (2009)

The first and perhaps most straightforward method of assigning solar rights is by express agreements between private parties, where these agreements have been implicitly or explicitly authorized by law. Express agreements are the most efficient means of allocating solar rights to the respective parties: each party understands her rights and has received compensation in some form or amount to which she has consented. Usually, the compensated parties are those who would have had the initial entitlement under the law— the burdened parties, and not the solar rights seekers. Used as devices to reassign these initial entitlements, express agreements come with significant transaction costs: bargaining is time-consuming and expensive, especially when attorneys must be hired, and formalities must be followed. Transaction costs may be particularly high in bilateral monopoly situations, where the possible parties to an express agreement are limited to a small number of individuals. These costs hinder the creation of express solar agreements.

Despite the costs, the law has allowed at least three types of express agreements to serve as the basis for a solar right....

a. Express Easements

... Easements allow one landowner (the dominant owner) to have certain rights over the real property of another landowner (the servient owner). These rights take one of two forms: affirmative rights that entitle the dominant owner to physical access of the servient parcel; and negative rights that encumber the servient owner's use of her property, usually preventing the servient owner from undertaking particular activities. An easement does not grant the dominant owner ownership rights, but rather allows the dominant owner to enforce the rights contained in the easement. These enforcement powers endure, and remain with the land for subsequent purchasers, until and unless some event or condition renders them unenforceable.

Solar easements, a kind of negative easement, can create solar rights between dominant and servient owners by burdening the servient owner's use of her property. More specifically,

Definitions — A contract for an **easement** gives the dominant holder a right-of-use or right-of-way over the land of the servient tenement. In contrast to a **license** that only conveys a personal privilege, an easement is an interest in the land and is often appurtenant to particular land so it passes from one owner to the next. Although the word **covenant** might be used for any agreement, it is now mostly employed in the restrictive covenant context to describe deed-mandated restrictions on building design, landscaping, or use requirements on several lots in a subdivided area.

a solar easement can prevent a servient owner from improving her property in a way that blocks sunlight from falling on all or part of the dominant estate (in effect, defining a solar skyspace). Although it is possible to argue that the common law contemplates solar easements, legislation allowing landowners to create express solar easements avoids ambiguity and has become popular....

Express solar easements have several benefits. Most obviously, each party to an easement has voluntarily bargained to a mutually agreeable result: the dominant owner receives a solar corridor, while the servient owner receives compensation to offset her burden. Dominant owners receive a property right that is usually permanent and irrevocable. Finally, private parties make and enforce solar easements, therefore obviating the need for unnecessary governmental bureaucracy. As a result of these benefits, as one scholar put it, solar easement statutes have become a popular and "inexpensive form of legislative cheerleading."

This form of "legislative cheerleading" has not, however, borne much fruit: a search of federal and state cases revealed not a single case dealing with express solar easements.... [This] likely ... reflects the fact that such easements are rare. Indeed, the primary benefit of the solar easement—its voluntary nature—may also prevent its widespread adoption. Potential obstructers might disagree on the terms of an easement or refuse to negotiate altogether. Even when all parties agree to negotiate, solar easements take time to formulate. Moreover, negotiations cost money—not just for attorneys' fees, recording fees, and other

Figure 2.4: Depiction of solar skyspace

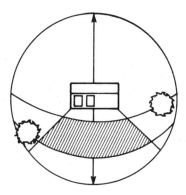
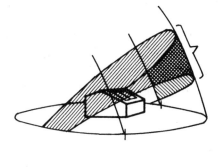

Martin Jaffe & Duncan Erley, U.S. Dep't of Hous.,
Protecting Solar Access for Residential Development 19 (1979)

administrative costs, but for the easement itself.... Servient owners may overcharge for easements, either because they overvalue their interests or because their relationships with the dominant owners function as a bilateral monopoly, each side being the only possible party to a transaction....

* * *

b. Covenants

In certain circumstances, covenants, a second type of express agreement, avoid the difficulties of express easements in promoting solar rights. Like express easements, covenants include conditions that run with the land and endure indefinitely.... A covenant must be recorded on the land records in sufficient detail to provide notice of the existence and substance of the covenant.... In this sense, covenants are efficient; purchasers with knowledge implicitly agree to incorporate the terms of the covenant in their purchase. Covenants also appear to be fair, because they often burden or benefit the owners of multiple parcels in the same way, with the same provisions applied to parcels in a geographically contiguous area, and because purchasers take land with notice of the covenant. The right to enforce (or the standing to overturn) covenants is shared between the owner of a covenanted property, other property owners burdened or benefited by the same covenant, and subsequent purchasers. Covenants appear most often, and function best, in residential neighborhoods with relatively homogenous lot sizes and structure types.

Some critics have called covenants a "two-edged sword because they can be used to either inhibit or enhance the use of solar systems." ... For example, [a developer] may ask her attorney to draft a covenant that prohibits the installation of "equipment" on rooftops. In the developer's mind, this covenant would ensure a uniform aesthetic and thereby preserve or enhance property values.... [H]owever the covenant precludes the possibility of locating a solar collector on the roof—often the most practical location for a collector....

To avoid such scenarios, some states have begun to legislate for covenants that promote, rather than hinder, solar collector use. Although courts will enforce covenants for solar access even if legislatures do not specifically authorize them, several states have made their authorization explicit. At least a dozen states go further, voiding restrictive covenants or deed conditions if they unreasonably restrict or increase the cost of a solar system. Three states have created special rules for condominiums or homeowners' associations, prohibiting certain restrictions on solar collectors. One state, Iowa, does not itself ban, but instead empowers localities to ban, covenants with unreasonable restrictions on solar collectors. Although one commentator has raised the issue of constitutional challenges to statutes voiding covenants that hinder solar collectors, no court has found, nor is one likely to find, such statutes unconstitutional.

The biggest barrier to covenants that promote solar collectors is ... practicality. Covenants are extremely difficult to impose retroactively on parcels in established neighborhoods, and therefore may only be practically useful in creating solar rights in new subdivisions. To impose a covenant on a new subdivision, a developer simply appends the covenant to the deed of each new parcel.... Covenants in new subdivisions therefore usually have low transaction costs. It is important to note that, although in theory the enactment of covenants in new subdivisions is easy, developers do not typically protect solar access voluntarily. Accordingly, several states have either allowed or required localities to consider solar access concerns when adopting subdivision regulations or approving subdivision requests from developers. When evaluated as a legal tool with the potential to create solar rights, such statutes have the same flaw as covenants—they apply prospectively only to those large-scale transactions which require subdivision review.

While enacting covenants in new subdivisions is relatively easy, enacting covenants in established neighborhoods requires significant involvement by individual parcel owners, some of whom may not want to permanently burden their properties for the sake of solar rights. Like an express easement, which requires a legal document separate from the deed that a property owner obtains upon transfer, a retroactive covenant requires an entirely new agreement. That agreement must address existing conditions (such as irregular lot sizes or unusually shaped structures), duration and termination issues, the substantive nature of the covenant, and any required financial exchanges.... Even if the property owners involved in a potential solar covenant could agree on all of the variables, the costs of bargaining for a covenant may equal or even exceed the transaction costs of express easements.... For these reasons, covenants—like express easements—show little promise in protecting solar rights, with the minor possible exception of the new residential subdivision.

c. Tenancy

Lessor-lessee arrangements, a third kind of express agreement that could create solar rights, suffer from some of the same deficiencies as express easements and covenants.... To give rise to a solar right, a lease must govern some unit of property through which the sun's rays must travel. Typically, solar leases involve airspace, known sometimes in the solar context as solar skyspaces. Airspace has long been recognized at common law as real property and may be legally distinct from ground or mineral estates. An individual who owns a piece of property in fee simple may sever the airspace from the ground parcel or craft a legal description which enables her to lease or burden just the airspace, without severance. A lease would give a solar user the ability to "occupy" the airspace without obstruction. Some states have tolerated leases that aim to provide solar access within existing landlord-tenant law. Only one state, Nebraska, explicitly recognizes leases for solar skyspaces and requires that such leases be in writing and recorded on the land records.

Most states, however, do not require such formalities for leases. People enter into leases far more frequently than they create easements and covenants. In non-complex transactions, leases involve only two parties, attorneys rarely participate, and negotiation may be minimal. Accordingly, of the three types of express agreements ... leases may have, on average, the lowest transaction costs. Many leases, however, are ultimately inefficient with respect to solar rights, because the duration of a tenancy limits the duration of the right, [which lessens the probability that a lessee will want to invest in a solar collector—Ed.].... Despite their low transaction costs, leases may be an impractical means of truly protecting solar rights.

———————

Notes and Questions

1. Consider who has the bargaining advantage in an easement transaction. Southside neighbors have little incentive to burden their properties by providing an easement to their neighbors to the north. To be of any benefit to the party seeking solar access, the easement must run with the land and bind future purchasers, which may impact the valuation of southern properties. Furthermore, even if the southern neighbors are willing to negotiate, there is no free market: they are the only party that has access to their neighbors' solar envelopes, and consequently they can demand prohibitive sums.

2. Several states have laws relating to solar and wind power access. http://dsireusa.org/incentives/index.cfm?SearchType=Access&&EE=0&RE=1. However, these laws vary widely, and few ban restrictive covenants prohibiting the installation of solar systems on roofs.

E.g., such a measure failed in Virginia in 2009, *see* http://www.ncsl.org/default.aspx ?tabid=16877.

3. Could the federal government play a role in this area? Under the authority of the Telecommunications Act of 1996, the FCC promulgated the Over-The-Air-Reception Device rule that explicitly restricted any private homeowner covenants that impaired the installation of certain satellite dishes. 47 C.F.R. § 1.4000. If the federal government nationally banned all restrictions against the installation of satellite TV dishes for the sake of competition, what rationales might support a similar law prohibiting restrictions on the installation of solar panels? Could national security be another rationale if having distributed solar energy generation makes the United States more resistant to terrorist attacks and provides backup for grid outages?

4. Solar panel leasing is becoming a popular option — *i.e.*, in June of 2011, Google invested $280 million in SolarCity, enough to build and lease solar power systems to between seven and nine thousand homeowners in the ten states in which SolarCity operates. A leasing company pays the upfront costs of the solar systems on individual buildings, and the building owners simply pay for power much in the way they previously paid the centralized utility. Do you think these leasing companies may provide more lobbying clout for protecting solar rights as they presumably are more organized and have more of an investment in ensuring solar access than do individual building owners? Should these leasing companies be subject to regulation by public utility commissions?

5. Who has the burden of taking to court a neighbor who is unwilling to comply with a contract? Once a case makes it to the court, neighbors seeking access to the sun would usually be the plaintiffs, which means they would usually be the party obligated to meet the burden of proof. Does it benefit society to place this burden primarily, or sometimes exclusively, on the parties trying to help solve our nation's energy problems by installing solar panels both to decrease their demands from the grid and to contribute excess power generated from their panels to the grid?

2. Tort Approaches

The previous section addressed protecting solar access through contracts between neighbors. When the parties cannot come to an agreement, the solar panel owner may have some recourse through tort law, such as (a) nuisance, (b) prescriptive easements, or (c) implied easements.

Sara C. Bronin, *Solar Rights*
89 B.U. L. Rev. 1217, 1250-65 (2009)

... [S]ome solar rights seekers may turn to the courts. In theory, court decisions result from careful analysis of law and balancing of equities, and courts apply precedent to adapt to new realities. Indeed, American courts have on numerous occasions created legal rights to advance innovations with broad social impacts. In 1946, for example, the Supreme Court accommodated the advent of the airplane era by limiting property owners' rights to only the airspace such owners could utilize. To have ruled otherwise, as Justice Douglas pointed out, would have exposed airline companies to so many private claims that travel by air would have been impossible. The Supreme Court's creation of a travelway for airplanes facilitated economic growth and transformed the way we live. Similarly, the scarcity of land and the proliferation of dense, high-rise condominium buildings gave rise to horizontal airspace as a unit of real property—a concept in property law, which had not existed before the advent of skyscrapers. The property right in airspace allowed property owners to maximize use of their land—much as a solar right would help to maximize the energy-saving technologies of the solar collector.

... [T]he vast majority of courts appear to be hostile to the creation of solar rights, despite the theoretical applicability of several strands of common law. Solar rights might be created, for example, under nuisance rules, whether private or public. They might also be created via court-assigned prescriptive easements, ... which property owners have enjoyed for some period of time, or under an implied easement theory, which would apply to certain property subdivisions. None of these theories has taken hold on a wide scale. To the contrary, they have been almost unanimously rejected.

... [E]ven if courts were receptive to solar rights theories, litigation will remain perhaps the least efficient and most expensive method of resolving solar rights. Court allocations can only assist with the protection of existing solar collectors, meaning that they are not useful in helping an individual decide whether to install a solar collector in the first place. More directly, the costs of litigation, borne by each party, exceed the costs of both express agreements and governmental allocations, and can be disproportionate to any anticipated benefit. Time also imposes a burden on solar rights seekers, as litigation can take months, and sometimes years. Uncertain outcomes and the existence of an adversary result in a stressful and complicated process, which at least one party will find unfair. Tracing each of these deficiencies through the judicial system demonstrates how courts' unwillingness to adapt to solar technology has severely limited solar rights and suggests that solar rights seekers should abandon the idea that courts will be willing allies in their cause.

a. Nuisance

Of the three possible court-made solar rights ... nuisance law seems the most capable of providing solar rights: well-developed and flexible, its balancing test methodology lends itself to the weighing of interests at stake in solar rights disputes. Despite this promise, however, nuisance law has not effectively been engaged to create solar rights....

Solar rights seekers have found only limited success in protecting access through private nuisance claims[, which allege harms against one discrete party or parties—Ed.]. The Restatement (Second) of Torts defines a private nuisance as "a nontrespassory invasion of another's interest in the private use and enjoyment of land" and requires that the invasion be intentional and unreasonable. The Restatement's rules instruct courts to weigh the harm and the utility of the activity; a private nuisance occurs if, on balance, the harm caused by the activity exceeds its benefit. Obstruction of a solar collector's access to light could therefore be considered a nuisance under Restatement principles if, on balance, the harm caused by the obstruction (say, rendering the solar collector defunct) is greater than benefits caused by the activity (say, erecting a tall structure that would shade the collector). Before 1982, in the few cases that reached the courts, the judiciary declined to find that obstruction of sunlight from reaching solar collectors was a private nuisance. In 1982, however, the Wisconsin Supreme Court decided *Prah v. Maretti*, which recognized a private nuisance claim for malicious obstruction of a solar collector under Restatement balancing principles....

Although solar collector proponents immediately lauded *Prah* as a sign that courts were finally beginning to recognize solar rights, the decision has attracted criticism [and has not had a significant impact on solar access law—Ed.]....

* * *

Given the paucity of relevant judicial activity, it seems unlikely that nuisance actions—whether private or public—will provide the solution to the challenge of allocating solar access rights. The unpredictability of outcomes may be the most significant deterrent: no matter how many cases courts decide, nuisance law always involves a highly individualized analysis of the applicable facts. Solar collector owners may be unwilling to bear the high costs of litigation for uncertain results. Nuisance litigation imposes not only private costs, but also the public cost on the courts, costs related to prosecution of public nuisances, and the consequences of erroneous judgments. The remedy granted may not necessarily mitigate these costs. On the one hand, a solar collector owner bringing a nuisance claim may want to receive damages and attorneys' fees to recoup out-of-pocket expenses and other losses. On the other, she may want an injunction to stop the nuisance itself—the only path to a secure right in solar access. Conflicting goals with respect to the remedy may further deter potential litigants.

b. Prescriptive Easements

Like nuisance law, the law of prescriptive easements provides a possible, but imperfect, means of securing solar rights through the courts. A prescriptive easement refers to a right of access "created from an open, adverse, and continuous use over a statutory period," which may be established without the consent of the property owner against whom the easement is claimed. Solar prescriptive easements date back to at least the reign of the Roman emperor Justinian, under whom codified laws prevented neighbors from blocking sunlight, which had previously been enjoyed by a property owner for light, heat, or sundial operation. A judge would decide the reasonableness of the expectation of sunlight one party could enjoy and the reasonableness of the amount of sunlight a neighbor might block. Similarly, in England, the common law included an "ancient lights" rule that granted a property owner the right to prevent a neighbor from blocking light that reached the interior of her building and that she had enjoyed continuously for twenty years. The amount of light protected was measured by the amount of indirect sunlight required to illuminate half of a room beyond the "grumble line"—the point beyond which a normal person might complain about lack of light....

* * *

According to reported cases and historical accounts, American courts at first embraced the ancient lights doctrine and its allowance of prescriptive easements in light. Treatises and courts confirmed, however, that by the late nineteenth century, the ancient lights rule had been rejected everywhere in this country, except in Louisiana. Courts justified this rejection on the grounds that settlement patterns differed in seventeenth-century England and nineteenth-century America, and that applying the rule in rapidly-growing cities and towns would impede development. By prioritizing land development over access to light, American courts boosted not only urban growth, but also individuals' rights to develop their properties without undue hindrance, as such individuals might have otherwise found it difficult to discover (and thereafter extinguish) their neighbors' continuing use of light. When deciding ancient lights rule cases, courts have often invited legislatures to set forth clear rules regarding prescriptive easements in light. Several legislatures have responded by prohibiting such prescriptive easements altogether.

Despite such an infertile judicial and legislative environment, the idea lingers that prescriptive easements may be an effective method of establishing lasting solar rights. Proponents of this view might believe that courts will begin to recognize and respond to the increasing importance of sunlight as a valuable economic commodity ... [and] establish a limited application of the ancient lights rules ... to protect solar collectors....

Although the possibility of granting prescriptive easements for solar collectors seems appealing, courts are unlikely to make such leaps. England's ancient lights rule, which requires a twenty-year occupancy period and which protects only a minimum amount of indirect light that reaches enclosed interior spaces, would hardly address the practical requirements of a solar collector owner. A potential solar collector user could not depend on a right vesting twenty years into the future, and might therefore decline to invest in expensive solar technology. Moreover, potential users might not want to gamble on courts' application of precedent meant to protect indirect lighting of building interiors to direct lighting required by solar collectors.

Irrespective of judges' attitudes towards prescriptive easements for the protection of solar rights, practical reasons militate against reliance on court-created prescriptive easements. Prescriptive easements may misallocate incentives, causing landowners to rush to develop their properties and file notices to extinguish possible claims by neighbors wishing to build solar collectors. If a good recording system is not in place, title searches may become extremely difficult and may reduce certainty in land purchases. And as described above, the inefficiencies of court actions will deter many solar rights seekers, and the outcomes will not satisfy all parties, leading to claims of unfairness.

c. Implied Easements

Less ink has been spilled over implied easements than prescriptive easements, perhaps because easements by implication occur only in very limited circumstances. A court may create an implied easement only if "an owner of two parcels of land uses one parcel to benefit the other to such a degree that, upon the sale of the benefited parcel, the purchaser could reasonably expect the use to be included in the sale." The court must therefore find unity of ownership prior to the conveyance of the new parcel, intent among the parties to create an easement, and a need for the easement. The rationale for implied easements rests in the notion that, given the facts, the parties did intend, or would have intended, to include the easement in the conveyance. Perhaps the most common example of an implied easement is a roadway on land conveyed to another, over which roadway the conveyor still requires access.

In the solar context, a solar collector owner who has sold a portion of her property might later seek an implied easement to prevent the buyer from doing something on the buyer's property (building a tall structure, for example) which would prevent sunlight from reaching the solar collector she used to meet her energy needs for years preceding the sale. After establishing unity of ownership, she would then have to argue that the parties intended to create an easement in light, but merely failed to do so in express terms. She would have to prove the intent of a party who—by virtue of being in court—firmly opposes her claim and would not admit to having such an intent. Finally, she would have to convince a court that she depends so heavily on the energy produced by a solar collector that it is rendered a "necessity" under common law precedent. With so much to prove, a solar rights seeker has a burden, which, in most cases, is extremely difficult to overcome.... The greatest barrier to implying an easement in solar collector cases appears to be the showing of necessity ... [although] [o]ther barriers, such as proving intent, also endure.

Courts should no doubt do more to weigh competing values, including public policy, when considering such cases. But even in the unlikely circumstance that courts begin to embrace the implied solar access easement, the limited circumstances in which such easements may occur would severely limit its utility. Implied easements in light have only been granted in three circumstances. First, they have been granted where the "light was so necessary to the trade use of a business premises that without it the property would be valueless." Under this standard, courts may be reluctant to find that access to solar collectors is necessary, so long as alternative forms of energy remain viable. Second, the easement seeker may have a claim if her access to light somehow related to a right of passage (the more common basis for an implied easement). It is difficult to see how this exception could be applied with respect to solar collectors. Third, other successful cases involve implied easements claimed by owners of private property that abuts public streets. Unless a solar collector owner asserts an implied easement over a public street, this exception is as unhelpful as the others. Only a few courts (and one state legislature) have allowed property owners to overcome the presumption against implied easements for light. This state of affairs seems unlikely to change in the immediate future.

After a review of the judicial developments with respect to nuisance, prescriptive easements, and implied easements, it is difficult to imagine that courts could ever become fully engaged with the development of a solar rights regime. Even if courts suddenly became receptive to solar rights, litigation would be a poor strategy for solar rights seekers for many reasons, including the uncertainty of the outcome and the related transaction costs. Rather than repeating "ancient" debates about ancient lights and other topics, modern scholars should shift their focus away from the courts ... [and] join a new debate about how jurisdictions might adopt an integrated approach, which addresses the concerns of both solar rights seekers and possible burdened parties.

Notes and Questions

1. Although French physicist Edmund Becquerel is credited with discovering the potential for generating electricity from sunlight in 1839, it was not until Bell Labs developed its power photocells in 1954 that photovoltaic generation of electricity became practical. Still, it took another two decades before PV panels were commercially available to the general public, and even then, their price made them prohibitive except for the wealthiest and most environmentally dedicated individuals. According to John Perlin in FROM SPACE TO EARTH (1999), solar panels cost approximately $300.00 per watt in 1954, and the price came down to approximately $15.00 per watt in the early 1970s. In 2009, the price of the silicon solar panels themselves (not the installed price) dropped below $1.00 per watt.

2. As in the case of common law contracts, consider who has the burden of proof in tort enforcement actions. In the late 1970s and early 1980s, the price of a solar PV system relative to the cost of litigation made it feasible to bring an action to enforce solar rights. Since that time, litigation costs have risen exponentially; in contrast, the price of solar panels has decreased dramatically. Now, the cost of just the first phases of litigation can easily exceed the entire cost of a typical solar array. Will this encourage the wasteful practice of simply walking away from the use of solar PV instead of seeking access through litigation?

3. What are the tests for nuisance, and how should a court resolve conflicts? The following case can provide some answers to these questions. Pay special attention to Footnote 5 of the majority opinion for some comments from Prosser on balancing competing property interests.

3. Prah v. Maretti

After the Arab oil embargo in 1973, renewable energy sources enjoyed a surge in development in the United States. Some courts began to articulate changing priorities affording more protection for solar rights than had previously been recognized since the industrial revolution. The Wisconsin Supreme Court's opinion in *Prah v. Maretti* reflects this shift of perspective.

Prah v. Maretti
321 N.W.2d 182 (Wis. 1982)

Abrahamson, J.

This appeal ... present[s] an issue of first impression, namely, whether an owner of a solar-heated residence states a claim upon which relief can be granted when he asserts that his neighbor's proposed construction of a residence (which conforms to existing deed restrictions and local ordinances) interferes with his access to an unobstructed path for sunlight across the neighbor's property. This case thus involves a conflict between one landowner (Glenn Prah, the plaintiff) interested in unobstructed access to sunlight across adjoining property as a natural source of energy and an adjoining landowner (Richard D. Maretti, the defendant) interested in the development of his land.

[The defendant purchased a lot south of the plaintiff and began constructing a home there that plaintiff alleged would substantially and adversely affect the integrity of plaintiff's solar system. The circuit court denied plaintiff's request for injunctive relief and granted summary judgment against the plaintiff—Ed.]

* * *

The plaintiff presents three legal theories to support his claim that the defendant's continued construction of a home justifies granting him relief: (1) the construction constitutes a common law private nuisance; (2) the construction is prohibited by sec. 844.01, Stats.1979-80; and (3) the construction interferes with the solar easement plaintiff acquired under the doctrine of prior appropriation.[4]

As to the claim of private nuisance the circuit court concluded that the law of private nuisance requires the court to make a comparative evaluation of the conflicting interests and to weigh the gravity of the harm to the plaintiff against the utility of the defendant's conduct. The circuit court concluded: "A comparative evaluation of the conflicting interests, keeping in mind the omissions and commissions of both Prah and Maretti, indicates that defendant's conduct does not cause the gravity of the harm which the plaintiff himself may well have avoided by proper planning." The circuit court also concluded that sec. 844.01 does not apply to a home constructed in accordance with deed and municipal ordinance requirements. Further, the circuit court rejected the prior appropriation doctrine as "an intrusion of judicial egoism over legislative passivity."

We consider first whether the complaint states a claim for relief based on common law private nuisance. This state has long recognized that an owner of land does not have an absolute or unlimited right to use the land in a way which injures the rights of others. The rights of neighboring landowners are relative; the uses by one must not unreasonably

---

4. Under the doctrine of prior appropriation the first user to appropriate the resource has the right of continued use to the exclusion of others. The doctrine of prior appropriation has been used by several western states to allocate water, and by the New Mexico legislature to allocate solar access, secs. 47-3-1 to 47-3-5, N.M.Stats.1978.

impair the uses or enjoyment of the other.[5] When one landowner's use of his or her property unreasonably interferes with another's enjoyment of his or her property, that use is said to be a private nuisance.

The private nuisance doctrine has traditionally been employed in this state to balance the rights of landowners, and this court has recently adopted the analysis of private nuisance set forth in the Restatement (Second) of Torts. The Restatement defines private nuisance as "a nontrespassory invasion of another's interest in the private use and enjoyment of land." Restatement (Second) of Torts sec. 821D (1977). The phrase "interest in the private use and enjoyment of land" as used in sec. 821D is broadly defined to include any disturbance of the enjoyment of property. The comment in the Restatement describes the landowner's interest protected by private nuisance law as follows:

> "The phrase 'interest in the use and enjoyment of land' is used in this Restatement in a broad sense. It comprehends not only the interests that a person may have in the actual present use of land for residential, agricultural, commercial, industrial and other purposes, but also his interests in having the present use value of the land unimpaired by changes in its physical condition. Thus the destruction of trees on vacant land is as much an invasion of the owner's interest in its use and enjoyment as is the destruction of crops or flowers that he is growing on the land for his present use. 'Interest in use and enjoyment' also comprehends the pleasure, comfort and enjoyment that a person normally derives from the occupancy of land. Freedom from discomfort and annoyance while using land is often as important to a person as freedom from physical interruption with his use or freedom from detrimental change in the physical condition of the land itself." Restatement (Second) of Torts, Sec. 821D, Comment *b*, p. 101 (1977).

Although the defendant's obstruction of the plaintiff's access to sunlight appears to fall within the Restatement's broad concept of a private nuisance as a nontrespassory invasion of another's interest in the private use and enjoyment of land, the defendant asserts that he has a right to develop his property in compliance with statutes, ordinances and private covenants without regard to the effect of such development upon the plaintiff's access to sunlight. In essence, the defendant is asking this court to hold that the private nuisance doctrine is not applicable in the instant case and that his right to develop his land is a right which is *per se* superior to his neighbor's interest in access to sunlight. This position is expressed in the maxim "cujus est solum, ejus est usque ad coelum et ad

---

5. In *Abdella v. Smith*, 34 Wis.2d 393, 399, 149 N.W.2d 537 (1967), this court quoted with approval Dean Prosser's description of the judicial balancing of the reciprocal rights and privileges of neighbors in the use of their land:

> "Most of the litigation as to private nuisance has dealt with the conflicting interests of landowners and the question of the reasonableness of the defendant's conduct: The defendant's privilege of making a reasonable use of his own property for his own benefit and conducting his affairs in his own way is no less important than the plaintiff's right to use and enjoy his premises. The two are correlative and interdependent, and neither is entitled to prevail entirely, at the expense of the other. Some balance must be struck between the two. The plaintiff must be expected to endure some inconvenience rather than curtail the defendant's freedom of action, and the defendant must so use his own property that he causes no unreasonable harm to the plaintiff. The law of private nuisance is very largely a series of adjustments to limit the reciprocal rights and privileges of both. In every case the court must make a comparative evaluation of the conflicting interests according to objective legal standards, and the gravity of the harm to the plaintiff must be weighed against the utility of the defendant's conduct." Prosser, *Law of Torts*, sec. 89, p. 596 (2d ed. 1971) (citations omitted).

In *United States v. Causby*, 328 U.S. 256 (1946), landowners were suing for trespass of airplanes in the skyspace directly over their properties under the "*ad coelum*" doctrine. The U.S. Supreme Court rejected the concept of *ad coelum* in the context of airplane sky space, stating, "[The ad coelum] doctrine has no place in the modern world." *Id.* at 260-61. However, the *Causby* court also stated, "[I]f the landowner is to have full enjoyment of the land, he must have exclusive control of the immediate reaches of the enveloping atmosphere. Otherwise buildings could not be erected, trees could not be planted, and even fences could not be run.... [Thus, a landowner] owns at least as much of the space above the ground as he can occupy or use in connection with the land ... [and invasions of that airspace] are in the same category as invasions of the surface." *Id.* at 264-65.

infernos," that is, the owner of land owns up to the sky and down to the center of the earth. The rights of the surface owner are, however, not unlimited. *U. S. v. Causby*, 328 U.S. 256, 260-1, 66 S.Ct. 1062, 1065, 90 L.Ed. 1206 (1946). *See also* 114.03, Stats.1979-80.

The defendant is not completely correct in asserting that the common law did not protect a landowner's access to sunlight across adjoining property. At English common law a landowner could acquire a right to receive sunlight across adjoining land by both express agreement and under the judge-made doctrine of "ancient lights." Under the doctrine of ancient lights if the landowner had received sunlight across adjoining property for a specified period of time, the landowner was entitled to continue to receive unobstructed access to sunlight across the adjoining property. Under the doctrine the landowner acquired a negative prescriptive easement and could prevent the adjoining landowner from obstructing access to light.

Although American courts have not been as receptive to protecting a landowner's access to sunlight as the English courts, American courts have afforded some protection to a landowner's interest in access to sunlight. American courts honor express easements to sunlight. American courts initially enforced the English common law doctrine of ancient lights, but later every state which considered the doctrine repudiated it as inconsistent with the needs of a developing country. Indeed, for just that reason this court concluded that an easement to light and air over adjacent property could not be created or acquired by prescription and has been unwilling to recognize such an easement by implication. (citations omitted).

Many jurisdictions in this country have protected a landowner from malicious obstruction of access to light (the spite fence cases) under the common law private nuisance doctrine. If an activity is motivated by malice it lacks utility and the harm it causes others outweighs any social values. VI-A Law of Property sec. 28.28, p. 79 (1954). This court was reluctant to protect a landowner's interest in sunlight even against a spite fence, only to be overruled by the legislature. Shortly after this court upheld a landowner's right to erect a useless and unsightly sixteen-foot spite fence four feet from his neighbor's windows, (citation omitted), the legislature enacted a law specifically defining a spite fence as an actionable private nuisance. Thus a landowner's interest in sunlight has been protected in this country by common law private nuisance law at least in the narrow context of the modern American rule invalidating spite fences. (citations omitted).

This court's reluctance in the nineteenth and early part of the twentieth century to provide broader protection for a landowner's access to sunlight was premised on three policy considerations. First, the right of landowners to use their property as they wished, as long as they did not cause physical damage to a neighbor, was jealously guarded. (citation omitted).

Second, sunlight was valued only for aesthetic enjoyment or as illumination. Since artificial light could be used for illumination, loss of sunlight was at most a personal annoyance which was given little, if any, weight by society.

Third, society had a significant interest in not restricting or impeding land development. (citation omitted). This court repeatedly emphasized that in the growth period of the nineteenth and early twentieth centuries change is to be expected and is essential to property and that recognition of a right to sunlight would hinder property development. The court expressed this concept as follows:

> "As the city grows, large grounds appurtenant to residences must be cut up to supply more residences.... The cistern, the outhouse, the cesspool, and the private drain must disappear in deference to the public waterworks and sewer; the terrace and the garden, to the need for more complete occupancy.... Strict limitation [on the recognition of easements of light and air over adjacent premises is] in accord with the popular conception upon which real estate has been and is daily being conveyed in Wisconsin and to be essential to easy and rapid development at least of our municipalities." (citations omitted).

Considering these three policies, this court concluded that in the absence of an express agreement granting access to sunlight, a landowner's obstruction of another's access to sunlight was not actionable. (citations omitted). These three policies are no longer fully accepted or applicable. They reflect factual circumstances and social priorities that are now obsolete.

First, society has increasingly regulated the use of land by the landowner for the general welfare. (citations omitted).

Second, access to sunlight has taken on a new significance in recent years. In this case the plaintiff seeks to protect access to sunlight, not for aesthetic reasons or as a source of illumination but as a source of energy. Access to sunlight as an energy source is of significance both to the landowner who invests in solar collectors and to a society which has an interest in developing alternative sources of energy.[11]

Third, the policy of favoring unhindered private development in an expanding economy is no longer in harmony with the realities of our society. (citation omitted). The need for easy and rapid development is not as great today as it once was, while our perception of the value of sunlight as a source of energy has increased significantly.

Courts should not implement obsolete policies that have lost their vigor over the course of the years. The law of private nuisance is better suited to resolve landowners' disputes about property development in the 1980's than is a rigid rule that does not recognize a landowner's interest in access to sunlight. As we said in *Ballstadt v. Pagel*, 202 Wis. 484, 489, 232 N.W. 862 (1930), "What is regarded in law as constituting a nuisance in modern times would no doubt have been tolerated without question in former times." We read *State v. Deetz*, 66 Wis.2d 1, 224 N.W.2d 407 (1974), as an endorsement of the application

---

11. State and federal governments are encouraging the use of the sun as a significant source of energy. In this state the legislature has granted tax benefits to encourage the utilization of solar energy. See Ch. 349, 350, Laws of 1979. *See also* Ch. 354, Laws of 1981 (eff. May 7, 1982) enabling legislation providing for local ordinances guaranteeing access to sunlight. The federal government has also recognized the importance of solar energy and currently encourages its utilization by means of tax benefits, direct subsidies, and government loans for solar projects....

of common law nuisance to situations involving the conflicting interests of landowners and as rejecting *per se* exclusions to the nuisance law reasonable use doctrine.

In *Deetz* the court abandoned the rigid common law common enemy rule with respect to surface water and adopted the private nuisance reasonable use rule, namely that the landowner is subject to liability if his or her interference with the flow of surface waters unreasonably invades a neighbor's interest in the use and enjoyment of land. Restatement (Second) of Torts, sec. 822, 826, 829 (1977). This court concluded that the common enemy rule which served society "well in the days of burgeoning national expansion of the mid-nineteenth and early-twentieth centuries" should be abandoned because it was no longer "in harmony with the realities of our society." *Deetz, supra,* 66 Wis.2d at 14-15, 224 N.W.2d 407. We recognized in *Deetz* that common law rules adapt to changing social values and conditions.[12]

Yet the defendant would have us ignore the flexible private nuisance law as a means of resolving the dispute between the landowners in this case and would have us adopt an approach, already abandoned in *Deetz,* of favoring the unrestricted development of land and of applying a rigid and inflexible rule protecting his right to build on his land and disregarding any interest of the plaintiff in the use and enjoyment of his land. This we refuse to do.[13]

---

12. This court has recognized "that the common law is susceptible to growth and adaptation to new circumstances and situations, and that courts have power to declare and effectuate what is the present rule in respect of a given subject without regard to the old rule.... The common law is not immutable, but flexible, and upon its own principles adapts itself to varying conditions." *Dimick v. Schiedt,* 293 U.S. 474, 487, 55 S.Ct. 296, 301, 79 L.Ed. 603 (1935), quoted with approval in *Schwanke v. Garlt,* 219 Wis. 367, 371, 263 N.W. 176 (1935).

In *Bielski v. Schulze,* 16 Wis.2d 1, 11, 114 N.W.2d 105 (1962), this court said: "Inherent in the common law is a dynamic principle which allows it to grow and to tailor itself to meet changing needs within the doctrine of *stare decisis,* which, if correctly understood, was not static and did not forever prevent the courts from reversing themselves or from applying principles of common law to new situations as the need arose. If this were not so, we must succumb to a rule that a judge should let others 'long dead and unaware of the problems of the age in which he lives, do his thinking for him.' Mr. Justice Douglas, Stare Decisis, 49 Columbia Law Review 735, 736 (1949).

"The genius of the common law is its ability to adapt itself to the changing needs of society." *Moran v. Quality Aluminum Casting Co.,* 34 Wis.2d 542, 551, 150 N.W.2d 137 (1967). *See also State v. Esser,* 16 Wis.2d 567, 581, 115 N.W.2d 505 (1962).

13. Defendant's position that a landowner's interest in access to sunlight across adjoining land is not "legally enforceable" and is therefore excluded *per se* from private nuisance law was adopted in *Fontainebleau Hotel Corp. v. Forty-five Twenty-five, Inc.,* 114 So.2d 357 (Fla.App.1959), *cert. den* 117 So.2d 842 (Fla.1960). The Florida district court of appeals permitted construction of a building which cast a shadow on a neighboring hotel's swimming pool. The court asserted that nuisance law protects only those interests "which [are] recognized and protected by law," and that there is no legally recognized or protected right to access to sunlight. A property owner does not, said the Florida court, in the absence of a contract or statute, acquire a presumptive or implied right to the free flow of light and air across adjoining land. The Florida court then concluded that a lawful structure which causes injury to another by cutting off light and air-whether or not erected partly for spite-does not give rise to a cause of action for damages or for an injunction. *See also People ex rel Hoogasian v. Sears, Roebuck & Co.,* 52 Ill.2d 301, 287 N.E.2d 677 (1972).

We do not find the reasoning of *Fontainebleau* persuasive. The court leaped from rejecting an easement by prescription (the doctrine of ancient lights) and an easement by implication to the conclusion that there is no right to protection from obstruction of access to sunlight. The court's statement that a landowner has no right to light should be the conclusion, not its initial premise. The court did not explain why an owner's interest in unobstructed light should not be protected or in what manner an owner's interest in unobstructed sunlight differs from an owner's interest in being free from obtrusive noises or smells or differs from an owner's interest in unobstructed use of water. The recognition of a *per se* exception to private nuisance law may invite unreasonable behavior.

Private nuisance law, the law traditionally used to adjudicate conflicts between private landowners, has the flexibility to protect both a landowner's right of access to sunlight and another landowner's right to develop land. Private nuisance law is better suited to regulate access to sunlight in modern society and is more in harmony with legislative policy and the prior decisions of this court than is an inflexible doctrine of non-recognition of any interest in access to sunlight across adjoining land.

We therefore hold that private nuisance law, that is, the reasonable use doctrine as set forth in the Restatement, is applicable to the instant case. Recognition of a nuisance claim for unreasonable obstruction of access to sunlight will not prevent land development or unduly hinder the use of adjoining land. It will promote the reasonable use and enjoyment of land in a manner suitable to the 1980's. That obstruction of access to light might be found to constitute a nuisance in certain circumstances does not mean that it will be or must be found to constitute a nuisance under all circumstances. The result in each case depends on whether the conduct complained of is unreasonable.

Accordingly we hold that the plaintiff in this case has stated a claim under which relief can be granted. Nonetheless we do not determine whether the plaintiff in this case is entitled to relief. In order to be entitled to relief the plaintiff must prove the elements required to establish actionable nuisance, and the conduct of the defendant herein must be judged by the reasonable use doctrine.

* * *

Because the plaintiff has stated a claim of common law private nuisance upon which relief can be granted, the judgment of the circuit court must be reversed [and the cause remanded]. We need not, and do not, reach the question of whether the complaint states a claim under sec. 844.01, Stats.1979-80, or under the doctrine of prior appropriation. (Citation omitted).

* * *

Callow, J. (dissenting).

The majority has adopted the Restatement's reasonable use doctrine to grant an owner of a solar heated home a cause of action against his neighbor who, in acting entirely within the applicable ordinances and statutes, seeks to design and build his home in such a location that it may, at various times during the day, shade the plaintiff's solar collector, thereby impeding the efficiency of his heating system during several months of the year. Because I believe the facts of this case clearly reveal that a cause of action for private nuisance will not lie, I dissent.

* * *

It is a fundamental principle of law that a "landowner owns at least as much of the space above the ground as he can occupy or use in connection with the land." United States v. Ausby, 328 U.S. 256, 264 (1946) (citations omitted). As stated in the frequently cited and followed case of *Fontainebleau Hotel Corp. v. Forty-Five Twenty-Five, Inc.*, 114 So.2d 357 (Fla.Dist.Ct.App.1959), *cert. denied*, 117 So.2d 842 (Fla.1960):

> "There being, then, no legal right to the free flow of light and air from the adjoining land, it is universally held that where a structure serves a useful and beneficial purpose, it does not give rise to a cause of action, either for damages or for an injunction under the maxim *sic utere tuo ut alienum non laedas*, even though it causes injury to another by cutting off the light and air and interfering with the view that would otherwise be available over adjoining land in its natural

state, regardless of the fact that the structure may have been erected partly for spite." *Id.* at 359 (emphasis in original).

See Venuto v. Owens-Corning Fiberglas Corp., 22 Cal.App.3d 116, 127, 99 Cal.Rptr. 350, 357 (1971). I firmly believe that a landowner's right to use his property within the limits of ordinances, statutes, and restrictions of record where such use is necessary to serve his legitimate needs is a fundamental precept of a free society which this court should strive to uphold.

As one commentator has suggested:

"It is fashionable to dismiss such values as deriving from a bygone era in which people valued development as a 'goal in itself,' but current market prices for real estate, and more particularly the premiums paid for land whose zoning permits intensive use, suggest that people still place very high values on such rights."

Williams, *Solar Access and Property Rights: A Maverick Analysis*, 11 Conn.L.Rev. 430, 443 (1979) (footnote omitted). *Cf.* Goble, *Solar Access and Property Rights: Reply to a "Maverick" Analysis*, 12 Conn.L.Rev. 270 (1980).

The majority cites two zoning cases, *Village of Euclid v. Ambler Realty Company*, 272 U.S. 365, 47 S.Ct. 114, 71 L.Ed. 303 (1926), and *Just v. Marinette County*, 56 Wis.2d 7, 201 N.W.2d 761 (1972), to support the conclusion that society has increasingly regulated private land use in the name of public welfare. *Supra*, at 189. The cases involving the use of police power and eminent domain are clearly distinguishable from the present situation as they relate to interference with a private right solely for the *public* health, safety, morals, or welfare. In the instant case, we are dealing with an action that seeks to restrict the defendant's private right to use his property, notwithstanding a complete lack of notice of restriction to the defendant and the defendant's compliance with applicable ordinances and statutes. The plaintiff who *knew* of the potential problem before the defendant acquired the land seeks to impose such use restriction to accommodate his personal, private benefit—a benefit which could have been accommodated by the plaintiff locating his home in a different place on his property or by acquiring the land in question when it was for sale prior to its acquisition by the defendant.

I know of no cases repudiating policies favoring the right of a landowner to use his property as he lawfully desires or which declare such policies are "no longer fully accepted or applicable" in this context. *Supra*, at 189.[2] The right of a property owner to lawful enjoyment of his property should be vigorously protected, particularly in those cases where the adjacent property owner could have insulated himself from the alleged problem by acquiring the land as a defense to the potential problem or by provident use of his own property.

The majority concludes that sunlight has not heretofore been accorded the status of a source of energy, and consequently it has taken on a new significance in recent years. Solar energy for home heating is at this time sparingly used and of questionable economic value

---

2. Perhaps one reason courts have been hesitant to recognize a cause of action for solar blockage is that such a suit would normally only occur between two abutting landowners, and it is hoped that neighbors will compromise and reach agreement between themselves. This has, undoubtedly, been done in a large percentage of cases. To now recognize a cause of action for solar blockage may thwart a policy of compromise between neighbors. *See* Williams, *Solar Access and Property Rights: A Maverick Analysis*, 11 Conn.L.Rev. 430, 441-42 (1979). *See also* S. Kraemer, *Solar Law*, 138 (1978) ("[a] deterring factor to the use of private nuisance to assure access to direct sunlight is the resultant litigation between neighbors").

because solar collectors are not mass produced, and consequently, they are very costly. Their limited efficiency may explain the lack of production.

Regarding the third policy the majority apparently believes is obsolete (that society has a significant interest in not restricting land development), it cites *State v. Deetz*, 66 Wis.2d 1, 224 N.W.2d 407 (1974). I concede the law may be tending to recognize the value of aesthetics over increased volume development and that an individual may not use his land in such a way as to harm the *public*. The instant case, however, deals with a *private* benefit.... While the majority's policy arguments may be directed to a cause of action for public nuisance, we are presented with a private nuisance case, which I believe is distinguishable in this regard.[3]

[The dissent (1) opines that what is desirable or advisable is not a question of fact, but instead a question of policy for the legislature. Chapter 354, Laws of 1981, enabled local governments to establish solar access ordinances, so the majority's decision is unwarranted judicial intrusion; (2) disagrees with the majority's interpretation of "invasion" under the Restatement (Second) of Torts § 821D (1977), saying that word generally means a "hostile entrance" not satisfied by an obstruction to sunlight; and (3) states that defendant's conduct of constructing his home despite plaintiff's notice to him that his construction would interfere with plaintiff's solar access was not "intentional and unreasonable" because "normal persons in that locality would not be substantially annoyed or disturbed by the situation." — Ed.]

* * *

I conclude that plaintiff's solar heating system is an unusually sensitive use. In other words, the defendant's proposed construction of his home, under ordinary circumstances, would not interfere with the use and enjoyment of the usual person's property....

Notes and Questions

1. Footnote 4 of the majority opinion observes that at least one state has applied prior appropriation, a first-in-time-first-in-right doctrine used primarily for water rights, for solar rights. What are some of the advantages or disadvantages of this approach?

2. What elements must a plaintiff show for nuisance? Must a neighbor's use be both intentional and unreasonable?

3. In his dissent, Judge Callow notes "plaintiff's solar heating system is an unusually sensitive use." Perhaps this was true in the late 1970s and early 1980s when solar systems were extremely costly and just becoming available to the public. However, the price of solar panels is decreasing, and as the Colorado Distributed Generation data in the box below illustrates, distributed solar installations are increasing at a dramatic rate in the

---

3. I am amused at the majority's contention that what constitutes a nuisance today would have been accepted without question in earlier times. Slip. op. at 10. This calls to mind the fact that, in early days of travel by horses, the first automobiles were considered nuisances. Later, when automobile travel became developed, the horse became the nuisance. Ellickson, *Alternatives to Zoning: Covenants, Nuisance Rules, and Fines as Land Use Controls*, 40 U.Chi.L.Rev. 681, 731 (1973). This makes me wonder if we are examining the proper nuisance in the case before us. In other words, could it be said that the solar energy user is creating the nuisance when others must conform their homes to accommodate his use? I note that solar panel glare may temporarily blind automobile drivers, reflect into adjacent buildings causing excessive heat, and otherwise irritate neighbors. Certainly in these instances the solar heating system constitutes the nuisance.

Colorado Distributed Generation—Distributed generation (DG) consists of small-scale electric generators such as rooftop or ground-mounted solar photovoltaic (PV) installations. According to the Governor's Energy Office REDI report of December 2009, Colorado has seen explosive growth in DG. In 2005, Colorado had less than 1 MW of installed solar PV; by 2009, that figure had grown to 45 MW of installed solar PV. Public Service Company of Colorado's 2010 REStandard Compliance Plan proposed to add an additional 260 MW of onsite solar PV.

United States. Do you think that solar panels will soon represent the automobile rather than the horse in Judge Callow's analogy in Footnote 3?

4. Consider what arguments might best provide a balance between the development of distributed solar power and a claim that property owners have an absolute right to use the space directly above their lots. Chapter 4 addresses how policy makers shifted the paradigm of absolute property rights in the context of water power by creating statutes that allowed a mill owner to flood upstream lands. One rationale for this shift was that mills provided a "public good." Distributed solar arrays provide a public good in at least two ways. First, by providing power to offset an individual home's demand, they can decrease the load demand on the entire public utility system—eliminating the need to construct more power infrastructure such as generation stations and transmission lines. Second, distributed solar panels can make the grid more secure by contributing excess power to the grid when there is an outage. Can you think of any additional arguments for solar power as a public good?

5. Solar PV represents an exemplary public-private partnership for the public good. Individuals contribute significantly, but the federal government also currently provides tax credits for up to 30 percent of the costs of solar panels. Rebates and credits from local governments and utilities for solar arrays connected to the grid also reflect society's recognition of their benefits to the public. As an example, Xcel Energy invested $212 million for its Colorado Solar Rewards program in just four years (90.4 MW between 2006 and 2010). For more information on public financial incentives, go to: www.dsireusa.org/incentives. Does the public good argument transform the battle over solar access away from a neighbor-against-neighbor tension as it was described in *Prah*?

C. Government Involvement

As renewable energy gains momentum, governments have recognized that leaving things up to the common law may not be an effective strategy. In addition, as we have seen above, governments—from federal to state to local—have provided financial incentives to develop solar, so there is a growing rationale for protecting that investment.

Section 1 will first explain reasons for government involvement and at which level. Next, Section 2 addresses some of the systems governments currently use to promote the use of solar, and Section 3 identifies some of the challenges in deciding what should be protected. Finally, Section 4 sets out a spectrum of legal strategies for promoting solar access illustrating how burden shifting and definitional choices might reflect changing energy supply priorities.

1. Reasons for Government Involvement and at Which Level

Financial incentives to encourage investment in rooftop solar panels have swelled in recent years. These incentives may come from the federal government, from state governments, from local utilities, or a combination of these sources. Aside from these financial incentives, at least thirty-four state and local governments have enacted some form of legislation to encourage the development of solar power. The following excerpt describes some of this growth in solar legislation.

Troy A. Rule, *Shadows on the Cathedral: Solar Access Laws in a Different Light*
2010 U. Ill. L. Rev. 851, 854-58 (2010)

a. Blistering Growth in Solar Energy Development

The U.S. solar energy industry has grown exponentially in recent years. The generating capacity of photovoltaic (PV) solar collector installations installed in the United States in 2008 was triple the amount installed in 2005 and more than ten times the amount installed in 2000, even though no utility-scale solar power plants came online in 2008. Such rampant expansion has been driven primarily by increases in grid-tied PV generating capacity—modest solar collector systems typically installed on the rooftops of homes and businesses. Globally, grid-tied PV generating capacity has increased by 600% since 2004.

Small-scale solar energy is a particularly attractive energy option. Landowners with installed PV systems usually use the solar-generated power on site, lessening the need for costly transmission facilities and reducing transmission-related energy losses. Unlike some other energy strategies, solar power generation does not emit greenhouse gases, threaten protected fish species, or create radioactive waste. Further, solar-generated electricity is most abundantly supplied on hot, sunny days when air conditioners are in use and electricity demands are at their peak.

Recent innovations in PV technology are improving solar panel efficiencies and thereby enhancing the economic viability of rooftop solar installations. Conventional crystalline PV solar panels are modules of small, connected cells comprised of copper, cadmium sulfide, silicon, and other materials. The cells are formulated to facilitate electricity-producing chemical reactions when struck by sunlight. Although most solar panels are still comprised of crystalline PV cells, new "thin film" solar panels are commanding a growing share of the PV market. Some predict that thin film technologies will ultimately make solar energy an economically competitive alternative to fossil fuel-based energy sources, even without government incentives.

Most of the recent increase in rooftop solar installations is not the result of innovation, however, but of generous state and federal government incentives and financing mechanisms aimed at spurring solar energy development. As of July 2009, forty-two states and the District of Columbia had adopted net metering programs that allow utility customers who generate power on their property (typically from small-scale solar or wind devices) to send excess power onto the electric grid and receive a credit on their electricity bills. In 2008, Congress extended the expiration date on the federal thirty percent investment tax credit for residential solar panel expenses and removed a $2,000 cap on the tax credit amount. State and local government programs offering additional cash and tax incentives, providing discounted financing for PV systems, and creating markets for renewable energy

> Virtually all forms of energy have been incentivized at one point or another in the United States. *See, e.g.,* http://tonto.eia.doe.gov/ftproot/service/emeu9202.pdf.

certificates have created further opportunities for cost savings in connection with solar energy systems. Depending on electricity rates, the combination of net metering and various incentives and programs can sometimes allow landowners to recoup the full amount of their investment in a PV system in just twenty years.

b. Renewed Attention to Solar Access Laws

Most state solar access laws existing today were enacted between 1978 and 1981. The oil embargos of the 1970s are credited with having catalyzed a period of legislative and scholarly interest in solar energy development during that period. Commentators at the time were boldly predicting that solar technologies would soon take a central role in U.S. energy policy. Dozens of state legislatures responded by enacting a wide spectrum of innovative solar access laws. When the energy crisis ended, conventional energy prices settled at lower levels, federal solar subsidies disappeared, and attention to solar energy issues faded. Left in the wake of the energy crisis were dozens of new solar access and solar rights statutes, many of which were inconsistent with each other and some of which seemed to deviate from existing law.

Almost thirty years later, another spike in energy prices and unprecedented government support for renewable energy have thrust solar power—and the legal issues associated with it—back into the spotlight. In 2007, the U.S. Department of Energy formed the Solar America Board for Codes and Standards (Solar ABCs) for the purpose of "resolving solar codes and standards issues." Solar ABCs promulgated a report in October of 2008 providing a comprehensive review of U.S. solar access laws, descriptions of "best practices," and a model statute. According to the report, the model statute was "intended to serve initially as a straw man for discussion among stakeholders and will be revised to reflect feedback based upon their needs." Although the model statute and best practices are a useful starting point for a discussion on solar access policy, they perpetuate certain aspects of existing solar access laws that diverge from prevailing common law principles and promote the suboptimal allocation of airspace rights.

Notes and Questions

1. Despite the insights in the *Prah* opinion, few states followed Wisconsin's lead, and *Prah* failed to create the watershed of solar access recognition that many solar proponents hoped. In Footnote 7 of his dissent in *Prah*, Judge Callow states, "I note that the federal government supports the plaintiff's position in the instant case. If solar energy is in the national interest, federal legislation should be enacted." *Prah v. Maretti*, 321 N.W.2d 182, 197 (Wis. 1982). Do you think federal laws are the best way to ensure solar access protection?

2. In the excerpt above, Rule mentions DOE's model solar access statute. Over the years, Congress has considered promoting solar access in a number of other ways. One example is a bill introduced in June of 2009. H.R. 2895, the Solar Opportunity and Local Access Right Act, would have established national net metering and interconnectivity standards for solar and would have prohibited a state or non-regulated utility from adopting or

enforcing any standard or requirement that restricted access to the electric power transmission or local distribution systems. This act would also have directed the Secretary of Housing and Urban Development (1) to issue regulations to prohibit any restriction impairing the ability of the owner or lessee of a one-family residential structure to install or use a solar energy system and (2) to issue regulations requiring that the application for approval of such a system: (a) be processed and approved in the same manner as an application for approval of an architectural modification to the property and (b) not be willfully avoided or delayed. http://www.govtrack.us/congress/billtext.xpd?bill=h111-2895. Although this bill died in committee, what parts do you think would be most likely to pass and what parts would be most problematic?

3. Senator Bernard Sanders of Vermont introduced the Ten Million Solar Roofs Act of 2011 (S. 1108) that would have streamlined permitting processes to help reduce the costs of installed solar PV. The bill was approved by the Senate Energy and Natural Resources Committee but never was considered on the Senate floor. Why do you think this was the result?

4. The approach used to protect solar access will vary depending on the level of government addressing the issue. While the federal government may encourage solar access for projects receiving federal funding, a nationwide mandate might not be appropriate for restricting buildings or vegetation because the way in which these obstructions interfere with sunlight will vary depending on a location's climate and latitude. A state might enact legislation as a matter of property law, but variations within a state might support delegating detailed decisions to local governments. The following subsection addresses some of the state and local mechanisms currently used for regulating solar rights.

2. Current Systems

As the public benefit of protecting solar rights gains increased recognition, some state and local governments have jumped in to enact legislation to modify common law allocations. Varieties of solutions have been tried—permits, zoning, and easement statutes are the most common. This section will examine each of these approaches.

a. Permit-Based Systems

A few state and local governments have permit-based systems. These can vary significantly depending on how they are structured.

Sara C. Bronin, *Solar Rights*
89 B.U. L. Rev. 1217, 1238-42 (2009)

Permits exemplify the benefits and flaws of governmental allocation of solar rights. To issue a permit, a state or local government agency must evaluate applications on a case-by-case basis while at the same time striving for consistency across decisions. A permit system might require several steps: a potential solar user must submit an application; neighbors must be notified and be given time to object and be heard; the relevant level of government must rule to grant or reject the permit; and, if issued, the permit must be registered. Applications must generally include descriptions of the real property on which the solar collector was located, dimensions needed for solar access over real property which would be affected by the right and present and future growth or structures which might interfere with the solar right.

In three states—New Mexico, Wyoming, Wisconsin—and the handful of cities where solar permit systems have been most fully realized, permit applications generally follow this pattern. New Mexico and Wyoming use a prior appropriation (first in time, first in right) approach similar to the approach sometimes used in water law. Both states allow the applicant-owner of a solar collector to attain rights to solar access if the owner used the collector before other uses that may block out such light, and if that use is beneficial. Successful applicants do not "own" the sunlight, but have a right to divert it for a beneficial use.... Once obtained, solar permits in both states, like water permits received through prior appropriation regimes, are freely transferable.

Wisconsin takes a different approach; it incorporates the reasonable use rule on private nuisance from the Restatement (Second) of Torts into the solar permit statute. The municipal agency which administers solar permits can only grant a permit if doing so would not unreasonably interfere with development plans, if no person has made substantial progress toward building a structure which would create an impermissible interference, and if the benefits to the public (including the applicant) will exceed the burdens of the grant. This weighing of the benefits and burdens on parties with competing interests reflects an approach grounded in nuisance law. The law allows permit holders to sue neighbors who interfere with the solar access granted by the permit, whether through vegetation or through construction, with remedies ranging from an injunction, to trimming vegetation, to damages.

Cities have created unique permitting regimes as well. Portland, Oregon has a solar collector permit system that exempts existing vegetation and solar friendly trees. The city of Ashland, Oregon, uses a system of solar access permits to protect solar collectors from shading by vegetation, but not from shading by buildings. Boulder, Colorado has established a permit system that protects existing or proposed solar collectors from being shaded by new construction or by vegetation. All of these municipal permits, like their state counterparts, create novel property rights for solar access.

The public creation of such property rights through permits has several significant flaws. As with other governmental allocations, permit systems require costly new bureaucracies, sometimes at both the state and municipal levels. In addition, they require individualized applications, the submission and review of which impose high costs on government, the applicant, and any affected third parties. Despite creating a time-consuming review process, the outcomes in permit decisions may not satisfy all of the affected parties, leading to claims that the permitting system is unfair. Moreover, because outcomes are unpredictable, benefited and burdened landowners may decline to enter into express agreements ... which would obviate the need for a permit. In the long term, property rights granted by permit might not be recorded on the land records, which could prevent subsequent purchasers (both benefited and burdened) from understanding their rights and duties.

In addition to the administrative challenges created by a permit program, many commentators have expressed concern that solar collector permits over-protect energy uses and thwart real estate development. Historically, American courts and legislatures resisted creating solar rights to avoid impeding development. While any solar rights regime might impede development, government-issued permits are more likely to impede development on a wide scale than express agreements between neighbors. In an urban setting, a solar permit owned by one landowner might prevent another landowner several blocks away from building a skyscraper that would shade the permit holder's property. In such a situation, the builder of the proposed skyscraper might petition the permitting agency for an exemption from the obligations of the permit. The denial of the petition would effectively prevent a skyscraper from being built; on the other hand, a grant of the exemption would erode the value of the permit system as a whole by introducing uncertainty into the entitlement process.

The would-be builder of the skyscraper, and others in similar situations, may have grounds for a takings claim against the permitting agency. Indeed, several scholars have argued that permit statutes, either as written or as applied, unconstitutionally take the property of burdened landowners. A takings challenge might succeed if a burdened property owner could prove that the permit reduced her property's value in violation of established takings precedent and that the government did not compensate her for this reduction....

Notes and Questions

1. The choice between a permit-based or zoning approach may depend on the level of government enacting the solar protection law. Zoning is usually a matter for local governments. If a legislature wishes to create a statewide system, permits may be a more palatable choice.

2. Zoning involves land-use regulation that impacts everyone within a district equally. Although some regulatory taking of private property for public use requires compensation under the Fifth and Fourteenth Amendments, as a general rule, equally-applied zoning regulation survives Takings Clause challenges. *See, e.g.*, *Kucera v. Lizza*, 69 Cal. Rptr. 2d 582, 588 (Cal. Ct. App. 1997) ("The preservation of sunlight has been recognized for nearly 40 years as a valid police-power purpose supporting height limitations.").

3. Do you think a case-by-case permit system for protecting solar access might be more vulnerable to constitutional challenges than a zoning regime? *See, e.g.*, Troy A. Rule, *Shadows on the Cathedral: Solar Access Laws in a Different Light*, 2010 U. ILL. L. REV. 851 (2010); Tawny L. Alvarez, Comment, *Don't Take My Sunshine Away: Right-to-Light and Solar Energy in the Twenty-First Century*, 28 PACE L. REV. 535 (2008); John W. Gergacz, *Legal Aspects of Solar Energy: Limitations on the Zoning Alternative from a Legal and Economic Perspective*, 3 TEMP. ENVTL. L. & TECH. J. 5 (1984).

b. Zoning

A permit system provides individual, lot-by-lot protection. In contrast, zoning provides area-wide protection.

Sara C. Bronin, *Solar Rights*
89 B.U. L. REV. 1217, 1242-50 (2009)

Like permit regimes, zoning ordinances require government decisions on individual applications—decisions that may be criticized on several grounds. Zoning refers to the regulation of uses, lot sizes, building characteristics, and other site features through a local body that has been publicly elected or appointed to uphold the map and text of the zoning ordinance. Localities that choose to address solar access through the enactment and application of their zoning ordinances must find grounds in state law. As a starting point, they may rely on the authority granted by enabling statutes in every state, which authorize them to provide for "safety, morals or general welfare" and "adequate light and air." To make localities' authority more explicit, at least thirteen states authorize localities to zone for solar access. In addition, a few states require that solar access be taken into account when designing zoning ordinances or comprehensive city plans, and various related initiatives have become law. Several other states explicitly prohibit localities from passing ordinances (zoning or otherwise) that would inhibit the operation of solar collectors.... The paucity of state laws relating to solar zoning confirms that more could be done at the state level to encourage this method, however flawed, of providing solar rights.

Whether solar zoning is specifically authorized, a zoning ordinance establishes a baseline from which property owners may request a deviation. Where authorized to do so, localities might zone to protect solar access in two ways—one that builds on the existing baseline, and one that resets the baseline. First, localities may allow solar rights seekers to obtain solar rights through existing processes for variances, special exceptions, and other flexibility rules common to zoning schemes across the United States. Alternatively, in the map and text of the zoning ordinance, or in a special solar zoning ordinance, localities may specify new "solar zones" which define how property owners in such zones may establish solar rights, either as of right or by individual petition.

The first method for establishing solar rights in a zoning scheme requires that a solar rights seeker follow established procedures to request specific relief from the zoning ordinance. To receive a favorable ruling for a variance, special exception, or other flexibility device from the zoning board, an applicant must submit plans for proposed construction and indicate how such plans comport with the zoning ordinance and, if applicable, the comprehensive plan.... The review process may be lengthy and expensive, and the board's ultimate decision may be at odds with either prior decisions or the ordinance itself.... If the zoning board granted her the variance, she would not receive a right to solar access, which is enforceable against others, but merely a right to establish access without the ability to change others' behavior. In other words, her variance would not allow her to prohibit a neighbor from erecting a skyscraper that shades her solar collector....

The second possible means of protecting solar rights—drafting new, comprehensive solar zoning provisions, or in other words, resetting the baseline—better serves solar rights seekers because such provisions may govern all properties within a neighborhood or neighborhoods, a situation which renders solar rights enforceable (at least in part) against others. At their most basic, such solar zoning ordinances could limit heights, restrict lot sizes, establish setback requirements (perhaps expanding setbacks for southern exposures—the preferred orientation for solar collectors in this country), and create other rules that would facilitate solar access. A more detailed ordinance might create an overlay zone to the zoning map or otherwise designate particular blocks as "solar blocks" and mandate solar access rights for parcels within that block. Perhaps the most sophisticated solar zoning ordinance in this country governs construction in Boulder, Colorado, which has created a system of "solar envelopes" and "solar fences," each of which function differently in different neighborhoods. The solar envelope, similar in concept to the solar skyspace, delineates a three-dimensional space over a parcel beyond which no construction or vegetation can occur without illegally interfering with the solar rights of neighbors. The solar fence represents a vertical plane along a property line that casts an imaginary shadow that cannot be exceeded in length by the shadows cast by any building or tree on the neighboring property. The Boulder solar ordinance divides the city into three zones, governed by area-wide rules establishing various solar envelope, solar fence, and other requirements. Commentators have lauded the envelope and fence elements of the Boulder system.

Boulder notwithstanding, local government experiments with solar zoning ordinances remain few and far between. Local governments may resist changing zoning ordinances because change requires money, time, expertise, and political capital which local officials may be unwilling to spend. However, because zoning occurs at the local level, zoning officials can enable solar access in a manner that responds to extant topography, vegetation, land uses, density, and building types. Moreover, unlike a statewide solar permit system, which would have to be created afresh, the boards, staff, and other administrative structures for a solar zoning ordinance already exist.... [A]ll zoning decisions are public documents and, especially if recorded on the land records, provide notice of solar access rights to third parties. Finally, zoning, if properly crafted, will likely avoid takings, equal protection, and due process challenges to which other systems (such as permits) may be subject.

Nonetheless, zoning presents concerns that cannot be overlooked. With respect to transaction costs, zoning applications and submissions consume time—with months required for appeals and public hearings, where applicable. Applicants with difficult cases may find themselves mired in bureaucracy.... Comprehensive ordinances that create building envelopes that enable the passage of light by segregating structures on individual large lots may, in effect, mandate sprawl ... a far worse problem arguably, than the low rate of solar collector utilization. Solar zoning should not rely on a large-lot solution.

In addition to these problems, solar zoning may raise fairness concerns. The text of a solar zoning ordinance may not account for variations in site conditions across the properties under its jurisdiction. Solar envelopes, for example, may be difficult to define on irregularly shaped parcels or in hilly areas, a situation that may lead to unequal application of the rules. Enforcement may also be arbitrary when zoning boards modify their interpretations of the zoning ordinance from case to case. In addition, the failure to compensate burdened parties may create severe inequities among landowners, and could also subject solar zoning ordinances to takings clause challenges....

Finally, zoning does not create a true vested property right. Even if zoning ordinances change relatively infrequently, changes to the scope of solar rights in the ordinance which are not accompanied by an exemption for nonconforming uses may mean that a property owner who used a solar collector under a previous ordinance must dismantle or otherwise modify her solar installation. Because it does not provide an enduring, secure property right, zoning is among the least effective means of securing solar access.

Notes and Questions

1. The City of Boulder's solar access code was drafted in 1991. The key drafters—Susan Osborne, (who was a city planner at the time the ordinance was drafted, and who became Mayor of Boulder in 2009), Alan Boles (an Assistant City Attorney for Boulder at the time he helped draft the plan), and Luke Danielson (currently an adjunct professor for the University of Denver Sturm College of Law)—believe it has been very successful. (Professor DuVivier has interviewed each of them.) *See also* Luke J. Danielson, *Drafting a Solar Access Ordinance: One City's Experience*, 3 Solar L. Rep. 911 (March/April 1982).

2. The City of Ashland, Oregon, has a comprehensive solar access zoning code that varies from Boulder's approach. John Fregonese and Gail Boyer Hayes collaborated in its creation and presented it as a model solar code in Gail Boyer Hayes, Solar Access Law (1979). Instead of employing a solar fence like Boulder, the Ashland ordinance establishes solar setbacks based on lot classifications as well as on northern lot line and slope calculations.

3. Assume you were drafting a solar access ordinance. Compare a lot-by-lot approach to area-wide protections. Which might best allow adaptation to specific circumstances? Which would be administratively easier to understand and apply? Which one would

Figure 2.5: Impact of setbacks and rooflines on solar access

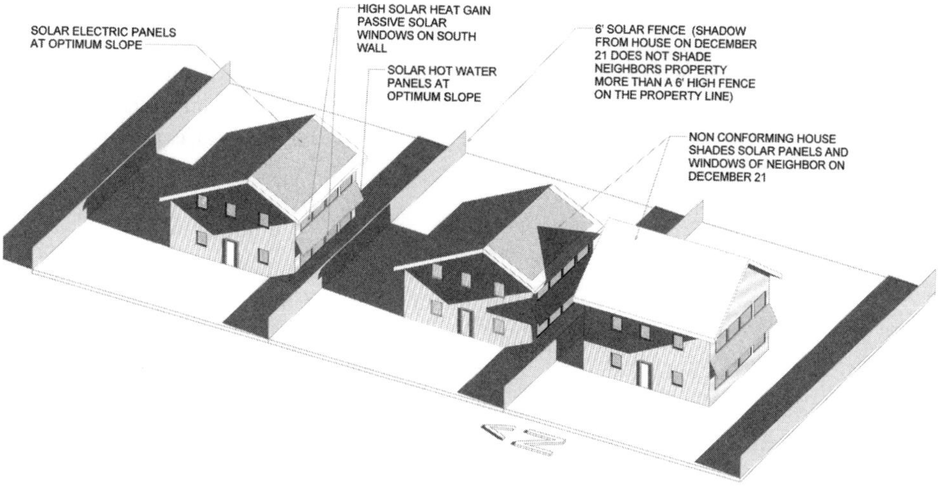

Image courtesy of Fuentesdesign.com

require a mechanism for keeping track of how many permits are issued, what property is affected, and the extent of the restrictions? Which provides more equal benefits and burdens and is more politically palatable? Which protects potential solar uses as well as current ones? Which provides incidental benefits—such as the aesthetic amenity of sunshine—in addition to solar energy uses?

4. As an alternative to zoning codes such as Boulder's or Ashland's, which set out explicit solar envelope protections, some municipalities have simply defined their building bulk planes so that the building to the south will have less impact on the building to the north. As an example, the City and County of Denver enacted a solar bulk plane ordinance in 1979 that provided for an asymmetrical roofline to protect the neighbor's access. In 2002, Denver replaced this solar bulk plane with one that provides only limited rooftop protection and is less favorable for solar rights than previous versions of its code. Figure 2.5 illustrates how building placement and rooflines can protect or obstruct a neighbor's use of solar resources.

5. In practice, both the Boulder and the Ashland systems have evolved from being universal zoning mechanisms to hybrid zoning and permit systems because in each situation, special permits are required for protection from tree shading. This issue will be discussed further in Section D of this chapter.

c. Legislative Solar Easements

A third way governments have become involved in solar allocation is through legislating a right for parties to enter into solar easement contracts.

Sara C. Bronin, *Solar Rights*
89 B.U. L. Rev. 1217, 1227-31 (2009)

... At least twenty-eight state statutes allow the creation and recording of express easements for solar access by private landowners. [These states include Alaska, California,

Colorado, Florida, Georgia, Idaho, Illinois, Iowa, Kansas, Kentucky, Maine, Maryland, Minnesota, Missouri, Montana, Nebraska, Nevada, New Hampshire, New Jersey, North Dakota, Ohio, Oregon, Rhode Island, Tennessee, Utah, Virginia, Washington, Wisconsin. Footnote 28 of the Bronin article provides statutory citations—Ed.] Solar easement statutes do not themselves create easements, but allow private entities and political subdivisions to create them. The majority of states require such easements to be in writing and contain detailed information about the size of the affected space, the manner of termination, and compensation. In most jurisdictions, the easement must also be recorded on the land records, to provide notice to individuals researching the dominant or servient estates.

* * *

... [The] costs may increase the already-high cost of solar energy systems and make them less attractive than cheaper forms of energy.

At least some of these costs stem from the assignment of initial entitlements; in the vast majority of jurisdictions, the initial entitlement rests with the potential obstructer, or the potential servient owner. The potential obstructer may never agree to an easement; even if she does, she has the power to set a high price on the easement. Because express easements often involve bilateral monopolies, an individual party can hold out or demand exorbitant compensation if she does not want to give up her entitlement. The assignment of the entitlement thus inhibits greater use of solar collectors.

One state, Iowa, assigns the initial entitlement in solar easements in a way that avoids at least some transaction costs. Like other states, Iowa allows users to create solar easements voluntarily. When a potential obstructer holds out, however, Iowa authorizes local regulatory boards to create easements without the burdened landowner's consent, provided that the burdened landowner receives just compensation. Local legislative bodies may establish "solar access regulatory boards" which govern applications for solar easements. An applicant must submit a statement of need, the legal description of the estates, a description of the solar collector, an explanation of the application's reasonableness, and a statement that the applicant has attempted to negotiate an easement. The law requires the review board to grant compensation for burdened property owners "based on the difference between the fair market value of the property prior to and after granting the solar access easement." Anecdotally, the statute has encouraged voluntary agreements. The Iowa approach reflects a sensible statutory solution to the holdout problem.

Notes and Questions

1. A state's recognition of an easement might be helpful. In what ways, if any, does an easement promote solar rights beyond a common law contract between neighbors?

2. Does having an easement statute change the bargaining positions of the parties or provide any additional incentive for a southern property owner to enter into such an agreement making solar agreements more likely?

3. Check your state's statute carefully. While these statutes may provide recognition for solar easements, some also have specific requirements for describing the solar right in writing, and some have eliminated common law remedies such as prescriptive easements

4. Iowa's solar easement statute is distinctive and will be discussed in more depth in Section C.4.b.i. below.

3. Deciding What to Protect

Significant public and private investment dollars are lost when existing solar thermal, solar PV, and passive solar systems are functioning below their potential or rendered useless because their access to the sun—their source of fuel—is obstructed. Yet in densely populated urban areas, this desire to access the sun can impact neighboring properties.

To clarify the discussion, let's first set out a few fundamentals. Current U.S. common law under the *ad coelum* doctrine recognizes a property owner's "exclusive control" of the airspace 90° up from the property lines on the surface to the height the surface owner "*can* occupy or use in connection with the land...." *United States v. Causby*, 328 U.S. 256, 265 (1946) (emphasis added).

Anywhere other than near the equator, the sun does not enter property at 90°. *See* Figure 2.6. Instead, as one moves further north of the equator, the angle of the sun in the winter, when it is needed most for passive heating or for PV electricity generation for heating or lighting, is steeply lower in the sky. Consequently, for solar access purposes, it is important that the sun remain unobstructed both over the surface of any property hosting a panel—labeled Solar Skyspace A—and over the unoccupied portion of a property to the south—labeled Solar Skyspace B in Figure 2.6.

Figure 2.6: Solar Skyspace B

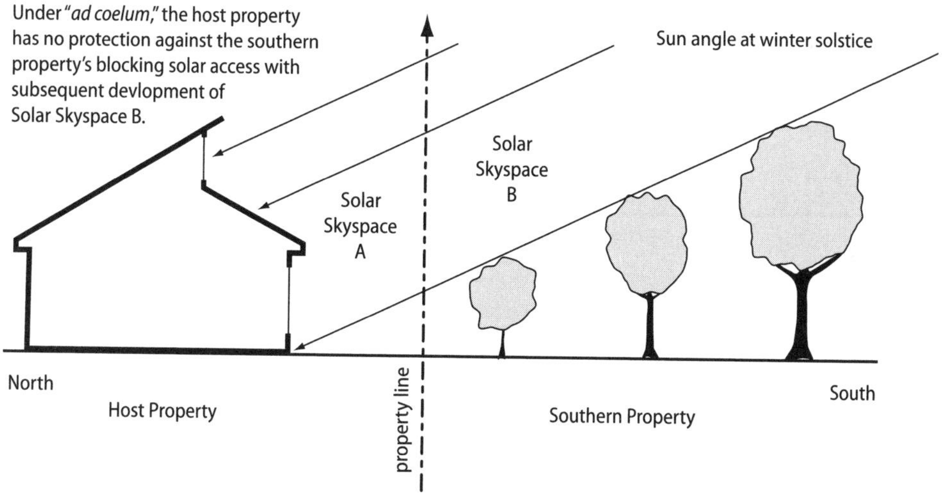

Notice the words "subsequent development" in Figure 2.6. Currently, few would suggest the elimination of existing structures or trees to create solar rights. Instead, in many instances, a utility will require an initial threshold of solar exposure before incentivizing installation of solar panels. The discussion here is focused on the status quo at the time solar rights are claimed. The word "can," highlighted in the quotation from the *Causby* case above, suggests that the common law currently protects the southern property owner's *possible future use* of Solar Skyspace B for any purpose over the host property's current use of the sun for energy purposes.

Society often reflects its priorities by determining which party bears the costs or will have the burden of proof in enforcing a right. This section provides a model for evaluating competing airspace rights.

Troy A. Rule, *Shadows on the Cathedral: Solar Access Laws in a Different Light*
2010 U. Ill. L. Rev. 851, 858-61 (2010)

The disparities among existing solar access laws and shortcomings of these laws are clearer when viewed within the "Cathedral Model"—a framework of entitlements, property rules, and liability rules set forth by Guido Calabresi and A. Douglas Melamed in 1972. The Cathedral Model can be a valuable device for comparing and analyzing resource allocation rules. Applying the Cathedral Model involves determining (1) which party should hold the scarce legal "entitlement" at issue and (2) whether to protect the entitlement with a "property rule" or a "liability rule." An entitlement is protected with a property rule if "other parties wishing to acquire the entitlement from its holder can do so only by purchasing it in a voluntary transaction at a price acceptable to" its holder. An entitlement is protected with a liability rule if a party other than the entitlement holder has a right to purchase it at a price equal to its objective value as determined by a (usually governmental) third party.

* * *

The Cathedral Model can be more easily understood through a simple example. Suppose that one party (polluter) discharges pollution into the air that causes injury to other parties (victims). Applying the Cathedral Model to the parties' conflict would involve first determining whether the polluter should be entitled to pollute or whether the victims should be entitled to pollution-free air. Once the entitlement is assigned, one must then determine whether to protect the entitlement by a property rule or a liability rule. Applying these two steps yields a total of four possible rules, conventionally enumerated as follows:

Rule One: The victims are entitled to pollution-free air, and their entitlement is protected by a property rule (the victims can obtain an injunction stopping the pollution without having to compensate the polluter);

Rule Two: The victims are entitled to pollution-free air, and their entitlement is protected by a liability rule (the victims are entitled compensatory damages from the polluter but cannot obtain an injunction stopping the pollution);

Rule Three: The polluter is entitled to pollute, and her entitlement is protected by a property rule (the victims can neither obtain an injunction stopping the pollution nor claim damages); and

Rule Four: The polluter is entitled to pollute, and her entitlement is protected by a liability rule (the victims have the right to purchase an injunction by paying the polluter its costs of stopping the pollution).

* * *

The Cathedral Model can be easily applied to the problem of solar access. A landowner whose trees or structures shade solar collectors on neighboring property is analogous to a polluter. A landowner whose solar collectors are shaded by a neighbor is a victim. The unwanted shade that damages solar collectors' productivity is analogous to pollution. Figure A is a table describing the four possible Cathedral Model rules in the solar access context. Interestingly, various state statutes currently in force correspond to all four rules.

Figure A

	Property Rule	Liability Rule
Entitlement to Solar User ("S")	Rule One: S may preclude N from shading S's solar panels (See statutes in New Mexico, Wyoming, Wisconsin (vegetation), and Massachusetts)	Rule Two: S is entitled to damages from N for the reduced productivity of S's solar panels caused by N's shading (See statutes in California (vegetation) and Wisconsin (structures))
Entitlement to Neighboring Airspace Owner ("N")	Rule Three: S has no claim against N for an injunction or for damages (See Fontainebleau; current law in most states)	Rule Four: S has a right to purchase an injunction or easement preventing N from having structures or trees on N's property that shade S's solar panels. (Current law in Iowa)

The Coase Theorem suggests that, if transaction costs are sufficiently low, any of the four rules described above would generate an efficient outcome because all four rules assign competing airspace rights to either one of the two parties and protect them by some legal rule. However, as Figure A shows, current solar access laws vary greatly across jurisdictions. This disparate statutory treatment and the infrequency of solar access cases can create uncertainty as to whom the rights are legally assigned. Even in jurisdictions where assignment of the entitlement is clear, the potential transaction costs of neighbor negotiations are too great for policymakers to expect Coasean bargaining to consistently and efficiently allocate competing airspace rights. A more careful examination of the unique attributes of solar access conflicts is thus required to determine which of the four Cathedral Model rules best promotes the efficient allocation of competing airspace rights.

Notes and Questions

1. Rule's analysis starts with consideration of who should be the "entitlement holder" for Solar Skyspace B. He argues, "Unlike water, oil, gas, or minerals, sunlight is not sufficiently 'scarce' to warrant property right protection." Rule, *supra*, at 861 (citing n. 66). What are some reasons to consider solar rights like water, oil, gas, or minerals? What are some reasons for distinguishing them? Does scarcity at a particular location—*i.e.*, only what is available in the solar skyspace surrounding one's solar panels—create sufficient scarcity to warrant property-right protection?

2. Instead of treating solar rights as property rights, Rule notes that "[b]ecause solar access conflicts are ultimately disputes over use of airspace, not sunlight, the entitlement in a Cathedral Model analysis of these conflicts must be defined accordingly. Should landowners who have installed or seek to install solar panels on their property (Solar Users) be legally entitled to an easement or other restriction across their neighbor's airspace to protect solar access? Or, should owners of properties near a Solar User (Neighbors) be entitled to exercise rights in the airspace above their property without liability for shading

nearby solar collectors?" *Id.* at 862. Consider the implications of Rule's distinction between solar access as an easement right as opposed to a right in property in the airspace above a neighboring lot.

3. Columbia Law School Professor Thomas W. Merrill posits that *Melms v. Pabst Brewing Company*, a Wisconsin Supreme Court opinion near the turn of the nineteenth century, sets up the property theorists' debate whether property is an individual right, existing primarily to protect the subjective expectations of owners, or whether it is a social institution, functioning to maximize the value that society ascribes to specific things. How might such a debate relate to the question of property rights to Solar Skyspace B?

4. Rule Three on Rule's chart illustrates the common law approach in most states. A few of the governmental solutions enacted to date shift the balance and give a solar user some claim to solar rights. Instead, the burden in most cases is still on the solar user to seek out a permit or enforce such a claim. The following section will address the spectrum of solar protection strategies and the impacts of shifting burdens.

4. Spectrum of Strategies

While the mechanism for protecting a solar right may vary—*e.g.*, an easement, a zoning ordinance, or a permit—perhaps more important is the definition of what is protected and the degree of protection. The Spectrum of Legal Strategies for Promoting Solar Access chart in Figure 2.7 shows how various government schemes promote the use of solar energy. Off the chart completely would be doing nothing, *i.e.*, allowing the common law to drive the development of solar energy systems in its haphazard way—looking back to precedents that did not take into consideration the technologies and policy considerations that might control today. We discussed these common law approaches above in Section B of this chapter.

At the least protective end of the spectrum are laws that remove impediments to installation of solar devices but that do nothing to truly encourage investment in solar energy because they do not provide protection for actual use of those devices. These

Figure 2.7: Spectrum of legal strategies for promoting solar power

Strongest Protections
- **Bulk plane** protections up to full solar skyspace.
- Restrictions on **vegetation**.
- Protection of solar skyspace through **(1) zoning or (2) permit systems that establish affirmative right** to access (burden on southern property to compensate if seeking change of status quo once panels are installed).

Middle Ground
- **Some rights, but burden on host property**, *i.e.*, Iowa statute gives host right to force easement on southern neighbor, but host must initiate action and compensate neighbor.

"Cheerleading"
- Grant the right to create **easements**.
- Grant **exceptions** to limiting provisions (*e.g.*, accessory structure limits, historic district regulations).
- **Prohibit** solar **restrictions**.

For a list of statutes, Rules, Regulations & Policies for Renewable Energy, Database of State Incentives for Renewables & Efficiency (DSIRE):
http://www.dsireusa.org/summarytables/rrpre.cfm.

For a good website for finding sample solar access codes nationwide:
http://en.openei.org/wiki/Solar_Access_Law/Guideline

"cheerleading" laws will be discussed in Section *a*. Section *b* will consider two middle-ground options that provide some protection but place the burdens for enforcing those protections almost exclusively on the developers of new sources of solar power for the grid, whether it is community solar garden groups, utilities, or individual users. Section *c* will flush out some of the strongest protection regimes.

a. Legislative Cheerleading

Because defining solar rights requires a delicate balancing act, legislators have been hesitant to make some of the tough choices. Instead, they enact statutes that appear to support solar, but in fact are nothing more than an "inexpensive form of legislative cheerleading." DONALD N. ZILLMAN, COMMON-LAW DOCTRINES AND SOLAR ENERGY, IN LEGAL ASPECTS OF SOLAR ENERGY, 32 (John H. Minian & William H. Lawrence, eds., 1981). Most solar easement and solar prohibition statutes might fit into the cheerleading category. They are an improvement over the common law because they recognize solar rights and cut back on some impediments, but they do not go to the next level of actually promoting solar uses.

A slight step up from simple cheerleading is a statute that eliminates restrictions on solar development. At least twenty-one states and some local governing bodies have laws voiding restrictive covenants that prohibit the installation of solar devices or interfere in other ways with solar access. Although these laws vary widely, they do provide some protection especially in areas with new developments.

Note that switching from an electric or gas clothes drier to the sun can save significant amounts of energy. Yet, many local governments and neighborhood associations ban clotheslines for aesthetic reasons. Consequently, some states have prohibited bans on clotheslines as another form of solar access law. *See, e.g.*, Florida Statutes § 163.04.

The following is an excerpt from one of the best summaries of our nation's current solar access laws.

Colleen McCann Kettles, *A Comprehensive Review of Solar Access Law in the United States: Suggested Standards for a Model Statute and Ordinance*
SOLAR AM. BD. FOR CODES AND STANDARDS at 6-8 (2008)

There are essentially two models that have perpetuated over the last two-plus decades that attempt to protect the right of homeowners to install solar energy systems. The first model addresses local government ordinances; the second model addresses private land use restrictions, such as covenants, conditions, and restrictions in deeds, as well as declarations in condominiums documents. Some states address both.

The typical language of a statute that protects solar rights at the state or local government level will contain language such as, "The adoption of an ordinance by a governing body which prohibits or has the effect of prohibiting the installation of solar collectors is expressly prohibited." The typical language of a statute that protects solar rights in the context of private land use restrictions is, "Any covenant, restriction, or condition contained in any deed, contract, security agreement, or other instrument affecting the transfer or sale of or any interest in real property which effectively prohibits the installation or use of a solar energy device is void and unenforceable." Some states distinguish their laws from others by defining solar energy device, providing or prohibiting retroactive effect, defining "effectively prohibiting" (usually by assigning a cost of compliance with a re-quirement). For the most part, the laws apply strictly to residential buildings, including condominiums.

* * *

Previous work has identified some of the shortcomings of traditional solar access laws (Starrs, Nelson, & Zalcman, 1999). The lack of awareness and understanding of solar rights statutes is one of the biggest obstacles to enforcement. The lack of awareness by homeowner associations and architectural review boards can lead to delays in processing applications and lawsuits that are expensive to defend and cost all parties, regardless of who prevails. Because, when a solar rights law awards the court costs and attorney fees to the prevailing party, and the homeowner is the prevailing party, they still end up paying since all homeowners in the community bear the common expenses, such as attorney fees. The lack of understanding of solar rights laws by homeowners and solar contractors can lead to missteps in the approval process. Most solar rights laws are not absolute; they still require that the homeowner apply to the architectural review board for approval, and the board has a degree of discretion in the approval process. Many homeowners and con-tractors believe that approval is not required and proceed with the installation without prior approval. This can lead to legal recourse by the association that has no bearing on the solar rights laws but rather pertain to the failure to follow association rules.

The following cases are examples of real events and represent the range of scenarios that occur on a daily basis.

Case 1: A homeowner purchases a solar energy system. The contractor arrives on site for installation. As neighbors notice the activity, they confront the homeowner and inquire as to the architectural review board's approval. The neighbor cites the solar rights law and says permission is not necessary. The association advises the homeowner to cease and desist work and to restore the premises to its original condition and levies a fine for every day they are in violation.

Case 2: A homeowner purchases a solar energy system. Approval from the architectural review board (ARB) is pending. The contractor applies for a permit from the local building agency, which refuses to issue the permit until a copy of the ARB approval is received. Alternately, the ARB requires a copy of the permit before approval is granted. The building permit process is so cumbersome, the contractor does not pull a permit, and ARB approval is denied.

Case 3: A homeowner considers purchase of a solar energy system. Deed restrictions require that the system not be visible from the street. The homeowner has a corner lot, and the only area not visible from the street faces north. The contractor devises a reverse mount for the collectors and runs afoul of local wind and structural codes.

Case 4: A homeowner/condominium association owns the exterior of the residence including the roof (common property). The request to install the solar energy system is denied, as they fear the roof warranty being voided, and question the liability for any damage to common property.

Case 5: A homeowner installs a solar energy system. A neighbor to the south has several very mature trees that are creeping into the solar window. The homeowner asks the neighbor to trim the trees, but the neighbor refuses, arguing that the shade of the trees reduces their air-conditioning load.

Case 6: A developer builds all homes in the community with a solar water heater and photovoltaic system. The solar window requires that a tree protected by the local landscape ordinance be removed. The developer is required to purchase and replant $20,000 [worth of] trees to compensate for the removal of the protected tree.

These are just a handful of the cases, all of which occurred in states with solar rights and solar access laws. The bottom line is that the law failed to protect the solar owner or cost the solar owner more than the value of the solar energy system to secure that protection.

Notes and Questions

1. Consider each of the real-case scenarios set out in the Kettles article. In what ways could solar installation be better promoted and conflict avoided?

2. Case 5 addresses the particularly thorny problem of protecting solar access from vegetation discussed more fully in Section D of this chapter.

3. One additional step for promoting solar that some state legislatures have proposed, and which has some federal support as well, is legislation to prevent state and local government agencies from charging excessive permit fees or plan review fees to customers who are installing solar energy systems. What do you think of such proposals and where would you place them on the spectrum of legal strategies for promoting solar power?

b. Middle-Ground Protections

Some state and local laws have moved up from cheerleading and beyond common law suits. What characterizes the strategies in this middle ground of the spectrum is that they provide some process for protecting solar rights, but as currently configured, all of the burdens for doing so—both legal and financial—fall on the host property owners. Two mechanisms will illustrate this point: Section i sets out Iowa's unique solar easement statute; Section ii describes the current state of California's Solar Shade Protection Act.

i. Iowa's Solar Easement Statute

In some ways, states that have enacted statutes recognizing solar easements have taken a step forward for solar rights. However, in another way, they are not providing any true support because they leave all of the negotiation up to the solar owner.

In most contract situations, there are some choices about the party with whom we will negotiate. Even in road easement situations, a landowner is rarely encircled by a single neighbor and may have the option of bringing a road in from directions other than the south. In contrast, solar users are usually captives to a single southern neighbor, and that sole neighbor may be demanding prohibitive sums or alternatively, refusing entirely to grant any easement.

Only the state of Iowa has an easement statute that shifts the balance to improve the chances of a fair negotiation and move this option into the middle ground. Below is the specific language of Iowa's interesting alternative.

Application for Solar Access Easement
Iowa Code § 564A.4 (2010)

1. An owner of property may apply to the solar access regulatory board designated under section 564A.3 for an order granting a solar access easement. The application must be filed before installation or construction of the solar collector. The application shall state the following:

a. A statement of the need for the solar access easement by the owner of the dominant estate.

b. A legal description of the dominant and servient estates.

c. The name and address of the dominant and servient estate owners of record.

d. A description of the solar collector to be used.

e. The size and location of the collector, including heights, its orientation with respect to south, and its slope from the horizontal shown either by drawings or in words.

f. An explanation of how the applicant has done everything reasonable, taking cost and efficiency into account, to design and locate the collector in a manner to minimize the impact on development of servient estates.

g. A legal description of the solar access easement which is sought and a drawing that is a spatial representation of the area of the servient estate burdened by the easement illustrating the degrees of the vertical and horizontal angles through which the easement extends over the burdened property and the points from which those angles are measured.

h. A statement that the applicant has attempted to voluntarily negotiate a solar access easement with the owner of the servient estate and has been unsuccessful in obtaining the easement voluntarily.

i. A statement that the space to be burdened by the solar access easement is not obstructed at the time of filing of the application by anything other than vegetation that would shade the solar collector.

2. Upon receipt of the application the solar access regulatory board shall determine whether the application is complete and contains the information required under subsection 1. The board may return an application for correction of any deficiencies. Upon acceptance of an application the board shall schedule a hearing. The board shall cause a copy of the application and a notice of the hearing to be served upon the owners of the servient estates in the manner provided for service of original notice and at least twenty days prior to the date of the hearing. The notice shall state that the solar access regulatory board will determine whether and to what extent a solar access easement will be granted, that the board will determine the compensation that may be awarded to the servient estate owner if the solar access easement is granted and that the servient estate owner has the right to contest the application before the board.

3. The applicant shall pay all costs incurred by the solar access regulatory board in copying and mailing the application and notice.

4. An application for a solar access easement submitted to the district court acting as the solar access regulatory board under this chapter is not subject to the small claims procedures under chapter 631.

Notes and Questions

1. At least one author has urged that Iowa's forced-easement statute is the best solution for solar access. Troy A. Rule, *Shadows on the Cathedral: Solar Access Laws in a Different Light*, 2010 U. ILL. L. REV. 851 (2010). As we saw above, state recognition of solar easements might do little to promote solar rights beyond what is possible under common law contract rules. How is Iowa's statute different?

2. What factors would you urge a solar access regulatory board to consider in determining compensation due to a neighbor whose right to build a structure is restricted? What factors should have weight in compensating for restrictions on planting vegetation that obstructs solar panels? How can you balance the value of these restrictions without likewise making the easement's cost prohibitive for the solar user?

3. Note that the Iowa statute describes the right recognized as an easement. In what ways is it similar to or different from a permit system?

4. The decision-maker under Iowa's solar easement statute is "the solar access regulatory board designated under [Iowa Code Annotated] section 564A.3." Section 564A.3 relies on existing entities, such as city council, the county board of supervisors, or even the district court to act as the "solar access regulatory board." Many court systems have specialty courts (such as juvenile, probate, or drug courts) for matters that require expertise and consistency of decisions. What arguments might you make for or against establishing a decision-making process in the solar arena that employed specialists?

ii. California's Solar Shade Control Act

"Shade pollution" can reduce the efficiency of any solar energy device, but it is particularly problematic for solar PV. According to a report from the National Renewable Energy Laboratory, because of circuitry in a typical crystalline silicon solar panel, "a shadow can represent a reduction in power over thirty times its physical size." CHRIS DELINE, NREL, CP-520-46001, PARTIALLY SHADED OPERATION OF A GRID-TIED PV SYSTEM (June 2009).

For thirty years, California had one of the strongest statewide solar shade protection statutes in the United States. As originally enacted in 1978, § 25983 of the California Solar Shade Control Act defined a violation as "an infraction" that would be enforced by a city attorney or district attorney as a public nuisance.

The only case ever brought under this section was in 2004, when a southern property owner refused to trim the Giant Redwoods that he had planted in a residential area and that were shading solar panels on the property to the north. When a court found the southern property owner guilty of an infraction and forced him to trim or remove his trees, he complained to California State Senator Joe Simitian about the statute and about the $37,000 in legal fees he had to pay. The California State Assembly passed Senator Simitian's bill to amend the Solar Shade Control Act, SB1399, and Governor Schwarzenegger signed it into law in the spring of 2008. Below are excerpts from the current version of the statute.

California Solar Shade Control Act
CAL. PUB. RES. CODE §§ 25980-82 (2011)

§ 25980. This chapter shall be known and may be cited as the Solar Shade Control Act. It is the policy of the state to promote all feasible means of energy conservation and all feasible uses of alternative energy supply sources. In particular, the state encourages the planting and maintenance of trees and shrubs to create shading, moderate outdoor temperatures, and provide various economic and aesthetic benefits. However, there are certain situations in which the need for widespread use of alternative energy devices, such as solar collectors, requires specific and limited controls on trees and shrubs.

§ 25981. (a) As used in this chapter, "solar collector" means a fixed device, structure, or part of a device or structure, on the roof of a building, that is used primarily to transform solar energy into thermal, chemical, or electrical energy. The solar collector shall be used as part of a system that makes use of solar energy for any or all of the following purposes:

(1) Water heating.

(2) Space heating or cooling.

(3) Power generation.

(b) Notwithstanding subdivision (a), for the purpose of this chapter, "solar collector" includes a fixed device, structure, or part of a device or structure that is used primarily to transform solar energy into thermal, chemical, or electrical energy and that is installed on the ground because a solar collector cannot be installed on the roof of the building receiving the energy due to inappropriate roofing material, slope of the roof, structural shading, or orientation of the building.

(c) For the purposes of this chapter, "solar collector" does not include a solar collector that is designed and intended to offset more than the building's electricity demand.

(d) For purposes of this chapter, the location of a solar collector is required to comply with the local building and setback regulations, and to be set back not less than five feet from the property line, and not less than 10 feet above the ground. A solar collector may be less than 10 feet in height only if, in addition to the five-foot setback, the solar collector is set back three times the amount lowered.

§ 25982. After the installation of a solar collector, a person owning or in control of another property shall not allow a tree or shrub to be placed or, if placed, to grow on that property so as to cast a shadow greater than 10 percent of the collector absorption area upon that solar collector surface at any one time between the hours of 10 a.m. and 2 p.m., local standard time.

* * *

§ 25983. A tree or shrub that is maintained in violation of Section 25982 is a private nuisance. . . .

§ 25984. This chapter does not apply to any of the following:

(a) A tree or shrub planted prior to the installation of a solar collector.

* * *

Notes and Questions

1. How does the amendment of § 25983, shifting the California statute from a public nuisance to a private nuisance model, change the dynamic of enforcement? Considering the significant public investment in many solar PV panels, should the burden of protecting that investment fall entirely on an individual host property owner?

2. How would a lawsuit under the current California Solar Shade Control Act diverge from a common law negligence suit as discussed in Section B.2 above? Even if the statute's definition of 10 percent shading as a nuisance might simplify the case, which party would have both the burden of bringing the case and the burden of proof in a trial? If a standard 3 kW solar PV system costs less than $37,000, wouldn't it be easier to write off the system than for a host property owner to put up an additional $37,000 or more in court and attorney's fees with no guarantee of prevailing in the lawsuit?

3. Unlike the Iowa forced-easement statute that requires a southern property owner to negotiate with a solar host, the revised California Shade Control Act tilts the balance back to the common law nuisance standard. If the southern property owner in the California case was willing to risk criminal charges and the cost of attorney's fees when the California Solar Shade Control Act established stronger protections, what incentives are there for current southern property owners to work things out with solar host properties to their north?

4. Look at the other provisions of the California Solar Shade Control Act. How does California define what types of solar energy are covered by the statute? Is passive solar covered in addition to solar PV? Is the full skyspace protected or just portions of it at certain times of day? How these definitions are crafted can reflect legislative policy decisions about the value of Solar Skyspace B for the production of solar energy as a public good. Can they also balance the public value against a southern neighbor's interest in landscaping or other personal uses of that airspace?

c. Strongest Protections

The "strongest protections" column of the spectrum chart is distinguished because each of the strategies listed provides substantial, affirmative protection for solar rights without significant additional burdens on those who are developing solar energy.

Addressing solar rights with these affirmative laws educates property owners about the interconnectedness of interests in Solar Skyspace B. For decades, smokers had no sensitivity about how their second-hand smoke impacted others. Similarly, many southern property owners have no sensitivity about how their placement or choice of a particular tree can detrimentally impact the energy equation for the entire grid and how simple some accommodations might be.

The strongest-protection laws are also visionary because they protect solar rights for future generations. The continuing strain on world energy resources will likely make options for solar PV or passive solar critical components for valuing future properties. Those without solar access may be substandard shade ghettos. Trees planted or buildings constructed today without sensitivity toward their shading impacts may create consequences that are costly or impossible to remedy for decades.

The spectrum chart includes three categories of strongest protections: bulk planes, restrictions on vegetation, and the establishment of affirmative rights through zoning or permit systems. Bulk planes were addressed above in Note 4 of Section C.2.b. and Figure 2.5. Vegetation issues will be addressed in Section D below. This section will, therefore, provide first an example of a successful solar access zoning regime and, second, some information about permitting systems.

i. Zoning Systems

Builders have never had an absolute right to construct structures to maximum square footages. Zoning codes have long mandated building design, setbacks, and other restrictions on new developments. Consistent area-wide rules also allow a balancing of property priorities without selectively disadvantaging particular southern property surface owners.

Protecting existing solar installations from new buildings or vegetation blockages maintains the status quo and prevents an infringement on the solar investment. Zoning and permit systems also provide a more cost-effective system for society than litigation to balance an absolute right of solar access with competing interests. The City of Boulder, Colorado, has had a zoning code that protects solar access since 1979 and that can serve as a model for other cities. Below are some of its key provisions.

Boulder Revised Code
Section 9-9-17 Solar Access

(a) Purpose: Solar heating and cooling of buildings, solar heated hot water, and solar generated electricity can provide a significant contribution to the city's energy supply. It is the purpose of this section to regulate structures and vegetation on property, including city-owned and controlled property, to the extent necessary to ensure access to solar energy, by reasonably regulating the interests of neighboring property holders within the city.

* * *

(d) Basic Solar Access Protection:

(1) Solar Fence: A solar fence is hereby hypothesized for each lot located in SA Area I and SA Area II. Each solar fence completely encloses the lot in question, and its foundation is contiguous with the lot lines. Such fence is vertical, opaque, and lacks any thickness.

* * *

(3) Insubstantial Breaches and Existing Structures: Insubstantial breaches of the basic solar access protection or of the protection provided by a solar access permit are exempt from the application of this section. A structure in existence on the date of establishment of an applicable solar access area, or structures and vegetation in existence on the date of issuance of an applicable solar access permit, are exempt from the application of this section. For purposes of this section, structures are deemed to be in existence on the date of issuance of a development permit authorizing its construction.

* * *

(f) Exceptions:

(1) Purpose: Any person desiring to erect an object or structure or increase or add to any object or structure, in such a manner as to interfere with the basic solar access protection, may apply for an exception.

* * *

(4) City Manager Action: The city manager may grant an exception of this section following the public notification period....

* * *

(5) Appeal of City Manager's Decision: The city manager's decision may be appealed to the BOZA [Boulder Board of Zoning Adjustments—Ed.]....

(6) Review Criteria: In order to grant an exception, the approving authority must find that each of the following requirements has been met.... [Specific requirements omitted—Ed.]

* * *

(I) All other requirements for the issuance of an exception have been met. The applicant bears the burden of proof with respect to all issues of fact.

(7) Conditions of Approval: The approving authority may grant exceptions subject to such terms and conditions as the authority finds just and equitable to assist persons whose protected solar access is diminished by the exception. Such terms and conditions may include a requirement that the applicant for an exception take actions to remove obstructions or otherwise increase solar access for any person whose protected solar access is adversely affected by granting the exception.

(8) Planning Board: Notwithstanding any other provisions of this subsection, if the applicant has a development application submitted for review that is to be heard by the planning board and that would require an exception, the planning board shall act in place of the BOZA, with authority to grant exceptions concurrent with other actions on the application, pursuant to the procedures and criteria of this section.

9-15-3 Administrative Procedures and Remedies.

(a) If the city manager finds that a violation of any provision of this title or any approval granted under this title exists, the manager, after notice and an opportunity for hearing under the procedures prescribed by chapter 1-3, "Quasi-Judicial Hearings," B.R.C. 1981, may take any one or more of the following actions to remedy the violation:

(1) Impose a civil penalty according to the following schedule:

(A) For the first violation of the provision or approval, $100.00;

(B) For the second violation of the same provision or approval, $300.00; and

(C) For the third violation of the same provision or approval, $1,000.00;

* * *

(4) Issue an order reasonably calculated to ensure compliance with the provisions of this title or any approval granted under this title.

9-15-4 Criminal Sanctions.

(a) The city attorney, acting on behalf of the people of the city, may prosecute any violation of this title or any approval granted under this title in municipal court in the same manner that other municipal offenses are prosecuted.

(b) The penalty for violation of any provision of this title is a fine of not more than $2,000.00 per violation. In addition, upon conviction of any person for violation of this title, the court may issue a cease and desist order and any other orders reasonably calculated to remedy the violation. Violation of any order of the court issued under this section is a violation of this section and is punishable by a fine of not more than $4,000.00 per violation, or incarceration for not more than ninety days in jail, or both such fine and incarceration.

* * *

9-15-5 Other Remedies.

The city attorney may maintain an action for damages, declaratory relief, specific performance, injunction, or any other appropriate relief in the District Court in and for the County of Boulder for any violation of any provision of this title or any approval granted under this title.

9-15-7 Private Right of Action.

Any person injured by a violation of any provision of this title or approval granted under this title may maintain an action for damages, declaratory relief, specific performance, injunction, or any other appropriate relief in the District Court in and for the County of Boulder against the person causing the violation. If plaintiff prevails, plaintiff shall be entitled to an award of attorney's fees. Upon filing such an action, plaintiff shall send notice thereof to the city, but nothing in this title authorizes the city or its employees or agents to be named as a defendant in such litigation.

Notes and Questions

1. Federal and local government incentives are currently more generous for solar PV, which generates electricity to the grid. Yet solar thermal and passive solar also benefit the grid by reducing overall demand. Looking at the Boulder ordinance above, consider what types of solar energy it protects. Does it cover solar PV systems? Solar thermal? Passive solar?

2. In addition to considering the rights protected above, consider to what degree the law protects each right. May the solar user claim protection for ground-mounted solar collectors or do the protections apply only to rooftop applications?

3. In contrast to the Iowa statute, who are the decision-makers under Boulder's ordinance? What do you think of having different departments making decisions — such as the building department to regulate the granting of building permits that satisfied the solar bulk plane requirements, the planning department to decide the waiver provisions, and the Board of Zoning Adjustment hearing appeals? Might some tensions arise if one department is more protective of solar rights and another favors developing larger structures? Under the City of Ashland, Oregon's solar access ordinance, the planning department retains jurisdiction for both granting and considering appeals for variances. City of Ashland Land Use Code § 18.70.060. Ashland has granted significantly fewer variances or exceptions for the standard solar access requirements than has Boulder.

4. Next, look at the Boulder ordinance to determine who is responsible for enforcement. What are the advantages and disadvantages of requiring a solar user to be responsible for bringing an enforcement action in district court? How would a citizen suit under this section vary from a standard common law nuisance action?

5. What is the government's enforcement role? What difference would it make if the rights were enforced by district courts or through municipal court systems? What are the advantages and disadvantages of allowing a solar user to file a complaint to ask a municipal official in the enforcement division to issue a citation and bring a neighbor to municipal court to enforce the solar access code in the same way as noise ordinances or dog barking ordinances are enforced?

ii. Permit Systems

This section will discuss the handful of statewide permit systems, some of which shift the balance entirely in favor of the solar user by allowing a first-in-time-first-in-right claim of prior appropriation.

Troy A. Rule, *Shadows on the Cathedral: Solar Access Laws in a Different Light*
2010 U. ILL. L. REV. 851, 876–78 (2010)

Rather than expressly assigning the Airspace Entitlement to either Solar Users or their Neighbors, statutes in New Mexico and Wyoming purport to use a "first-in-time" rule analogous to the prior appropriation doctrine in water law. Because the effect of these statutes is to assign competing airspace rights to Solar Users, the statutes are versions of Rule One [of Rule's Cathedral model. *See* Section C.3, *supra*—Ed.]

In New Mexico and Wyoming, a landowner can unilaterally acquire solar access rights across Neighbors' airspace, without compensating Neighbors, by being the first to make "beneficial use" of the airspace. A landowner who installs a qualifying solar collector, records a valid solar right instrument or declaration with the county clerk, and satisfies statutory neighbor notice requirements under these statutes acquires "solar rights." Solar rights acquired under these statutes are not rights in sunlight itself or in some other scarce resource for which private property rights did not previously exist. The New Mexico and Wyoming statutes define a "solar right" as a property right "to an unobstructed line-of-sight path from a solar collector to the sun, which permits radiation from the sun to impinge directly on the solar collector." In essence, a solar right is an easement across a Neighbor's airspace for the specified purpose of solar access. New Mexico's statute even has language requiring that a solar right "be considered an easement appurtenant."

In solar access disputes in New Mexico and Wyoming, "priority in time" supposedly "[has] the better right." Unfortunately, solar access conflicts are rarely disputes over competing solar access easements in which one Solar User erects a solar collector in the solar access path of another Solar User. Instead, such disputes are almost always between Solar Users seeking to obtain or enforce solar access rights and Neighbors with no interest in installing solar collectors who seek only to preserve existing airspace rights. In nearly every circumstance, Neighbors were "first-in-time" with respect to the Airspace Entitlement because they hold title to the surface estate directly below the airspace at issue. Although the New Mexico and Wyoming statutes are a well-intended effort to innovatively promote solar access, they ignore Neighbors' existing airspace rights and misapply the prior appropriation doctrine. The statutes seem based on the presumption that neither Solar Users nor their Neighbors already possess rights in the airspace at issue. In truth, Neighbors of Solar Users do hold such rights under common law.

The New Mexico and Wyoming statutes are not the first-in-time rules they purport to be, but they do adjust or reallocate existing property rights among landowners based on priority in time of use. They can thus generate many of the same unintended consequences associated with first-in-time rules.... [T]he Wisconsin and Massachusetts permit-based solar access statutes are like the New Mexico and Wyoming statutes in this regard. All of these statutes enable Solar Users to unilaterally acquire rights in or impose restrictions on Neighbors' airspace, but only to the extent that the airspace is not already occupied. Such approaches promote solar energy development by motivating Solar Users to install solar collectors quickly before Neighbors make use of the airspace needed for solar access. They may also, however, encourage opportunistic landowners to install solar panels with ulterior motives of acquiring a view easement across Neighbors' property or of preventing or delaying Neighbors' more productive uses. The rules might also motivate Neighbors to overdevelop their properties with trees or structures to avoid forfeiting their airspace rights to new Solar Users. Because they impose individualized burdens based on the needs of individual private landowners and without compensation, the rules are also more vulnerable to constitutional challenge.

Notes and Questions

1. The basis for Rule's assumption that southern property owners may claim Solar Skyspace B is that "in nearly every circumstance," the southern property owners are "'first-in-time'... because they hold title...." The U.S. Supreme Court may have created this assumption about airspace title by using wording that suggested protection of the unused space for future uses: "Otherwise buildings *could not be erected*, trees *could not be planted*, and even fences *could not be run*...." *United States v. Causby*, 328 U.S. 256, 264 (1946) (emphasis added). Could legislatures modify this common law recognition of airspace title? Could they create a different property regime that inverted the timing when title to Solar Skyspace B would vest, *i.e.*, vesting for first-in-time use of that specific airspace, not when title to the land surface below it was conveyed?

2. In a companion article to *Solar Rights*, Bronin proposes looking to water law regimes to establish a system of permits as the best solution to address solar access problems. Sara C. Bronin, *Modern Lights*, 80 U. Colo. L. Rev. 881 (2009). She advocates setting an entitlement to the sun to produce "socially beneficial results" and to "adequately compensate burdened landowners." *Id.* at 882. What do you see as the advantages and disadvantages of this proposal?

3. Rule raises concerns about solar rights delaying a southern property's "more productive uses." Strict first-in-time-first-in-right regimes have caused waste in the water context. Could a permit system address this by providing an objective determination of what constitutes "more productive uses"? Similarly, could a permit system encourage development alternatives or mitigation measures? Currently many developers are required to mitigate the impacts of the changes they seek by building roads, sewer systems, parks, or schools. Could they similarly be required to accommodate existing solar power systems through set-back and roof-obstruction adjustments or by providing new, unblocked locations for solar panels? Would this at least address the host property's lost investment which, under the common law, the southern property owner currently has no obligation to consider?

4. Rule suggests that a host property owner might have ulterior motives for installing panels, such as creating a view easement. Could the law be drafted to curb such opportunistic motives?

5. Does creation of a solar permitting system require an expensive new bureaucracy for enforcement? Would each state system require a distinct solar permitting agency? What existing agencies might be best suited for such a task? What do the Iowa and Boulder laws do?

D. Trees and Distributed Solar Power

A recurring problem with renewable energy is that some of the strongest proponents of alternative or "green" energy production are surprisingly unwilling to compromise in allowing that production to go forward. Perhaps one of the most puzzling illustrations of this tension is the conflict between proponents of distributed solar generation and tree advocates.

Government and civic organizations have long encouraged the planting of trees for windbreaks around homes and erosion prevention as well as for aesthetic reasons. Many government agencies still do so today (*e.g.*, The ABA's "One Million Trees by 2014" and Denver's "Mile High Million"). Unfortunately, many of these programs may be contributing to the conflicts down the road between solar access and vegetation by encouraging the planting of taller trees without consideration of their shade pollution potential.

Trees can cool homes, shade streets, and help avoid the "heat island" effect of asphalt. These benefits make trees an appealing ingredient in the energy mix. However, when scrutinized more closely, the benefits of trees in many parts of the United States cannot come close to the benefits of solar panels in the energy equation. For example, one study cited by the Colorado Tree Coalition, which focused on examples in Arizona and southern California, showed that shade trees can reduce energy use for air conditioning by 214 to 642 kWh per house per year. In comparison, the typical solar array of approximately 5 kW can produce over 6,000 kWh per house per year—over ten times the electricity potentially saved by tree shading.

While there may be reasons to preserve a particularly aesthetic or historic tree, increasing the number of solar panels that can contribute electricity to the grid might warrant an adjustment to any automatic default toward retaining offending vegetation. The topic appears to be so emotional that few governmental entities have attempted to address it. Compare the following two approaches as well as the Boulder ordinance in the previous section. Some issues to consider as you review these ordinances are the rights protected, the mechanisms for enforcement, and efforts to balance competing interests.

The Right Trees—Understory trees and bushes that have a mature height of less than twenty-five feet can provide sufficient shade for energy savings without impacting neighboring properties or blocking one's own rooftop solar panels.

In the Right Place—Trees should be planted on the east and west side of buildings to protect against unwanted heat entering in the summer. Avoid planting any trees to the south. Increasing evidence shows that even deciduous trees on the south negatively impact a building's energy use because their limbs provide significant shade during the critical winter months, increasing the heating load and interfering with PV electricity production.

Other Considerations—Although trees can be valuable for absorbing CO_2, recent studies have shown that not all trees are equally good from an air pollution perspective. Some trees emit more VOCs than they absorb, contributing to ozone problems. http://www.i4es.org/treeproject.html.

Ordinance—Gainesville, Florida
Ch. 30—Land Development Code

§ 30-251. Elements of compliance.

All property within the city shall be subject to the following regulations.... No parcel within the city may be cleared, grubbed, filled or excavated, nor shall any building be altered or reconstructed in a manner which changes the site plan, site use or increases the impervious surface area except in compliance with this article. Requirements of these sections do not exempt property owners from compliance with any other section of this chapter.

* * *

(7) *Design principles and standards.* All landscaped areas required by this article shall conform to the following general guidelines:

a. The preservation of native trees and shrubs is strongly encouraged to maintain healthy, varied and energy-efficient vegetation throughout the city, and to maintain habitat for native wildlife species.

b. The landscaping plan should integrate the elements of the proposed development with existing topography, hydrology and soils in order to prevent adverse impacts such as sedimentation of surface waters, erosion and dust.

c. The functional elements of the development plan, particularly the drainage systems and internal circulation systems for vehicles and pedestrians, should be integrated into the landscape plan. The landscaped areas should be integrated, especially to promote the continuity of on-site and off-site open space and greenway systems, and to enhance environmental features, particularly those features regulated by the environmental overlay districts (article VIII).

d. The selection and placement of landscaping materials should maximize the conservation of energy through shading of buildings, streets, pedestrian ways, bikeways and parking areas. The use of wind for ventilation and the effect on existing or future solar access shall be considered.

e. Landscaping design should consider the aesthetic and functional aspects of vegetation, both when initially installed and when the vegetation has reached maturity. Newly installed plants should be placed at intervals appropriate to the size of the plant at maturity, and the design should use short-term and long-term elements to satisfy the general design principles of this section over time. The natural and visual environment should be enhanced through the use of materials which achieve a variety with respect to seasonal changes, species of living material selected, textures, colors and size at maturity.

§ 30-254. Permits for tree removal.

* * *

(e)*Permit approval criteria.* Removal or relocation of regulated trees shall be approved by the city manager or designee upon a finding that the trees pose a safety hazard; have been weakened by disease, age, storm, fire or other injury; or prevent the reasonable development of the site, including the installation of solar energy equipment. Regulated trees shall not be removed, damaged or relocated for the purpose of locating utility lines and connections unless no reasonably practical alternative as determined by the city manager or designee is available.

* * *

Figure 2.8: Depiction of tree heights for solar access

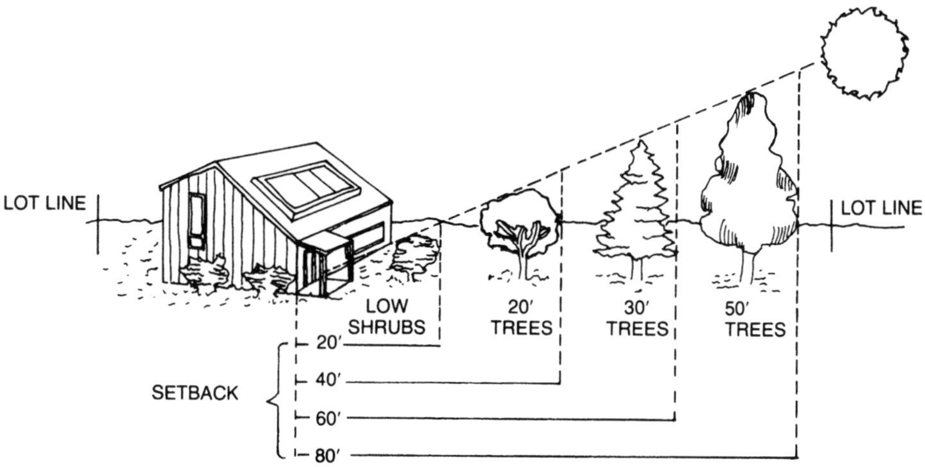

Duncan Erley & Martin Jaffe, U.S. Dep't of Hous.,
Site Planning for Solar Access 98 (1979)

Ordinance — Ashland, Oregon

Ch. 18 — Land Use

§ 18.70.010 Purpose and Intent.

The purpose of the Solar Access Chapter is to provide protection of a reasonable amount of sunlight from shade from structures and vegetation whenever feasible to all parcels in the City to preserve the economic value of solar radiation falling on structures, investments in solar energy systems, and the options for future uses of solar energy.

§ 18.70.070 Solar Access Permit for Protection from Shading by Vegetation.

A. A Solar Access Permit is applicable in the City of Ashland for protection of shading by vegetation only. Shading by buildings is protected by the setback provisions of this Ordinance.

B. Any property owner or lessee, or agent of either, may apply for a Solar Access Permit from the Staff Advisor. The application shall be in such form as the Staff Advisor may prescribe but shall, at a minimum, include the following:

1. A fee of Fifty ($50.00) Dollars plus Ten ($10.00) Dollars for each lot affected by the Solar Access Permit.

* * *

6. The tax lot numbers of a maximum of ten (10) adjacent properties proposed to be subject to the Solar Access Permit. A parcel map of the owner's property showing such adjacent properties with the location of existing buildings and vegetation, with all exempt vegetation labeled exempt.

7. The Solar Access Permit height limitations as defined in Section 18.70.050 of this Ordinance, for each affected property which are necessary to protect the solar energy system from shade during solar heating hours. In no case shall the height limitations of the Solar Access Permit be more restrictive than the building setbacks.

C. If the application is complete and complies with this Ordinance, the Staff Advisor shall accept the solar access recordation application and notify the applicant. The applicant is responsible for the accuracy of all information provided in the application.

D. The Staff Advisor shall send notice by certified letter, return receipt requested, to each owner and registered lessee of property proposed to be subject to the Solar Access Permit....

E. If no objections are filed within thirty (30) days following the date the final certified letter is mailed, the Staff Advisor shall issue the Solar Access Permit.

F. If any adversely affected person or governmental unit files a written objection with the Staff Advisor within the specified time, and if the objections still exist after informal discussions among the objector, appropriate City Staff, and the applicant, a hearing date shall be set and a hearing held in accordance with the provisions of Section 18.70.080.

* * *

18.70.110 Effect and Enforcement.

A. No City department shall issue any development permit purporting to allow the erection of any structure in violation of the setback provisions of this Chapter.

B. No one shall plant any vegetation that shades a recorded collector, or a recorded collector location if it is not yet installed, after receiving notice of a pending Solar Access Permit application or after issuance of a permit. After receiving notice of a Solar Access Permit or application, no one shall permit any vegetation on their property to grow in such a manner as to shade a recorded collector (or a recorded collector location if it is not yet installed) unless the vegetation is specifically exempted by the permit or by this Ordinance.

C. If vegetation is not trimmed as required or is permitted to grow contrary to Section 18.70.100(B), the recorded owner or the City, on complaint by the recorded owner, shall give notice of the shading by certified mail, return receipt requested, to the owner or registered lessee of the property where the shading vegetation is located. If the property owner or lessee fails to remove the shading vegetation within thirty (30) days after receiving this notice, an injunction may be issued, upon complaint of the recorded owner, recorded lessee, or the City, by any court of jurisdiction. The injunction may order the recorded owner or registered lessee to trim the vegetation, and the court shall order the violating recorded owner or registered lessee to pay any damages to the complainant, to pay court costs, and to pay the complainant reasonable attorney's fees incurred during trial and/or appeal.

D. If personal jurisdiction cannot be obtained over either the offending property owner or registered lessee, the City may have a notice listing the property by owner, address and legal description published once a week for four (4) consecutive weeks in a newspaper of general circulation within the City, giving notice that vegetation located on the property is in violation of this Ordinance and is subject to mandatory trimming. The City shall then have the power, pursuant to court order, to enter the property, trim or cause to have trimmed the shading parts of the vegetation, and add the costs of the trimming, court costs and other related costs as a lien against that property.

E. In addition to the above remedies, the shading vegetation is declared to be a public nuisance and may be abated through Title 9 of the Ashland Municipal Code.

F. Where the property owner or registered lessee contends that particular vegetation is exempt from trimming requirements, the burden of proof shall be on the property owner or lessee to show that an exemption applies to the particular vegetation.

Notes and Questions

1. Ashland and Boulder both have a hybrid solar access system. There is standard protection of a solar envelope to prevent the construction of buildings or other structures that interfere with solar access. However, a special permit is required for protection from vegetation. The drafters of Boulder's solar ordinance claim they wanted a uniform zoning system, but defaulted to a hybrid one requiring special permits for trees in response to taunts that otherwise they were "tree mutilators." Do you think such a hybrid system makes sense in terms of resources? What arguments can you make that trees should require this special treatment?

2. Notice that § 18.70.110 (E) of Ashland's ordinance declares that "shading vegetation … [is] a public nuisance…." What is the significance of this declaration? Does it mean that a public official, instead of a private homeowner, would be responsible for enforcement? Compare Ashland's regime to the current version of California's Solar Shade Control Act discussed in C.4.b.ii above.

3. Virtually all current solar access laws protect existing vegetation at the time a solar device is installed. At a minimum, utilities usually will not provide credits if the area where solar panels are to be installed is already shaded. Is it reasonable to expect landowners with trees to keep them trimmed to maintain the status quo at the time of installation? Should the utility that is using the grid-connected PV generation be responsible just as it is for trimming vegetation below its utility lines?

4. As the country's energy needs become more pressing, might this provide more incentive to eliminate provisions grandfathering in existing vegetation and to allow the trimming of existing trees? What rationales are most compelling—national security, safety, the public good of contributing to or stabilizing the electricity grid?

5. Interviews by the author with Ashland and Boulder officials in 2010 suggest the vegetation permit portions of their ordinances are not used frequently. For example, the Ashland planners said that they had fewer than five vegetation permits in the thirty plus years since its ordinance was passed in 1981. But these officials opined that the shift to protection for solar owners made permits and litigation unnecessary because southern neighbors are more likely to accommodate when they know they, not the solar users, have the burden of proof. Do you think this is a persuasive rationale for enacting more burden-shifting regulation in the solar access arena?

Chapter 3

Wind

This chapter addresses wind power. Like solar, wind is infinitely renewable as long as variations in temperatures of the earth's surface cause air to circulate when warmer air rises above descending cooler air.

In recent years, wind has been a favorite for meeting renewable energy mandates. Although intermittent, wind power can be more reliable than solar photovoltaics, which stop generating electricity whenever a passing cloud shades the panels. In contrast to random cloud cover, wind patterns are more predictable because they are driven by the interaction of large-scale atmospheric flow with fixed landforms. Some of these wind events, such as nocturnal low level jets in the Great Plains or valley breezes in the Columbia River Valley, recur frequently. Also, variations in these wind flows can be more easily predicted than can generation interruptions caused by cloud cover in the PV context.

Wind power also has been one of the most cost effective sources of renewable electricity. The price of wind power has risen because of increased demand for turbines and global price increases for steel, cement, copper, and other key components for the development of a wind farm. Yet wind's wholesale price can be competitive with natural gas generation, and it even stands up well against coal generation if one accepts the conclusions of the economists cited in Chapter 1.A.1 who conclude the true cost of coal generation is between nine and twenty-seven cents per kWh. *See also* Figure 1.4.

Wind power development has progressed at breakneck speed in recent years. The installed wind capacity in the United States increased almost fourteen times between 2000 and 2009. U.S. wind capacity increased 50 percent in 2008 alone when 8,500 new MW of wind power generation were constructed nationwide. The new total installed wind capacity of 25,300 MW allowed the United States to surpass the previous world leader, Germany, which had approximately 24,000 MW in 2008. By 2009, the U.S. total installed capacity increased to approximately 35,000 MW. But world competition is intense, and China took

> **Will climate change impact wind power?** Ironically, some research shows that global warming may diminish the potential for developing wind power because there will be more sultry hot days without breezes to propel turbine blades.

Figure 3.1: U.S. total installed wind energy nameplate capacity and generation

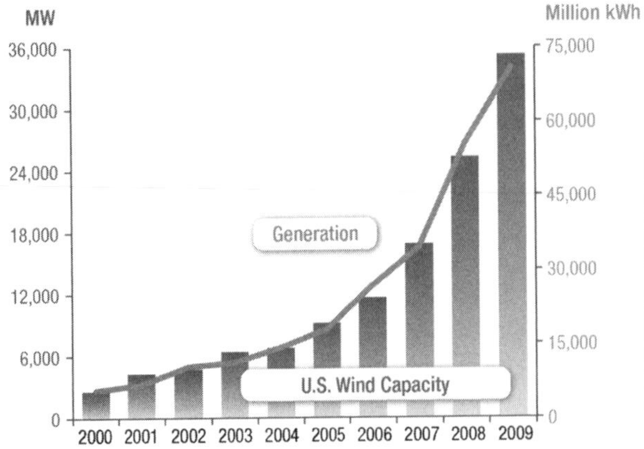

	U.S. Wind Energy Generation (Million kWh)	U.S. Wind Energy Capacity and Percent Increase from Previous Year	
		Total (MW)	% Increase
2000	5,593	2,578	2.6%
2001	6,737	4,275	65.8%
2002	10,354	4,686	9.6%
2003	11,187	6,353	35.6%
2004	14,144	6,725	5.9%
2005	17,811	9,121	35.6%
2006	26,589	11,575	26.9%
2007	34,450	16,812	45.2%
2008	55,363	25,237	50.1%
2009	70,761	35,159	39.3%

National Renewable Energy Lab., 2009 Renewable Energy Databook 58 (2010).

the lead in annual installed wind capacity by adding 13,800 MW that same year. Figure 3.1 shows the growth in U.S. wind energy capacity since the year 2000.

Figure 3.2 lays out a chart of the most common technologies for converting wind to electricity and some of the legal and land use issues. While the previous chapter focused primarily on distributed solar generation, this chapter will concentrate on the mode of wind development with which Americans have had the most experience: terrestrial wind on private lands.

Significant wind power resources are available on public lands, many of them offshore. Because the development of resources on public lands raises significantly different legal issues from private land development, offshore and public land wind will be addressed in Chapter 8. As of 2010, the United States had yet to build a single offshore wind turbine, but in April of that year, the Obama administration approved a 130-turbine wind farm in Nantucket Sound, which for years had been opposed by residents of Cape Cod, Nantucket, and Martha's Vineyard.

Much of the nation's best wind resources and greatest electricity needs are along the coasts, so the United States is moving toward following Europe's example of investing extensively in this option. Currently, additional offshore wind projects other than Cape Wind have also been approved by the federal government, and Google has announced plans to build a backbone transmission system to carry wind-generated electricity up and down the East Coast. Because the cost of offshore wind is several times higher per kWh than terrestrial wind, some industry participants are less confident and have scaled back on earlier plans for offshore development.

As wind power is poised to become more than a fringe source in the electricity generation pool, we are beginning to see more pushback against it. This chapter explores the background of wind power, the property status of wind, and then various areas of legal conflict between wind and wildlife, wind and humans, and wind and the development of other resources. Finally, most wind farm siting is currently controlled at the local level, so this chapter will address the ramifications of the local-control construct.

Figure 3.2: Wind power types

Land type & scale	Technologies	Land use issues	Legal
Private land: terrestrial utility-scale	Current state of the art is 1.5 to 3.5 MW turbines from 216 feet (66m) to 328 feet (100m) tall.	According to awea.org, utility scale wind requires about sixty acres per megawatt of installed capacity. Turbines located on ridgelines in hilly terrain can require as little as two acres per megawatt.	Accumulation of leases, easements, or severed wind deeds for a "farm." A buffer zone is required to prevent upwind turbines from deflecting downwind resources.
Private land: terrestrial distributed wind	Mostly rural applications; relatively rare and inefficient in urban areas because of obstructions.	Rooftop or yard installation.	For landowners' own electricity use.
Public land	Similar to Terrestrial Utility-Scale	Acquisition & land issues to be discussed in Chapter 8.	On state land, leases from states. On federal level, the Bureau of Land Management uses rights of way (ROW) or rights of use and easement (RUE).
Offshore wind	Generally larger turbines than for terrestrial wind, current proposals up to 7.5 MW.	Acquisition & land issues to be discussed in Chapter 8.	Permitting through federal government for turbines more than three miles from shore. States generally regulate the area between, so state permits or leases will be required for facilities closer to shore and for connection to shore for federal offshore turbines.

A. History & Context

This first section briefly sets out the history of wind power and some context for the discussion of legal issues related to terrestrial wind power development.

K.K. DuVivier, *Animal, Vegetable, Mineral — Wind?* *The Severed Wind Power Rights Conundrum*
49 WASHBURN L. J. 69, 72-73 (2009)

The use of wind as a power source has ancient origins. As early as 3,000 B.C.E., humans harnessed the wind to sail boats up the Nile River. Recognizing the potential of wind power, the Chinese are credited as being the first to erect land-based windmills to pump water around 200 B.C.E. Next, the Persians automated the task of grinding grain with vertical-axis windmills between 500 and 900 A.D. Historians believe that merchants and

Crusaders brought the concept from the Middle East to Europe where windmills became widely accepted. The Dutch are credited with refining the tower windmill in the sixteenth and seventeenth centuries and using it to drain lakes, marshes, and even the sea.

Windmills were "the 'electrical motor' of pre-industrial Europe." Yet, wind power is not constant, so during the Industrial Revolution the steam engine and then electricity were introduced as more reliable energy alternatives. Wind turbines remained prevalent on farms and ranches in rural portions of the United States until the 1950s when these areas were connected to the electric power grid through efforts of the Rural Electrification Administration and its successor agencies.

As electricity became the currency for power in most non-transportation engines in the United States, scientists explored the use of wind turbines to generate electricity, starting as early as the 1880s. Yet, the popularity of using wind to generate electricity "has always fluctuated with the price of fossil fuels." When fossil fuel prices have been low, wind power has not been able to compete. However, at times when the United States has been faced with oil embargoes and calls for energy independence from foreign oil, the interest in wind power has gained support and wind generation technologies have made significant progress. Nevertheless, a lack of steady research and development funding from the U.S. government and a lack of consistent tax treatment by Congress have, until recently, put U.S. wind generators competitively behind their European counterparts.

Most of the excerpts in this chapter focus on community-scale or commercial-scale wind farms on private lands instead of individual small-scale turbines. This is for two reasons. First, small-scale wind turbines are relatively rare in developed areas because the high density of buildings obstructs wind patterns, preventing the wind from being predictable or intense enough to provide a reliable power source. Second, small-scale wind in rural areas rarely evokes the types of ownership or access problems discussed in this chapter because landowners locate them on their own property for their own benefit. The excerpt below addresses small-scale wind briefly.

Carol Sue Tombari, POWER OF THE PEOPLE: AMERICA'S NEW ELECTRICITY CHOICES
93-95 (2008)

We hear a lot about utility-scale wind, but small-scale wind (10 kWh or less) can also be quite cost-effective. It can be either grid-connected and placed on buildings, or off-grid, to offset the need to string wires to a remote location. Humans have used wind for hundreds, if not thousands of years. Before electricity, windmills captured the power of the wind to turn a shaft which turned a millstone to grind grain or drive a pump to lift water.

We don't call them windmills any more. They're wind turbines. They have three blades mounted on a shaft, forming the rotor. The propeller-like blades act like airplane wings. When the wind blows over a blade, a pocket of low-pressure air forms on the downwind side. The low-pressure pocket pulls the blade toward it (lift), causing the rotor to turn. The lift is actually stronger than the wind blowing against the front side of the blade (drag). The combination of lift and drag cause the rotor to spin. The turning shaft spins a generator located at the top of the tower (in a box called the nacelle) and makes electricity.

The Good

If you live in a windy area, and especially if you need electricity in remote locations not served by the grid, small wind is for you. Stand-alone small wind turbines are especially useful for pumping water and for communications. In remote areas of China, small wind and PV are often used in tandem. Grid-connected small wind can also help cut electric bills.

New models that are particularly suited for rooftop installation recently hit the market. Residential and commercial customers alike may now take advantage of "urban" wind.

The Bad

The power of prevailing winds is a direct function of distance off the ground. The higher you go, the better the wind resource. This is an issue if you mount a small turbine on an urban or suburban roof. If the tower should fall down, there's greater risk of injury or damage in a populous area than there is in a sparsely populated place. In urban and suburban applications, building codes and homeowner association covenants are likely to prohibit small wind on buildings. In addition, insurers raise liability issues. Consequently, the urban and suburban residential and commercial applications may be limited. Aesthetics may also be an issue to some.

... [T]he wind doesn't always blow. The intermittent nature of the resource is another negative, though this is not as critical a factor as with utility- scale wind turbines....

The Balance

On balance, rural applications of small wind are probably easiest to achieve at present. This may change over time. Rural residents are accustomed to the look of wind turbines through their history with windmills (some of which are still standing). Importantly, building codes and covenants are not as likely to present barriers in rural areas as they are in urban and suburban communities.

The following excerpt provides an overview of some of the benefits and drawbacks of commercial-scale terrestrial wind power generation.

Ronald H. Rosenberg, *Diversifying America's Energy Future: The Future of Renewable Wind Power*
26 Va. Envtl. L.J. 505, 523-33 (2008)

1. Benefits of Wind Power Electricity Generation

Wind energy development spans a lengthy time period and encompasses a number of phases. The process runs from site monitoring to facility construction to plant operation and electricity generation to decommissioning of the development. As a process, wind power involves different advantages and disadvantages at varying stages of the development timeline....

a. Eliminating Fuel Costs in Electricity Generation

Proponents of wind power technology emphasize a range of reasons to support the rapid expansion of wind-generated electricity and motive power. First and foremost, wind power is a renewable and indigenous form of non-fossil fuel electricity. Once a wind turbine is installed, there is no fuel cost for the generation of power and consequently no fuel cost volatility. Second, the wind follows predictable patterns, yet its kinetic energy is available without cost to the turbine owner solely because of the siting location of the turbine in a windy area. Third, wind power is an inexhaustible supply without raw material or fuel costs, thereby making the inflationary characteristics of coal, natural gas, and oil irrelevant to the economic calculus of the project. As a result, the geopolitical complications of fossil fuels supplied from non-domestic sources cease to be a concern for the wind power electricity generator. Finally, since the fuel is naturally-occurring wind, there are no adverse impacts on workers, the environment, or local communities from fuel extraction. By comparison, coal—the nation's largest electricity supply fuel—imposes serious societal costs through air pollution, water pollution, water resource use, solid waste generation, and land contamination.

b. Zero Air Pollution & Global Warming Emissions

Arguably the strongest advantage of wind power is the fact that it does not create significant air pollution or GHG emissions at any point in its life cycle. Fossil fuel combustion is the largest source of carbon dioxide emissions in the United States, and electricity generation comprises nearly half of that large source of emissions. Conventional coal, natural gas, and oil-fired power plants annually emit thousands of tons of emissions of sulfur dioxide, nitrogen oxides, carbon monoxide, particulate matter, hydrocarbons, mercury, and other pollutants, while wind power produces zero emissions. With increased domestic and international emphasis on the elimination of GHGs, the substitution of fossil fuel-generated electricity with non-combustion-produced electricity will help alleviate climate change in the future. As U.S. policymakers embrace more rigorous GHG reduction goals, wind power will be emphasized as a viable alternative energy source.

c. No Water Use for Cooling

Wind power generation requires minimal amounts of water during operation, in stark contrast to the water use of conventional thermoelectric fossil fuel plants. As a result,

---

Does wind power cause increased GHG emissions?

In 2010, the Independent Petroleum Association of Mountain States (IPAMS) commissioned BENTEK Energy to prepare a study, which concluded increases in wind generation would harm Colorado's air quality. IPAMS claimed that inefficiencies from changing the cycle of coal-fired power plants used to back up the intermittency of wind power resulted in more air pollution. A representative of Colorado's largest utility refuted the IPAMS conclusion. "[Xcel Energy must— Ed.] ramp up and down other power plants as the wind changes. Generally, we prefer to ramp gas-fired plants because they respond quickly to sudden system changes. If we ramp coal-fired units, the plant's efficiency may decline, causing its emission rate to increase for short periods.... The IPAMS study correctly points to this fact, but then carries its conclusions too far. The study implies that small, short-term emission increases associated with ramping result in significant increases in the total emissions. This is simply wrong. Since 2007, we have added hundreds of megawatts of wind generation, and our overall emissions have declined." Frank Prager, *Setting the Record Straight on Wind Energy*, Denver Post, May 28, 2010, at 11B.

In addition, utilities are developing high-efficiency, quick-response natural gas plants, like GE's FlexEfficiency 50 CC , to specifically improve the efficiencies involved with ramping up and down for fluctuations in solar or wind energy production.

thermoelectric power plants use enormous quantities of water nearly equal to irrigation, which is mostly withdrawn from U.S. fresh water supplies. Both fuel-cycle and consumptive (evaporative) water use for coal and nuclear-generated electricity range in the billions of gallons per year. This intensive water use is often the most serious limiting factor in the permitting of these plants, especially in arid areas where water is scarce. As competition for fresh water becomes more intense, non-water using energy technologies like wind power boast an additional advantage. Wind power does not use water because it employs kinetic, not thermal, energy to spin the turbines in its generators.

d. Elimination of Solid and Hazardous Wastes Resulting from Fuel Preparation and Pollution Control

After the construction of a wind power facility, there is no solid or hazardous waste requiring disposal as a byproduct of generation. By contrast, the DOE estimates that the preparation of coal prior to power plant combustion generates solid waste at ten percent of the coal mined, resulting in millions of tons of coal wastes in need of disposal as part of the process of electricity generation. After combustion, large amounts of additional solid waste result from coal burning in the form of boiler slag, fly ash, and scrubber sludge produced by sulfur dioxide and particulate removal equipment. The lack of solid waste disposal problems is yet another significant environmental advantage of wind power, since the left-over residue of coal combustion must be disposed of in landfills.

e. Community and Regional Economic Benefit of Wind Energy Facility

The development of wind farms often occurs in rural communities experiencing depressed or reduced economic conditions, and wind power projects often have a positive economic impact on the employment in a construction area. The assembly of pre-fabricated wind turbines and towers employs construction workers at an estimated average rate of 4.8 job years (direct and indirect employment) per one MW of wind power construction. Using this ratio, a fifty MW wind farm would produce 240 job years of employment for workers who construct the facility. One estimate of employment impacts suggests that by 2015, wind energy projects in California alone would produce 2690 construction jobs and 121 million dollars in income. It is also estimated that between nine and ten full-time service personnel would be need[ed] to maintain a one hundred MW wind farm. This continual employment benefit would occur in rural areas and be distributed over a large area. In addition to these employment benefits, state governments would collect sales and income taxes from new construction, and local governments would benefit from increases in real estate tax bases due to the presence of the new wind farm equipment.

f. Supplementary Income to Rural Landowners

Wind farms use leased land, or land upon which royalties or land fees must be paid to the landowner. In these rural areas, there are often few leasing alternatives and none that pay the high level of lease or royalty payments of between 3000 and 4000 dollars per turbine per year. Depending on the amount of land leased and the number of turbines,

The lease and royalty payments for wind turbines are a boon to many farmers and ranchers. However, wind royalties are not as lucrative as mineral royalties that might also be available for the land. Section E of this chapter addresses some of the conflicts between the development of wind and the development of these other resources.

Decreasing Variability by Dispersing Sources — NREL studies have shown that by coordinating operations over a wider coverage area, western states can tap wind and solar energy for 35 percent of their electricity needs by 2017 without building additional infrastructure or back-up gas-burning power plants. "When you coordinate the operations between utilities across a large geographic area, you decrease the effect of variability of wind and solar energy sources, mitigating the unpredictability of Mother Nature." NREL News Release NR-2210, *Western Wind and Solar Integration Study*, (May 20, 2010), www.nrel.gov/wwsis (quoting Dr. Debra Law, NREL project manager for the study).

the lease payments would constitute much-needed income for rural land owners with few economic alternatives. New wind power lease payments supplement rural incomes, potentially allowing farmers and ranchers to remain on the land to continue traditional agricultural or ranching activities. This would maintain the rural life and culture that is rapidly disappearing in many areas. In addition, wind energy development would generally be compatible with other existing land uses, including livestock grazing, recreation, wildlife habitat, and oil, gas, and geothermal production.

2. Potential Drawbacks of Wind Power Generation

While there are many advantages to wind power, some disadvantages exist. Every energy-producing technology involves pros and cons that must be evaluated by government policymakers, private investors, and the general public. Some disadvantages are inherent in the nature of the wind energy technology itself, while others relate to the use of the technology at particular locations....

a. Consistency of the Wind Resource

One potential disadvantage of wind power relates to the nature of the wind resource itself: the blowing of wind is intermittent and occurs according to atmospheric conditions, not human energy needs. Wind does not always blow when energy is required and, in general, cannot be stored for later use. Wind speed and availability—and consequently the amount of electricity generated—often varies day-to-day. It is feared that utilities relying on wind power will need to develop or purchase costly reserve capacity to fill in when wind power is not available. Advocates of wind power believe that wind resources, while not consistent, are predictable and can be connected into the electrical grid with small cost penalties. The high level of private investment in wind power suggests that wind-generated electricity can provide a valuable flow of power that can be integrated into the utility transmission system. Further research will undoubtedly address this important question.

b. The Availability of Optimal Wind Power Sites

Good wind sites with the highest wind power classifications are often located in remote places, far from the high-density metropolitan areas with the greatest energy demands. An examination of the U.S. Wind Energy Resource Atlas reveals that many of the highest potential class six and seven wind areas are located in the upper Midwest, hundreds of miles from the closest population source. Since many remote locations where wind energy resources exist are often not located proximate to high capacity utility transmission lines, power connections must be built to link the wind electricity generators to the utility power grid. The high costs of building this necessary connective infrastructure can create serious obstacles for wind power projects. In addition, frequent popular opposition to the construction of high voltage lines can make it difficult, and in some cases impossible, to obtain the necessary governmental support for new construction. Even if remotely-located wind power sources are able to connect, they may be charged high access fees to use existing transmission lines. Furthermore, these lines may have limited transmission capacity allocated on a first-in-time principle, with a discriminatory effect on new power generators like wind farms. With the improvement in wind power-operating efficiencies, however, less desirable sites can economically produce electricity, especially if transmission lines and capacity are well situated.

There are also large, high-potential wind sites in offshore locations in coastal waters and in the Great Lakes that represent potentially huge amounts of electrical generation. In Europe, siting constraints for land-based wind farms have resulted in the construction of eighteen projects located in the North Sea providing 804 MW with over eleven GW of new offshore projects planned by 2010. Offshore wind is identified as an attractive alternative for a number of reasons, including: first, the most forceful and consistent winds exist offshore; second, offshore sites exist within reasonable distances from the major urban load centers, especially in the mid-Atlantic and New England; and third, underwater transmission line siting and distant turbine location can minimize aesthetic and land use objections. While offshore wind project construction costs can range between forty and seventy-five percent higher than land-based projects, they also boast compensating productivity advantages because the wind capacity factor is considerably greater than that of most on-shore facilities. Offshore activities within three miles of the coast come under state regulatory authority, while those beyond the three mile limit are the responsibility of the federal government. The federal Energy Policy Act of 2005 allocated jurisdiction over the development of alternate energy-related uses (including wind power) on the Outer Continental Shelf (OCS), including the power to grant permission to use the OCS for such purposes, to the Minerals Management Service (MMS) of the Department of the Interior. Once in place, the final program rules will likely open up large tracts of offshore areas to wind farm development.

[Offshore wind will be discussed further in Chapter 8.—Ed.]

c. Wind Power Technology and the Cost of Electricity

If wind power is to be widely adopted, then the cost of production must be reduced to a level near that of fossil fuel plants. Fortunately, the economics of wind-generated electricity have changed considerably over the last quarter century, with costs dramatically lower than they once were. Improvements in turbine design and electronic controls have led to significant production efficiencies. Technological improvements, however, must continue in order to assure competitive wind power pricing.

Production costs must also be reduced at secondary or sub-optimal wind power sites. The cost of wind energy varies greatly depending upon the wind speed at the site. Most existing wind projects are located at the best sites (class six and seven) with the lowest generation costs. Recent DOE estimates put wind power electricity costs at these locations between three and six cents per kWh, making wind power cost-competitive when compared to fossil fuel plants. However, prime sites will be exploited first, and wind power plants will eventually need to be constructed at secondary, less- desirable sites with lower wind speeds and higher generating costs. Either government subsidy programs or higher consumer prices would be necessary to assure that class four and five site electricity remains competitive. Federally-supported research is currently seeking ways to advance the technology in order to bring down the costs at sites with lower wind speeds. Cost considerations must be kept in mind in order to spread the advance of wind power, especially at remote locations with high transmission costs.

d. Competition for Land for Wind Energy Projects

Wind power development must compete with other land uses that might be more highly prized or valued. While farming and ranching activities are generally compatible with the generation of wind energy, other types of uses might be considered incompatible with the installation of large wind turbines. Several preservation policies expressed in BLM documents exclude wind development construction from special public lands and recreation areas. This kind of land competition pits renewable wind power energy goals against land preservation. As the controversy surrounding the Cape Wind offshore wind project near Cape Cod demonstrates, land use compatibility questions can be central to the objections raised against the siting of a wind power facility, even if the objectors approve of wind power generally. Land use considerations are often the heart of objections to large wind power projects and require that project siting choices be made in a decision-making process designed to identify and balance multiple values. Even in England, where strong government and popular support for renewable wind power exists, disagreements occasionally surface.

e. Potential Adverse Land Use Impacts of Wind Power Projects

The construction and operation of wind farms or utility-sized projects present a number of potential conflicts with neighboring residents and land uses. As with any large-scale energy generation project, background land use patterns will be changed by the new energy development. The construction phase of a wind power facility carries with it the potential of interference with a number of different interests, including wildlife habitat, water quality, cultural resources, geologic features, air quality, vehicular traffic, and occupational safety. Once a wind generation project is operational, prominent concerns include the aesthetic impact of a large number of wind turbines, interference with communications, shadow flicker, noise produced by rotating blades, impact on aircraft communications and navigation systems, ice throws from the blades of turbines, and effects on resident or migrating bird and bat populations. Some also criticize wind power for potential adverse effects on adjacent property values, although recent analysis has not borne this out. As research and experience with wind power technology becomes increasingly available, it is possible to separate verifiable claims of harm from those without basis in fact. Additionally, research is likely to indicate useful methods of planning turbine locations so as to mitigate some of the potentially harmful effects of wind power siting.

Notes and Questions

1. Although U.S. wind power capacity increased by a record 50 percent in 2008, the economic recession set back wind development. Wind turbine deliveries declined by approximately 50 percent in the early part of 2010 due to competition from natural gas (which dropped 25 percent in price, down to $4.20 per million BTUs). Utilities have been hesitant to sign long-term power purchase agreements with wind developers when natural gas prices are low and electricity demand has declined because of the weak economy. Could Congress help guarantee a more steady development of wind power by enacting a federal renewable mandate? Do you think this is a good idea?

2. Although small wind generators currently make sense in rural areas, to date, they have not become a standard fixture in urban areas. If small generators become increasingly

efficient, what legal problems might arise as they are installed in urban areas? What parallels and distinctions do you see between a right to solar access and a right to wind access?

B. Property Rights & Wind Severance

As developers put together land packages for wind farms, thousands of ranchers and farmers have begun to deed away their wind rights separately from their surface ownership rights, in the same way they might have severed mineral rights. This section addresses some of the problems with treating wind in the same way as minerals and how some states have responded by prohibiting severance.

K.K. DuVivier, *Animal, Vegetable, Mineral—Wind? The Severed Wind Power Rights Conundrum*
49 WASHBURN L. J. 69, 75-98 (2009)

From the perspective of a celestial being or an alien spaceship, a property right in the United States consists of more than a flat postage stamp on the surface of the earth. Instead, a property owner has a three-dimensional right, which when pivoted to a cross-section view, reveals that private ownership extends above and below the line at the earth's surface. This vision of property ownership is labeled the *ad coelum* or "unified fee" doctrine. Under this doctrine, the owner of the soil, or surface, also has ownership rights in everything from the center of the earth to the skies. Commentators cite the *ad coelum* or unified fee doctrine as justification for wind rights severance.

Borrowing from early English law, the courts "of practically every state" in the United States have at one point adopted the *ad coelum* maxim. [The "*ad coelum*" doctrine was discussed in the context of solar rights in Chapter 2—Ed.] Although it has been modified or rejected in some jurisdictions, the concept of property ownership as three dimensional—a pillar or cone from the center of the earth to the heavens—is frequently the starting point for the severed-estate analysis.

Early commentators viewed division of the unified fee estate as a "derogat[ion] from the general rights of property." The following ... explores the evolution of mineral severance and the justifications for it, including the belief that severance facilitates development of the subsurface for the public good and that concurrent use of the surface and the subsurface estate could occur with minimal disturbance.... [T]hese rationales for severance do not support its application in the context of wind rights. Instead of facilitating development, the severance of both the wind and the mineral estates creates new obstacles to a party attempting to create a wind farm. Also, the perpetual footprint of wind turbines, and the accompanying spider web of transmission, collection, and distribution lines, interferes with many concurrent uses of the surface.

1. The Evolution of Mineral Severance

The primary rationale for severing wind estates from surface estates is drawn from the analogy that mineral estates can be severed from the surface....

How did the concept of separate ownership of minerals first arise? It appears that one of the rationales for severance was the wondrous and divine nature of metals. The word "royalty" means not only a person of royal rank, but also a right or prerogative of the sovereign to receive a percentage of mining proceeds. In the first period of Greek mining, many believed that a royalty was owed to the gods. For example, Greek miners sent one tenth of the production from the mines of Siphnos to the shrine of the god Apollo at Delphi. Later, when payment was stopped and the mines were flooded by the sea, "this disaster was ascribed to the wrath of Apollo at being deprived of his divine royalty!" In England, the rationale was that "gold and silver are the most excellent things which the soil contains [so] the law has appointed them, as in reason it ought to, to the person most excellent, and that is the king."

Aside from the beauty of metals, conquering sovereigns claimed right to them for pragmatic reasons as well—they were necessary for warfare. Pliny called iron "the most deadly fruit of human ingenuity" because it was forged to create weapons of war. Historians argue that empires have been founded based on the advantages some conquerors enjoyed from using superior metals as opposed to those available to the vanquished civilizations.

* * *

While the Greek and Roman systems allowed prospectors to retain a portion of the fruits of their efforts, the concept of granting an entirely separate estate below the surface seems to have originated in Western Europe in the area that now encompasses modern Germany. Reasoning that the subterranean art of mining was "carried on neither within the same boundaries nor by the same persons as agriculture," it is not surprising that these peoples created the fable of the mining dwarf and established separate mining laws "not dependent upon the ordinary laws of property." Thus arose the principle of Bergbaufreiheit or "free mining," which recognized the "existence of an estate in minerals, entirely independent of the estate in soil." In contrast to serfs who were tied to the land of a particular lord, free miners were permitted unrestricted exploration on the government's or others' lands and, because of their special talents, were allowed to participate in creating the rules that controlled how they extracted the minerals they found. Eventually, the German government entirely surrendered its claim to rights in the minerals within its borders, instead placing itself on the same footing as private citizens with respect to ownership.

Other countries did not widely embrace the German concept of mineral severance. Spanish law, which is the source of the mining law in Mexico and most South American countries, retained ownership of mineral rights in the sovereign. In France, the monarch owned the mineral rights until that country no longer had a monarch. In 1791, the law changed to make mineral deposits in France the property of the nation, and the government granted concessions for them. Later, under Napoleon's law of 1810, ownership of minerals went to the owner of the land's surface. However, the French government retained the right to grant a separate mineral right, even in perpetuity, to someone other than the surface owner, so long as the grantee paid tribute to the state.

International ownership of wind—In most countries, the national or federal government owns minerals and grants leases or concessions to develop them. Some might also consider wind a nationally-owned natural resource, however, the province of Chubut in southern Argentina passed legislation stating that the wind is a natural resource that is locally owned by the province. http://www.legischubut2.gov.ar/index.php?option=com_content&view=article&id=265:sesion-1236-28-12-10&catid=59:ano-2010&Itemid=196 (defining wind power as renewable energy and a natural resource, which under the Argentine Constitution is owned by the province in which it is located).

As English law developed in a common law system, mining law was "complicated with many local regulations and 'immemorial customs.'" Generally, precious metals, and in some cases copper and tin, belonged to the crown. Aside from these claims by the crown, the fee owner of the land enjoyed a prima facie holding of the mineral rights to the land.

Later attempts to charge free miners with trespassing failed after British courts determined that free miners had the legal basis for claiming their own rights in mining fields. The courts then, and now, have struggled with the nature of the mineral right. Some characterize it as simply a right to dig and a property right in the minerals themselves once separated from the soil. The alternative created a property interest in the minerals in place, a separate and enduring real estate interest that coexisted with the surface estate.

Significantly, British cases that permitted severance noted that allowing a separate mineral estate was an abrogation of the common law concept of absolute ownership by the surface owner. Consequently, the right to sever an estate out of the surface owner's right is often strictly construed.

This "derogat[ion] from the general rights of property" seems to be justified by two concepts. First, the severed minerals are so far below the surface that the surface owner normally would not want, or be skilled enough, to exploit them. For example, even under the German code, surface deposits are left to the surface owner. Likewise, in Saxony— where the brown coal, or lignite, is often less than 150 feet below the surface and mined by surface methods, such as "opencast" or open pit mining—coal belongs to the surface owner. Thus, only the deeper minerals were severed—with the expectation that they could be exploited with minimal disturbance to the surface. The well-recognized mining law concepts of subjacent support and compensation for the surface owner further reinforce that a system endorsing severance does not anticipate significant conflicts between mineral and surface estate owners.

The second justification for abrogating the common law and allowing severance of minerals as a separate estate is the "benefit to the public secured thereby in the extraction of the mineral from the bowels of the earth." Not only did the public benefit from the availability of more minerals as a resource, but the royalties also provided an "advantage … to the coffers of the state."

* * *

2. Problems Applying Traditional Mineral Severance Rationales to Wind

As property law has evolved, it has created increased flexibility for land owners to divide their estates. Consequently, landowners appear not only to have authority over the wind that flows across their surface estates, but also authority to sever the wind rights from those surface estates. Despite this authority, the question remains not whether wind can be severed, but whether it is in society's best interest to allow wind severance.

Traditional mineral severance rationales of minimal surface disturbance and encouraging development of resources seem to support non-severance in the wind context. First, wind power development requires more extensive use of the surface than most mineral development. Much mineral development is through underground mining, *in situ* leaching, or extraction drilling, all of which have a fairly small surface footprint in comparison to wind. In contrast, wind ties up much of the surface for roads, substations, operations and maintenance facilities, and laydown yards. While some surface uses, such as grazing or farming, may coexist with wind development, spider webs of subsurface and overhead transmission, collection, and distribution lines can interfere with many other uses of the land.

Furthermore, even if mineral development is through strip or open-pit mining, which requires removal of the entire surface, the use is temporary and the surface is reclaimed after the mineral is extracted. In contrast, wind power is renewable and never depleted, thus, wind generation facilities might require perpetual surface use.

The second traditional justification for allowing severance of minerals is that mineral extraction and exploitation benefits the public by encouraging skilled workers to develop the resource, thus, making more minerals available for public consumption. Yet, instead of encouraging wind development, severing wind rights impedes it for at least three reasons.

First, separating wind rights from the surface estate removes the surface owner from the negotiating table. Because wind development requires extensive, long-term surface use, the surface owners are the parties most impacted. Taking them out of the equation seriously complicates surface access and damages negotiations.

Second, landowners who retain control over both the mineral and wind rights can serve important roles as mediators in disputes between competing developer interests. Landowners who receive royalties from both mineral and wind development have a financial incentive to see both enterprises coexist. This incentive is eliminated when mineral and wind rights are severed and the owners of these separate estates seek only to maximize their own distinct interests.

Finally, the first two points above create such serious surface rights issues that many commercial-scale wind investors are hesitant to work with landowners who have severed their wind rights. Alternatively, these investors hold up financing until they are provided with surface use agreements from all interested parties, which can sometimes be an insurmountable prerequisite.

In summary, while property law may permit the severance of wind rights, the traditional rationales for mineral severance do not support severance as the most effective method for encouraging the development of wind power. . . .

* * *

Only a handful of state legislatures have addressed wind ownership issues, and even fewer have recognized and explicitly restricted wind severance. Without legislative guidance, courts logically turn to precedents defining the status of other resources, such as oil or water, for analogies. Yet, defaulting to traditional models is unlikely to encourage the best development of our country's wind resources.

Without a legislative restriction, property law appears to allow severance of wind in a manner comparable to mineral severance. While some mineral models may be appropriate for resources developed underground, time has proved false the assumption that everything that happens below ground can stay below ground without impacting the surface. Wind farms require long-term and extensive use of land, not only on the surface, but also immediately above and below the surface for transmission, distribution, and collection lines. As a result, wind development needs careful consideration to coordinate it with other surface and subsurface uses.

Furthermore, a specific analogy with oil extraction is problematic because applying the common law rule of capture to wind may cause wasteful development as it did with oil until legislatures introduced mechanisms to adjust the deficiencies of the common law model. An analogy to water also carries with it the baggage and deficiencies of the prior appropriation system. Progress toward treating wind rights in a manner similar to solar rights may be a step in the right direction. However, few states have effective solar right regimes, and the distinctions between wind and solar suggest that wind rights are best addressed by legislation specifically catered to wind development.

South Dakota Codified Laws § 43-13-19
Severance of Wind Energy Rights Limited

No interest in any resource located on a tract of land and associated with the production or potential production of energy from wind power on the tract of land may be severed from the surface estate as defined in Section 45-5A-3, except that such rights may be leased for a period not to exceed fifty years. Any such lease is void if no development of the potential to produce energy from wind power has occurred on the land within five years after the lease began. The payment of any such lease shall be on an annual basis.

Notes and Questions

1. Many states have statutes recognizing that a property owner may create a wind easement. Are these statutes any different from the "cheerleading" statutes discussed in the context of solar rights in Chapter 2?

2. To help address some of the issues raised by wind development, some states have gone beyond simply recognizing a right to enter into a wind easement. In 1996, South Dakota enacted the above statute limiting wind severance. One of the original proponents of the South Dakota provision said that his concern was for farmers in the state who were accepting leases for potential wind development without realizing that it could ultimately destroy all surface use. In the last five years, other states, such as North Dakota and Nebraska, have followed South Dakota's lead and prohibited severance of wind rights from surface rights. Wyoming and Montana are also considering the possibility. What do you see as the pros and cons?

3. As the leading state for wind power development, Texas most likely has the highest acreage in the country of wind rights that are already severed. Yet, just as it has resisted unitization in the context of oil and gas development, Texas appears to be resisting efforts to restrict wind severance. At least one author has suggested that the market may solve the problem: commercial-scale wind investors are hesitant to work on properties where the wind is severed from the surface estate. Lisa Chavarria, *The Severance of Wind Rights in Texas*, University of Texas School of Law's Wind Energy Institute, 5 (Jan. 2009). What do you think of this market approach?

4. Each wind turbine creates a "wake" behind it that can impact the productivity of downwind turbines. Troy Rule, *A Downwind View of the Cathedral: Using Rule Four to Allocate Wind Rights*, 46 San Diego L. Rev. 207, 208 (2009). Consequently, there is an art in positioning wind turbines to maximize production in an area. Figure 3.4 dramatically

Wind statutes by state—For lists of the wind statutes in each state, you can look to National Wind Coordinating Committee & National Conference of State Legislatures, *State Siting and Permitting of Wind Energy Facilities* (April 2006), http://www.nationalwind.org/assets/publications/Siting_Factsheets.pdf.

Figure 3.3: Horns Rev 1 Windfarm

Horns Rev 1 owned by Vattenfall. (Photographer Christian Steiness).

illustrates the wake effect. One of the problems with a strict prior appropriation model is that a first-in-time, first-in-right regime ensures that the older uses, but not the best uses, must be fully satisfied before newer uses. Is there some way to allocate rights so that older turbines do not impact the downwind use of more efficient and more productive windfarms? Should there be?

5. Like oil, wind is fugacious in that it flows across multiple properties and is not valuable until reduced to possession. In the context of oil, this is called the "rule of capture," and this rule has caused owners on either side of a property line to throw additional money and resources into dueling production wells in the race to be the first to extract the resource. Uncontrolled extraction also cuts back on the productivity of oil fields, causing waste not only of additional unnecessary infrastructure but of the oil resource itself. Many states have attempted to mitigate the negative effects of the rule of capture by imposing spacing, pooling, and unitization requirements. What do you think about the possibility of states enacting comparable rules of cooperative development to substitute for the competitive rule of capture in the context of wind development? Would this encourage more efficient production with fewer turbines?

6. Currently those landowners with turbines located on their properties can receive significantly higher royalties than adjacent owners who do not have a turbine. Even under a unitization scheme, can the landowners with turbines make an argument that the hassle of having turbines on their properties justifies additional compensation for the noise, the traffic, or their inability to use the surface in that area?

Can wind power learn from the history of waste and tribulations during the evolution of oil and gas development law? *See, e.g.,* Thomas A. Mitchell, *The Future of Oil and Gas Conservation Jurisprudence: Past as Prologue*, 49 Washburn L.J. 379 (2010); Jacqueline Weaver, *The Tragedy of the Commons from Spindletop to Enron*, 24 J. Land Resources & Envtl. L. 187 (2004). Several state legislatures, including California, Colorado, Montana, and Texas have explored enacting or revising their wind power statutes.

C. Wind v. Wildlife

Some environmental groups that might support green energy development have aligned themselves against wind power because of its impacts on wildlife, especially birds and bats. The noise and dust of construction and operation of a site can disturb species and can affect a much larger area than the immediate development site because of fragmentation of habitat and disruption of foraging, mating, and nesting. Even the increased human presence after construction can cause some species to permanently abandon an area.

One species currently affected by the development of renewable energy projects is the greater sage-grouse, whose population has declined from 45 to 80 percent since 1965. In the context of wind, the major problems for sage-grouse are human presence, which disrupts habitat, and "shadow flicker" from rotating blades, which simulates the approach of avian predators.

The following excerpt and case explain some of the current concerns related to wind and wildlife and relevant laws to address them.

1. Background

Roger L. Freemen & Ben Kass, *Siting Wind Energy Facilities on Private Land in Colorado: Common Legal Issues*
39 Colo. Law. 43, 46-48 (May 2010)

The development of wind energy facilities has the potential to impact aquatic, terrestrial, and airborne wildlife. The most common impacts are to bird and bat species and include collisions, electrocution, habitat removal, habitat fragmentation, and displacement. It is estimated that approximately 10,000 to 40,000 birds are killed each year by the wind turbines currently operating in the United States. In comparison, approximately 60 to 80 million bird deaths each year are attributed to motor vehicles, and 40 to 50 million deaths are attributed to collisions with communication towers. Although these numbers tend to downplay the impact of wind turbines on birds, wind energy projects still are much more likely to kill threatened or endangered species than vehicle collisions. Despite the threat to birds posed by wind energy development, several national environmental organizations, including the Audubon Society, have stated their support for the expansion of the wind energy industry under certain conditions.

The national debate continues about the policy tradeoffs of expanding wind development despite avian mortality; in this article, the focus is on the overall legal constraints faced by a wind developer. The following is an overview of [federal] wildlife laws and regulations most relevant to wind energy development in [Colorado].

Endangered Species Act (ESA)

Arguably, the most stringent of the federal environmental laws, the ESA provides for the conservation and protection of endangered and threatened species and their ecosystems.... Congress intended that endangered species be given the highest priority under the ESA and, thus, courts readily grant injunctive relief for violations of the ESA, notwithstanding the potential costs involved.

The ESA has three provisions most applicable to the construction and operation of wind energy facilities. Section 9 of the ESA prohibits any person from "taking" endangered or threatened species, which includes harming a listed species, or harassing a listed species by significantly disturbing normal behavior patterns, such as breeding, feeding, or sheltering. Thus, where development of a wind energy project site or operation of the turbines harms' or harasses a protected species, wind energy developers may be subject to stiff penalties and possibly to an injunction halting the project. Although Section 10 of the ESA allows the U.S. Fish and Wildlife Service (FWS) to issue "incidental take" permits, allowing certain activities to proceed despite the potential for harm to a species, these permits are not easily obtained.

Section 7 of the ESA requires that all federal agencies ensure that their actions not jeopardize the continued existence of any endangered or threatened species or result in the destruction or adverse modification of designated critical habitat. For example, wind energy projects are affected where the project requires a federal approval, uses federal funding, or involves a connection to a federal transmission line.

Although the definition of "federal action" under the ESA is subject to varying court interpretations, agency consultation typically is warranted even for smaller-scale wind energy projects. As mentioned, Section 9 of the ESA is a strict liability provision that does not require intent or knowledge of a violation. Further, ... [some state] regulations require consultation with the [state divisions of wildlife — Ed.] for all wind projects seeking to sell power to regulated utilities. Thus, consultation with the [federal and state wildlife services] prior to commencing a project is the most effective way for wind project developers to avoid costly delay and coordinate the sequence and timing of its activities, a pivotal part of the planning process.

Migratory Bird Treaty Act (MBTA)

The MBTA establishes protections for birds migrating through U.S. airspace. The MBTA protects more than 800 species of birds, many of which are found in Colorado, and makes it illegal to take, capture, or kill migratory birds and imposes fines up to $15,000 or imprisonment on "any person, association, partnership, or corporation" who violates its provisions. The MBTA is distinct from the ESA because it protects migratory bird species that are not necessarily threatened or endangered. Like the ESA, knowledge or intent is not required for liability to attach under the MBTA. Thus, the MBTA is a strict liability statute and does not provide for incidental take permits to cover accidental impacts from a wind energy project.

Several aspects of wind development may be subject to the MBTA provisions, including site clearing, construction of towers to gather metrological data, and wind turbine operation. Because the MBTA is implemented and enforced by the FWS, consultation regarding the ESA and MBTA often can be concurrently completed. Consultation typically will include an analysis of potential impacts of proposed wind projects on bird species protected under the MBTA.

Bald and Golden Eagle Protection Act (BGEPA)

The BGEPA was passed in 1940 to prevent the extinction of the bald eagle, and was amended in 1962 to include protection of golden eagles. The act makes it unlawful to "take" any bald or golden eagle, their parts, nests, or eggs by shooting at, poisoning, wounding, killing, capturing, trapping, collecting, molesting, or disturbing the eagles. Although in July 2007 the bald eagle was removed from endangered and threatened species list, it still receives protection under the BGEPA. The BGEPA imposes substantial fines or imprisonment for violations.

Like the ESA and the MBTA, provisions of the BGEPA apply to nearly all aspects of a wind energy project, including site clearing, gathering of meteorological data, and construction and operation of wind turbines. Although a take permit is available to move a golden eagle nest to prevent harm to the nest or eggs, the BGEPA is a strict liability statute and does not provide exceptions for accidental impacts from wind projects. The BGEPA is implemented and enforced by the FWS, and consultation regarding the impacts of a particular wind project can be completed concurrently with the ESA and MBTA.

---

2. Animal Welfare Institute v. Beech Ridge Energy LLC

A 2005 study by the U.S. Forest Service, specifically examined bird mortality from collisions. This study concluded that 58.2 percent of annual bird mortalities in the United States were caused by collisions with buildings and less than .01 percent were caused by collisions with wind turbines. WALLACE P. ERICKSON, ET AL., A SUMMARY AND COMPARISON OF BIRD MORTALITY FROM ANTHROPOGENIC CAUSES WITH AN EMPHASIS ON COLLISIONS, USDA FOREST SERVICE GEN. TECH. REP. PSW-GTR-191, 1029, 1039 (2005). This same study also concluded that the number of bird deaths attributed to wind turbines, approximately 28,000, was dwarfed by the number of birds killed by cats each year, approximately 100 million. *Id.*

Furthermore, now that wind developers are aware of bird issues, they have modified their siting and equipment designs to reduce opportunities for turbines to harm birds. The towers are slimmer and without rungs to make them less inviting as roosts. To make it less of a refuge for prey animals, land at the base of turbines is not allowed to go wild. Turbine speeds and blade designs also have been adjusted. Figure 3.4 shows successive generations of wind turbines in the San Gorgonio region of California, approximately

Figure 3.4: Turbines in the San Gorgonio region of California

Photo by K.K. DuVivier.

one hundred miles east of Los Angeles. Some of these turbines were erected before 1994, and some are more recent. Note at least three different turbine designs in this one patch of the wind farm.

While bird fatalities from direct collisions with turbines range from about 1.8 to 2.2 per turbine per year, the fatalities for bats are much higher, ranging from .8 to 53.3 bats per year depending upon the location. Most bats are not killed by direct collisions but instead by "barotrauma," a hemorrhage in the lungs due to a sudden drop in air pressure. Some speculate that the bats are drawn into the danger zone of one to two meters of the rotor blades because turbines simulate roosting sites. Developers are looking for technological solutions. For example, increasing the cut-in speed, the speed at which turbines begin to produce power, from 8 to 11 miles per hour has been shown to decrease bat fatalities up to 93 percent.

The following case describes the consequences of not addressing potential bat fatalities when initially siting a wind farm.

Animal Welfare Institute v. Beech Ridge Energy LLC
675 F. Supp. 2d 540 (D. Md. 2009)

Titus, J.

This is a case about bats, wind turbines, and two federal polices, one favoring protection of endangered species and the other encouraging development of renewable energy resources. It began on June 10, 2009, when Plaintiffs Animal Welfare Institute ("AWI"), Mountain Communities for Responsible Energy ("MCRE"), and David G. Cowan (collectively, "Plaintiffs") brought an action seeking declaratory and injunctive relief against Defendants Beech Ridge Energy LLC ("Beech Ridge Energy") and Invenergy Wind LLC ("Invenergy") (collectively, "Defendants"). Plaintiffs allege that Defendants' construction and future operation of the Beech Ridge wind energy project ("Beech Ridge Project"), located in Greenbrier County, West Virginia, will "take" endangered Indiana bats, in violation of Section 9 of the Endangered Species Act ("ESA"), 16 U.S.C. § 1538(a)(1)(B).

* * *

Wholly-Future Violations Under the ESA

Defendants argue that the ESA's citizen-suit provision bars actions alleging "wholly-future" violations of Section 9 of the statute, where there is no past, current, or continuing "take." This is an issue of first impression in the Fourth Circuit.

At first glance, a superficial reading of the text of the ESA would appear to lend some support to Defendants' position. The citizen-suit provision employs the present tense, allowing a private party to commence a civil action against anyone "who is alleged to be *in violation of* any provision of this Act...." 16 U.S.C. § 1540(g)(1)(A) (emphasis added). Defendants note that the Supreme Court and the Fourth Circuit have interpreted identical language in the citizen-suit provision of the Clean Water Act ("CWA"), 33 U.S.C. § 1365, and argue that these cases stand for the proposition that *"there is no jurisdiction over claims of wholly future violations."* Defs.' Surreply and Pre-Trial Br. at 2-3 (emphasis in original) (citations omitted).

Defendants' reliance on the CWA cases is misplaced. In *Gwaltney of Smithfield, Ltd. v. Chesapeake Bay Found., Inc.,* the issue before the Supreme Court was whether the CWA confers jurisdiction over citizen suits for wholly-*past* violations. 484 U.S. at 54-56, 108 S.Ct. 376. Correlatively, in *American Canoe Association v. Murphy Farms,* the Fourth Circuit held that to establish jurisdiction under the CWA, a plaintiff must either prove violations that continue on or after the date the complaint is filed or show a likelihood of future recurrence of violations. 412 F.3d at 539. These CWA cases clearly do not address claims of wholly-future violations.

Moreover, the ESA's citizen-suit provision provides for injunctive relief which by design prevents *future* actions that will take listed species. Congress explained that citizen-suit actions allow any person "to seek remedies involving injunctive relief for violations or *potential* violations of the Act," H.R. Rep. 93-412 (1973) (emphasis added), suggesting that a historic violation is not necessary. The Court therefore concludes that the citizen-suit provision includes within its scope wholly-future violations of the statute.

The text of Section 9 and its legislative history also indicate that Congress intended that the "take" provision be expansive in scope. By prohibiting any "attempt" to harm, wound, kill, or harass a listed species, 16 U.S.C. § 1532(19), Congress clearly manifested an intent that Section 9 was designed to include claims of future injury. Furthermore, the Senate confirmed that the term "take" is defined "in the broadest possible manner to include every conceivable way in which a person can 'take' or attempt to 'take' any fish or wildlife." S.Rep. No. 93-307, at 7 (1973), *reprinted in* 1973 U.S.C.C.A.N. 2989, 2995. Protecting against the threat of imminent future harm is clearly consistent with Congress' broad definition of the term "take."

In addition, the Court finds that Defendants' interpretation of the ESA's citizen-suit provision as precluding claims for wholly-future violations is inconsistent with the very purpose of the Act. As discussed in *supra* Part I, Congress' intent when enacting the ESA was to protect and conserve threatened and endangered species, whatever the cost. *Tenn. Valley Auth. v. Hill,* 437 U.S. 153, 184, 98 S.Ct. 2279, 57 L.Ed.2d 117 (1978). Requiring that a listed species be harmed, wounded, killed, or harassed before conferring jurisdiction would thwart this central goal of the Act.

Accordingly, the Court holds that the ESA's citizen-suit provision allows actions alleging wholly-future violations of the statute, where no past violation has occurred. The Court's holding is consistent with the text of the citizen-suit provision, the legislative history, the purpose of the ESA, as well as decisions from the Ninth Circuit squarely addressing the issue. (Citations omitted).

Requisite Degree of Certainty Under the ESA

Neither the Supreme Court nor the Fourth Circuit has yet had the opportunity to decide whether under Section 9 of the ESA, a plaintiff must establish by a preponderance of the evidence that the possibility of a take is likely or certain, or something in between. Plaintiffs urge the Court to apply ordinary principles of tort causation, which would require that they demonstrate that a take is merely more likely than not. Defendants contend that Plaintiffs must prove by a preponderance of the evidence that the challenged activity is certain to harm, kill, or wound Indiana bats.

Although the Act is silent as to the requisite degree of certainty for establishing a take under Section 9, the FWS regulations implementing the ESA suggest that the standard for "harm" is higher than for "harassment." The regulations define the term "harass" as "an intentional or negligent act or omission, which creates the *likelihood* of injury to wildlife by annoying it. . . ." 50 CFR § 17.3 (emphasis added). However, the term "harm" means "an act which *actually* kills or injures wildlife." *Id.* (emphasis added). The omission of the word "likelihood" and the insertion of the word "actually" in the latter definition suggest that a plaintiff must prove that harm is more than merely "likely" to occur.

The explanatory commentary to this regulation indicates that harm cannot be speculative. The FWS stated that it inserted the term "actually" before "kills or injures" because "existing language could be construed as prohibiting the modification of habitat even where there was no injury." (Citations omitted). The FWS further opined that the "redefinition sufficiently clarifies the restraints of Section 9 so as to avoid injury to protected wildlife due to significant habitat modification, while at the same time precluding a taking where no actual injury is shown." 46 Fed.Reg. 54,748, 54,749 (Nov. 4, 1981).

Similarly, *Sweet Home* appears to suggest that mere likelihood of harm is insufficient under Section 9. In *Sweet Home,* the Court held that the Secretary of the Interior did not exceed his authority when including *habitat modification and degradation* in the afore-mentioned regulation defining the term "harm." 515 U.S. at 707-8, 115 S.Ct. 2407. Throughout the majority opinion, the Court, quoting the regulation, repeatedly stated that "actual" injury is required. (Citations omitted). By underscoring the need for actual injury, the Court implied that harm cannot be hypothetical.

Courts outside of the Fourth Circuit addressing the issue of the requisite degree of certainty of harm have articulated varying standards, and have not always distinguished between harm, kill, wound, and harass. (Citations omitted).

The First Circuit, for example, held in *American Bald Eagle v. Bhatti* that "[t]he proper standard for establishing a taking under the ESA, far from being a numerical probability of harm, has been unequivocally defined as a showing of 'actual harm.'" 9 F.3d 163, 165 (1st Cir. 1993) (rejecting the notion that "a one in a million risk of harm is sufficient to trigger the protections of the ESA"). The case involved a claim that American Bald Eagles would be harmed by a controlled deer hunt in a public forest because some of the wounded deer would not be recovered ("cripple-loss deer"), that they would die within the feeding area of the birds, and that bald eagles might be harmed by consuming lead in the deer carcasses. *Id.* at 164. Both the district court and the First Circuit found that the speculative risk of harm was insufficient to assert a claim under Section 9 of the ESA.*Id.* at 166; *see also id.* at 166 n. 4 ("Appellants have not shown that bald eagles have ingested lead slugs nor fragments thereof during past hunts or will ingest lead slugs or fragments thereof during future hunts....").

Because the risk of harm was highly speculative in *American Bald Eagle,* the First Circuit's observations regarding the degree of certainty of harm required by the ESA were not necessary to the decision. However, the Ninth Circuit, where most Section 9 actions involving land-use activities have been brought, has squarely addressed the issue.

In *Marbled Murrelet v. Pacific Lumber Co.,* the Ninth Circuit required that a plaintiff establish a "reasonable certainty of imminent harm." 83 F.3d 1060, 1068 (9th Cir.1996) ("The district court did not clearly err in finding marbled murrelets were nesting in Owl Creek and that there was a reasonable certainty of imminent harm to them from Pacific Lumber's intended logging operation.") Two years later, in *Defenders of Wildlife v. Bernal,* the court appeared to raise the standard, holding that plaintiffs "had the burden of proving by a preponderance of the evidence that the proposed construction *would* harm a pygmy-owl by killing or injuring it, or would more likely than not harass a pygmy-owl by annoying it to such an extent as to disrupt its normal behavioral patterns." 204 F.3d 920, 925 (9th Cir.2000) (emphasis added). However, the Ninth Circuit did not state that it was departing from *Marbled Murrelet,* but instead clarified that in its previous decision it had held that "a *reasonably certain threat of imminent harm* to a protected species is sufficient for issuance of an injunction under Section 9 of the ESA." *Id.* at 925 (emphasis added).

* * *

The Court agrees with the standard adopted in *Marbled Murrelet,* and holds that in an action brought under Section 9 of the ESA, a plaintiff must establish, by a preponderance of the evidence, that the challenged activity is reasonably certain to imminently harm, kill, or wound the listed species. To require absolute certainty, as proposed by Defendants, would frustrate the purpose of the ESA to protect endangered species before they are injured and would effectively raise the evidentiary standard above a preponderance of the evidence. The reasonable certainty standard, in combination with the temporal component, is consistent with the purpose of the Act, its legislative history, the implementing regulations, and Supreme Court precedent.

* * *

The fact that there are no caves within five miles of the project site known to currently contain Indiana bats makes it less likely that Indiana bats are present at the site in large numbers during fall swarming and spring staging than if there were hibernacula within this area. However, the absence of hibernacula within five miles does not eliminate the possibility that Indiana bats are present at the site. For example, Indiana bats have been found more than five miles from hibernacula during fall swarming. (Citations omitted).

Moreover, the five mile distance has no bearing on the question of the presence of Indiana bats during *migration.* (Citations omitted). In fact, Robbins testified that he has captured Indiana bats at other wind project sites where the closest hibernaculum was approximately 100 miles away (a Priority 4 cave). Trial Tr. 205:25-206:8, Oct. 21, 2009. Robbins also opined that hibernacula within 150 miles of the Beech Ridge Project site, including Hellhole Cave (a Priority 1 cave), would be within the migratory range of Indiana bats. Trial Tr. 206:23-207:11, Oct. 21, 2009.

* * *

Likelihood of a Take of Indiana Bats at the Beech Ridge Project Site

It is uncontroverted that wind turbines kill bats, and do so in large numbers. Defendants contend, however, that Indiana bats somehow will escape the fate of thousands of their less endangered peers at the Beech Ridge Project site.

Defendants argue that Indiana bats do not fly at the height of the turbine blades. Lacki and Tyrell stated that Indiana bats are "edge foragers," meaning they tend to forage for food directly below or at the tree canopy. (Citations omitted). Lacki opined that Indiana bats are not going to be in locations, such as the area above the tree canopy, where "their foraging approach is likely to render them vulnerable." Trial Tr. 225:8-13, Oct. 23, 2009. Tyrell speculated that the tree canopy at the Beech Ridge Project site is sixty to eighty feet above the ground, Trial Tr. 20:1-20:4, Oct. 29, 2009, which is below the lowest part of the rotor swept area.

However, Plaintiffs' expert Kunz, one of the leading bat biologists in the country, stated that with the development of acoustic technology and thermal cameras, there is growing research that bats can fly as high as a kilometer or more above the ground, and that Indiana bats may also fly at these altitudes. Trial Tr. 49:1-18, Oct. 22, 2009. Kunz explained that bats fly above the tree canopy as warm air carries insects high above the surface of the earth. *Id.* at 50:1-19 (stating that insects can be carried as high as 2.5 km above the ground). Kunz opined that "the fact that Indiana bats were detected at ground level ... suggests that they would also would also [sic] equally likel[y] be detected higher up in the rotor swept region." Trial Tr. 77:10-14, Oct. 22, 2009. Moreover, the height at which Indiana bats forage has no relation to how high they fly during migration. *See, e.g.,* Trial Tr. 84:11-17, Oct. 29, 2009 (Tyrell).

Defendants also point out that no Indiana bat has been confirmed dead at any wind power project in the country, which they contend supports a conclusion that Indiana bats, unlike other bat species, are somehow able to avoid harm caused by wind turbines.

However, other *Myotis* species have been reported killed at wind power projects. (Citations omitted). Plaintiffs' experts opined that biologically, Indiana bats are no less vulnerable than other *Myotis* species to turbine collisions and barotrauma. (Citations omitted).

In addition, post-construction mortality studies are generally inefficient (for example, due to scavenging), thus making the chances of finding the carcass of a rare species even smaller. (Citations omitted). At trial, Gannon criticized those mortality studies-like those proposed at the Beech Ridge Project site, Trial Tr. 60:24-61:14, Oct. 23, 2009 (Groberg)-that survey only a subset of the turbines: "[i]f you've got a haystack, and you're only looking at a very small portion of that haystack, what's the odds that you're going to find something rare in the haystack?" Trial Tr. 64:5-65:1, Oct. 21, 2009.

* * *

The Court agrees with [Plaintiffs'] very credible expert opinions. The Court finds that there is no reason why Indiana bats would not fly at a height of 137 to 389 feet above the ground, within the rotor swept area of the turbines at the Beech Ridge Project site. Plaintiffs have presented compelling evidence that Indiana bats behave no differently than other *Myotis* species that have been killed by wind turbines and Defendants have failed to rebut this fact. Furthermore, the Court is not surprised that no dead Indiana bat has yet been found at any wind project because few post-mortality studies have been conducted, mortality searches are generally inefficient, and Indiana bats are rare.

Based on the evidence in the record, the Court therefore concludes, by a preponderance of the evidence, that, like death and taxes, there is a virtual certainty that Indiana bats will be harmed, wounded, or killed imminently by the Beech Ridge Project, in violation of Section 9 of the ESA, during the spring, summer, and fall.

* * *

Because entirely discretionary adaptive management will not eliminate the risk to Indiana bats, the Court has no choice but to award injunctive relief.

Injunctive Relief

Because the Court has found that the Beech Ridge Project will take Indiana bats, injunctive relief is appropriate under Section 11 of the ESA. The question, then, is what form that injunctive relief should take. The ITP [Incidental Take Permit] process is available to Defendants to insulate themselves from liability under the ESA and, while this Court cannot require them to apply for or obtain such a permit, it is the only way in which the Court will allow the Beech Ridge Project to continue.

The Court sees little need to preclude the completion of construction of those forty turbines already under construction, but does believe that any construction of additional turbines should not be commenced unless and until an ITP has been obtained. The simple reason for this is that the ITP process may find that some locations for wind turbines are entirely inappropriate, while others may be appropriate.

There is, by the same token, no reason to completely prohibit Defendants from operating wind turbines now under construction once they are completed. However, in light of the record developed before this Court, that operation can only occur during the periods of time when Indiana bats are in hibernation, i.e., from November 16 to March 31. (Citations omitted). Outside this period, determining the timing and circumstances under which wind turbine operation can occur without danger of the take of an Indiana bat is beyond the competence of this Court, but is well within the competence of the FWS under the ITP process.

Accordingly, the Court will enjoin all operation of wind turbines presently under construction except during the winter period enumerated above. However, the Court invites the parties to confer with each other and return to the Court, if agreement can be reached, on the conditions under which the wind turbines now under construction would be allowed to operate, if at all, during any period of time outside of the hibernation period of Indiana bats.

Conclusion

As noted at the outset, this is a case about bats, wind turbines, and two federal policies, one favoring the protection of endangered species, and the other encouraging development of renewable energy resources. Congress, in enacting the ESA, has unequivocally stated that endangered species must be afforded the highest priority, and the FWS long ago designated the Indiana bat as an endangered species. By the same token, Congress has strongly encouraged the development of clean, renewable energy, including wind energy. It is uncontroverted that wind turbines kill or injure bats in large numbers, and the Court has concluded, in this case, that there is a virtual certainty that construction and operation of the Beech Ridge Project will take endangered Indiana bats in violation of Section 9 of the ESA.

The two vital federal policies at issue in this case are not necessarily in conflict. Indeed, the tragedy of this case is that Defendants disregarded not only repeated advice from the FWS but also failed to take advantage of a specific mechanism, the ITP process, established by federal law to allow their project to proceed in harmony with the goal of avoidance of harm to endangered species

* * *

This Court has concluded that the only avenue available to Defendants to resolve the self-imposed plight in which they now find themselves is to do belatedly that which they should have done long ago: apply for an ITP. The Court does express the concern that any extraordinary delays by the FWS in the processing of a permit application would frustrate Congress' intent to encourage responsible wind turbine development. Assuming that Defendants now proceed to file an application for an ITP, the Court urges the FWS to act with reasonable promptness, but with necessary thoroughness, in acting upon that application.

The development of wind energy can and should be encouraged, but wind turbines must be good neighbors. Accordingly, the Court will, albeit reluctantly, grant injunctive relief as discussed above.

Notes and Questions

1. The court required Beech Ridge Energy to have an Incidental Take Permit under Section 9 of the ESA even though no bats were sighted on the property and no bats had been injured. What was the court's reasoning?

2. Beech Ridge Energy was expected to anticipate bat issues, prepare a Habitat Conservation Plan, and apply for an Incidental Take Permit all *prior to* construction. This failure created a potentially lengthy delay in project implementation after most of the capital was already invested. Does this timing make sense?

3. The USGS has studied the impact of wind turbines on wildlife. *See, e.g., http://www.fort.usgs.gov/products/publications/pub_abstract.asp?PubID=22170*. In February of 2011, the USFWS published Draft Land-Based Wind Energy Guidelines to address wind and wildlife issues. http://www.fws.gov/windenergy/docs/Wind_Energy_Guidelines_ 2_15_2011FINAL.pdf ["Guidelines"]. The Guidelines' intent is to "guide a developer's decision process as to whether or not a selected location is appropriate for wind development." *Id.* at 2. The Guidelines set out five tiers of review, specifically: (1) a preliminary evaluation; (2) site characterization; (3) preconstruction field monitoring and assessments to predict impacts; (4) post-construction fatality monitoring; and (5) other post-construction research to evaluate direct and indirect effects of development and to assess how these effects might be addressed. Do you think this process will give developers enough information at each stage of the project to determine whether to proceed or to abandon it if the costs are too high?

4. The iterative review process proposed by the Guidelines may mean years of preliminary study and assessment before a wind project can become operational. Some predict it may take up to seven years, in contrast to just a matter of days or months for an oil well to be approved and drilled on federal lands. Could this extra time and cost jeopardize investments in wind development?

5. The Guidelines also note, "The Service will regard such voluntary adherence and communications as evidence of due care with respect to avoiding, minimizing, and mitigating adverse impacts to species protected under the MBTA and BGEPA, and will take such adherence and communication fully into account when exercising its discretion with respect to any potential referral for prosecution related to the death of or injury to any such species." *Id.* If you were a wind developer, would this assurance be enough?

6. In 2008, the Bureau of Land Management prepared a programmatic environmental impact statement (PEIS) addressing the development of wind energy on federal lands within its jurisdiction. Under this PEIS, the BLM has issued rights-of-way for approximately 500 MW of wind power. What do you think are some of the measures the BLM identified for mitigating wildlife impacts of wind projects? After four years of deliberations, the USFS also issued guidelines for permitting wind turbines on Forest Service lands, including wildlife monitoring requirements. U.S. FOREST SERV., WILDLIFE AND FISHERIES PROGRAM MGMT HANDBOOK 2609.13_80 (effective Aug. 4, 2011); U.S. FOREST SERV., SPECIAL USES HANDBOOK 2709.11_70 (effective Aug. 4, 2011).

D. Wind v. Humans

Wind turbines have been criticized for disrupting the lives of birds and bats, but now they also are increasingly under attack for disturbing humans. Noise and shadow flicker are the most common concerns, with some correlation between the most vigorous complaints and parties who do not receive economic benefit from nearby installations. For those not near enough to experience noise or flicker, objections often focus on aesthetic issues such as views. The following excerpt and case explore the wind-and-human interface conundrum.

1. Background

This excerpt describes concerns related to shadow casting or shadow flicker and to the noises wind turbines can cause. Also, for safety reasons, aviation laws require wind towers to sport flashing lights and to be colored white or red, stymieing efforts to reduce their visual impact.

Bent Ole Gram Mortensen, *International Experiences of Wind Energy*
2 Envtl. & Energy L. & Pol'y J. 179, 189-92 (2008)

People in densely populated areas demonstrate a growing opposition to the location of nearby wind turbines, especially when it comes to siting larger wind farms. A major rationale for this backlash against land-based wind farms stems from perceived negative aesthetic effects on the natural landscape. Yet another rationale relates to potential nuisances created by wind turbines, including (1) the noise and vibrations that echo from the rotor blades, gearbox, and generator, and (2) the shadows and reflections created by the rotor blades and the tower. Of course, the perceived negative impacts of wind turbines on the landscape and creation of public nuisances are also highly subjective.

During standard operation, a wind turbine causes both mechanical and aerodynamic noise. The noise level depends on the design of the tower and rotor blades, as well as the rotational speed of the rotor blades. Technological innovations have reduced mechanical noise such that the noises alone are not considered a problem in a properly designed wind turbine. While the nature of wind turbines likely mean that a certain measure of aerodynamic noise will remain, even that type of noise has been reduced with newer turbines, and the amount of noise created by wind turbines is often considered low when compared with other common sources. However, low-frequency noise remains a problem, even if the actual impact of low-frequency noise on human beings is not yet very well understood.

Moving rotor blades also project flickering shadows onto land and buildings that can result in a common yet predictable nuisance known as "shadow casting." Shadow casting represents a potential nuisance to those living or working close to wind turbines, and the negative impacts associated with this nuisance are considered greatest in the context of larger wind turbines, whose flickering shadows can sweep great distances. In spite of the nuisance effect of shadow casting, the severity of the problem can depend on location....

Another turbine-related nuisance occurs when sunrays reflect directly from moving turbine rotor blades, causing a disorienting and inevitable flashing effect in the area of the turbine. The magnitude of this nuisance, also known as "blade glint," may depend on various factors ... [and] may be mitigated by innovations in wind turbine design and cautious turbine siting. For instance, the use of non-reflective paint has proven particularly effective in managing blade glint.

Additionally, national aviation legislation often requires cautionary lighting on turbine towers.... At night, such lighting, intended to be highly visible, can dot the landscape, thereby presenting another aesthetic impact that some people find offensive. Nuisances are normally regulated either by public law, especially environmental law, or by local laws that limit the level of disturbances permitted by neighboring activities including wind turbines. Turbine-related nuisances are most problematic in densely populated areas, in which it can be difficult to site wind turbines at a suitable distance from communities and homes.

Wind turbines are highly visible elements in the landscape, and in order to ensure the best possible wind access, turbines often must be sited in a manner that retains their visibility. In some countries, even mountain ridges are used for turbines. Large modern wind turbines are sometimes the largest structures in the vicinity of certain urban and rural areas, and the combined effects of rotor blade motion and the expansive areas of land required for optimal wind energy production serve to draw further attention to the wind turbines.

The acceptance of such structures in the landscape may differ significantly from area to area. Many people find wind turbines intrusive from an aesthetic point of view, and changes to the landscape may meet opposition from aesthetic, preservation-oriented environmentalists and neighboring groups, especially in developed countries. Planning laws normally contain regulations that preclude wind turbines and other structures from being sited within certain areas, such as nature reserves.

Furthermore, special regulations regarding minimum distances to neighboring houses exist in some countries. Modern wind turbines are generally taller and have longer rotor blades than older wind turbines. At present, the largest designs involve rotor diameters in excess of one hundred meters and a capacity of more than five MW. Longer rotor blades enable wind farms to operate with fewer turbines, which also provide indirect advantages such as lower maintenance costs and higher production. Additionally, the lower rotational speed of longer turbine blades results in reduced visual distraction in comparison with faster moving blades of smaller turbines.

––––––––––

Notes and Questions

1. Some California utilities are importing wind power generated from outside the United States. Matthew Fleischer, *Cross (Border) Winds*, HIGH COUNTRY NEWS, Feb. 15, 2010, at 6. Incentives for locating wind farms in Mexico instead of the United States include lower costs and a quicker licensing process. Could this result in fewer U.S "green" jobs?

2. What about social justice or environmental justice concerns? Is this solution simply exporting U.S. environmental concerns with wind power production to another country that may provide fewer protections for its citizens?

3. A technical analysis of wind energy facilities' impacts on the property values of nearby residences found no evidence that prices of homes surrounding wind facilities are consistently, measurably, or significantly affected by either the view of wind facilities or the distance of the home to those facilities. http://www1.eere.energy.gov/windandhydro/impacts_siting.html. Does this market data refute the aesthetics argument against wind?

2. Rankin v. FPL Energy, LLC

The following case addresses the aesthetics issues in more depth as the Texas Court of Appeals determines whether the Horse Hollow Wind Farm could be considered a nuisance under Texas law.

Horse Hollow II, with a power capacity of 299 MW, is the largest wind farm in Texas. According to the American Wind Energy Association, there are three "Horse Hollow Wind Farms" in Texas. *http://www.awea.org/la_usprojects.cfm.*

Rankin v. FPL Energy, LLC

266 S.W.3d 506 (Tex. App. 2008), *pet. denied*

Strange, J.

Several individuals and one corporation (Plaintiffs) filed suit against FPL Energy, LLC; FPL Energy Horse Hollow Wind, LP; FPL Energy Horse Hollow Wind, LP, LLC; FPL Energy Horse Hollow Wind GP, LLC; FPL Energy Callahan Wind Group, LLC; and FPL Energy Callahan, LP (FPL). Plaintiffs sought injunctive relief and asserted public and private nuisance claims relating to the construction and operation of the Horse Hollow Wind Farm in southwest Taylor County. FPL filed a motion for partial summary judgment directed at Plaintiffs' nuisance claims, and the trial court granted it in part dismissing Plaintiffs' claims to the extent they were based on the wind farm's visual impact. Plaintiffs' remaining private nuisance claim proceeded to trial. The jury found against Plaintiffs, and the trial court entered a take-nothing judgment. [This court affirmed the finding of no nuisance, and the Texas Supreme Court denied certiorari.—Ed.]

* * *

Analysis

FPL's Motion for Partial Summary Judgment

FPL asked the trial court to dismiss Plaintiffs' public and private nuisance claims contending that Plaintiffs could not assert a nuisance claim based upon the wind farm's aesthetical impact and that Plaintiffs' deposition testimony precluded their remaining nuisance claims. The trial court granted the motion in part and dismissed "Plaintiffs' claims of public and private nuisance asserted in whole or in part on the basis of any alleged aesthetic impact of [FPL's] activities." The trial court later included an instruction in the jury charge that excluded their consideration of the wind farm's aesthetic impact.

* * *

Texas Nuisance Law

Texas law defines "nuisance" as "a condition that substantially interferes with the use and enjoyment of land by causing unreasonable discomfort or annoyance to persons of ordinary sensibilities." (Citations omitted). Nuisance claims are frequently described as a "non-trespassory invasion of another's interest in the use and enjoyment of land." (Citations omitted).[4] But despite this exclusionary description, in some instances an action can be both a trespass and a nuisance. *See, e.g., Allen v. Virginia Hill Water Supply Corp.*, 609 S.W.2d 633, 636 (Tex.Civ.App.-Tyler 1980, no writ) (continuing encroachment upon the land of an adjoining owner by either erecting or maintaining a building without any right to do so is a trespass and a private nuisance).

In practice, successful nuisance actions typically involve an invasion of a plaintiff's property by light, sound, odor, or foreign substance. For example, in *Pascouet*, floodlights that illuminated the plaintiffs' backyard all night and noisy air conditioners that interfered with normal conversation in the backyard, that could be heard indoors, and that interrupted plaintiffs' sleep constituted a nuisance. 61 S.W.3d at 616. In *Bates*, the court noted that

---

4. Trespass to real property occurs when a person enters another's land without consent. *Wilen v. Falkenstein*, 191 S.W.3d 791, 797 (Tex.App.-Fort Worth 2006, pet. denied). To recover damages, a plaintiff must prove that (1) the plaintiff owns or has a lawful right to possess real property; (2) the defendant entered the plaintiff's land and the entry was physical, intentional, and voluntary; and (3) the defendant's trespass caused injury to the plaintiff. *Id.* at 798.

foul odors, dust, noise, and bright lights could create a nuisance. 147 S.W.3d at 269. In *Lamesa Coop. Gin v. Peltier,* 342 S.W.2d 613 (Tex.Civ.App.-Eastland 1961, writ ref'd n.r.e.), a cotton gin's operations were a nuisance because of its loud noises and bright lights that could be seen and heard on plaintiff's property and because of the dust, lint, and cotton burrs that would be carried there.

Texas courts have not found a nuisance merely because of aesthetical-based complaints. In *Shamburger v. Scheurrer,* 198 S.W. 1069 (Tex.Civ.App.-Fort Worth 1917, no writ), the defendant began construction of a lumberyard in a residential neighborhood. Neighboring homeowners filed suit and contended that the lumberyard would be unsightly, unseemly, and have ugly buildings and structures. The court held that this did not constitute a nuisance, writing:

The injury or annoyance which warrants relief against an alleged nuisance must be of a real and substantial character, and such as impairs the ordinary enjoyment, physically, of the property within its sphere; for if the injury or inconvenience be merely theoretical, or if it be slight or trivial, or fanciful, or one of mere delicacy or fastidiousness, there is no nuisance in a legal sense. Thus the law will not declare a thing a nuisance because it is unsightly or disfigured, because it is not in a proper or suitable condition, or because it is unpleasant to the eye and a violation of the rules of propriety and good taste, for the law does not cater to men's tastes or consult their convenience merely, but only guards and upholds their material rights, and shields them from unwarrantable invasion. *Id.* at 1071-72. In *Dallas Land & Loan Co. v. Garrett,* 276 S.W. 471, 474 (Tex.Civ.App.-Dallas 1925, no writ), the court found that a garage being built for residents of an apartment complex was not a nuisance because "[m]atters that annoy by being disagreeable, unsightly, and undesirable are not nuisances simply because they may to some extent affect the value of property." In *Jones v. Highland Mem'l Park,* 242 S.W.2d 250, 253 (Tex.Civ.App.-San Antonio 1951, no writ), the court held that the construction of a cemetery on adjacent property did not constitute a nuisance, noting: "However cheerless or disagreeable the view of the cemetery in question may be to appellees, and no matter what unpleasant or melancholy thoughts the same may awaken, no reason is thereby shown why appellants should be restrained from making such use of their property."

Plaintiffs' Nuisance Claim

Plaintiffs advance several arguments why this case law does not preclude their private nuisance action. First, they argue that aesthetics may be considered as one of the *conditions* that creates a nuisance. Plaintiffs concede that, if their only complaint is subjectively not liking the wind turbines' appearance, no nuisance action exists. But, they contend that the jury was entitled to consider the wind farm's visual impact in connection with other testimony such as: the turbines' blinking lights, the shadow flicker affect they create early in the morning and late at night, and their operational noises to determine if it was a nuisance. Second, they note that nuisance law is dynamic and fact-specific; therefore, they contend that older case holdings should not be blindly followed without considering intervening societal changes. Third, nuisance claims should be viewed through the prism of a person of ordinary sensibilities and case law involving unreasonable plaintiffs asserting subjective complaints should be considered accordingly.

FPL responds that the trial court ruled correctly because no Texas court has ever recognized a nuisance claim based upon aesthetical complaints and notes that, in fact, numerous courts have specifically rejected the premises behind such a claim. *See, e.g., Dallas Land & Loan,* 276 S.W. at 474; *Shamburger,* 198 S.W. at 1071 ("the law will not declare a thing a nuisance because ... it is unpleasant to the eye.") FPL argues that sound public policy supports such a rule because notions of beauty or unsightliness are necessarily subjective in nature and that giving someone an aesthetic veto over a neighbor's use of his land would be a recipe for legal chaos. Finally, FPL argues that the wind farm does not prevent any of the plaintiffs from using their property but at most involves an emotional reaction to the sight of the wind turbines and contends that an emotional reaction alone is insufficient to sustain a nuisance claim.

When FPL moved for summary judgment, Plaintiffs presented affidavits from the plaintiffs to establish that the wind farm was a nuisance. Plaintiffs' affidavits personalize individual objections to the wind farm's presence and to the use of wind turbines for generating electricity commercially. They also express a consistent theme: the presence of numerous 400-foot-tall wind turbines has permanently and significantly diminished the area's scenic beauty and, with it, the enjoyment of their property. Some Plaintiffs, such as Linda L. Brasher, took issue with the characterization of her complaint as just aesthetics. She acknowledged not liking the turbines' looks but contended that they had a larger impact than mere appearance. Brasher stated that she and her husband had purchased their land to build a home and to have a place "for strength, for rest, for hope, for joy, for security—for release."[sic] They had plans for building and operating a small bed and breakfast but cancelled those plans in response to the wind farm. Brasher characterized the presence of the wind farm as "the death of hope."

Plaintiffs' summary judgment evidence makes clear that, if the wind farm is a nuisance, it is because Plaintiffs' emotional response to the loss of their view due to the presence of numerous wind turbines substantially interferes with the use and enjoyment of their property. The question, then, is whether Plaintiffs' emotional response is sufficient to establish a cause of action. One Texas court has held that an emotional response to a defendant's lawful activity is insufficient. *Maranatha Temple, Inc. v. Enterprise Products Co.,* 893 S.W.2d 92 (Tex.App.-Houston [1st Dist.] 1994, writ den'd), involved a suit brought by a church against the owners and operators of companies involved in an underground hydrocarbon storage facility. The church's claims included a nuisance action. The trial court granted summary judgment against the church. *Id.* at 96. The question before the Houston First Court was whether a nuisance action could exist when the only claimed injury was an emotional reaction to the defendants' operations. The court found that a nuisance could occur in one of three ways: (1) by the encroachment of a physically damaging substance; (2) by the encroachment of a sensory damaging substance; and (3) by the emotional harm to a person from the deprivation of the enjoyment of his or her property, such as by fear, apprehension, offense, or loss of peace of mind. *Id.* at 99. The court noted that nuisance claims are subdivided into nuisance per se and nuisance in fact.[7] *Id.* at 100. Because the operation of the storage facility-just like FPL's wind farm-was lawful, it could not constitute a nuisance per se. This last factor was critical. The court recognized that no case or other authority specifically gives a nuisance-in-fact cause of action based on fear, apprehension, or other emotional reaction resulting from the lawful operation of industry and affirmed the summary judgment. 893 S.W.2d at 100 & n. 6.

Plaintiffs do not contend that FPL's operations are unlawful but minimize this factor by arguing that even a lawful business can be considered a nuisance if it is abnormal and out of place in its surroundings. Plaintiffs are correct that several Texas courts have recited

---

7. A nuisance per se is an act, occupation, or structure that is a nuisance at all times, under any circumstances, and in any location. A nuisance in fact is an act, occupation, or structure that becomes a nuisance by reason of its circumstances or surroundings. *Freedman v. Briarcroft Prop. Owners, Inc.,* 776 S.W.2d 212, 216 (Tex.App.-Houston [14th Dist.] 1989, writ denied).

this general principle; but, in each of the cases cited by Plaintiffs, the nuisance resulted from an invasion of the plaintiff's property by flooding, flies, or odors. We cannot, therefore, agree with Plaintiffs that merely characterizing the wind farm as abnormal and out of place in its surroundings allows a nuisance claim based on an emotional reaction to the sight of FPL's wind turbines.

We do not minimize the impact of FPL's wind farm by characterizing it as an emotional reaction. Unobstructed sunsets, panoramic landscapes, and starlit skies have inspired countless artists and authors and have brought great pleasure to those fortunate enough to live in scenic rural settings. The loss of this view has undoubtedly impacted Plaintiffs. A landowner's view, however, is largely defined by what his neighbors are utilizing their property for. Texas case law recognizes few restrictions on the lawful use of property. If Plaintiffs have the right to bring a nuisance action because a neighbor's lawful activity substantially interferes with their view, they have, in effect, the right to zone the surrounding property. Conversely, we realize that Plaintiffs produced evidence that the wind farm will harm neighboring property values and that it has restricted the uses they can make of their property. FPL's development, therefore, could be characterized as a condemnation without the obligation to pay damages.

Texas case law has balanced these conflicting interests by limiting a nuisance action when the challenged activity is lawful to instances in which the activity results in some invasion of the plaintiff's property and by not allowing recovery for emotional reaction alone. Altering this balance by recognizing a new cause of action for aesthetical impact causing an emotional injury is beyond the purview of an intermediate appellate court. Alternatively, allowing Plaintiffs to include aesthetics as a condition in connection with other forms of interference is a distinction without a difference. Aesthetical impact either is or is not a substantial interference with the use and enjoyment of land. If a jury can consider aesthetics as a condition, then it can find nuisance because of aesthetics. Because Texas law does not provide a nuisance action for aesthetical impact, the trial court did not err by granting FPL's motion for partial summary judgment and by instructing the jury to exclude from its consideration the aesthetical impact of the wind farm. Issue One is overruled.

Notes and Questions

1. What do you think of the *Rankin* court's rationale? Do you agree that "notions of beauty or unsightliness are necessarily subjective in nature and that giving someone an aesthetic veto over a neighbor's use of land would be a recipe for legal chaos"? If not, how should a court weigh aesthetic claims?

2. While the *Rankin* court decided that a nuisance action could not be based on aesthetic concerns, the Supreme Court of Appeals of West Virginia seems to have reached the opposite result in *Burch v. Nedpower Mount Storm*, 647 S.E.2d 879 (W. Va. 2007). In *Rankin*, the plaintiffs were attempting to shut down the Horse Hollow Wind Farm. In contrast, the *Burch* case involved a pre-construction effort to enjoin the construction of a 200-turbine wind power facility. Although the circuit court granted judgment on the pleadings for the defendant, the Supreme Court of Appeals of West Virginia reversed, holding that the plaintiff's allegations of nuisance were legally sufficient to survive an attempt at injunction. How would you reconcile the two approaches?

3. Beyond aesthetics, health concerns might be an alternative basis for a nuisance claim. Broken turbine blades and "ice throw" (large chunks of ice thrown off turbine blades)

> **Wind power and radar** — Because current wind turbine technology employs towers that reach heights of over 300 feet, it is subject to regulation by the Department of Defense (DOD), the Federal Aviation Administration (FAA), and the Federal Communication Commission (FCC). In 2006, the DOD found that wind turbines could impair radar systems. Another concern is that large wind farms can distort weather forecasts, appearing on Doppler radar as tornados. The FAA has also determined that some wind towers cannot be located near airports because they interfere with air traffic.

have mostly been eliminated by newer turbine designs. However, at least one doctor has diagnosed "wind turbine syndrome," described as "the cluster of symptoms—sleeplessness, headaches, depression, dizziness and nausea—that she has identified in people ... who live within a mile of industrial-size wind turbines." Kristin Choo, *The War of Winds*, A.B.A. J. 54, 56 (Feb. 2010). Consider how recognition of this syndrome might change the potential for a wind nuisance lawsuit.

4. Noise appears to be one of the biggest complaints against wind turbines. EPA guidelines recommend 55 dBA (A-weighted decibels) for daytime outdoor noise limits in urban areas and 45 dBA for rural areas. http://www.epa.gov/history/topics/noise/01.htm. Yet lower decibel noises can disturb night sleep, and "rhythmic pulsing" as well as "low-frequency noise—sound that vibrates relatively slowly and is pitched low on the scale of sounds audible to the human ear"—may be especially hard to ignore. Choo, *supra*, at 57. Could one argue that noise, as with aesthetics, has a subjective component?

5. Can wind developers proactively address some of the most common complaints? Some companies are paying landowners up front in exchange for a waiver of the right to sue on the basis of noise. Do you think this is a good idea or might it be interpreted as an admission that noise complaints are legitimate concerns?

E. Wind v. Other Resources

Even when wind development is located in areas that do not create wildlife, health, or aesthetic issues, it still confronts opposition—sometimes by developers of competing interests. This section focuses on the potential conflicts between wind and mineral development. It also addresses some lease provisions that may help anticipate and minimize conflicts.

K.K. DuVivier and Roderick E. Wetsel, *Jousting at Windmills: When Wind Power Development Collides with Oil, Gas, and Mineral Development*
55 Rocky Mtn. Min. L. Inst. 9-2 to 9-30 (2009)

Success for the renewable energy economy rides on wind power. Although wind currently accounts for only 1 percent of the total electricity generation in the United States, the Obama administration hopes to leverage it to 25 percent by 2030. In 2007, the United States' installed wind power capacity totaled 16,515 megawatts (MW). Approximately 15 percent of this total came from California, the birthplace of the modern wind energy industry. Four other key states for wind generation are Iowa, which now has surpassed California for the number two slot; Minnesota; Washington; and Colorado. But currently Texas reigns as king, with more than a quarter of the U.S. total generating capacity.

As fate would dictate, wind companies have constructed their new projects in vast rural areas within or adjacent to the Texas oilfields, making conflict between the two industries inevitable. Not only does Texas hold the top rank for wind generating capacity, it also is number one for oil production in the continental United States. In fact, potential for clashes have erupted across the country as wind resources seem to have an uncanny knack for overlapping existing mineral-rich areas. [Figure 3.5 shows U.S. wind resources and their concentration in many parts of the country that also have significant oil and gas or mineral potential—Ed.]

* * *

The Wind Boom

… There were no commercial-scale wind turbines in Texas prior to 1995, but by the first quarter of 2009, the exponential growth of the wind industry evoked memories of the oil booms during the early part of the twentieth century. Soon after 2000, small towns in West Texas and the Texas Panhandle became hives of activity as landmen descended in droves, presenting unprepared landowners with leases written in fine print and the promise of riches rivaling those of the speculators who first brought "big oil" to this part of the country in the last century.

In the years between 2003 and 2009, the West Texas town of Sweetwater evolved from being "The Home of the World's Largest Rattlesnake Roundup" to being "The Wind Energy Capital of the World," with three of the world's largest wind farms. The Sweetwater area

Figure 3.5: U.S. wind resource potential

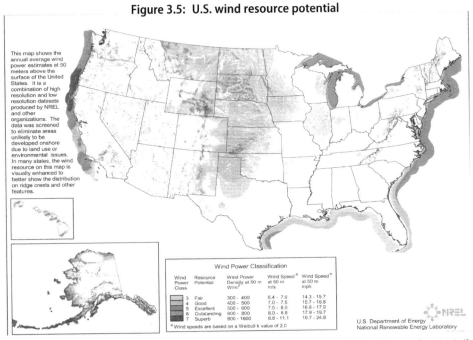

National Renewable Energy Lab., http://www.nrel.gov/gis/pdfs /windsmodel4pub1-1-9base200904enh.pdf (last visited July 17, 2011)

was attractive because it provided three ideal conditions for development of a wind farm: (1) adequate wind capacity of approximately 38 percent to 42 percent; (2) proximity to existing high-voltage transmission lines; and (3) plenty of wide-open space less than 200 miles from the metropolitan load centers where the electricity can be sold.

In order to reap the benefits of this new industry, landowners in many Texas counties formed wind associations and selected "steering committees" to hire attorneys to attract wind developers and to negotiate wind leases. Many county governments offered attractive ad valorem tax. Promoters began putting together packages of leases in order to "flip" deals to interested developers. In 2003, the mayor of Sweetwater formed the West Texas Wind Energy Consortium in order to educate landowners and organize political support for wind development in Texas....

In 2007, the Texas Public Utilities Commission (PUC) reviewed 25 areas nominated for transmission development and designated a handful from West Texas and the Texas Panhandle to be Competitive Renewable Energy Zones (CREZ). This designation, and the hearings and publicity that preceded it, set off a further land rush by developers. Throughout the last half of 2007 and all of 2008, landmen descended like Biblical locusts upon many rural towns in the Texas Panhandle.

Although Texas experienced the fastest growth, similar wind booms have spread to other states. With close to 1,000 new units statewide in 2008 alone, wind turbines seemed to be popping up almost as fast as corn stalks in some Iowa counties. Many states with Renewable Portfolio Standards—such as California, Colorado, Iowa, Kansas, New Mexico, Oklahoma, and Wyoming—have seen significant increases in wind development....

Wind power has been "the fastest growing source of new power generation" over the last few years. Wind energy generating capacity in the United States increased by 27 percent in 2006, 45 percent in 2007, and another 50 percent in 2008....

Although the wind boom was slowed somewhat by the national recession at the end of 2008, no bust is in sight. Statements of support for wind power from President Barack Obama and Secretary of the Interior Ken Salazar suggest the wind industry will continue to be robust for years to come.

Impact of Wind Energy on the Mineral Industry

The spread of the wind boom across mineral-producing states has set off an alarm among mineral owners because of the large swaths of land needed for wind development. The first concern is the enormous size of today's turbines, which have power ratings ranging from 250 watts to five MW. For example, the General Electric 1.5 MW turbine rises to a height of eighty meters (262 feet) at its hub and has a rotor radius of approximately thirty-eight meters (125 feet). Although the surface footprint for each turbine is relatively small in relation to its height, wind development requires more extensive surface use than traditional oil and gas development because of additional factors: (1) turbine spacing, (2) buffer zones, (3) other surface uses such as for roads, substations, operations and maintenance facilities, and laydown yards, and (4) overhead and underground transmission, collection, and distribution lines.

First, and most obviously, land is needed for the placement of the turbines themselves. A variety of factors determines turbine spacing, including terrain, wind speed, wind direction, turbine size, and access to an electric grid....

Second, when putting together a wind farm, developers must include land not only for the placement of the turbines themselves, but also for buffer zones to prevent obstructions upwind. Many wind sites include leases or non-obstruction easements for land at least ten rotor-lengths (approximately one-half to one mile) away from the turbines to avoid "waking" or wind disturbance to the nearby generators. Depending upon the topography and number of turbines installed, a typical wind farm can cover anywhere from 3,000 to 150,000 acres, which may or may not include the additional acreage needed for buffer zones.

Third, wind companies must utilize significant portions of the surface for (1) roads, (2) O&M [operations & maintenance] facilities and substations, and (3) laydown yards [locations where the turbines and other equipment are set out flat before being erected]. Wind farm roads are huge in comparison with oilfield roads and may be as much as sixty feet wide prior to turbine construction in order to accommodate the large cranes needed to erect the turbines. Service roads connect each of the turbines; access roads provide ingress and egress to and from public roads and adjoining properties. Each wind farm has one or more O&M facilities and substations. These facilities include power stations and company offices and are located on tracts of three to five acres each.

Fourth, and most significantly, the turbines are linked by a spiderweb of underground and overhead transmission, collection, and distribution lines. Although these lines take up little surface space, the fact that they must be crossed or avoided can interfere with concurrent use of the same land for oil, gas, and mineral exploration and development. Large overhead lines containing many strands of wires up and down the poles are of particular concern, as it is often difficult for drilling rigs to be moved underneath them. [Figure 3.6 shows a representation of the spiderweb of subsurface lines running beneath a typical wind farm—Ed.]

This plethora of surface and subsurface activities required to develop wind power is made possible by a broad and extensive "purpose clause" in wind leases. The broad powers granted to wind companies in some wind leases have raised tensions with oil, gas, and mineral developers. In recent years, disputes have arisen between mineral companies and

Figure 3.6: Subsurface imprint of wind farms

Illustration by Christopher Van Essen

wind companies about the conduct of seismic operations, location of drilling rigs and tank batteries, use of roads, and ingress and egress to properties. Although there has not yet been any reported litigation regarding these issues, the storm is on the horizon.

Common Law Approach I—Dominant-Servient Estate and Accommodation

The Industrial Revolution made the search for and supply of fossil fuels and metals to feed factories a priority. When landowners severed estates to allow for development of these resources, the mineral estate earned nearly uncontested dominance over the surface. Consequently, courts have upheld the principle that ownership of minerals includes an implied right to interfere with the surface owner's activities and to use as much of the surface "as necessary" in accessing and extracting the minerals. Thus, a dominant owner is liable to the servient owner only for damages inflicted negligently.

The hardship the "unidimensional" dominant-servient estate doctrine imposed on surface owners has been mitigated in some situations by the "multidimensional" accommodation doctrine. Under this doctrine, courts require the mineral owner to accommodate a surface owner's use if the mineral owner has a reasonable alternative for accessing and extracting its minerals. Yet, the accommodation doctrine does not restrict mineral development altogether.

* * *

Altering Common Law Approaches through Express Agreements

Wind developers, particularly those from Europe, have been surprised to learn that under U.S. law their wind interest might be servient to dominant mineral estates. They have been rightfully concerned about investing hundreds of millions of dollars in a wind project that could be subject to interference by the owners of the mineral estate. Before providing financing, some investors require a title search and a mineral endorsement. These are available, however, only if the title company finds that there is little or no likelihood of mineral development.

When mineral leases currently exist on the property, or if there is any future potential for mineral development, most wind investors require a more proactive approach. Instead of relying on a judge's resolution of potential conflicts, they seek instead to alter the common law regimes through express agreements.

* * *

If Wind Rights Are First In Time

If the wind rights grantor owns the surface and mineral estates, and has not previously sold or leased any part of the estate, then wind developers have been able to negotiate clauses in their leases that greatly restrict oil, gas, and mining activities on the surface. Some clauses in early wind leases even attempted to reverse the dominant estate doctrine and make the mineral estate servient to the wind estate.

Additionally, wind companies have mandated that all future oil and gas leases entered into by the surface owner contain provisions referencing the wind lease and requiring the oil and gas lessee to enter into a surface use or accommodation agreement with the wind lessee. A surface accommodation agreement makes provision for any concurrent surface operations (including required distances from facilities), notice prior to the commencement of drilling or construction, use and maintenance of roads, indemnity for surface damages and personal injuries, and insurance. An accommodation agreement is now customary for an oil company which desires to drill on a wind farm.

As a further impediment, wind companies in their leases have sought to impose restrictions on surface use for oil and gas development. These clauses are very broadly written so as to prohibit the location of drilling rigs or other oil and gas facilities within a specified number of feet of any existing wind turbine, substation, or transmission line. Such clauses also provide that in any future oil and gas or mining lease, the surface owner must provide that the mineral company will not conduct any activities within the areas specified and will not otherwise unreasonably interfere with the wind company's rights under its lease....

Many wind leases also contain a broad "no-interference clause," which provides that the surface owner and its lessees shall not currently or prospectively disturb or interfere with the construction, installation, maintenance, or operation of the wind power facilities or the undertaking of any other activities permitted under the lease....

If Mineral Rights Are First In Time

In conflict areas, it is more likely that the mineral estate has been severed and perhaps leased before a wind developer enters the scene. In these situations, wind developers first provide the mineral interest owner with notification.

Next, as a first line of defense when the mineral estate beneath a wind lease is severed from the surface, wind companies have sought to obtain a surface waiver or non-interference agreement from the mineral interest owners who did not also own the surface estate....

These efforts have often proved futile. With the assumption that they have the common law advantage of dominant estate ownership and have no obligation to accommodate the servient surface use of the wind lessee, some mineral owners have hindered development of wind projects by refusing to negotiate reasonable non-disturbance agreements or have requested exorbitant sums as compensation for them.

The Role of the Grantor

Concurrent wind and mineral development is more likely when the grantor can act as referee between these separate interests. The grantor can try to negotiate clauses in the lease agreements that put pressure on lessees to work together....

However, the grantor can also be caught in the middle of battles between wind and mineral developers. For example, oil companies have fought back against wind development leases by requesting promises of their own from the grantors. Oil leases now frequently require that payment of the bonus consideration is contingent upon and subject to execution of an accommodation agreement by any wind lessee on the property.... Demands from wind lessees or mineral lessees that the grantor make their rights dominant can put the grantor in an untenable position, inviting litigation.

Furthermore, tensions between wind and mineral developers can be heightened if the grantor is not positioned to intervene. This occurs in at least two situations. First, the federal government's standard form lease reserves the right to lease different resources to different parties because the government believes that wind and mineral development are compatible....

Second, if a private grantor severed the mineral estate before executing the wind lease, a wind developer might be required to work with a mineral lessee who has interests that do not align with the wind grantor. The potential conflicts increase significantly with severance of the wind from the surface estate. When the wind rights are owned by one party and the mineral rights by another, there is little incentive for any of the parties to work together.

The situation is further exacerbated if the surface owner, who is most impacted by both wind and mineral surface operations, receives no royalty or other benefit from the development of either resource.

Common Law Approach II — Multiple Mineral Development

There is a certain irony about disputes between wind and mineral interests, especially when the conflict is between oil and gas companies and wind companies. First, some of the large wind power developers are divisions of oil and gas companies, so one division may be fighting with another in the same company. Second, wind is considered intermittent power; it can produce electricity only when the wind is blowing. Consequently, our nation can develop wind as an alternative renewable power source only if there is a back up, usually from fossil fuel plants run with oil or, more often, with natural gas.

Although some have stated that a wind lease is "incontestably not a transfer of mineral rights," the first of only two courts in the United States that have addressed the severance of wind analogized wind rights to oil and gas interests. [*Contra Costa Water Dist. v. Vaquero Farms, Inc.*, 68 Cal. Rptr.2d 272 (Cal. Ct. App. 1997).] That decision addressed wind severance in a backhanded way through condemnation, but other courts may use the rationale that the wind estate should be treated in the same way as a mineral estate.

Thus, the initial question in approaching conflicts between mineral lessees and wind lessees is the status of the wind estate. Under the "ad coelum" doctrine, the owner of the soil, or surface, also has ownership rights in everything from the center of the earth to the skies. Application of the ad coelum doctrine may justify characterizing wind flowing across a piece of land as a severable wind power estate. If such a wind power estate is viewed as part of the surface estate, then traditional notions of the dominant-servient estate and accommodation doctrines may apply.

However, the development of wind is comparable to the development of other mineral commodities and, arguably, wind estates are closer to mineral rights. If courts decide to treat wind as a "mineral," then the common law rules that apply to conflicts between mineral and surface lessees may not control. Instead, courts may prefer to look to alternative common law models, such as those controlling multiple mineral development, including (1) avoidance; (2) first in time, first in right; and (3) equal dignity.

Avoidance is one strategy employed in multiple mineral development contexts. For example, after concluding that joint development of potash and oil and gas was "unworkable," the United States and the State of Utah withdrew certain lands in the Cane Creek area from oil and gas leasing. Similarly, in New Mexico, oil and gas operations that conflict with potash development may be prohibited. Some investors are eyeing abandoned mine sites as potential locations for renewable energy development because they may receive incentives for making use of the site, and if a deposit is d[e]pleted, there should be no competition of use of the surface.

When avoidance is not an option, however, alternative methods for resolving conflicts must be addressed. Although there is no well-defined system for resolving conflicts, the traditional approach appears to be one of "first in time, first in right." For example, in the Powder River Basin of Wyoming, the government had issued several leases before it considered withdrawal from leasing to avoid conflict between coal and oil and gas development. To address the problem there, some of the subsequent leases include special stipulations prohibiting coal operations that might unreasonably interfere with preexisting oil and gas leases.

> **Is siting renewables on abandoned mines or other brownfields a good solution?**
>
> One approach that addresses the land use issue of wind (and sometimes solar development) is the placement of renewable resources in brownfields, areas that have been used for industrial projects in the past and are currently inappropriate for housing or other needs because they are polluted and need reclamation. This concept is discussed in more depth in Chapter 5.

While first in time, first in right may be the current approach of the U.S. government in multiple mineral development contexts, it is better as a default procedure. An alternative that is "consistent with the balancing mechanisms of multiple use philosophies," is an equal dignity of estates approach. If mineral estates have equal dignity, a court may value interference with a competing mineral right more highly than it might value interference with use of the surface. A coal mining case from the eastern U.S. can provide an example.

Although the rationales for upholding a right of access to develop underlying strata vary, a leading coal case on the topic is *Chartiers Block Coal Co. v. Mellon*. In this case, a coal lessee sought to restrain oil and gas operations by a subsequent lessee, alleging that the drilling was a hazard to its coal mining operations. The Pennsylvania Supreme Court denied the injunction. The landowner's initial grant retained the underlying strata and a right of access to it; otherwise the reserved mineral estate below the coal would be inaccessible and valueless. The majority in *Chartiers* conditioned the oil and gas lessee's right of access on indemnification to the coal operator for damages.

The *Chartiers* decision included a concurring decision basing the right of access on a reciprocal servitude theory. The reciprocal servitude theory did not rest on priority of possession or indemnification alone. Instead, the concurrence resolved the conflict through an approach similar to the accommodation doctrine: giving the trial court discretion to impose terms for the right of access, for the precautions each lessee must observe, and for compensation. Ultimately, the concurrence urged the trial court to "exercise its equitable powers to adjust and balance the competing interests."

From a landowner's perspective, it might be more profitable to develop the traditional mineral estate instead of the wind estate. However, a multiple mineral development framework based on equal dignity of the estates might at least provide indemnification for the wind developer without having to prove negligence on the part of the mineral lessee. Regardless of the common law model used, it seems preferable for both parties to participate in good faith negotiations for a joint use agreement instead of litigating and leaving their fate within a judge's discretion.

Notes and Questions

1. Federal mineral leases presume that multiple minerals can be developed. Yet, some commentators have concluded that multiple mineral development has not worked well in the mineral context. Is accommodation a reasonable alternative?

2. Although first-in-time is often the default in a multiple development context, should other balancing criteria be considered? Does first-in-time make sense if wind power precludes development of more monetarily lucrative reserves of oil and gas or minerals?

Also, should wind power development come first if extracting another deposit could be completed in less than five years, when a wind farm might tie up other development of the land for twenty to fifty years?

3. Another area of tension between wind power and other resources is the use of transmission facilities. During periods of high water and high winds in 2011, the Bonneville Power Administration, the federal agency that operates hydropower dams on the Columbia River, ordered wind farms to shut down for several hours each day. The BPA claims it cannot shut off the dams instead because water over the spillways would cause turbulence that could injure juvenile salmon. The closures have cost wind energy companies millions of dollars because they cannot deliver the power and resulting tax and energy credits they have committed to California utilities. On June 13, 2011, five wind companies filed a claim with FERC alleging BPA used its transmission marketing power to curtail competing generators. Can you think of ways to address these conflicts?

4. The DuVivier-Wetsel article included specific examples of agreements as appendices. One sample follows here. As you review it, consider whether there is anything you would add or delete.

K.K. DuVivier and Roderick E. Wetsel, *Jousting at Windmills: When Wind Power Development Collides with Oil, Gas, and Mineral Development*
55 Rocky Mtn. Min. L. Inst. 9-37 to 9-44 (2009)

Appendix III: Sample Accommodation Agreement

NOTICE OF CONFIDENTIALITY RIGHTS UNDER SECTION 11.008 OF THE TEXAS PROPERTY CODE: IF YOU ARE A NATURAL PERSON, YOU MAY REMOVE OR STRIKE ANY OF THE FOLLOWING INFORMATION FROM THIS INSTRUMENT BEFORE IT IS FILED FOR RECORD IN THE PUBLIC RECORDS: YOUR SOCIAL SECURITY NUMBER OR YOUR DRIVER'S LICENSE NUMBER.

ACCOMMODATION AGREEMENT

DATE: _____

PARTIES: _____
         ("WIND LESSEE" or sometimes "Party")

and _____
         ("O&G LESSEE" or sometimes "Party")

SUBJECT PREMISES: See Exhibit A attached hereto and incorporated herein.

1. <u>Purpose</u>. WIND LESSEE has acquired, or has an option to acquire, by surface lease, the right to install and operate facilities for the generation and transmission of electric power derived from wind energy on the Subject Premises. The parties acknowledge that WIND LESSEE intends to use the Subject Premises for the construction, installation, operation and maintenance of large wind turbines, overhead transmission, collection, and distribution electric and communication lines, substations, switching stations, roads, operations buildings, roads and related equipment and facilities (all of which are collectively referred to herein as "Wind Facilities") for the conversion of wind energy to electricity and for the collection and transmission of wind-generated electric power. O&G LESSEE

is or will be the owner of a lease of all or an interest in the oil, gas and other gaseous substances that can be produced from a well or wells in and under the Subject Premises, or a portion thereof. The parties acknowledge that O&G LESSEE intends to use the Subject Premises, or portions thereof, to explore for, drill wells for, mine, produce and transport oil, gas and other gaseous substances that can be produced from a well or wells from the Subject Premises and to install and maintain wells, tanks, tank batteries, roads, pipelines, flow lines, power lines, compressors, and other permanent and semi-permanent equipment and facilities for the production, handling, storage, treatment, transportation and marketing of such production (all of which are collectively referred to herein as the "O&G Facilities"). WIND LESSEE and O&G LESSEE desire to mutually agree, in the respects set forth herein, for themselves and for their respective heirs, successors and assigns, on the manner in which they will exercise the rights associated with their respective estates so that each will accommodate, and interfere as little as reasonably practical with, the use of the Subject Premises by the other.

2.   Oil and Gas Operations. O&G LESSEE agrees that all exploration for and development and production of oil, gas and other gaseous substances that can be produced from a well on the Subject Premises will be conducted in a manner that will reasonably accommodate WIND LESSEE's activities, and, insofar as is reasonably possible and without increasing the cost or risk (including economic risk) of such oil and gas development activities, will not interfere with the operation of the Wind Facilities. Without limiting the generality of the foregoing, and in addition to all other covenants and obligations imposed by law, O&G LESSEE agrees as follows:

(a)   Unless otherwise agreed to by WIND LESSEE, no O&G Facilities for O&G LESSEE's operations will be located within three hundred (300) feet of the center point of any wind turbine tower, other than roads as otherwise provided for herein.

(b)   O&G LESSEE will not place O&G Facilities on the Subject Premises without first consulting with WIND LESSEE in order to determine a location for such items that will not (or will as little as reasonably possible without increasing O&G LESSEE's cost or risk) materially disrupt the flow of wind currents over and across the land, such that the wind disruption or any other circumstance resulting from the placement of such equipment or facilities reduces the capacity of the Wind Facilities to generate electricity from the wind; provided that drilling rigs and workover rigs may be erected temporarily without such consultation and remain in place so long as necessary for the drilling, reworking, or recompletion of the well for which they have been engaged. WIND LESSEE will notify O&G LESSEE within thirty (30) days after receipt of O&G LESSEE's notice, as hereinafter provided, if any equipment or facilities O&G LESSEE proposes to place on the Subject Premises will cause such a disruption, and will provide such information and documentation as may be necessary to demonstrate the disruptive effect of O&G LESSEE's proposed placement, whereupon O&G LESSEE will be obligated to take WIND LESSEE's notice into account and accommodate WIND LESSEE's reasonable requests concerning such O&G Facilities to the extent above provided. O&G LESSEE agrees not to construct or erect any building or structure on the Subject Premises higher than forty (40) feet, other than a temporary drilling or workover rig, without the prior written approval of WIND LESSEE.

(c)   O&G LESSEE and its agents, employees and contractors will not enter upon the area within fifty (50) feet immediately surrounding WIND LESSEE's turbines and other Wind Facilities without the permission of WIND LESSEE and will not tamper with any Wind Facilities, other than roads or at crossings of underground or overhead electric transmission or collection Wind Facilities (as otherwise provided herein).

(d)   Unless otherwise agreed to by WIND LESSEE, in conducting geophysical exploration or construction of O&G Facilities, O&G LESSEE will perform no blasting within two hundred (200) feet of any Wind Facility. O&G LESSEE will notify WIND LESSEE of any intended blasting operations more than twenty (20) days prior to commencement of same. If due to subsurface conditions, greater distances from Wind Facilities are required by WIND LESSEE to avoid damage to the Wind Facilities or subsurface support thereof, WIND LESSEE will notify O&G LESSEE within ten (10) days after such notice of additional setback requirements and the location thereof, and O&G LESSEE will comply with any reasonable requests by WIND LESSEE for such additional setback. O&G LESSEE will take all available precautions to shield the Wind Facilities against blasting debris. O&G LESSEE agrees to monitor any seismic surveying to ensure compliance with this Agreement.

(e)   O&G LESSEE will give WIND LESSEE at least twenty (20) days notice (i) prior to commencement of the drilling of any well, which notice will advise WIND LESSEE of the proposed location, a description, with approximate dimensions, of the drilling rig and other equipment that will be used, and the estimated time for the drilling and completion of the well, and (ii) prior to the construction or installation of any O&G Facilities, including the location at which each is proposed to be placed and their approximate dimensions. O&G LESSEE agrees to consult with WIND LESSEE and to provide such further information regarding proposed operations as WIND LESSEE may reasonably request including maps or plats depicting the location of such O&G Facilities, if prepared for O&G LESSEE. O&G LESSEE agrees to consult with WIND LESSEE concerning any matter in which WIND LESSEE perceives the possibility of conflict between the parties' respective uses of the Subject Premises.

3. Wind Operations. WIND LESSEE agrees to conduct its operation and development of the Subject Premises for the conversion of wind energy to electricity and for the collection and transmission of wind-generated electric power and operation of the Wind Facilities in a manner that will reasonably accommodate O&G LESSEE's said activities, and without increasing the cost or risk (including economic risk) of such wind development activities, will not interfere with the operation of the O&G Facilities. Without limiting the generality of the foregoing, and in addition to all other covenants and obligations imposed by law, WIND LESSEE agrees as follows:

(a)   Unless otherwise agreed to by O&G LESSEE, WIND LESSEE will not locate any wind turbine structure, or any part thereof, within three hundred (300) feet of an existing well capable of producing oil, gas and other gaseous substances.

(b)   WIND LESSEE agrees not to obstruct O&G LESSEE's ingress and egress to and from any existing wells and facilities used in the production of oil, gas and other gaseous substances and will not (without O&G LESSEE's consent) locate its wind turbines so closely to each other as to prevent the safe and orderly passage of vehicles and equipment between them.

(c)   WIND LESSEE will not enter upon the area within fifty (50) feet immediately surrounding O&G LESSEE's wells, tank batteries, and other surface facilities without the permission of O&G LESSEE and will not tamper with any equipment or other property of O&G LESSEE, other than roads or at crossings of underground or overhead electric lines, pipelines, or flowlines (as otherwise provided herein).

(d)   Unless otherwise agreed to by O&G LESSEE, in conducting construction of Wind Facilities, WIND LESSEE will perform no blasting within two hundred (200) feet of any O&G Facility. WIND LESSEE will notify O&G LESSEE of any intended blasting operations more than twenty (20) days prior to commencement of same. If due to subsurface conditions, greater distances from O&G Facilities are required by O&G LESSEE to avoid damage to the O&G Facilities or subsurface support thereof, O&G LESSEE will notify WIND LESSEE within ten (10) days after such notice of additional setback requirements and the location thereof, and WIND LESSEE will comply with any reasonable requests by O&G LESSEE for such additional setback. WIND LESSEE will take all available precautions to shield the O&G Facilities against blasting debris. WIND LESSEE agrees to monitor any seismic surveying to ensure compliance with this Agreement.

(e)   WIND LESSEE will provide O&G LESSEE at least twenty (20) days notice of any new construction or installation of Wind Facilities, including the nature and location of each item thereof, and notice upon completion of any such construction or installation. WIND LESSEE will provide O&G LESSEE such further information concerning the Wind Facilities as O&G LESSEE may reasonably request and as may be relevant to O&G LESSEE's operations on this Agreement, including maps or plats depicting the location of such Wind Facilities if prepared for WIND LESSEE. WIND LESSEE agrees to consult with O&G LESSEE concerning any matter in which O&G LESSEE perceives the possibility of conflict between the parties' respective uses of the Subject Premises.

4. Safe and Legal Operation. O&G LESSEE and WIND LESSEE each agree to conduct its respective operations on the Subject Premises in a safe and prudent manner, which specifically includes travel at safe speeds (forty-five (45) miles per hour during the day and thirty-five (35) miles per hour at night, or less) along roads across the Subject Premises, and in a manner that does not pose a danger to the property or personnel of the other or risk of contamination or pollution of the surface or subsurface or of water resources, and in full compliance with all applicable laws, rules, regulations and orders of any governmental authority having jurisdiction.

5. Road Use. O&G LESSEE and WIND LESSEE each agree that the other may use all roads located on or serving as access to the Subject Premises that are constructed or maintained by either Party, provided such Party has the right to grant such right to the other. Each Party agrees to repair all damage caused by its use to any jointly used road, and each of the parties agrees to bear and pay a proportionate share of the cost of maintaining such roads in good condition and repairing damage that is not directly attributable to use by one Party or other, according to the amount of each respective Party's use. In the event that either party (the "Responsible Party") fails to repair any damage to a road caused by it or its agents or contractors, and such damage prevents or impedes the other party's (the "Non-Responsible Party") use thereof within a reasonable period of time (not to exceed ten (10) days) after notice to the Responsible Party), the Non-Responsible Party shall have the right to repair such damage and the Responsible Party shall reimburse the Non-Responsible Party for all costs incurred to repair such damage; provided, however, in the event the damage occurs during construction of Wind Facilities (in the case of damage caused by the O&G LESSEE), or drilling operations for oil or gas (in the case of damage caused by the WIND LESSEE), the Non-Responsible Party shall have the right to repair such damage if not repaired within forty-eight (48) hours after notice to the Responsible Party if the damage to such road impedes the Non-Responsible Party's construction or drilling operations, and the Responsible Party shall reimburse the Non-Responsible Party for all costs incurred to repair such damage.

6. Variances. Notwithstanding the distances proscribed in Paragraphs 2 and 3 above, in the event that either party believes that such party's operations or facilities can be

conducted or placed within such proscribed distances without causing damage to or impacting the use of the other party's facilities or such parties' operations, such party (the "Requesting Party") may provide to the other party (the "Responding Party") notice of a requested variance from the proscribed distances, which notice shall include engineering analysis and data to support the Requesting Party's position that no detriment to the other party's facilities or operations will occur. In the event that the parties reach an agreement to allow the variance, the Requesting Party can proceed with such operations or facilities. If the Responding Party refuses to agree to such variance, the dispute shall be submitted to an independent engineer selected by the parties, whose determination shall be final. The fees and expenses of the independent engineer shall be paid by the Requesting Party if the independent engineer refuses to grant the variance and by the Responding Party if the independent engineer grants the variance.

7. <u>Crossings</u>. In constructing the Wind Facilities and O&G Facilities, the applicable Party (the "Crossing Party") may cross existing or proposed locations of pipelines, above or below ground electric and communication lines, and roads of the other Party (the "Crossed Party") with the same facilities of the Crossing Party, unless such crossings will unreasonably interfere with the operation or maintenance of such facilities. The Crossing Party will notify the Crossed Party of the approximate locations of any crossings and if any such crossings require the relocation or reconstruction of any of the Crossed Party's Facilities. If relocation or reconstruction of any facilities of the Crossed Party is required, after consulting with the Crossed Party, the Crossing Party will relocate or reconstruct such facilities and will be responsible for any costs incurred by the Crossed Party in connection with such relocation or reconstruction.

8. <u>Delivery Notices</u>. All notices required or permitted under this Agreement shall be in writing and may be delivered personally, by mail, by commercial courier or delivery service, or by facsimile or other electronic transmission, and shall be deemed given when actually received by the recipient or delivered at the address of the receiving Party set forth below, or at such other address of which either Party may notify the other from time to time. Each Party agrees to notify the other as promptly as reasonably possible of any damage to the other caused by its operations. For such purpose, each Party agrees to furnish the other the name of a person or persons who will be available at all times to contact. Each Party further agrees to make reasonable efforts to notify the other Party as soon as reasonably possible in case of an emergency involving the other Party's operations or if it becomes aware of circumstances involving the other Party's operations or the parties' mutual rights and obligations that require prompt action to avoid damage, loss or liability. Notices hereunder shall be addressed as follows:

If to WIND LESSEE: address on page 1

If to O&G LESSEE: address on page 1

9. <u>Severability</u>. If any term or provision of this Agreement, or the application thereof to any person or circumstance shall, to any extent, be determined by judicial order or decision to be invalid or unenforceable, the remainder of this Agreement or the application of such term or provision to persons or circumstances other than those as to which it is held to be invalid, shall be enforced to the fullest extent permitted by law.

10. <u>Governing Law</u>. Except as otherwise provided herein, this Agreement be governed by the applicable laws of the State of Texas, and _____ County, Texas, shall be considered the proper forum or jurisdiction for any disputes arising in connection with this Agreement.

11. <u>Counterparts</u>. This Agreement may be executed in multiple counterparts, each of which shall be deemed the original, and all of which together shall constitute a single instrument.

12. <u>Authority</u>. The signatories hereto warrant that each has the authority to execute this Agreement on behalf of any entities which are Parties to this Agreement and that each such entity has executed this Agreement pursuant to its organizational documents or a resolution or consent of its governing body.

13. <u>Counterpart Execution</u>. This Agreement may be executed in any number of counterparts, and each counterpart shall be deemed to be an original instrument, but all such counterparts shall constitute but one Agreement. Signature and acknowledgment pages of all counterparts may all be attached to one counterpart for recording purposes.

14. <u>Successors and Assigns</u>. This Agreement shall be binding on the parties hereto and their respective heirs, successors and assigns, and the covenants and obligations expressed herein shall be covenants running with the ownership of the respective parties' interests in the Subject Premises. No Assignment of this Agreement shall be effective until the Party assigning this agreement furnishes the other Party a copy of the fully executed assignment.

Notes and Questions

1. What do you think of the sample Accommodation Agreement? How does it address the wind lessee's needs? How does it help from the oil and gas lessee's perspective?

2. Wind lessees frequently must assign their agreements to lenders for collateral. Note that Section 14 of the sample Accommodation Agreement above does not mention assignment to lenders specifically, but it does restrict assignment to being effective only when a copy of the agreement is furnished to the other party. What do you think of language common in other agreements that provides a wind lessee with the absolute right to assign to a lender and that also requires the other party to execute a form of consent to such assignment "in a form customary to the industry"?

3. Drafting a wind lease involves several considerations. In addition to considering what rights are to be granted, a lease would normally set terms for each phase—evaluation (typically 1-10 years), construction (1-2 years), and operations (20-50 years). What other clauses might you recommend if you represented a wind lessee? For wind lease drafting suggestions from the landowner perspective, go to www.windustry.org/leases.

4. Some wind leases are drafted to state that after the operations phase (20-50 years), the turbines revert to the ownership of the landowner. Do you think this is a good idea? What if the power purchase agreement has expired—where would the landowner market the power? What if the amortization period for the turbines has run—can the landowner afford decommissioning costs? Some local authorities are beginning to step in to require decommissioning as part of any wind energy development plan. Should this be required? If not, will our hillsides of the future be pocked with defunct transmission lines and wind turbines because most rancher and farmer lessors are not equipped to dismantle tons of concrete and steel? Does this emphasize the need for local control as addressed in the next section?

F. Local Controls

Some areas have greater potential for large-scale wind power development, but any locality with sustained wind speeds of greater than 10 to 15 mph can support a wind turbine or wind farm. Paul Komor, Renewable Energy Policy 33 (2004). Forty-five states have wind development potential based on this 10-to-15-mph threshold. Furthermore, as wind technology continues to advance, harvesting of low wind will become more efficient and consequently more prevalent.

In the absence of federal or state law preempting control, the task of regulating this wind development falls on hundreds of local counties and municipalities. The following case addresses one county's effort to ban completely any wind power construction within its borders.

Zimmerman v. Board of County Commissioners of Wabaunsee County
218 P.3d 400 (Kan. 2009)

Nuss, J.

Procedural and Factual Background

Plaintiffs are owners of land in Wabaunsee County who have entered into written contracts for the development of commercial wind farms on their properties. Intervenors are the owners of wind rights concerning other properties in the county.

Defendant is the three-member Board of County Commissioners of Wabaunsee County. The county is roughly thirty miles long and thirty miles wide, containing approximately 800 square miles and 7,000 people. It is located in the Flint Hills of Kansas, which contain the vast majority of the remaining Tallgrass Prairie that once covered much of the central United States.

In October 2002, the county zoning administrator told the Board that he had been contacted by a company desiring to build a wind farm in the county. At that time, the county had no zoning regulations relating specifically to wind farms. The next month, the Board passed a temporary moratorium on the acceptance of applications for conditional use permits for wind farm projects until the zoning regulations could be reviewed. The moratorium was extended on at least five occasions.

The following month, December 2002, the county planning commission conducted its first public meeting to discuss amending zoning regulations regarding commercial wind farms.

* * *

[The Board held a number of meetings and hearings and issued its decision through Resolution No. 04-18 on July 12, 2004. Among other things, the resolution amended Article 31-112 of the zoning regulations—Ed.] to include a new paragraph (5):

"5. *No Commercial Wind Energy Conversion System, as defined in these Regulations, shall be placed in Wabaunsee County.* No application for such a use shall be considered." (Emphasis added.)

Plaintiffs sued the Board in district court, seeking a judicial declaration that the Board's action in passing Resolution No. 04-18 be null and void. Plaintiffs also sought damages under a number of different theories.

Community projects

Sometimes community ownership of a wind or solar project can help overcome opposition that a private development by large commercial interests or a neighbor-against-neighbor development might encounter. Community solar in the form of solar gardens was discussed in Chapter 2. For more information about community wind, see http://www.windustry.org/communitywind.

Yet, community ownership is no guarantee that a project will avoid opposition. The City of New Ulm, Minnesota decided to develop its own municipal wind farms and acquired wind leases to put up five wind turbines as a power source for the city. Landowners across the Minnesota River opposed the project and refused to give the necessary buffer zone rights. In response, the City of New Ulm threatened to use eminent domain to acquire wind rights on nearby farmland. The state public utilities commission de-escalated the standoff when it rejected New Ulm's project permit.

In February of 2011, Columbia Law School's Center for Climate Change posted a Model Wind Energy Ordinance for Municipalities. The current version is designed for New York municipalities and provides a framework for addressing turbine placement, setbacks, noise levels, and aesthetic concerns.

Without filing an answer, the Board filed a motion to dismiss for failure to state a claim and for lack of jurisdiction. In decisions dated February 23 and July 22, 2005, the district court, Judge Klinginsmith, held the Board followed the proper procedures under K.S.A. 12-757(d) in adopting Resolution 04-18 and dismissed Count II. It also dismissed four more of Plaintiffs' claims. These were Count I: state preemption; Count IV: violation of the Contract Clause of the United States Constitution; Count V: violation of the Commerce Clause of the United States Constitution; and Count VI: federal preemption. It reserved judgment on the remaining Count III (unconstitutional taking) and Count VII (42 U.S.C. § 1983 [2000]) holding that their consideration was premature until the court could determine the reasonableness under K.S.A. 12-760 of the Board's adoption of Resolution No. 04-18.

* * *

Aesthetics

As the court held in *Gump*, Kansas appellate courts have long allowed aesthetics to be considered in zoning matters. 35 Kan.App.2d at 509-10, 131 P.3d 1268; see, *e.g.*, *Ware v. City of Wichita*, 113 Kan. 153, 157, 214 P. 99 (1923) (recognizing in a zoning case that "[t]here is an aesthetic and cultural side of municipal development which may be fostered within reasonable limitations. (Citations omitted). Such legislation is merely a liberalized application of the general welfare purposes of state and federal constitutions."). As our court acknowledged 60 years later: "[T]he current trend of the decisions is to permit regulation for aesthetic reasons. (Citations omitted).

* * *

Judge Ireland found that "[t]here is no doubt the County looked at the aesthetics of having the wind generators as a compatible or incompatible use with the Flint Hills area." We agree, particularly when the Board has cited its finding to the record for our review. We also agree that these Board's findings could reasonably have been found to justify its decision: that the commercial wind farms would adversely, if not dramatically, affect the aesthetics of the county and for that reason should be prohibited. See *Golden v. City of Overland Park*, 224 Kan. 591, 596, 584 P.2d 130 (1978) ("'[A] court is not free to make findings of fact independent of those explicitly or implicitly found by the city governing body, but is limited to determining whether the given facts could reasonably have been found by the zoning body to justify its decision.'").

Nonconformance with comprehensive plan and other considerations

More than aesthetics considerations are alleged in the instant case. The Board's submitted Findings of Fact first referenced specifics of the county's Comprehensive Plan 2004:

"9. The final adopted Plan, Wabaunsee County Comprehensive Plan 2004, includes the following goals and objectives, which were developed as a direct result of a county-wide survey and focus groups by the Plan Preparation Class:

* * *

"b. *Maintain the rural character of the county with respect to its landscape, open spaces, scenery, peace, tranquility and solitude.*

* * *

"d. Develop realistic plans to protect *natural resources such* as the agricultural land, *landscape, scenic views, and Flint Hills* through regulatory policy.

* * *

"h. *Develop tourism programs involving* historic properties, *nature of rural character, and scenic landscape.*" (Emphasis added).

In the Board's Conclusions, it then determined that the commercial wind farms were not in conformance with the Comprehensive Plan 2004 for numerous reasons....

* * *

In sum, the Plaintiffs and Intervenors have not established that the Board acted unreasonably in amending its zoning regulations to prohibit commercial wind farms in its county.

* * *

Issue 4: *The district court did not err in dismissing the claim alleging that the zoning amendment violated the Contract Clause.*

Plaintiffs and Intervenors next briefly argue that Judge Klinginsmith erred in dismissing as a matter of law their claim alleging violation of the Contract Clause of the United States Constitution, U.S. Const., Art. 1, § 10, cl. 1. Plaintiffs contend that they entered into wind leases prior to the adoption of the Board's ban and that the county cannot pass a law that interferes with the enforcement of a contract. They set forth the standard expressed in *Energy Reserves Group v. Kansas Power & Light*, 459 U.S. 400, 103 S.Ct. 697, 74 L.Ed.2d 569 (1983), and summarily conclude their contractual rights were impaired. They do not discuss how.

* * *

Here, even without further factual development, we hold for the Board. First, when considering whether a contractual relationship is substantially impaired, we focus on whether the subject matter of the contract is regulated. We observe that the land use field is heavily regulated in Kansas. See, *e.g.*, K.S.A. 12-701 *et seq.* Additionally, as more fully discussed later on the issue of state preemption, Intervenors themselves argue that commercial power, *e.g.*, electric power, is also regulated by the Kansas Corporation Commission. When regulation already exists, it is foreseeable that changes in the law may alter contractual obligations. See *Energy Reserves Group,* 459 U.S. at 416, 103 S.Ct. 697. Accordingly, we can find no substantial impairment. See also *Schenck v. City of Hudson,* 997 F.Supp. 902, 907 (N.D.Ohio 1998)*aff'd without opinion,*208 F.3d 215 (6th Cir.2000) ("Land use and building regulation have long existed in [the town]. Plaintiffs entered their contracts aware that government regularly affected zoning and building issues."); *Kittery Retail Ventures, LLC v. Town of Kittery,* 856 A.2d 1183, 1195 (Me.2004) (Contractual relationships not substantially impaired in rezoning case because "[i]n Maine, land use is an area that has traditionally been regulated by the state and municipalities.").

We also note that according to the dates provided in the briefs of Plaintiffs and Intervenors, all of their relevant contracts were entered into after the Board had declared its first moratorium in November 2002. Accordingly, they were on notice that Board action could be taken contrary to their future contracts. *Cf. Energy Reserves Group,* 459 U.S. at 416, 103 S. Ct. 697 (Contract provision suggested "that ERG knew its contractual rights were subject to alteration by state price regulation. Price regulation existed and was foreseeable as the type of law that would alter contract obligations."); *Alliance of Auto. Mfrs.,* 430 F.3d at 42 (Contracts entered into after the law was passed were not substantially impaired because they were "executed with the knowledge and expectation of pervasive state regulation.").

Second, even assuming that the Resolution substantially impaired the contractual interests, we hold that it served significant public purposes. These purposes included "aesthetics" and consistency with the county's comprehensive plan and, although not relied upon by the district court, also the environmental, ecological, and surface and subsurface water concerns expressed by the Board as bases for its Resolution. See *Northwestern Nat. Life Ins. Co. v. Tahoe Regional,* 632 F.2d 104, 105 (9th Cir.1980) (under a Contract Clause analysis, court noted that restrictive zoning to discourage urbanization of open areas is legitimate exercise of state's police power.); *Schenck,* 997 F.Supp. at 907 ("Land use is a legitimate matter of concern for [town].'"); *cf. Keystone Bituminous Coal Assn.,* 480 U.S. at 505, 107 S.Ct. 1232 (The State's "strong public interest" in the environment "transcend[ed] any private agreement between contracting parties.").

Third, the means chosen by the Board to implement these significant public purposes are not deficient, *i.e.*, they are reasonable and necessary. See *U.S.D. No. 443,* 266 Kan. at 79, 966 P.2d 68; see also *Keystone Bituminous Coal Assn.,* 480 U.S. at 505, 107 S.Ct. 1232. In forming this conclusion, we are guided by the Supreme Court's acknowledgment that "courts should '"properly defer to legislative judgment as to the necessity and reasonableness of a particular measure." [Citation omitted.] 480 U.S. at 505, 107 S.Ct. 1232. The evidence in the record reveals that the Board has drawn a reasonable line-based in part upon size, power, and use-between those wind farms that are allowed and those that are not. See, *e.g., Houlton Citizens' Coalition v. Town of Houlton,* 175 F.3d 178, 191 (1st Cir.1999) (Court deferred to town council's judgment that the waste management system created by the ordinance was a "moderate course designed to achieve the permissible purposes stated in the ordinance's preamble."). Like the United States Supreme Court in *Keystone Bituminous Coal Assn.,* 480 U.S. at 505, 107 S.Ct. 1232, "[w]e refuse to second-guess" the Board's "determinations that these are the most appropriate ways of dealing with the problem."

* * *

Affirmed in part, and cross-appeal denied; several issues stayed pending receipt of supplemental briefs and oral argument.

Notes and Questions

1. In Section D of this chapter, we saw that aesthetics alone were not enough in the *Rankin* case to support a nuisance claim in Texas. What does the *Zimmerman* court say about aesthetics as the basis for Wabaunsee County's complete prohibition against wind farms?

2. Because much of wind power development is dependent on siting— which is a land use function traditionally controlled by local governments— counties and cities have significant influence over the future of wind development. Yet, we see in *Zimmerman* that local control can result in countywide bans preventing a wind farm from developing even after it already had acquired permission from individual landowners. What issues do you see with the Kansas Supreme Court's holding?

3. Some states, including Wisconsin, Washington, and Ohio have enacted statewide statutes to promote wind power development by restricting local control. *E.g.*, http://legis.wisconsin.gov/2009/data/acts/09Act40.pdf. Do you think that such statewide laws might conflict with the traditional powers of local governments to protect the health and safety of their residents? Do you think that states with a strong tradition of home rule for cities and local governments might have a more difficult time centralizing the regulation of wind development?

4. The Telecommunications Act of 1996 barred local governments from considering the environmental impact of radio frequency radiation emissions when regulating the placement of cell towers. Could the federal government similarly step in to protect wind development? *See, e.g.*, Patricia E. Salkin, director of Government Law Center at Union University's Albany Law School— past chair of ABA Section of State and Local Government Law, *quoted in* Kristin Choo, *The War of Winds*, A.B.A. J. 54, 60 (Feb. 2010).

Chapter 4

Hydropower

Hydropower was one of the oldest forms of energy harnessed before the industrial revolution and currently accounts for the largest percentage of renewable energy generation in the United States. Despite hydropower's significant role in the renewable energy mix, its influence has diminished over the years, and its share of the overall electricity generation pie is decreasing. In the 1940s, 40 percent of the electricity generated in the United States came from hydroelectric plants; in 2009, conventional hydropower's share of the U.S.'s net generation of 3.95 GWh of electricity was less than 7 percent. *See* Figure 4.1

Hydropower still reigns as the largest source of renewable energy in the United States, and in 2010, accounted for 70 percent of all of U.S. renewable energy production. The vast majority of hydropower generation is concentrated in just three states: Washington,

Figure 4.1: U.S. electricity generation by source (%) 2000–2009

	Coal	Petroleum Liquids	Petroleum Coke	Natural Gas	Other Gases	Nuclear	Hydro	Renew-ables	Hydro Pumped Storage	Other	Total Generation (million kWh)
2000	51.7%	2.7%	0.2%	15.8%	0.4%	19.8%	7.2%	2.1%	-0.1%	0.1%	3,802,521
2001	50.9%	3.1%	0.3%	17.1%	0.2%	20.6%	5.8%	1.9%	-0.2%	0.3%	3,737,052
2002	50.1%	2.0%	0.4%	17.9%	0.3%	20.2%	6.8%	2.1%	-0.2%	0.4%	3,858,919
2003	50.8%	2.6%	0.4%	16.7%	0.4%	19.7%	7.1%	2.1%	-0.2%	0.4%	3,883,783
2004	49.8%	2.5%	0.5%	17.9%	0.4%	19.9%	6.8%	2.1%	-0.2%	0.4%	3,970,782
2005	49.6%	2.5%	0.6%	18.8%	0.3%	19.3%	6.7%	2.2%	-0.2%	0.3%	4,056,199
2006	49.0%	1.1%	0.5%	20.1%	0.3%	19.4%	7.1%	2.4%	-0.2%	0.3%	4,065,762
2007	48.5%	1.2%	0.4%	21.6%	0.3%	19.4%	6.0%	2.6%	-0.2%	0.3%	4,158,267
2008	48.2%	0.8%	0.3%	21.4%	0.3%	19.6%	6.2%	3.1%	-0.2%	0.3%	4,121,184
2009	44.6%	0.7%	0.3%	23.3%	0.3%	20.2%	6.9%	3.6%	-0.1%	0.3%	3,953,898

Source: EIA

Note: Electricity generation from hydro pumped storage is negative because more electricity is consumed than generated by these plants.

National Renewable Energy Lab., 2009 Renewable Energy Databook 12 (2010)

Figure 4.2: Dams by primary purpose from NID report

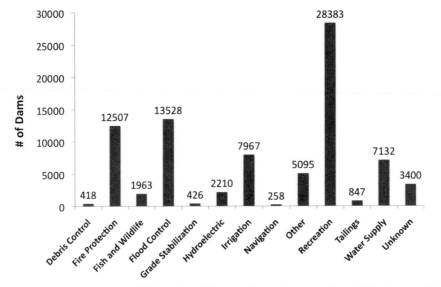

U.S. Army Corps of Engineers, National Inventory of Dams, NID National, *available at* http://geo.usace.army.mil/pgis/f?p=397:5:3152048205233717:NO.

California, and Oregon, which together account for more than half of the U.S. total. *See* Figure 1.8.

Hydropower's contribution of almost 7 percent of the U.S.'s total electricity needs is even more impressive when we consider that fewer than 3 percent of the nation's total of over 82,500 dams were built primarily for hydroelectric generation. According to a National Inventory of Dams prepared by the U.S. Army Corps of Engineers in 2007, the majority of U.S. dams were built for recreation, flood control, fire protection, irrigation, water supply, and purposes other than electricity generation. *See* Figure 4.2. Although electricity generating turbines can be, and in some instances have been, added to existing dams, the majority are still not equipped for hydropower.

A large part of the hydroelectric power story is the involvement of the U.S. government in leasing, building, and operating dams. From an electricity standpoint, the government's involvement in hydropower makes sense: today's plants are about 90 percent efficient, making them the most efficient means of producing electric energy. Despite its efficiency, hydropower has significant environmental consequences, many discussed in this chapter, which have resulted in a movement toward dismantling dams.

While the reclamation of these dams means some hydroelectric capacity is being eliminated, there is significant potential for new hydropower generation. A 1998 Department of Energy resource assessment identified 5,677 sites in the United States with undeveloped hydroelectric capacity of 30,000 MW. Much of this potential comes not from traditional hydroelectric dams, but instead from pumped storage and alternative hydrokinetic generators, using low-head, low-power technologies that allow us to capture electricity from waves, tides, or in-stream flows.

While pumped storage is covered in this chapter, hydrokinetic generation will be addressed in Chapter 8.B.2. Figure 8.5 also describes some of the technologies used to generate electricity from water. A handful of these technologies are unique to one form of hydropower generation (such as the wave devices). Others — such as the horizontal or vertical axis turbines — can be employed in dams, run-of-river, small hydro, or pumped storage applications.

The following section explains some of the basics about how electricity can be created by water.

A. How Hydropower Works

This chapter will focus on what are currently the three most prevalent types of hydropower plants: impoundment, diversion, and pumped storage.

Impoundment facilities — typically large-scale hydropower systems — are currently the most common. They use dams to store river water in an upstream reservoir. Gravity creates electricity: in response to demand, dam managers route the reservoir water into intakes called penstocks. As the flow of the water turns a mechanical turbine, electricity is created by the attached generators. Sections B through D of this chapter will focus primarily on the legal issues related to impoundments.

Diversion facilities channel a portion of a river, stream, or irrigation canal into a turbine to generate electricity. Although some diversion facilities, also sometimes called run-of-river facilities, require dams and encounter many of the same issues as impoundments, smaller-scale diversions can avoid many of these stream-flow and habitat disruptions. Small hydro will be discussed further in Section E of this chapter.

Pumped storage is a mechanism for expanding the nation's peak power generation at times of high demand. When demand is low, such as in the night, water is pumped against gravity to an upper reservoir where it is stored. When a utility needs extra generation capacity to meet peak loads, the water in the upper reservoir is released to generate hydropower. Some of the legal and environmental issues raised by pumped storage will be discussed further in Section E of this chapter.

The following excerpt explains the basic science of hydroelectric power generation.

U.S. Dep't. of Interior, Bureau of Reclamation, Power Resources Office, RECLAMATION: MANAGING WATER IN THE WEST — HYDROELECTRIC POWER
1-6 (July 2005)

Hydroelectric power comes from water at work, water in motion. It can be seen as a form of solar energy, as the sun powers the hydrologic cycle, which gives the earth its water. In the hydrologic cycle, atmospheric water reaches the earth's surface as precipitation. Some of this water evaporates, but much of it either percolates into the soil or becomes surface runoff. Water from rain and melting snow eventually reaches ponds, lakes, reservoirs, or oceans where evaporation is constantly occurring.

Moisture percolating into the soil may become ground water (subsurface water), some of which also enters water bodies through springs or underground streams. Ground water may move upward through soil during dry periods and may return to the atmosphere by evaporation.

Water vapor passes into the atmosphere by evaporation then circulates, condenses into clouds, and some returns to earth as precipitation. Thus, the water cycle is complete. Nature ensures that water is a renewable resource.

1. Generating Power

In nature, energy cannot be created or destroyed, but its form can change. In generating electricity, no new energy is created. Actually one form of energy is converted to another form.

To generate electricity, water must be in motion. This is kinetic (moving) energy. When flowing water turns blades in a turbine, the form is changed to mechanical (machine) energy. The turbine turns the generator rotor, which then converts this mechanical energy into another energy form—electricity [using a Faraday motor—Ed.]. Since water is the initial source of energy, we call this hydroelectric power or hydropower for short.

At facilities called hydroelectric powerplants, hydropower is generated. Some powerplants are located on rivers, streams, and canals, but for a reliable water supply, dams are needed. Dams store water for later release for such purposes as irrigation, domestic and industrial use, and power generation. The reservoir acts much like a battery, storing water to be released as needed to generate power.

The dam creates a "head" or height from which water flows. A pipe (penstock) carries the water from the reservoir to the turbine. The fast-moving water pushes the turbine blades.... The water's force on the turbine blades turns the rotor, the moving part of the electric generator. When coils of wire on the rotor sweep past the generator's stationary coil (stator), electricity is produced.... When the water has completed its task, it flows on unchanged to serve other needs.

[Figure 4.3 shows the parts of a traditional dam and turbine. For animations of how electricity is generated by hydropower, *see* www.fwee.org—Ed.]

2. Transmitting Power

Once the electricity is produced, it must be delivered to where it is needed.... Dams are often in remote locations and power must be transmitted over some distance to its users.

Vast networks of transmission lines and facilities are used to bring electricity to us in a form we can use. All the electricity made at a powerplant comes first through transformers, which raise the voltage so it can travel long distances through powerlines. (Voltage is the pressure that forces an electric current through a wire). At local substations, transformers reduce the voltage so electricity can be divided up and directed throughout an area.

Transformers on poles (or buried underground, in some neighborhoods) further reduce the electric power to the right voltage for appliances and use in the home....

* * *

3. How Power is Computed

... The actual output of energy at a dam is determined by the volume of water released (discharge) and the vertical distance the water falls (head).... The head and the discharge at the power site and the desired rotational speed of the generator determine the type of turbine to be used.

The head produces a pressure (water pressure) and the greater the head, the greater the pressure to drive turbines. This pressure is measured in pounds of force (pounds per square inch). More head or faster flowing water means more power.

Figure 4.3: Traditional dam and turbine

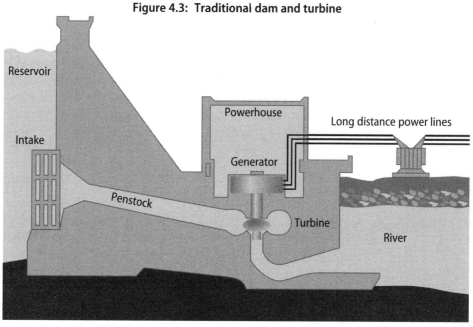

Tenn. Valley Auth., http://www.tva.gov/power/hydroart.htm (last visted July 17, 2011)

Notes and Questions

1. Originally, the power of water was used directly onsite, without converting it to electricity. Some argue that large dam projects in the western United States were primarily motivated by the desire to control water for land development and irrigation, not for power generation. However, President Franklin Delano Roosevelt and western congressmen recognized that they could use electricity as an excuse and a subsidy for these massive, and otherwise uneconomical, dam projects. *See, e.g.*, MARC REISNER, CADILLAC DESERT: THE AMERICAN WEST AND ITS DISAPPEARING WATER 140 (2d ed. 1993).

2. Once the large dam projects were built, utilities encouraged consumers to increase demand for the electricity produced. Hoover Dam has produced an average of 4.2 TWh/year, and Nevada, a sparsely populated state, receives almost one quarter of that production. Do you want to guess why there are so many lights on the Las Vegas strip?

Reliability of hydropower — Although it is often touted as being more available on demand than wind or solar power, even hydropower can be intermittent: Venezuela has imposed rolling blackouts to prevent water levels in the Guri Dam from falling to critical lows amid droughts like the one in January of 2010. The dam feeds hydroelectric plants supplying 73 percent of Venezuela's electricity.

B. Early State Control and Property Issues

Long before humans discovered electricity or fossil fuels, they learned to harness the energy in water to work for them. Over 2,000 years ago, the Greeks employed water wheels to grind wheat. Since that time, water has been used not only for grist (or grain-grinding) mills, but also for sawing wood, and for carding and fulling fibers.

Section 1 describes how humankind used water first as a general energy source and then to supply electricity. The *Bean* case in Section 2 addresses water power before electricity generation and how society chose to balance the right to produce that power against the rights of other riparian landowners.

1. History

This excerpt describes the evolution of hydropower from feudal times to the 19th Century.

Sarah C. Richardson, Note, *The Changing Political Landscape of Hydropower Project Relicensing*
25 Wm. & Mary Envtl. L. & Pol'y Rev. 499, 501-03 (2000)

a. Feudal Origins

The monumental hydropower dams of the twentieth century are technological descendants of the humble watermills that were one of humankind's earliest methods of transferring energy from nature to run machinery. Gristmills powered by water, common in medieval times, were often erected by feudal lords and granted to the miller for public use. The early stages of the Industrial Revolution were powered to a great extent by milldams that made it possible to run larger, more efficient sawmills and textile factories.

The growth of waterpower occurred hand-in-hand with, and was a partial cause of, changing conceptions of property rights. As the landscape of the United States became industrialized (primarily along its rivers), economic development became more highly valued than traditional, agrarian land uses.

Evidence of the changing legal view of water rights can be found in a series of "Mill Acts" passed in the late eighteenth century that "were, more than any other legal measure, crucial in dethroning landed property from the supreme position it had occupied in the eighteenth century world view, and ultimately, in transforming real estate into just another cash-valued commodity." As a commodity, land—and the water that flows over it—becomes more susceptible to cost-benefit analyses and governmental constraints on its use than under the traditional conception of private property as sacrosanct.

b. Industrial Development in the Nineteenth Century

Although the new industrial uses for rivers in the early nineteenth century were private enterprises, their benefits were seen as redounding to the good of society, as one Virginia judge stated in 1816: "[T]he property of another is, as it were, seized on, or subjected to injury, to a certain extent, it being considered in fact for the public use." Another state judge spoke of the Massachusetts Mill Act as being

> designed to provide for the most useful and beneficial occupation and enjoyment of natural streams and watercourses, where the absolute right of each proprietor

Figure 4.4: History of hydropower

B.C.	Hydropower used by the Greeks to turn water wheels for grinding wheat into flour, more than 2,000 years ago.
Mid-1770s	French hydraulic and military engineer Bernard Forest de Bélidor wrote Architecture Hydraulique, a four-volume work describing vertical- and horizontal-axis machines.
1775	U.S. Army Corps of Engineers founded, with establishment of Chief Engineer for the Continental Army.
1880	Michigan's Grand Rapids Electric Light and Power Company, generating electricity by dynamo belted to a water turbine at the Wolverine Chair Factory, lit up sixteen brush-arc lamps.
1881	Niagara Falls [illuminated—Ed.] by hydropower.
1882	World's first hydroelectric power plant began operation on the Fox River in Appleton, Wisconsin.
1886	About forty-five water-powered electric plants in the U.S. and Canada.
1887	San Bernardino, CA, opens first hydroelectric plant in the west.
1889	Two hundred electric plants in the U.S. use waterpower for some or all generation.
1901	First Federal Water Power Act.
1902	Bureau of Reclamation established.
1907	Hydropower provided 15 percent of U.S. electrical generation.
1920	Hydropower provided 25 percent of U.S. electrical generation. Federal Power Act establishes Federal Power Commission authority to issue licenses for hydro development on public lands.
1933	Tennessee Valley Authority established.
1935	Federal Power Commission authority extended to all hydroelectric projects built by utilities engaged in interstate commerce.
1937	Bonneville Dam, first Federal dam, begins operation on the Columbia River. Bonneville Power Administration established.
1940	Hydropower provided 40 percent of electrical generation. Conventional capacity tripled in United States since 1920.
1980	Conventional capacity nearly tripled in United States since 1940.
2003	About 10 percent of U.S. electricity comes from hydropower. Today, there is about 80,000 MW of conventional capacity and 18,000 MW of pumped storage.

U.S. Dep't. of Energy, http://www1.eere.energy.gov/windandhydro/hydro_history.html

to use his own land and water privileges, at his own pleasure, cannot be fully enjoyed, and one must of necessity, in some degree, yield to the other.

The Mill Act's controversial constraints on the "absolute right" of landowners are echoed nearly two centuries later by recent controversies over the functions, benefits, and regulation of hydropower dams....

---

2. Bean v. Central Maine Power Co.

The following case illustrates how New England states used statutes to reallocate rights to stream and river water—as well as rights to adjacent lands and the rights to the flow of the river—in order to promote the "public good" of encouraging mill development. The plaintiffs were seeking damages for a loss "of the enjoyment of the flow of water in a swift current through [their] lands...." These plaintiffs argued that the downstream dams prevented them from also using their upstream waters for mill purposes. Determine who the court said was entitled to damages and who was not. Also, consider the court's rationale for its conclusions.

Bean v. Central Maine Power Co.
173 A. 498 (Me. 1934)

Barnes, J.

The erection and operation of Wyman Dam, between Moscow and Pleasant Ridge, in Somerset county, occasioned a material change in the surface level of the Kennebec river above the dam, and flowing of riparian lands on each side, including such lands of both plaintiffs, the Bingham Company land, at the southerly bound of Carrying Place Plantation, west of the river, and the Bean land, in Carratunk, east of the river and far above the Bingham Company land.

[Plaintiffs in this case were not seeking compensation for their flooded lands but instead for the loss of river current through their lands. They argued that the swift current, which they also could have used to drive a mill wheel, was "a valuable incorporeal hereditament incident to their lands...."

The defendant's position was that even if an upstream landowner had potential to develop a millsite, which was lost when the downstream landowner flooded those upstream lands when he developed his mill, he did not owe the upstream owner any compensation for "render[ing] useless as a power privilege the upper site ... [even if the loss is] based on inability to make a profit from development of the upper power privilege."—Ed.]

* * *

Industrial development had not advanced in England, at the time of first New England settlement, to the stage of construction of dams for sawing timber or grinding grain by water power. It is said that sawmills driven by water power were in successful operation in New England more than thirty years before an attempt was made to build such in the mother country.

Definition of "flowing" — Justice Barnes uses the term "flowing" not only to describe the movement of water in a stream, but also to describe what we might now call "flooding" of lands adjacent to the stream. The Oxford English Dictionary supports Justice Barnes' characterization, as one possible definition of flowing is "an overflowing; a flood." The Oxford English Dictionary 1095 (2d ed. 1989).

Permanent settlements in the area, now the state of Maine, were established before enactments of the Massachusetts Bay Colony were accepted and recognized as the law of this locality.

* * *

Such rules of English common law as the early colonists adopted became the common law of the land of the colonists, together with other laws deemed by them to be of prime importance and adapted to the needs of the inhabitants of the new land.

Under the common law of England, the bed of a river was the property of the state; a riparian proprietor owned only to low-water mark on the shore of a river. At the time of the first settlements in the new world, the chief service of a river was as a highway.

Obstructions on a river bed were abatable if proven a nuisance to the public.

In England there was recognized the exception that an obstruction erected by the sovereign was not abatable.

This exception was adopted in New England, with the further exception that dams might be erected, and mills driven by water power might be maintained, as of public use and benefit; hence the expression "mill privilege."

Under the common law as recognized by Massachusetts Bay Colony, a proprietor's land, bounded on a stream, extended to the midthread of the current.

If one owned the banks on both sides of a river, above the reach of the tide, he owned the bed of the stream, and his dam, on his land, could not be prostrated [cast down] unless by order of court for the abatement of a public nuisance.

Under the doctrine of reasonable use, common-law rights and duties protected and restricted those who would develop a mill privilege; for examples, they had the right, as against the public, to convert a current, valuable to timber men, to a still pond, and the duty not to obstruct a river below the mark to which the tide of ocean flowed.

Experience showed that raising a head of water sufficient for reasonable operation of a mill frequently flowed river banks and adjoining lands beyond the bounds of what the mill man owned or could control by virtue of grant, and controversies and lawsuits arose; wherefore the mother colony, in 1714, enacted legislation, the first Mill Act, so far as Maine is concerned, "That where any person or persons have already, or shall hereafter set up any water-mill or mills, upon his or their own lands, or with the consent of the proprietors of such lands legally obtained, whereupon such mill or mills is or shall be erected or built, that then such owner or owners shall have free liberty to continue and improve such pond, for their best advantage, without molestation."

Then, in harmony with the common-law rule that, if one man's property is taken, to another's advantage, the taker shall make good the loss, the act provided for an impartial "appraisal of the yearly damage done to any person complainant by flowing his or their land as aforesaid...."

A similar act was passed after the establishment of the commonwealth of Massachusetts, and by the first Legislature of Maine, Laws 1821, c. 45.

Then, by R. S. 1841, c. 126, §1, our Legislature provided: "Any man may erect and maintain a water mill, and a dam to raise water for working it, upon and across any stream that is not navigable, upon the terms and conditions and subject to the regulations hereinafter expressed"; and in the regulations provided by Section 2, "No dam shall be erected to the injury of any mill lawfully existing either above or below it, on the same stream; nor to the injury of any mill site, on which a mill or mill dam shall have been lawfully erected and used, unless the right to maintain a mill on such last mentioned site, shall have been lost or defeated by an abandonment, or otherwise."

Subsequent amendments, not vital here, have been made, and the present law, R. S. c. 106, prescribes, "Any man may on his own land, erect and maintain a water-mill and dams to raise water for working it, upon and across any stream, not navigable; * * * upon the terms and conditions, and subject to the regulations hereinafter expressed" (section 1); retains the clause of exception, section 2; by subsequent section provides, "Any person whose lands are damaged by being flowed by a mill-dam * * * may obtain compensation for the injury, by complaint to the superior court," etc. (section 4); and, if injury compensable in damages is established, by section 9 provides, "The court shall appoint three or more disinterested commissioners of the same county, who shall go upon and examine the premises, and make a true and faithful appraisement, under oath, of the yearly damages, if any, done to the complainant by the flowing of his lands * * * described in the complaint. * * * They shall also ascertain, determine, and report what sum in gross would be a reasonable compensation for all the damages, if any, occasioned by the use of such dam;" and makes provision for collection of such compensation.

* * *

A mill privilege, as the term is used here, presupposes a mill site, understood when the first Mill Act was passed as a place on a stream where a dam might be seated to furnish power for grist, saw, carding, and fulling mills, and it may be mills of other sorts, "serviceable for the public good, and benefit of the town, or considerable neighborhood, in or near to which they may have been erected." Mill seat, now mill site, and mill privilege have been household words of the people served by power dams on streams since the mudsill of the first dam was seated in the territory now the state of Maine, for full three hundred years.

The terms are synonymous, used interchangeably to name a location on a stream where by means of a dam a head and fall may be created to operate water wheels.

The property right in a mill site has been recognized, and protected by legislation, as an incorporeal hereditament attached to the land of the riparian owner, and since 1841 a proprietor of an upper mill privilege, in this state, cannot be deprived thereof if his privilege has been developed and not clearly abandoned, defeated, or lost.

Is this incorporeal hereditament, when no dam or mill has been erected, a property right that may not be taken from the riparian owner by the filling and maintaining of a pond for operating a lower water mill, without compensation in damages?

Riparian owners have been deprived of certain rights in rivers and streams as American history has been written, as in New England, where exclusive right to the taking of food fish has been granted to towns, unquestioned to this day in certain Maine towns, or in mineral bearing states where the very water of the stream is appropriated by a first taker

---

Definition of "fulling" — The Oxford English Dictionary defines *fulling* as "[t]he process of cleansing and thickening cloth by beating and washing; also called *milling*." The Oxford English Dictionary 253 (2d ed. 1989). It later defines a *fulling mill* as "a mill in which cloth is fulled or milled by being beaten with wooden mallets, which are let fall upon it (or in modern use, by being pressed between rollers) and cleansed with soap or fuller's earth." *Id.*

for the furtherance of mining, or where irrigation projects are of public benefit, and in all states where for a public use a water district includes a stream.

A riparian owner on a floatable stream has not a monopoly of the use of the stream or its banks. He must yield to the rights of others, at reasonable times, to float timber down the stream, and allow necessary use of his banks by the owners of the timber and their servants, as travel up and down the banks is called for; he must allow the passage of boats.

In these and other ways the right of the owner in his mill privilege is limited. To erect a dam and mill thereon, when thereby no owner above or below is injured, is his right, but he must so operate his dam as to let the natural volume of the stream pass through, as well as the logs of the river driver. Further, it was declared in Massachusetts, when our present state was a part of the former, that, if a lower proprietor on a stream shall erect and maintain a dam for furnishing power to water wheels, and the pond created by such dam shall flow a mill site above, never improved, or improved and abandoned, the upper owner cannot recover damages of the lower, although so long as he maintains his dam he deprives the upper proprietor of any right to use his privilege to work a mill. This follows as a result of the nature and extent of the right in the upper owner. His right is defeasible, and, if it is not asserted and availed of by him, he must submit to lower development, on a scale commensurate with the needs of the section benefited, and he may not have damages for the right of which he is deprived, a right which he shared with other riparian owners and lost when such other made prior appropriation of his site. The lower "owners shall have free liberty to continue and improve such pond, for their best advantage, without molestation." Colonial Act of 1714. The lower owners may erect a water mill, and if, "in so doing any land shall be flowed not belonging to the owner of such mill, it shall be lawful for the owner or occupant of such mill to continue the same head of water to his best advantage, in the manner and on the terms hereinafter mentioned." Laws of Maine, 1821, c. 45.

These, however, only assure to the lower owner his common-law right to flow so far as necessary for reasonable use. The rule that appropriation of an unimproved or abandoned mill site is damnum absque injuria originated in Massachusetts, and is known as the Massachusetts rule: "for the owner of a mill site, who first occupies it by erecting a dam and mill, will have a right to water sufficient to work his wheels, if his privilege will afford it, notwithstanding he may, by his occupation, render useless the privilege of any one above or below him upon the same stream." Hatch v. Dwight, 17 Mass. 289, 296, 9 Am. Dec. 145.

It is important to note that this case was tried upon issues raised before the separation of Maine from Massachusetts.

"The usefulness of water for mill purposes depends as well on its fall as its volume. But the fall depends upon the grade of the land over which it runs. The descent may be rapid, in which case there may be fall enough for mill sites at short distances; or the descent may be so gradual as only to admit of mills at considerable distances. In the latter case, the erection of a mill on one proprietor's land may raise and set the water back to such a

---

Damnum absque injuria refers to "loss or harm that is incurred from something other than a wrongful act and occasions no legal remedy." Black's Law Dictionary 450 (8th ed. 2004).

distance as to prevent the proprietor above from having sufficient fall to erect a mill on his land. It seems to follow, as a necessary consequence from these principles, that in such case, the proprietor who first erects his dam for such a purpose has a right to maintain it, as against the proprietors above and below; and to this extent, prior occupancy gives a prior title to such use. It is a profitable, beneficial, and reasonable use, and therefore one which he has a right to make. If it necessarily occupy so much of the fall as to prevent the proprietor above from placing a dam and mill on his land, it is damnum absque injuria. * * * Such appears to be the nature and extent of the prior and exclusive right, which one proprietor acquired by a prior reasonable appropriation of the use of the water in its fall; and it results, not from any originally superior legal right, but from a legitimate exercise of his own common right, the effect of which is, de facto, to supersede and prevent a like use by other proprietors originally having the same common right." Cary v. Daniels, 8 Metc. (Mass.) 466, 477, 41 Am. Dec. 532.

"This priority of first possession necessarily arises from the nature of appropriation; where two or more men have an equal right to appropriate, and where the actual appropriation by one necessarily excludes all others, the first in time is the first in right." Gould v. Boston Duck Co., 13 Gray (Mass.) 442, 451.

"To the extent to which the descent or fall of water in a stream is taken up and occupied by the erection of dams for the purpose of carrying mills, the right of other owners on the same stream, who have not improved their sites for the creation of water power and the driving of mills, is abridged and taken away. In such case, prior occupancy gives priority of title. Although the right to the use of water is inherent in or appurtenant to land, it is nevertheless in a certain sense a right publici juris, and subject to the rule of law, which regards the erection of a dam for the purpose of creating mill power a profitable, beneficial and reasonable use of the stream, of which riparian proprietors on the same stream, who have not appropriated the force and fall of the water on their own land, cannot complain. It is damnum absque injuria. * * *

"It is in view of this well established doctrine of the common law of this state, that the provisions of the mill act, so called, are to be construed and administered. By the first section of the Rev. Sts. c. 116, which is substantially a reenactment of St. 1795, c. 74, § 1, full power is given to any person to erect and maintain a water mill and dam to raise water upon any stream not navigable, according to the terms and conditions and subject to the regulations therein expressed. The only limitation on this power, so far as the rights of other owners of mill sites or water powers on the same stream may be affected by its exercise, is found in the second section of the same chapter, and in St. 1841, c. 18, which provides that no such dam shall be erected to the injury of any existing mill or of any mill site which shall have been previously used or occupied. But no provision is made to protect unoccupied or unimproved mill sites. Nor are they included specifically as a subject of damages in the fourth section of the statute, which provides for a compensation to parties 'whose land is overflowed or otherwise injured' by the erection and maintenance of a dam. The great purpose of these statutes, as declared in the preamble to St. 1795, c. 74, was to prevent the erection and support of mills from being 'discouraged by many doubts and disputes.' They were not intended to confer any new right, or to create an additional claim for damages, which did not exist at common law. They only substituted, in the place of the common law remedies, a more simple, expeditious and comprehensive mode of ascertaining and assessing damages to persons whose lands were overflowed or otherwise injured by the erection and maintenance of dams on the same stream, for the purpose of creating a water power and carrying mills. It follows that, as a riparian proprietor could recover at common law no damages occasioned to an unimproved or unappropriated mill site by the erection of a dam and mill on the same stream below, he cannot maintain a complaint under the mill acts to recover similar damages." Fuller v. Chicopee Mfg. Co., 16 Gray (Mass.) 43.

And the court says, in the same opinion: "This is the first case, so far as we know, in which an attempt has been made by a complaint under Rev. Sts. c. 116, or under the previous statutes enacted for the erection and regulation of mills, to claim damages for injury done to an unoccupied mill site. The fact that there is no precedent for such a claim is not conclusive, but it is strong evidence against the existence of any such right as the complainant sets up in the present case."

Residents in the province of Maine, before separation from the mother state, are conclusively held to have adopted the common law, as expressed by the courts of that state and Massachusetts Bay Colony.

The declaration of the common law in Hatch v. Dwight, supra, is as effective, if not repealed, in Maine, as if it were a declaration of our court, because the plaintiff in that suit acquired an interest in the privilege under litigation in 1807, took possession in 1817, and the writ was brought before Maine became a separate state.

* * *

In that case [Hatch] the question at issue was the quantity of water which the proprietors of an upper mill privilege, improved by a dam, were allowed to discharge by means of a penstock from their dam to a trench which diverted the water from the natural channel of the river and returned it thereto at a point below the dam of proprietors of a lower privilege, also improved.

In the opinion, the learned jurist gives expression to some of the many principles of law then limiting the rights of riparian owners whose lands extend to the thread of the same stream.

Several of his observations were but dicta, and, in the hundred years that have followed the decision in Hatch v. Dwight, supra, the Massachusetts court has not abandoned that decision.

Owners of riparian lands on any river, from its source to its mouth, have rights in common. They may make reasonable use of its current over rips and falls not appropriated by the local owner and over or through the obstructions caused by reasonable appropriation by the local owner.

* * *

... [T]he Mill Act makes the appropriation by construction on the lower site, before any development is begun on the upper site, a rightful appropriation, "known and admitted by the law." There is no reason to suppose that Judge Story conceived that his findings on the facts before him, the right by grant or long-established user to divert water from a lower proprietor, would be asserted as restricting the right given by the common law to flow the lands of an upper proprietor.

In Maine, litigation over rights in water powers began soon after the establishment of the state, and the principle was announced by our court in 1832 that the right of the owner of an undeveloped mill site is not complete. As against the owner of a lower site, the right to develop and use the upper is suspended, if the lower is first developed and flows the upper site, suspended so long as by the use of the lower site the other is submerged.

* * *

> **More Latin?** "Qui prior est in tempore, potier est in jure" means "first in time, first in right." Black's Law Dictionary 1865 (9th ed. 2009).

In 1868, in a case for damages for flowing an improved upper mill site, these words were used, "The plaintiff's dam was originally erected before the defendant's. This is not controverted. In cases of this description qui prior est in tempore, potior est in jure," and the authority given is Cary v. Daniels and Gould v. Boston Duck Company, cases hereinbefore cited. Lincoln v. Chadbourne, 56 Me. 197.

From the date of that decision, the principle has stood unattacked and in reports and students' texts Maine is considered as having adopted the Massachusetts rule.

"A mill owner can at any time appropriate for raising and maintaining a head of water for working his mill so much space in the river valley as has not already been appropriated by some other mill owner for his own mill." National Fibre Co. v. Electric Co., 95 Me. 318, 49 A. 1095, 1096.

As against all the world except riparian proprietors, one who owns a mill site may seek damages if deprived of his right.

But because of the right common to riparian proprietors, publici juris, to further the public good, the doctrine of appropriation of a mill privilege grew up as naturally as the doctrine of appropriation of the water of a stream for mining grew and established itself over this country from the mountains of the West to the plains. It was founded on necessity, based on the conditions of the watershed of Massachusetts and Maine, at a time when a twelve-foot head of water was a monstrous power and what would now be tiny mills were necessities of domestic and industrial life.

In the present era of industrial development, in the few states that have not coal, but have streams in volume and character like ours, there is ever more insistent demand for the development of water power sites, not in separate independent units, however, but in aggregate of head, as the topographical features of the watershed dictate. So that what may have never suggested profitable development as a power site, until a great enterprise was begun, now demands the changing of the river from a stream of strong swift current to a pond, with the consequence that a recognizable but unprofitable mill site may be flowed by a lower riparian owner without damages for the appropriation or change.

It is conceivable that on any half mile of the river along Carratunk a dam might be erected, though at such expense as to be an unprofitable venture, if its pond were filled, but none of any economic value if all were built upon.

> **Encouraging "use-it-or-lose-it" for mill sites** — A streamside owner of land that could make a good mill site, but who has not yet developed that site, can seek recovery from anyone "except riparian proprietors" if that site is damaged. The Mill Acts encouraged riparian proprietors to develop mills at the expense of upstream lands by exempting those riparian proprietors from damages for the upstream flooding. Thus, the upper owner who never built a mill (or who built one and abandoned it) cannot recover damages because the upstream owner's right is defeasible and lost when the downstream owner made reasonable use of the area by flooding it to create enough "head" for the downstream mill.

> **Questions about the court's rationale** — What is the nature of the public good that the Mill Acts protect? What if one mill site prevents another from being profitable? If the law does not allow the injured mill sites to recover damages for changes in their water flows making their sites unprofitable, does it recognize a claim by the injured mill site owners to a share of the "great dam" downstream? If not, does this interpretation of the Mill Acts favor the larger mill developers over the smaller ones?

Construction at the strategic point flows out many possible sites, and the law as understood in this state favors the erection of the great dam, for the good of the greater number.

Flowing the lands of another for the purpose of working mills is a right recognized in this jurisdiction, not as an exercise of the eminent domain, for our mills are not of public use, as the term is understood in law, and our Constitution does not authorize taking for the benefit of the public as does that of Massachusetts. Brown v. Gerald, 100 Me. 351, 370, 61 A. 785, 70 L. R. A. 472, 109 Am. St. Rep. 526; Murdock v. Stickney, 8 Cush. (Mass.) 113; Bates v. Weymouth Iron Co., 8 Cush. (Mass.) 548, 553; Lowell v. Boston, 111 Mass. 454, 464, 15 Am. Rep. 39; Turner v. Nye, 154 Mass. 579, 28 N. E. 1048, 14 L. R. A. 487.

Flowing of riparian lands is an adjustment and regulation to assure development of reasonable use of such lands among riparian owners. See cases cited in Brown v. DeNormandie, 123 Me. at page 541, 124 A. 697.

In that adjustment we do not recognize, in theory or in fact, that the owner of land flowed by a pond for a water mill is a part owner in the developed lower privilege. He still owns his flowed land, and may still use it on which to sink a pier or in which to drive piling, Jordan v. Woodward, 40 Me. 317, 324, or submit it to any reasonable use not detrimental to the maintenance of the pond.

But he does not participate in the ownership of the dam and mill below. He is not entitled to share in the profits of the lower development simply from the fact that his unimproved mill site or the rocky course of the bed of the river on his land does its part in upholding the impounded water.

Items of alleged damage for changing the current to still pond water are not to be included in the evidence for consideration by the commissioners; their statement is not pertinent to process under the Mill Act.

Notes and Questions

1. The Mill Acts shifted several significant aspects of property law. These acts allowed mill owners to flood upstream property and eliminated or restricted the ability of affected upstream property owners to sue for trespass or nuisance, which would have been permitted under the common law. The acts also eliminated the upstream owners' ability to seek a permanent injunction or to recover punitive damages. Finally, before the Mill Acts were enacted, upstream owners could and did dismantle mill dams under the common law right to use self-help to abate a nuisance. This too, the Mill Acts prohibited. Consider the significance of each of these shifts.

2. Justice Barnes suggests that the basis for supporting the right of a mill owner to flood others' lands or even to destroy flows of an "unprofitable mill site" is "to further the

public good...." He explains later in the *Bean* decision that flooding "of riparian lands is an adjustment and regulation to assure development of reasonable use of such lands among riparian owners." Could a similar argument be made to support solar or wind access statutes? Look back at Chapters 2 and 3 to consider similar adjustments to common law property rights that might assure reasonable use of lands for the development of these alternative energy sources. For example, should a southern property owner's interest in Solar Skyspace B be defeasible if solar power generation is a first-in-time reasonable use of that space?

3. Justice Barnes is careful to note that mill rights are not "an exercise of eminent domain" and that mills "are not a public use...." While early grist mills may have been accessible to all of the public, subsequent mills for commodities such as cotton or paper were generally private enterprises. This evolution led courts to justify the Mill Acts on a broader definition of public good: "the interest which the community at large has in the use and employment of mills...." *Fiske v. Framingham Mfg. Co.*, 29 Mass. (12 Pick.) 68, 72 (1831). What do you think of this rationale?

4. The *Fiske* court also noted that the Mill Acts were "designed to provide for the most useful and beneficial occupation and enjoyment of natural streams and water-courses, where the absolute right of each proprietor to use his own land and water privileges, at his own pleasure, cannot be fully enjoyed, and one must of necessity, in some degree, yield to the other." *Id.* at 73. Could this rationale support a shift in common law property rights to provide a better balance in developing other alternative energy sources for the public good?

5. The court in the *Fiske* case also stated that the Mill Acts were justified to give the riparian owners full use of their properties: "the nature of the property ... is often so situated, that it could not be beneficially used without the aid of this power." *Id.* at 72. Does this argument that a particular property right is valueless without associated rights of necessity or access sound similar to arguments wind or mineral owners might make as discussed in Chapter 3?

C. Federal Control of Hydropower Dams

While authority over waterways first resided with states, such as Maine's control of mill sites in the *Bean* case above, the balance of this authority shifted around the turn of the 20th century. The federal government, which had controlled navigable waters under the Commerce Clause since the early 1800s, began to assert its power over any obstructions in navigable waters, including dams.

As the U.S. government encouraged westward expansion, it became obvious that water was critical to settlement. In 1902, Congress created the Reclamation Service to control waterways and distribute water to places in the arid West that had none by constructing dams and irrigation works. The U.S. Army Corps of Engineers' National Inventory of Dams illustrates the magnitude of this undertaking. About 97 percent of the current dams in the United States were built since 1900. The number built from 1900 to 1940 is impressive: 10,380, but the number built between 1950 and 1980 is staggering: 43,000. *See* Figure 4.5. This section addresses first the rise of federal control over dams and hydropower production and then the decline in dam building since its peak in the 1960s.

Figure 4.5: Dams by completion date

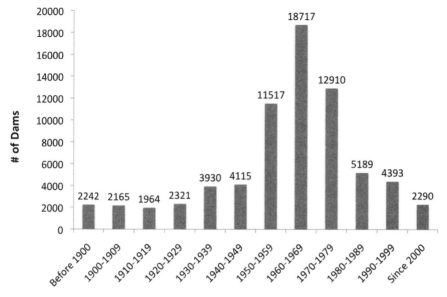

U.S. Army Corps of Engineers, National Inventory of Dams, NID National, *available at* http://geo.usace.army.mil/pgis/f?p=397:5:3152048205233717:NO.

1. Rise of Federal Control

The following excerpt and statute sections illustrate the federal government's role in licensing, building, and, in many cases, operating some of our nation's largest hydroelectric generating facilities.

Sarah C. Richardson, Note, *The Changing Political Landscape of Hydropower Project Relicensing*
25 Wm. & Mary Envtl. L. & Pol'y Rev. 499, 502-07 (2000)

The history of federal hydropower-dam regulation begins almost a century before any such regulation was actually enacted, when the U.S. Supreme Court decided in 1824 that Congress had regulatory power over navigation grounded in the Commerce Clause of the U.S. Constitution. In the well-known case of *Gibbons v. Ogden*, the Court stated that

> All America understands, and has uniformly understood, the word "commerce," to comprehend navigation.... The word used in the constitution, then, comprehends, and has been always understood to comprehend, navigation within its meaning; and a power to regulate navigation, is as expressly granted, as if that term had been added to the word "commerce."

For most of the nineteenth century, however, Congress took a laissez-faire attitude to the regulation of waterpower. Not until the very end of the century was Congress' navigation power given statutory formulation in the Rivers and Harbors Appropriation Act of 1899, Section 10 of which requires Congressional approval for the "creation of any obstruction ... to the navigable capacity of any of the waters of the United States...."

* * *

The first decades of the twentieth century saw a fierce and prolonged debate over the appropriate form of federal approval for one type of obstruction to navigable capacity: hydropower dams. In 1908 and 1909, President Theodore Roosevelt vetoed Congressional attempts to allow dams on the Rainy and James Rivers because the legislative authorizations were not time-limited. In one veto statement, Roosevelt said:

> The public must retain the control of the great waterways. It is essential that any permit to obstruct them for reasons and on conditions that seem good at the moment should be subject to revision when changed conditions demand.... Provision should be made for the termination of the grant or privilege at a definite time, leaving to future generations the power or authority to renew or extend the concession in accordance with the conditions which may prevail at the time.

Roosevelt, a dedicated conservationist, was determined that the country's waterways should remain in public control, because "actual experience of what happens with indeterminate public-utility franchises proves that they are in the vast majority of cases practically perpetual. Each right should be issued to expire on a specified day without further legislative, administrative, or judicial action."

Roosevelt's position was part of a battle between the conservationists he represented and a coalition of private hydropower developers and states-rights advocates who wanted to curtail federal control." Both groups brought tremendous pressure to bear on Congress, and the struggle for water power legislation lasted for fifteen years." It is ironic that, almost a century later, private hydropower developers, rather than supporting states' rights to condition dam licenses, argue vociferously that the federal authority eventually vested in FERC should not be eroded....

Roosevelt's insistence on a fixed expiration date for hydropower licenses was unwavering, and in the end he prevailed. When the Federal Water Power Act ("FWPA," amended in 1935 as the Federal Power Act ("FPA")) was passed in 1920, it included a fifty-year limit on license terms. Half a century was deemed long enough to protect hydropower owners' investment by allowing them to recoup their construction costs and achieve profitability, secure in the knowledge that the terms of a project license can be changed only by mutual consent of the Commission and the licensee.

When the FWPA was enacted in 1920, hydroelectric power comprised a significant portion of the nation's electrical generating capacity (about 30 percent—a much greater percentage than the 10 to 12 percent it represented at the end of the twentieth century). To encourage and coordinate hydropower development, the FWPA created the Federal Power Commission, precursor of the Federal Energy Regulatory Commission, with the exclusive power to license hydropower projects on the nation's "navigable waters."

This entrustment of power to a federal agency (an unusual precursor of the other President Roosevelt's New Deal, when such delegations became common) was intended to remedy jurisdictional conflicts created by the General Dam Acts of 1906 and 1910, which had required hydropower developers to submit their plans to the Secretary of War and the Chief of Engineers. The Dam Acts were widely seen as a practical failure, since (as a Congressman noted at the time) "all waterpower development under Federal control [had] practically ceased." To remedy that failure, the Federal Water Power Act created an independent commission, made up of five members appointed by the President with the advice and consent of the Senate.

Consolidation of Federal Authority Over Hydropower Projects

The FWPA, and later the FPA, provides a grant of broad authority to the Commission, enabling it to "prescribe, issue, make, amend, and rescind such orders, rules, and regulations as it may find necessary or appropriate to carry out the provisions" of the Act, in awarding licenses to projects that are "best adapted to a comprehensive plan for improving or developing a waterway ..." The Constitutional power over navigation is incorporated expressly and comprehensively, with navigable waters defined as "those parts of streams or other bodies of water over which Congress has jurisdiction under its authority to regulate commerce ... and which either in their natural or improved condition ... are used or suitable for use for the transportation of persons or property in interstate or foreign commerce." Judicial interpretations of Congress' navigation power have reinforced its breadth to include nonnavigable sections of navigable waterways, nonnavigable tributaries that affect the navigability of connected waterways, and even projects on nonnavigable waters that affect commerce through transportation of hydroelectricity in interstate commerce.

In its noteworthy 1940 decision of *United States v. Appalachian Elec. Power Co.*, the U.S. Supreme Court held that Congress had delegated broad authority to FERC over rivers that were either navigable themselves or whose use would affect national commerce. In the Court's opinion, "It cannot properly be said that the constitutional power of the United States over its waters is limited to control for navigation.... In truth the authority of the United States is the regulation of commerce.... That authority is as broad as the needs of commerce." This expansive authority brought an even wider range of hydropower projects under federal authority, increasing FERC's role as a licenser, and leading to licenses for hundreds of dams in the ensuing decades.

Another significant hydropower-project case heard by the Supreme Court in the 1940s, *First Iowa Hydro-Electric Cooperative v. Federal Power Commission*, involved the state of Iowa's attempt to impose conditions on a FERC-approved project that was to divert most of the waters of the Cedar River in order to run an enormous electricity generator. Iowa's laws required that diverted water be returned "to the nearest practicable place without being materially diminished in quantity or polluted or rendered deleterious to fish life." The Court reinforced FERC's authority over hydropower licensing by finding that the Commission's congressionally delegated power preempted any state-imposed conditions.

To require the petitioner to secure the actual grant to it of a State permit ... as a condition precedent to securing a federal license for the same project under the Federal Power Act would vest in ... Iowa a veto power over the federal project. Such a veto power easily could destroy the effectiveness of the Federal Act. It would subordinate to the control of the State the "comprehensive" planning which the Act provides shall depend upon the judgment of the Federal Power Commission or other representatives of the Federal Government.

Section 27 of the Federal Power Act provides that the Act is not intended to affect "laws of the respective States relating to the control, appropriation, use, or distribution of water used in irrigation or for municipal or other uses, or any vested right acquired therein." The *First Iowa* Court interpreted this "saving clause" as limited primarily to the "proprietary rights" of the state in its waters, for the use of its citizens, which did not include streamflow for fishes. As a consequence, FERC's authority over hydropower projects was unquestioned for almost fifty years, until conflicts between the states and FERC that involved hydropower license conditions took a new direction in the last decade of the century....

Federal Power Act
16 U.S.C. § 797

* * *

(e) To issue licenses to citizens of the United States, or to any association of such citizens, or to any corporation organized under the laws of the United States or any State thereof, or to any State or municipality for the purpose of constructing, operating, and maintaining dams, water conduits, reservoirs, power houses, transmission lines, or other project works necessary or convenient for the development and improvement of navigation and for the development, transmission, and utilization of power across, along, from, or in any of the streams or other bodies of water over which Congress has jurisdiction under its authority to regulate commerce with foreign nations and among the several States, or upon any part of the public lands and reservations of the United States (including the Territories), or for the purpose of utilizing the surplus water or water power from any Government dam, except as herein provided:.... In deciding whether to issue any license under this subchapter for any project, the Commission, in addition to the power and development purposes for which licenses are issued, shall give equal consideration to the purposes of energy conservation, the protection, mitigation of damage to, and enhancement of, fish and wildlife (including related spawning grounds and habitat), the protection of recreational opportunities, and the preservation of other aspects of environmental quality.

———————

Colorado River Basin Project Act
43 U.S.C. § 1501

(a) It is the object of this chapter to provide a program for the further comprehensive development of the water resources of the Colorado River Basin and for the provision of additional and adequate water supplies for use in the upper as well as in the lower Colorado River Basin. This program is declared to be for the purposes, among others, of regulating the flow of the Colorado River; controlling floods; improving navigation; providing for the storage and delivery of the waters of the Colorado River for reclamation of lands, including supplemental water supplies, and for municipal, industrial, and other beneficial purposes; improving water quality; providing for basic public outdoor recreation facilities; improving conditions for fish and wildlife, and the generation and sale of electrical power as an incident of the foregoing purposes.

* * *

———————

Notes and Questions

1. What is the legal basis for federal authority over hydropower dams? What do you see as the pluses and minuses of federal involvement?

2. Eighteen years after the National Reclamation Act of 1902, Congress passed the Federal Power Act, 16 U.S.C. § 791-828. The Federal Power Act created the Federal Power Commission, predecessor to the current Federal Energy Regulation Commission (FERC). FERC has the authority to license non-federal hydroelectric power plants, and it is unlawful to operate any hydroelectric generation facility without such a license. What do you see as the benefits of requiring federal licensing? What are some of the drawbacks?

Note about the Colorado River Basin Project Act — The Colorado River Basin Project Act began with the negotiation of the Colorado River Compact in 1922, the first in a series of negotiations, agreements, and court cases that collectively comprise the "Law of the Colorado River." In 1922, California was the primary user of the river water, as Wyoming, Arizona, Nevada, and New Mexico were largely uninhabited, and Colorado and Utah had barely begun expanding. Early negotiations could not reach a compromise on the amount to apportion for each state, but the states could agree to divide the Colorado River into two artificial basins. California, Arizona, and Nevada took the lower basin; Colorado, Wyoming, Utah, and New Mexico took the upper basin. The upper basin was allotted seven million acre feet, the lower basin eight million acre feet, with roughly one million acre feet reserved for Mexico. Interestingly, the Colorado River's annual flow does not typically exceed eighteen million acre feet, so virtually the entire annual flow is allocated. Although all of the states signed the Compact in 1922, the voters and legislatures of the individual states did not authorize ratification.

The Colorado River's annual flow is highly erratic, and in order to ensure that both basins received their apportioned amount, the states began to consider projects to store water. Such storage projects were particularly important to the Upper Basin states because they could not yet fully utilize their apportioned amounts and feared that California would start to "borrow" their water. However, the expense of developing the projects was daunting, so using the electrical power generation capability of the dam to offset the cost was proposed as an alternative revenue generator.

3. Some have argued that FERC's jurisdiction in this area impedes the development of hydropower — especially for small-scale and offshore projects as discussed further in Section E of this chapter. What do you think about FERC's role?

4. In addition to being the licensing authority for all dams, the federal government (through the Army Corps of Engineers) controls 609 dams and maintains and operates seventy-five hydroelectric facilities that generate 24 percent of the nation's hydropower and 3 percent of the U.S.'s total electricity. Do you think that being directly involved in the hydropower business is an appropriate role for the federal government? Do federally-run hydropower dams have an unfair advantage? Look back to the discussion in Chapter 3 about how the Bonneville Power Administration has required wind farms to shut down during high-wind, high-water periods. Is there a way for other renewable energy sources to receive federal backing similar to what hydropower has enjoyed?

5. Again review the above excerpts from the FPA and the Colorado River Basin Project Act. What is the basis of authority for each act? How do the purposes vary and what role does electricity generation play in each?

2. Remaining State and Local Influence

The vast majority of dams in the United States are owned by private parties or local governments. *See* Figure 4.6. However, the federal government asserted expansive power over all dams through the FPA licensing authority. The FPA does not limit its federal licensing authority to persons or corporations; it explicitly requires licenses from states and municipalities as well.

Despite this federal assertion of power, Section 27 of the FPA also recognizes that FERC's licensing authority was not intended to eliminate all state or local authority over

Figure 4.6: Dams by owner type

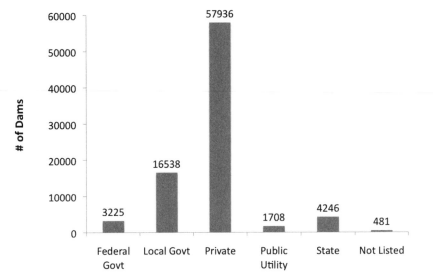

U.S. Army Corps of Engineers, National Inventory of Dams, NID National, *available at*
http://geo.usace.army.mil/pgis/f?p=397:5:3152048205233717:NO.

waterways. Consequently, the waterways that support hydropower dams are regulated by a patchwork of federal, state, local, and sometimes tribal laws.

Even when the federal government operates a hydroelectric facility, there are a number of additional players whose interests are impacted. Federal law may preempt the others when explicit, yet even explicit language may not easily resolve some disputes. An alternative mechanism for avoiding litigation when there are conflicts is a group approach as in the adaptive management plan for Glen Canyon discussed in the following excerpt.

Lawrence Susskind, Alejandro E. Camacho & Todd Schenk, *Collaborative Planning and Adaptive Management in Glen Canyon: A Cautionary Tale*
35 Colum. J. Envtl. L. 1, 7-29 (2010)

I. The Colorado River's Enduring Resource Conflict and the Glen Canyon Dam

The Colorado River is the lifeblood of much of the western United States, providing water to seven American states and Mexico. What was once a wild river, flowing from the Rocky Mountains through parched deserts and the Grand Canyon into the Gulf of California, is now heavily utilized and highly regulated. The Law of the River—a collection of statutes, agreements, regulations, and numerous court decisions—dictates how the river will be managed, including how water will be allocated among the various users and territories.

To improve management and storage of water from the river, the Bureau of Reclamation constructed the Glen Canyon Dam above Lee's Ferry, Arizona and created Lake Powell between 1956 and 1963.... The Bureau of Reclamation can store water in Lake Powell—and Lake Mead downstream—and release it when necessary to smooth out the Colorado's significant year-over-year variability in flow and ameliorate the impacts of droughts.... [I]n light of the various stakeholders with diverging interests in the Dam's operation, as well as the wide range of often conflicting laws that influence the management of the Dam and the surrounding natural resources, Congress established the Glen Canyon Adaptive Management Program (AMP) as an innovative experiment in resource management.

* * *

A. The Stakeholders

In addition to operating the Dam to ensure that obligations under the Law of the River are met, other competing considerations influence how the Dam is operated. First, while the primary purpose of constructing the Glen Canyon Dam was to regulate the flow of the river, a secondary objective has always been to generate hydroelectric power. Revenue from power sales is paying off the Dam's construction debt, albeit slowly. Operating the Dam optimally for hydroelectric power generation requires fluctuating water releases throughout the course of each day, depending on demand. Second, some advocate for operating the Dam in a manner that alleviates environmental impacts. Over time, concerns arose regarding the Dam's impacts on the environment and endangered species such as the humpback chub. Traditionally, the Colorado swelled in the spring with sediment-laden snowmelt, then receded in the summer, depositing the sediment and replenishing sandbars in the process. Species indigenous to the area, including the humpback chub, adapted to these conditions over time. The operations regime favored by hydroelectric interests and used in practice disrupts these natural conditions: water is impounded, making it cooler and allowing the sediment to settle, then released through turbines in fluctuations defined by electricity needs. Conservationists have, therefore, called for changes in water releases aimed at ameliorating these impacts, including slower ramping rates and high-flow releases in spring. Finally, groups with other interests—such as sport fishing, whitewater rafting, other recreational interests, and protecting cultural sites—influence how the Dam should be operated.

The Bureau of Reclamation and the states are interested primarily in the water storage services the Dam provides.... The Bureau of Reclamation, the states, the Western Area Power Administration, and the contractors that purchase power are also concerned about maximizing power generation. The Dam is an important source of power for the region, producing approximately 4.5 billion kilowatt-hours annually, which offsets about 2.5 million tons of coal or eleven million barrels of oil. The "controlled floods" advocated by some conservation and recreation interests lower the Dam's power generating potential. Any changes to the permitted ramping rate (i.e., the speed at which releases change) or seasonal and/or daily restrictions also undercut power generation.

Environmental organizations, including the Sierra Club, opposed the initial plan for what was to become the Colorado River Storage Project Act of 1956, which called for a series of dams along the Colorado River, including two in Dinosaur National Monument. The Sierra Club's primary concern at the time was that "no major scenic resource should be sacrificed for a power project." The Sierra Club eventually dropped its opposition to the Colorado River Storage Project Act, including the Glen Canyon Dam, in exchange for project modifications that canceled the two upstream dams at Echo Park and Split Mountain in Dinosaur National Monument. Many environmentalists came to regret this acquiescence. Indeed, in light of concerns about the impact of the Dam on the environment and endangered species, environmental groups have challenged the Dam's existence and management ever since.

There are eight endangered and three threatened species in the area: four of the endangered species—the southwestern willow flycatcher, humpback chub, razorback sucker, and Kanab ambersnail—have been adversely affected by dam operations. The humpback chub and razorback sucker are of particular interest because they are found only in the Colorado River Basin. Despite a recent stabilization in the estimated chub population, their current numbers are much lower than they were before the river was heavily modified. According to conservationists, a number of changes caused by building the Dam present challenges to these endangered species, including decreased sediment load, cooler and more constant temperatures, more constant flows rather than natural seasonal variation, beach and bar erosion, and the arrival of invasive species.

Environmentalists also argue that the water storage services the dam provides are not particularly valuable, given that sufficient storage capacity exists elsewhere in the system and that a staggering volume is lost to evaporation from Lake Powell annually. Recognizing that outright removal of the Dam is unlikely, environmental organizations and other conservation interests, including the U.S. Fish and Wildlife Service, advocate flow modifications, like controlled flood releases and restricted ramping rates. Such modified flow regimes would be designed to restore and maintain the habitat and other conditions essential for species like the humpback chub. Modified flow regimes, however, often conflict with water supply and power interests.

The area around Glen Canyon remains only sparsely populated; with no roads and a harsh landscape, the area contained even fewer residents when the Dam was proposed. It is the traditional territory of the Havasupi, Hopi, Hualapi, Navajo, Pueblo, and Southern Paiute tribes, who attach great religious and cultural significance to sites within the canyons and along the river. It appears, however, that there was little opposition from the tribes at the time of construction, perhaps because the Dam brought tangible economic benefits in the form of employment opportunities, and the Navajo Nation was compensated for the land it lost.

Overall, the impacts of the Dam on tribes have been mixed. The Dam and associated tourism are a major source of income for the Navajo Nation and other tribes; however, the flooding of the canyon, the erosion resulting from the modified downstream flow, and tourism have harmed important sacred and historical sites. Beyond specific places of historical and cultural significance that have been impacted, various zones, vistas, and the general attributes of the region are considered culturally important, and even sacred, by both Native American groups and non-native groups in the area, and these vistas and attributes have been altered as a result of the Dam's construction.

* * *

The animosity among these stakeholders has increased over time, as their positions regarding releases have hardened and each has felt increasingly threatened by the demands of others. Though perhaps popular when approved, large dams like the Glen Canyon Dam have become controversial and politically unattractive.... Additionally, serious conflicts have arisen regarding management of the Colorado River more generally as population growth, economic development, and climate change have exacerbated water scarcity, increased electricity demand, and compounded environmental impacts. Climate change threatens to magnify the problem in the longer term if it reduces stream flow as predicted.

Table 1 summarizes the primary interests of the stakeholder groups involved in the AMP as reflected in their legal mandates and stated interests. Each stakeholder group or

Table 1

Stakeholder group/agency	Mandate
U.S. Bureau of Reclamation	Hydroelectric power generation and water extraction
U.S. Bureau of Indian Affairs	Provide services to and manage land in trust for American Indian tribes
U.S. Fish and Wildlife Service	Natural resource management
U.S. National Park Service	Natural resource management
Western Area Power Administration	Hydroelectric power generation
Arizona Game and Fish Department	Natural resource management
Tribes (X 6)	Protect the interests and enhance the wellbeing of tribe members, including fostering economic opportunities, protecting cultural tradition, and maintaining a healthy environment
States (X 7)	Water extraction and hydroelectric power generation
Environmental groups (X 2)	Nature conservation
Recreation groups (X 2)	Recreation
Power purchasers (X 2)	Hydroelectric power generation

agency gets one vote unless otherwise noted (i.e., each of the seven states gets a vote, as do each of the environmental groups). These general views may vary from issue to issue, and certain stakeholder groups may split internally on a specific issue (e.g., though the states often agree, their interests on a particular matter may conflict in important respects).

B. *The Regulatory Setting*

The Bureau of Reclamation operates the Glen Canyon Dam in accordance with the Colorado River Storage Project Act of 1956. This law authorized construction of the Glen Canyon Dam—along with other dams, reservoirs, power plants, and transmission infrastructure in the upper Colorado basin—and enumerates the Dam's water management and power generation goals. The Bureau of Reclamation, a division of the DOI, was created in 1902 to promote settlement and economic development in the West by facilitating the capture and delivery of water to meet the needs of farmers and communities. Today, it is the largest water wholesaler in the country, and the second largest producer of hydroelectric power in the Western states. The Bureau's mission has evolved to recognize the various benefits and costs associated with its work of regulating rivers. Today, its declared goal is "to manage, develop, and protect water and related resources in an environmentally and economically sound manner in the interest of the American public." Fulfilling this mission involves making difficult choices regarding how dams like Glen Canyon Dam should be operated to balance a variety of interests and comply with numerous regulations.

The regulations governing the Dam's management have changed over time, reflecting both shifting interests among stakeholders and increased scientific understanding. Perhaps as a result, the multiple, and often conflicting, laws and directives governing the operation of the Dam establish no clear prioritization among the various competing usage demands. The only cultural or environmental stipulation in the Colorado River Storage Project Act of 1956 is that the Secretary of the Interior must "take adequate protective measures to preclude impairment of the Rainbow Bridge National Monument." Various environmental and cultural preservation acts passed in subsequent years—particularly the National Historic Preservation Act (1966), the National Environmental Policy Act (1969), and the Endangered Species Act (1973)—have had major implications for the Dam's operation. For example, the Endangered Species Act explicitly protects the humpback chub, which the Dam has impacted adversely. This statutory protection has been the foundation of numerous lawsuits and biological opinions filed over the past few decades.

In 1992, Congress enacted the Grand Canyon Protection Act ("GCPA") in an effort to consolidate the body of regulations governing the Dam's operations. Rather than clarifying priorities and sorting out conflicting regulations, the GCPA confused matters. While allowing for a decrease in power generation, the GCPA reinforced the water management and hydroelectric priorities the Dam was initially meant to serve. At the same time, it stated that the dam and water resources should be managed in "such a manner as to protect, mitigate adverse impacts to, and improve the values for which Grand Canyon National Park and Glen Canyon National Recreation Area were established, including natural and cultural resources and visitor use." Thus, the GCPA does not set priorities among cultural, environmental, and recreational interests; nor does it mandate how they should be reconciled with water management objectives when the interests conflict. In fact, the GCPA seems to suggest that all demands can be met, and that the GCPA should in no way affect water allocations or conflict with any federal environmental laws.

… The GCPA asked the Secretary [of the Interior] to take responsibility for long-term monitoring of the Dam's impact so that operations could be adjusted over time to account for new information or changed circumstances. Presumably, long-term monitoring would determine the impacts that management has on "the natural, recreational, and cultural resources of Grand Canyon National Park and Glen Canyon National Recreation Area." Furthermore, the GCPA requires that such monitoring be conducted in consultation with various stakeholders, ranging from the governors of the affected states to the recreation industry.

* * *

II. The Persistence of Problems at Glen Canyon

Since its creation a decade ago, the AMP has received praise from various agency officials and scholars who maintain that the Glen Canyon Dam AMP is a successful model of collaborative, adaptive regulation and management. Despite these accolades and considerable funding, a growing number of observers have concluded that the Glen Canyon Dam AMP has been far from successful.

The Glen Canyon Dam AMP should not be considered a success because it has failed to address effectively the concerns that led to its creation in the first place, including: (1) developing a stakeholder-supported operating plan responsive to increased understanding; (2) averting litigation and other attempts to resolve conflict outside of the AMP context; and (3) protecting the downstream ecology, including endangered species.…

A. There Has Been Little Progress on Formulating a Long-term Plan to Operate the Dam

Despite more than fifteen years of research and negotiations, the Dam operates under the same "modified low fluctuating flows" regime as it did in 1996. This lack of progress is discouraging given the commitment of the AMP and its stakeholders to ongoing adaptive management.... In particular, the Glen Canyon Dam AMP has yet to resolve how power generation should be reconciled with ecological and other uses that compete for the Dam's resources. For example, evidence from three, well-publicized controlled flood experiments indicates that vulnerable species, particularly the humpback chub, greatly benefit from seasonal flow changes, yet no subsequent changes have been made to long-term operations to incorporate such information.

The strongest opposition to flow regime change has come from power generation interests. The Colorado River Storage Project Act of 1956 mandates the maximization of power generation revenues, provided that operations do not impinge on the Colorado River Compact or other relevant compacts. This mandate gives power interests authorization to operate the Dam in a manner most beneficial to them, subject to other laws. Controlled floods represent lost revenues to the power industry—an estimated four million dollars in the case of the 2008 experiment. It is still not clear whether power interests will be compensated for this loss. It is unsurprising, given these losses, that power interests are opposed to changes in the Dam's operation.

* * *

B. The AMP Has Been Unable to Avert Unproductive Extra-Programmatic Conflict

Lawsuits filed as early as 1973, only ten years after the Dam was completed, challenged various resource management decisions. Indeed, it was a legal victory won by environmental groups—National Wildlife Federation v. Western Area Power Administration—that led to the creation of the AMP in the first place. The AMP was created to facilitate conflict resolution without resorting to litigation. Under an effective collaborative adaptive management program, stakeholders would reflect jointly on what they had learned and engage in collaborative problem solving to improve the Dam's operations. Unfortunately, under the Glen Canyon AMP, stakeholders hold fast to their positions and continue to spend time and resources challenging each other. As a result of the lack of progress, [Adaptive Management Working Group] (AMWG) members have turned to litigation rather than reliance on the AMWG to resolve disputes over dam operations.

In 2006, five environmental organizations sued the Bureau of Reclamation over the impacts the Dam continues to have on endangered species like the humpback chub. This suit was settled when the Bureau of Reclamation agreed to conduct a new study of native fish and habitats in concert with the Fish and Wildlife Service. The Grand Canyon Trust, an environmental group and member of the AMWG, filed a lawsuit against the Bureau of Reclamation in December 2007 accusing the agency of managing water releases to benefit power generators at the expense of the downstream fish habitat. In March 2008, the Grand Canyon Trust and Earthjustice filed a complaint against the Bureau of Reclamation and the Fish and Wildlife Service, alleging Endangered Species Act violations. United States District Judge David Campbell ruled against the Fish and Wildlife Service in May 2009, requiring the agency to reconsider its approach to evaluating the Dam's impacts on humpback chub.

C. The Downriver Ecology is Still in Jeopardy

In 2008, the Fish and Wildlife Service reiterated that the ecosystem below the Dam has been heavily modified from its pre-dam state. Federal agencies are attempting to ameliorate the situation by making further flow modifications and removing nonnative species, but the changing stream flow (particularly coldwater releases and unnatural flow regimes caused by the Dam) and land use changes have greatly diminished the species' habitat. The humpback chub thrive in warm, sediment-rich flows that create fast moving currents, eddies, and associated beach formations. The Fish and Wildlife Service postulates that, historically, humpback chub were found throughout the Grand Canyon, while today they are largely confined to a few sections and tributaries that remain largely undisturbed by human intervention. According to the Fish and Wildlife Service, "[m]any of the physical changes in the post-dam Colorado River are believed to have contributed to eliminating spawning and recruitment of humpback chub in the mainstem river."

The precarious state of the downriver ecology is particularly disconcerting because anticipated stressors, such as climate change, are likely to strain the ecosystem even further. Fish and Wildlife acknowledges that the effects of climate change should factor into how the Dam is operated, as the low reservoir levels associated with droughts from 2004 to 2006 demonstrate the potential for climate change to impact humpback chub. Perhaps more disturbingly, recent findings by University of Colorado researchers suggest that climate change and population growth could dry up the Colorado River's reservoir by 2057. This would profoundly impact human settlements, agriculture, and the riverine environment.

Recent evidence suggests that the humpback chub may have temporarily benefitted from recent temporary high-flow releases. These releases are byproducts of AMP experiments with various flow regimes used to assess the impacts on species populations and ecosystem health starting in 1996. The U.S. Geological Survey ("USGS") reported in April 2009 that humpback chub populations increased by fifty percent between 2001 and 2008, a significant recovery after steady declines in the 1990s. The USGS acknowledges the difficulty of determining why the population rebounded, but argues that the experimental water releases are probably one factor.

One might consider the humpback chub's recovery to be evidence that the AMP is doing its job. After years of research, however, debates continue regarding whether or not flow regimes should be permanently modified to protect the health of the chub population. Furthermore, the AMP's reluctance to adopt a modified flow regime or even to continue with high flow tests suggests that any successes attributable to the experimental water releases are only temporary and could be erased by the cessation of controlled flooding. Though the DOI recently directed the development of a protocol for conducting even more high-flow experiments, the fact that ongoing dam operations have never been formally changed to incorporate the apparent benefits of the experimental releases on downriver ecosystems indicates the AMP's limited commitment to adaptive management and jeopardizes the ancillary ecological benefits obtained through experimentation.

Notes and Questions

1. The Glen Canyon experience exemplifies problems with dam sites throughout the United States. While there may have been relative consensus about the purpose and benefit of a dam at the time it was built, subsequent experience reveals some of the negative impacts, and competing interests are now raising their voices. What were some of the different interests involved in the Glen Canyon project? How do they overlap or conflict?

2. The excerpt notes that the Grand Canyon Protection Act of 1992 did not set priorities among cultural, environmental, and recreational interests. Do you think it is Congress's

role to establish such priorities? Do you think the collaborative planning and adaptive management approach discussed in the excerpt might be a better mechanism? Why or why not?

3. Although some consider the Glen Canyon Adaptive Management Plan to be a success, the article authors disagree. How has the AMP failed in (1) developing a successful long-term plan; (2) averting litigation; or (3) protecting downstream ecology?

D. Environmental Balance

Depending on one's perspective, the reservoirs that result from an impoundment hydropower plant may be a benefit. They can help with water supply and flood control as well as offering public recreational opportunities such as boating, fishing, and swimming. Yet the damming of water also has its downsides such as displacing humans, plants, and animals from submerged habitats or changing ecosystems through diversion or insufficient stream flows. These environmental concerns mark much of the focus in hydropower law since the 1980s. This section will address efforts by states and environmental groups to take greater control of environmental protection in the context of dam relicensing. The second part of this section addresses the dam decommissioning process.

1. Relicensing Procedures

The limited terms of federal dam projects and the imposition of environmental requirements gave states and other groups new ammunition to contest dams. In some instances, the added requirements are so restrictive that they effectively block relicensing. Congress attempted to limit environmental pushback through provisions in the Energy Policy Act of 2005 as addressed in the following excerpt.

Rick Eichstaedt, Rebecca Sherman & Adell Amos, *More Dam Process: Relicensing of Dams and the 2005 Energy Policy Act*
50-Jul Advocate (Idaho) 33, 33-36 (June/July 2007)

Every hydropower facility in the nation not operated by the federal government must obtain an operating license from the Federal Energy Regulatory Commission (FERC) if that facility affects interstate commerce, navigable waterways, or meet other criteria. In Idaho, dams such as Idaho Power Company's Hells Canyon Complex on the Snake River and Avista Corporation's Post Falls dam on the Spokane River are subject to FERC's jurisdiction.

License renewal, called "relicensing," occurs only once every thirty to fifty years. Each time a license is issued or renewed the process provides an opportunity for any interested stakeholder to earn legal standing and influence the terms of the next license. Because a license term lasts for decades, dams seeking license renewal today were effectively grandfathered from complying with existing environmental laws and standards. Thus, relicensing is a significant opportunity to address a hydropower dam's environmental footprint and to weigh the commitment of watershed resources for another thirty to fifty years.

Many authorities in the hydropower licensing kitchen

While FERC administers hydropower licenses, many federal and state agencies share the authority to craft license conditions. Under Section 18 of the Federal Power Act, federal fisheries agencies, such as the United States Fish and Wildlife Service or National Oceanic and Atmospheric Administration (NOAA, formerly the National Marine Fisheries Service), may require a licensee to construct a fish passage device around a dam. FERC has no authority to deny or alter the prescription, even if it is a reservation of authority. Section 4(e) of the Federal Power Act empowers any manager of a federal reservation upon which the hydropower project lies to place license conditions for the "adequate protection and utilization of the reservation." Again, FERC has no authority to alter these conditions. Nor can FERC deem submitted Section 4(e) conditions untimely or inappropriate and ignore them; only a court may review these conditions. The United States Forest Service (Forest Service), the Bureau of Indian Affairs, and the Bureau of Land Management (BLM), among others, frequently utilize this Section 4(e) conditioning authority.

A state agency may also place conditions in its certification that will ensure that water quality is protected. These conditions are mandatory, and certification may include minimum instream flow requirements. In May 2006, the United States Supreme Court in *S.D. Warren Co. v. Maine Bd. of Environmental Protection* issued a unanimous decision confirming that water quality certifications are required for federal licensure of hydropower dams. The combination of Supreme Court reinforcement and the Energy Policy Act amendments discussed below have emphasized the importance of the Clean Water Act certification authority.

The Federal Power Act grants certain agencies the authority to provide recommendations that FERC must consider. Section 10(j) of the Federal Power Act allows state and federal fisheries agencies to provide recommendations for the protection, mitigation, and enhancement of fish and wildlife. These recommendations must receive expert agency deference at FERC. The licensing process may also trigger requirements for formal review and consultation under the Endangered Species Act if the dam impacts threatened or endangered species.

The Energy Policy Act of 2005: Federal Power Act Amendments

Signed into law in August 2005, the Energy Policy Act of 2005 (EPAct) established an interim process to challenge Sections 18 and 4(e) prescriptions and conditions of relicensing.

Under the new amendments, once an agency files its proposed Section 4(e) conditions or Section 18 fishway prescription, a challenger has thirty days to respond with a request for hearing and/or an alternative....

The EPAct also established a strict requirement for consideration of submitted alternative conditions. By the terms of the EPAct, the secretary of the department in which the agency is housed must accept an alternative to a Section 18 fishway prescription if the alternative provides equal or greater protection than the original condition and either costs less to implement or generates more power. For Section 4(e) conditions, the standard is looser: the secretary must accept the alternative if the condition is adequate for the protection and utilization (but does not necessarily provide equal or greater protection) of the federal reservation, and costs less to implement or generates more power. In either instance, if an alternative does not meet these criteria, the department must still consider the alternative against several criteria, including energy supply, air quality, and navigation....

* * *

Use of the EPACT amendments in individual proceedings

The regulations established a December 19, 2005 deadline for any retroactive EPAct challenges in pending licensing proceedings, or proceedings in which no license had issued but preliminary terms and conditions had been filed. Licensees for fifteen hydropower projects filed challenges by that deadline....

The federal departments began consideration of retroactive hearing requests and alternatives at the same time new requests and alternatives were filed. As more hearings were requested, ALJ actions resolved outstanding questions about the hearing process. First, the ALJs provided varying responses to determining whether an issue subject to an EPAct challenge is factual or material. Some ALJs suggested that the party requesting the hearing should have the opportunity to develop the facts to demonstrate materiality, while others have dismissed issues that raise legal, non-material, and policy questions. However, ALJ opinions have consistently held that the party requesting the hearing bears the burden of proof.

One striking feature of the hearing requests is that they have almost exclusively led to settlements between the conditioning agency and the hearing requestor, who in all instances outside of the Klamath proceeding have been the licensee. These settlements typically do not resolve the stated factual issues, rather they result in the agency either revising its underlying conditions and issuing wholly new license conditions or abandoning its license conditions altogether. Intervenors in the hearing typically do not have access to these settlement discussions, and once a settlement is complete, the hearing request is usually withdrawn. In some instances, hearing requestors have asserted their rights to reinitiate a hearing if license conditions change or have amended the original hearing request or alternative. Conservation and recreation interests note that these settlements result in reduced resource protection.

The Hells Canyon Settlement

* * *

... Hells Canyon was the first non-retroactive application of the EPAct challenge process. On February 27, 2006, Idaho Power filed challenges against preliminary conditions filed by the Forest Service and the BLM. Idaho Power's filing against the Forest Service alone consisted of more than 200 pages challenging twenty-six issues of material fact and included more than twenty alternative conditions for agency review.

Two conservation groups, the states of Oregon and Idaho, and NOAA all intervened in the Forest Service's EPAct proceeding. At the same time, the state of Oregon, two conservation groups, and the Shoshone-Bannock Tribes filed alternatives to the NOAA's Section 18 fishway prescription, a reservation of authority in lieu of an actual fish passage measure.

The BLM reached a settlement with Idaho Power and filed revised Section 4(e) conditions within three months of the hearing request. Despite repeated requests by the interveners to participate in the Forest Service's settlement discussions, the Forest Service and Idaho Power reached settlement....

... The settlement resulted in significantly weaker mitigation measures and land management. For example, instead of requiring Idaho Power to acquire 1,522 acres to address the depletion of riparian habitat, the Forest Service required the company to acquire only 56.3 acres. Finally, Idaho Power successfully excluded a suite of intervenors and submitted revised alternatives that mirrored the settlement after the statutory deadline had expired.

The first hearing: Klamath River

In fall 2006, the Klamath Hydroelectric Project became the first project subject to a completed trial-type hearing. The Klamath Project includes four dams in Oregon and California that many licensing parties are actively working to remove. PacifiCorp, the owner and operator of the Klamath Project, requested an EPAct hearing on April 28, 2006, challenging prescriptions and conditions of NOAA and agencies within the Department of the Interior.

The Departments of Commerce and Interior consolidated the hearing requests and conducted one hearing adjudicated by a Coast Guard ALJ, Judge Parlen McKenna. The California Department of Fish and Game, Klamath Tribes, Hoopa Valley Tribes, and several conservation groups all filed notices of intervention. A five-day hearing with fifty-seven available witnesses was held in Sacramento, California in late August 2006.

On September 27, 2006, Judge McKenna issued a decision. Of the fourteen disputed material facts, PacifiCorp prevailed on issues related to the recreational use of the river and on lamprey habitat and survival. On the central factual disputes, including whether anadromous fish occurred above the project facilities and whether current habitat and water quality conditions above and through the facilities would support repopulation of these fish, Judge McKenna confirmed the original positions taken by the resource agencies, the Tribes, and the conservation groups: that "PacifiCorp failed to prove its version of the facts."

Although the result of the hearing was an incremental loss from the factual position underlying the preliminary conditions, the federal resource agencies, the conservation groups, and the Tribes achieved positive precedent for all future EPAct cases. The order's factual conclusions form a strong record that will substantiate agency conditions and prescriptions under any future legal review. The Klamath proceeding also demonstrated the scale of time, expense, and human resources required to challenge and defend mandatory conditions under the EPAct hearing.

* * *

Conclusion

At a May 2006 hearing, Senator Larry Craig described FERC hydropower licensing as "12 and 14 year processes that cost millions and millions of dollars." The hydropower provisions of the EPAct were billed as a mechanism to streamline the process and increase its affordability. However, in practice it is clear that the provisions have not streamlined the process or reduced the cost for licensing participants. Instead, the EPAct, if carried out, makes the process more complex, litigious, and expensive. A more common outcome of hearing requests is not a confirmed set of facts but an exclusive supra-licensing settlement proceeding that revises the underlying agency mandatory conditions. This threatens only to further complicate the relicensing process and undermine the intent of the Federal Power Act.

Notes and Questions

1. As Figure 4.5 shows, the heyday of dam construction is over. From a peak of 18,717 dams completed in the 1960s, dam construction dropped off to 4,393 in the 1990s and to only 2,290 since 2000. Is this decrease simply a consequence of a time-consuming and bureaucratic licensing process? Does the fact that the process is complex reflect a balancing of other values?

2. What do you think about the 2005 EPAct's provisions to streamline the relicensing process? Does the excerpt above suggest it has been effective?

3. The hydropower story perhaps illustrates the conflict of any large-scale renewable energy development. At 90 percent efficiencies, hydropower is one of the most efficient ways of producing electricity. Hydropower does not produce Greenhouse Gases and is available in large quantities on demand. Yet, any large-scale power source cannot be constructed without some negative impacts on the environment—most notably habitat disruption. Unlike newer renewable resources that are just now attempting to gain a foothold in the market, hydropower made significant inroads in the United States during an era that did not require as much public input about aesthetic and environmental concerns. How can the United States attempt to meet its unrelenting demand for more energy without some environmental tradeoffs? Do you have suggestions for how to reach a balance that allows new energy sources to be developed?

2. Dam Decommissioning

The first of the excerpts in this section explains the movement behind decommissioning some hydropower dams. It tells the story of how Americans placed hydropower above wildlife in developing dams along the Columbia River and of efforts in the last few decades to restore the lost salmon runs. The second excerpt addresses some of the legal nuts and bolts of dam decommissioning.

Michael C. Blumm, Erica J. Thorson, & Joshua D. Smith, *Practiced at the Art of Deception: The Failure of Columbia Basin Salmon Recovery Under the Endangered Species Act*
36 ENVTL. L. 709, 711–24 (Summer 2006)

I. Introduction

For at least a quarter century, national policy has been to restore the Columbia Basin's salmon runs. Once the world's largest, the Columbia's salmon runs were decimated first by over-fishing and later by water project development, which transformed the basin into the largest interconnected hydroelectric system in the world and created a seaport in Idaho, some 465 miles inland.

After unsuccessfully experimenting with massive reliance on hatcheries to substitute for salmon habitat lost to water project development, Congress ordered modifications in the operations of Columbia Basin dams in an innovative 1980 statute, the Northwest Power Act. Although the drafters of that statute were quite optimistic that those operational changes and other modifications to the dams would reverse the salmon's decline, the measures instituted under the 1980 statute were unable to prevent the listing of several salmon species under the Endangered Species Act (ESA) in the early 1990s.

The ESA era ushered in by the listings began with great anxiety among the electricity, navigation, and other river-dependent industries that so-called draconian ESA measures would elevate salmon protection over hydropower generation or barge transport of agricultural goods. But over a dozen years after the initial listings, the issuance of several biological opinions (BiOps)—designed to avoid jeopardy to listed salmon—produced no such reallocation of Columbia Basin hydrosystem priorities.

* * *

II. The Relationship Between Hydropower and Salmon

Between the 1930s and the 1970s, hydropower development reconstructed the mighty flows of the Columbia River and its principal tributary, the Snake River. By the middle of the 1970s, the completion of the four dams on the Lower Snake River—Ice Harbor, Lower Monumental, Little Goose, and Lower Granite—created a series of deep, slackwater pools that transformed Lewiston, Idaho into a deepwater port.

In the rush to develop the Columbia Basin, the federal government did not entirely ignore the plight of salmon, however. In 1945, when Congress authorized the McNary Dam—in the same statute that sanctioned the Lower Snake Dams—it pledged that "adequate provision shall be made for the protection of anadromous fishes by affording them free access to their natural spawning grounds." Despite this directive suggesting that salmon conservation was a federal priority, hydropower operations have always remained the dominant use of the rivers in the Columbia Basin, even though in 1980 Congress passed the Northwest Power Act (NPA), which called for "parity" between salmon conservation and hydropower production. A dozen years after enactment of the NPA, the listing of salmon under the ESA eclipsed the NPA as the primary tool for salmon conservation. The ESA did not, however, stem the decline of the Columbia Basin salmon populations—largely because NOAA has continued to preserve the hydropower status quo over the survival and recovery needs of salmon....

A. The Legal Framework: The Northwest Power Act and the Endangered Species Act

Congress first expressed concern about the potential effects of Columbia River hydropower operations on salmon as long ago as 1937, when it enacted the Bonneville Power Act. But it was not until the passage of the NPA in 1980 that Congress seriously attempted to protect and restore the Columbia Basin's salmon runs. The NPA directed the Northwest Power Planning Council (Council) to create a program to "protect, mitigate, and enhance" damaged salmon runs "to the extent affected by the development and operation" of the hydropower system. The ensuing program—the Columbia Basin Fish and Wildlife Program—was once touted by the Council as "the most ambitious effort in the world to save a biological resource," and aimed to achieve parity between salmon and hydropower. Despite years of efforts to fulfill this promise, however, the program failed to restore the Columbia Basin's decimated salmon runs.

Although Congress passed the modern ESA in 1973, after hydropower had imperiled the Columbia Basin's salmon runs for most of the twentieth century, the ESA did not become the central player in salmon legal protection until the 1990s. Citizens began the ESA era by invoking the public petitioning process for listings after the NPA failed to achieve its objective of putting salmon on par with hydropower. In 1991, NOAA listed the Snake River sockeye as endangered, and the next year listed two species of Snake River chinook as threatened. By 2005, NOAA had listed thirteen Columbia Basin salmon runs under the ESA.

* * *

A species's listing implicates the ESA's protections, including bans on sales, imports, exports, and "takes" of endangered species and, in some circumstances, threatened species. Section 7 of the ESA also prohibits federal agencies from proceeding with an action that is likely to jeopardize a listed species or adversely modify its critical habitat.

* * *

The action agency must initiate a formal consultation with the appropriate agency to determine whether the proposed action is likely to jeopardize the species or result in adverse critical habitat modification only if the biological assessment indicates that the proposal is likely to affect adversely either the listed species or its critical habitat. Otherwise, the action may proceed without formal consultation....

The Columbia Basin hydropower operations have been the subject of numerous BiOps over the last dozen years....

B. Current Status of the Columbia River Basin's Listed Salmon Runs

At the time NOAA published its 2004 BiOp, twelve Columbia Basin salmon species were listed under the ESA. Most of these species were listed as threatened; only two were listed as endangered—the Snake River sockeye and the Upper Columbia River spring chinook....

Current Snake River salmon runs are a mere shadow of their former abundance and vigor. The Hells Canyon Dam complex, completed in 1967, and the four Lower Snake River dams, the last of which was completed in 1975, have had a devastating impact on Snake River chinook species, effectively eliminating nearly 50% of their historical spawning habitat....

* * *

Of all the Columbia Basin runs, none have fared worse than those of the Upper Columbia River. The Grand Coulee Dam, completed in 1941, is a major barrier to spawning habitat for salmon, excluding Upper Columbia River runs from nearly fifty percent of their historic spawning grounds. Both Upper Columbia River spring chinook and steelhead continue to spawn precariously in drainages between Rock Island Dam and Chief Joseph Dam (immediately below Grand Coulee), the limit of upstream passage today. In 1998, however, fewer than one-hundred wild Upper Columbia River spring chinook returned. Despite increasing returns during the period of 2001 to 2003, this species faces a high rate of extinction, primarily because hatchery fish account for approximately seventy to ninety percent of the returns....

All of the Lower Columbia River listed species—chinook, steelhead, and coho—have sustained losses of thirty-five to forty percent of historic habitat due to impassable dams....

* * *

The Willamette River runs listed under the ESA consist of the Upper Willamette River chinook and the Upper Willamette River steelhead. Both species have experienced significant loss of spawning grounds due to the thirty-seven dams that occupy their historic habitat....

* * *

Although ocean conditions during 2000-2003 produced a boon in salmon productivity, as evidenced by increases in returning spawner numbers for almost all populations, all listed salmon runs in the Columbia Basin face the likelihood of endangerment, if not extinction, within the foreseeable future.... The high returns reported at the turn of the twenty-first century were the result of favorable ocean conditions, not improvements in hydropower operations.

———————

Sarah C. Richardson, Note, *The Changing Political Landscape of Hydropower Project Relicensing*

25 Wm. & Mary Envtl. L. & Pol'y Rev. 499, 507-16 (2000)

1. Growth of Environmental Awareness after the Mid-Century

The latter half of the twentieth century has seen a profound change in our society's perception of the environmental costs of dams. "[T]he balance of power struck during the Progressive era in favor of centralized federal authority over the uses of the Nation's navigable waters" has begun to wobble because of changes in public policy goals that have resulted in a markedly altered view of the public interest in hydropower and in natural resources. The growth of environmental awareness in the 1960s and '70s brought with it a new perspective on the balance of costs and benefits represented by dams.

The costs can be severe: since dams are created in order to barricade rivers, they also obstruct the fish swimming in those rivers, which drastically alters the life cycle of indigenous migratory fish species. By slowing rivers and reducing downstream water levels, dams increase water temperatures and reduce oxygen levels. The obstructions cause silt to collect on upstream riverbeds, which also destroys habitat and kills many kinds of fish. Fish swimming downstream may be killed directly by being drawn into and cut up by power turbines.

Hydropower dams also cause severe changes in water levels by withholding and then releasing water to generate power for peak periods; fluctuating water levels interrupt the natural growth and reproduction cycles of many species. "Thus, although a source of renewable energy, hydropower consumes another valuable natural resource: free-flowing rivers and the many ecological, recreational, aesthetic, and economic benefits that rivers provide."

During the first sixty years of the Commission's existence, it turned down only one proposed license on aesthetic or recreational grounds. Well into the 1980s, the agency gave short shrift to environmental factors in licensing decisions and continued to act as "a friend of the hydroelectricity industry and a nemesis of environmentalists." In fact, FERC didn't even promulgate regulations implementing NEPA until 1987, seventeen years later than the statutory requirement, even though one of the first modern environmental-law cases, *Scenic Hudson Preservation Conference v. Federal Power Commission*, established clearly the Commission's duty to consider environmental factors in the licensing process.

As another momentous sign of changes to come, in 1967 the Supreme Court held in *Udall v. Federal Power Commission* that FERC had failed to consider fishery-resource impacts of the project under consideration. In the Court's view, section 10(a) of the FPA requires FERC to balance power generation against environmental impacts, because the "public interest" provision of the Act encompasses "preserving reaches of wild rivers and wilderness areas, the preservation of anadromous fish for commercial and recreational purposes, and the protection of wildlife." The Court's championing of fish against hydropower signaled a profound change:

> The importance of salmon and steelhead in our outdoor life as well as in commerce is so great that there certainly comes a time when their destruction might necessitate a halt in so-called "improvement" or "development" of waterways. The destruction of anadromous fish in our western waters is so notorious that we cannot believe that Congress through the present Act authorized their ultimate demise.

By 1986, Congress had decided that FERC should be giving more attention to nonpower interests, so it amended the Federal Power Act by enacting the Electric Consumers Protection Act (ECPA), which provided for stronger environmental protection. ECPA added Section 10(j) to the FPA, which requires FERC to include in a new license any terms and conditions pertaining to fish and wildlife that are recommended by state and federal fish and wildlife agencies. The ECPA compels FERC to give equal consideration in relicensing procedures to energy conservation, fish and wildlife preservation, recreational opportunities, energy conservation, and protection of environmental quality.

This "equal consideration" mandate requires FERC to consult with federal, state and local resource agencies, including fish, wildlife, recreation and land management agencies, in order to assess more accurately the impact of a hydropower dam on the surrounding environment. In its evaluation of environmental impacts, FERC is obligated to prepare an Environmental Impact Statement (EIS) or Environmental Assessment (EA), investigative reports which assess the environmental consequences of a proposed hydropower project and compare the impacts with those of alternatives to the suggested action. [One adaption that is made to mitigate fish declines is a fish ladder like the one in Figure 4.7, which allows fish to access a reservoir above a dam— Ed.]

* * *

2. Dam Decommissioning

a. FERC's 1994 Policy on Dam Decommissioning

A combination of factors, including greater understanding of the environmental costs of dams, increased value given to nonpower river uses, and the swelling flood of relicensing applications, led FERC to formulate a new dam-decommissioning policy in 1994. The new policy relied heavily on Section 10(a) of the FPA....

* * *

In its policy statement, FERC concluded that, in order to "satisfactorily protect the public interests involved," the Commission has "the legal authority to deny a new license at the time of relicensing if it determines that, even with ample use of its conditioning authority, no license can be fashioned that will comport with the statutory standard under section 10(a)...." Outright denial will be rare, according to the Policy Statement, as licensing conditions will provide the required balance between power and environmental safeguards. Decommissioning is more likely to occur when "the licensee of an already marginal project is confronted with additional costs at relicensing that render the project uneconomic." Those costs will arise when FERC imposes environmental or other conditions, and decommissioning will be the *de facto* result even if not expressly commanded by FERC.

In response to industry commentators' objections that, if license conditions make the project uneconomic, then the conditions must be rejected as unreasonable, FERC quotes a Seventh Circuit opinion: "[T]here can be no guarantee of profitability of water power projects under the Federal Power Act; ... values other than profitability require appropriate consideration."

The Policy Statement notes that frequently conditions placed on a project's license come, not from FERC, but from the state in which the dam is located. The Clean Water Act (CWA) empowers states to approve or deny water-quality certification for hydropower projects, and the Supreme Court has ruled that FERC must include the state's CWA conditions in a license. Moreover, when the Energy Consumer Protection Act was passed, Congress mandated that FERC licenses "shall be subject to and contain such conditions

Figure 4.7: Fish ladder at Bonneville Dam on the Columbia River

Photo by K.K. DuVivier

as the Secretary of [federal land management agencies] shall deem necessary for the adequate protection and utilization of such [federal land] reservation...."

b. Funding Dam Decommissioning

The crucial question of how decommissioning will be funded is of particular concern to the states (or, if the dam is on federal land, other federal agencies) who may find themselves with a defunct and "deadbeat" (non-power-producing) dam left on their hands by an owner who could not afford the decommissioning costs. The 1994 Policy Statement briefly considers requiring hydropower projects to institute decommissioning funds, but rejects that idea because it "could mean unnecessarily tying up substantial amounts of the capital of financially sound licensees in less than optimum investments for extensive periods." FERC leaves open the possibility that a licensee might act on its own to set up a fund if the risk of decommissioning looms ahead, and could recover the costs of the decommissioning fund in its rates.

The Commission also discarded a second funding option, establishing an industry-wide decommissioning fund financed by annual charges, because it found inadequate evidence of the need for such an administratively challenging fund. Although the Commission rejected the adoption of such long-term funding requirements, it made clear its policy that "[t]he licensee has the responsibility for project retirement," not the federal government. FERC's rationale for this position is that "the licensee created the project and benefited from its operations."

c. The Decommissioning Policy in Action

The first test of FERC's new policy was presented three years after the Policy Statement by the relicensing process for the Edwards Dam, built on the Kennebec River in Maine's capital, Augusta. The dam, one of the hundreds built to provide mechanical power for sawmills in the early nineteenth century, had been a mainstay of industrial development in the area, and later was converted to generate electricity. By the time its license came up for renewal, however, the dam was generating only one-tenth of one percent of Maine's electricity, and many advocacy groups were calling for restoration of the historic—and remunerative—shad, sturgeon, sea bass, and salmon fisheries on the river, devastated by the dam for more than 160 years.

In 1997, FERC denied the renewal of the Edwards Dam license. "The potential for fisheries restoration was so great, the electricity generated so minimal, that the consensus for removal was almost inevitable." Although the dam operator at first threatened to fight the denial, the parties reached a settlement in which the dam was turned over to the state. Upstream hydropower dams and a downstream shipbuilding company jointly funded the removal and fish-restoration programs. In this agreement, the consortium of hydropower dams was given a longer time before it had to install required fish ladders, and the shipbuilder was allowed to fill in seventeen acres of the river to build new dry docks.

d. Negotiated Decommissioning Settlements: The Future

The "many-back scratcher" settlement reached in Maine provides a glimpse of the likely future for dam-relicensing controversies. Similarly, in September 1999, a voluntary agreement to remove the Condit Dam, located in southwestern-Washington state, was signed by the Yakima Indian Nation, PacifiCorp, environmental groups, state and federal fishery agencies, and the Columbia Inter-Tribal Fish Commission, after two years of negotiations. The agreement provided that the dam, a 14-megawatt project on the White Salmon River, will be removed to open up a salmon run from the Columbia River, but that before removal it could continue operating for seven years in order to generate funds that will offset the decommissioning costs.

Because of the agreement, PacifiCorp will be able to avoid carrying out relicensing conditions (primarily fish passages) that had been spelled out in FERC's Environmental Impact Statement for the project. Fulfilling those conditions could have cost $30 million or more, whereas removing the dam and contributing to fishery restoration will cost an

For an interactive graphic of how the Glines Canyon dam in Washington state will be dismantled, go to "Dam removal begins, and soon the fish will flow" (Sept. 17, 2011) on the Los Angeles Times website.

estimated $17.15 million. Just as FERC foresaw in its 1994 Policy Statement, decommissioning for this project resulted from environmental conditions that rendered the hydropower project uneconomic.

Such environmental/economic trade-offs may become a regular part of the hydropower field, as hydropower projects coming up for relicensing find themselves in a new regulatory landscape where they face FERC- or state-imposed environmental conditions, and they know that FERC will not necessarily shy away from decommissioning. If, as in the Edwards and Condit Dam situations, restoration programs would cost far more than removal of dams that do not produce highly profitable amounts of power, hydropower operators are likely to turn to their surrounding municipalities and states in search of innovative solutions.

Notes and Questions

1. "The largest dam removal project ever undertaken in the United States" began on September 17, 2011, as bulldozers ate away at the 108-foot-high Elwah Dam in Washington State. Kim Murphy, *Dam Removal Begins, and Soon the Fish Will Flow*, L.A. TIMES, Sept. 17, 2011. The Elwha Dam, and its 210-foot-high companion, the Glines Canyon Dam, were built in 1913 to provide electricity for industry in Port Angeles. Don Laford, the project's construction manager, estimated that removal of the two dams would take three years and cost approximately $27 million. Reuters reported that the entire river restoration project—"including new power sources, water treatment plants, re-vegetation, and other improvements"—would have a price tag of $325 million. Finding sources of funding is a serious issue because decommisioning was not considered at the time these and other dams in the past were constructed. The excerpt above discussed possible mechanisms to fund decommissioning? What were they? Can you think of other options?

2. In our history, Americans have focused on present production of resources without considering the cost to future generations of reclaiming or decommissioning a resource development or generating site. As a result, we have left a legacy of ruined dam sites and abandoned mines that continue to pollute water resources for generations. Can we use the dam decommissioning story as a cautionary tale that the end game should be considered upfront for all energy projects?

3. Unfortunately, there is no guarantee that decommissioning dams will mean that recreational activities or wildlife can return to pristine habitat. Immediately after the Gold Ray dam was dismantled on the Rogue River in Oregon, prospectors rushed in to dredge the riverbed searching for gold. The removal of one priority of the past—the building of dams—just made way for another past priority—the exploitation of minerals. Is there a way to make sure current priorities prevail?

E. Future Development of Hydropower Resources

Despite a decline in the enthusiasm for large-dam hydropower projects, new technologies promise to preserve hydropower's stature as a significant player in the mix of renewable energy resources for the United States. The National Hydropower Association is bullish about hydropower's potential, predicting that the U.S.'s current hydropower capacity of approximately 96 MW could be doubled. Here are a few ways the National Hydropower Association, the Bureau of Reclamation, and others have set out for increasing the percentage of hydropower in the U.S. energy mix.

1. Retrofitting and Uprating Existing Dams: According to a 2006 study from the Idaho National Laboratory, the United States has only 192 large (over 30 MWa) hydroelectric dams. Furthermore, of the thousands of large and small dams in the country, under 3 percent produce electricity. Some of these existing dams can be retrofitted to add hydroelectric turbines. In addition, many of the existing hydroelectric dams have antiquated turbine technologies. Uprating existing hydroelectric generator and turbine units at power plants is one of the most immediate, cost-effective, and environmentally acceptable means of developing additional hydroelectric power. From 1978 to 2005, the Bureau of Reclamation added more than 1,600,000 kW of capacity at an average cost of $69 per kilowatt.

2. Developing Smaller, Low-Head Hydropower Plants: A 2006 nationwide mapping survey by the Idaho National Laboratories ("2006 INL Study") identified a handful of sites suitable for large, high-head dams, but determined that none of these were feasible for development because of environmental concerns. However, this same survey identified over 127,000 sites where electricity could be generated from smaller, low-head hydropower in running rivers, streams, or irrigation canals. If fully developed, these small hydro sites could produce over 29,000 MW per year (or MWa—See Figure 1.1). Figure 4.8 shows how a small hydro site might be configured.

While the Department of Energy defines large hydropower as facilities that have a capacity of more than 30 MWa, there is no consistent definition of small-scale or micro-hydropower. DOE considers the 100 kWa to 30 MWa range as small-scale, but the Bureau

Figure 4.8: Major components of a small hydro system

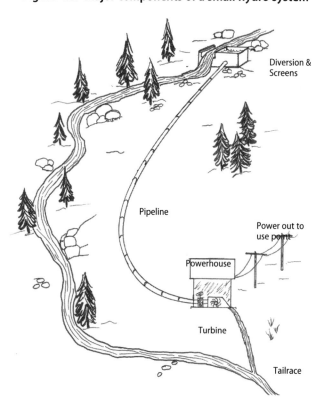

> **Micro-hydropower**—The smallest generators can be installed almost anywhere and cause no significant impacts on the environment. A small or micro-hydroelectric power system can produce enough electricity for a home, farm, ranch, or, in developing countries, a village. For photos of examples, search online for Colorado Small Hydro Association or Hydroscreen.com.

of Reclamation puts locations with a drop of less than sixty-five feet and a generating capacity of less than 15,000 kW in the smaller, low-head category. The INL study defined a "small hydro" facility as any site that produced between 1 and 30 MWa and a "low power" facility as one that can produce less than 1 MWa.

As of the time of the 2006 INL study, there were only 2,186 small and low power hydro facilities in the United States. FERC's complicated regulatory process may be one of the main reasons the small hydro figures are so low; in the past thirty years, FERC has only issued permits for twenty-four Small Hydro projects. The Small Hydro Federal Permitting Program Study by the Colorado Governor's Energy Office (GEO) in 2010 concluded that "the resources needed today to obtain a hydropower permit from FERC represent a disproportionate burden for developers of small projects. As a result, the development of this renewable resource is stifled nation-wide." The National Hydropower Association agreed with this assessment, and in 2011, successfully lobbied members of Congress to introduce H.R. 795, the Small-Scale Hydropower Enhancement Act, which would exempt hydropower projects of less than 1.5 MW from FERC licensing requirements.

Colorado took an alternative approach, entering into a Memorandum of Understanding (MOU) with FERC in August of 2010 that attempts to streamline and simplify small scale hydropower permitting. Through the MOU, Colorado's GEO takes a proactive approach. The GEO assumes the duties of potential small hydro applicants to contact and submit project applications to resource agencies, affected tribes, and the public. Under the MOU process, FERC has not waived or eliminated any of its requirements. Instead, FERC simply has agreed to allow the GEO to do the work of the applicant by producing written waivers from affected parties to meet the first two stages of FERC's consultation process. Colorado and FERC are touting the MOU as a potential model nationwide.

3. Pumped storage: Hydropower would seem to be the ideal renewable resource to back up intermittency problems with other renewable resources such as wind or solar. Hydropower generation can be started and stopped on demand by simply opening or closing the penstocks. Yet sometimes hydropower dams have less flexibility because there are additional considerations they must address such as maintaining reservoir levels or avoiding turbulence in spillways that can degrade fish habitat. Pumped storage simply requires pumping supplies to an uphill reservoir at times of low demand and releasing that water through the hydropower turbines and into a lower reservoir during times of high electricity demand. See Figure 4.9.

Pumped storage is the one form of hydroelectric power that is specifically designed to meet peak load demands or to complement intermittent renewable energy development. In fact, pumped storage is currently the leading form of energy storage in the United States at approximately 21 GW of pumped storage generating nameplate capacity. All other forms of energy storage—including batteries, chemicals, hydrogen, pitch, purchased steam, sulfur, tire-derived fuels, and miscellaneous technologies—accounted for only 1 GW of U.S. generator nameplate capacity in 2009. http://www.eia.gov/cneaf/electricity/epa/epaxlfile1_2.pdf

Figure 4.9: Pumped-storage hydro

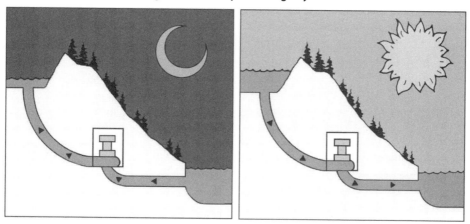

Tenn. Valley Auth., http://www.tva.gov/power/pumpstorart.htm (last visited July 19, 2011)

While it may be appealing to marry one renewable electricity source with others, this match has its drawbacks. Most notably is the Energy Return on (Energy) Investment (EROI or EROEI), a concept which will be discussed in more detail in the following chapter on Biomass. In short, while pumped storage represented 1.8 percent, or approximately 20 GW, of the nation's total electricity generating *capacity* in 2009, it represented a negative number (−0.1 percent) of U.S. electricity *generation* that year. *See* the NREL chart in Figure 4.1. This negative return results because the water that is being pumped to the upper reservoir is working against gravity, and more energy is required to get it uphill than the same water can generate as it flows downhill through the hydropower turbines. In fact, the return of energy for pumped storage is approximately 0.7:1. According to Dr. Robert B. Schainker, Senior Technical Executive with the Energy Storage Program for the Electric Power Research Institute (EPRI), the EROEI for battery storage (0.75:1) is in the same range as pumped hydro, but Compressed Air Energy Storage (CAES) is much better at 1.429:1. Other concerns that have arisen about pumped storage include use of the land to create the reservoirs and water loss from the upper reservoir by leakage or evaporation as it is being held for controlled release.

In addition to the above efforts, the Department of Energy has funded manufacturers designing more environmentally friendly hydropower turbines for both the United States and worldwide markets. Some of DOE's work on encouraging these new, more efficient hydropower technologies is described in the following excerpt.

U.S. Dep't of Energy, Energy Efficiency and Renewable Energy Wind and Water Program—Technologies

http://www1.eere.energy.gov/windandhydro/hydro_rd.html

1. Hydropower Research and Development

The United States faces many challenges as it prepares to meet its energy needs in the twenty-first century. Electricity supply crises, fluctuating natural gas and gasoline prices, heightened concerns about the security of the domestic energy infrastructure and of foreign sources of supply, and uncertainties about the benefits of utility restructuring are all elements of the energy policy challenge. Hydropower is an important part of the diverse energy portfolio that is needed for a stable, reliable energy sector in the United States.

Responding to these national energy issues, DOE recently restructured hydropower R&D, which is now organized around two primary areas:

- Enhancing the viability of hydropower—developing new, cost-effective, advanced technologies that will have enhanced environmental performance and greater energy efficiencies. When implemented, these technologies will enable a 10 percent growth in hydropower generation at existing plants.

- Expanding the application of hydropower—providing supporting research in power systems integration, resource assessment, innovative technology characterization, valuation and performance metrics, industry support, and technology acceptance.

2. Current R&D

While hydropower turbine manufacturers have incrementally improved turbine technology to improve efficiencies, the basic design concepts haven't changed for decades. These late 19th and early 20th century designs did not consider environmental effects, since little was known about environmental effects of hydropower at the time.

During the 1980s, the environmental concerns in the United States became more important in hydropower projects, both existing and planned. This trend is slowly spreading across the globe.

The hydropower industry recognizes that hydropower plants have an effect on the environment. The industry also recognizes that there is a great need to bring hydro turbine designs into the 21st century. The industry visualizes innovative hydro turbines designed from a new perspective. This perspective would look at the "turbine system" (which could include everything except the dam and powerhouse) to balance environmental, technical, and economic considerations.

a. Environmental Challenges

Although hydroelectric power plants have many advantages over other energy sources, the potential environmental impacts are also well known. Most of the adverse impacts of dams are caused by habitat alterations. Reservoirs associated with large dams can cover land and river habitat with water and displace human populations. Diverting water out of the stream channel (or storing water for future electrical generation) can dry out streamside vegetation. Insufficient stream flow degrades habitat for fish and other aquatic organisms in the affected river reach below the dam. Water in the reservoir is stagnant compared to a free-flowing river, so water-borne sediments and nutrients can be trapped, resulting in the undesirable growth and spread of algae and aquatic weeds. In some cases, water spilled from high dams may become supersaturated with nitrogen gas and cause gas-bubble disease in aquatic organisms inhabiting the tailwaters below the hydropower plant.

Hydropower projects can also affect aquatic organisms directly. The dam can block upstream movements of migratory fish such as salmon, steelhead, American shad, sturgeon, paddlefish, and eels. Downstream-moving fish may be drawn into the power plant intake flow and pass through the turbine. These fish are exposed to physical stresses (pressure changes, shear, turbulence, strike) that may cause disorientation, physiological stress, injury, or death.

R&D is currently underway that will help fishery biologists and turbine designers better understand what is happening in the turbine passage. Biological tests are being conducted that will quantify the physical stresses that cause injury or death to fish. In addition to these tests, tools are being developed to help both the engineers and biologists. These tools include developing a Sensor Fish, which is a "crash dummy fish." It will be able to measure the physical stresses in a turbine passage and can be used instead of live fish to gather information. Another tool is the development of a computational fluid dynamics program that models potential fish behavior in the turbine passage. The test results and tools will help turbine manufacturers design a more environmentally friendly turbine, which will reduce the physical stresses to which fish are exposed. New products such as greaseless bearings eliminate the possibility of petroleum products being released in the water.

b. Hydro Turbine Development

In the mid-1990s, the U.S. Department of Energy began research into advanced hydropower technology. The goal is to develop systems that generate more electricity with less environmental impact....

Notes and Questions

1. What do you think of the ways the government proposes to expand U.S. hydropower resources? Which sound most promising to you? What do you think might be some of the legal issues involved with each?

2. What do you think of the process for encouraging small hydro created through the FERC-Colorado MOU? If FERC is not eliminating any of its permitting requirements, are the problems of development simply shifted to Colorado? If this is the case, is the state a better entity to handle these hurdles than individual small hydro developers? Could a similar streamlining model be used to expedite the development of other renewable resources?

3. Pumped hydro can never have an Energy Return on Energy Investment (EROEI) of greater than one because of the physics of the process. Is the value of having energy where-you-want-it-when-you-need-it worth this extra cost in terms of energy invested?

4. Although there might be an argument for pumped storage to meet peak demand, should the negative EROEI be a basis for discouraging its use as back-up for intermittent renewables? What if, despite this negative EROEI, the cost of pumped storage is lower than the other storage alternatives such as compressed air storage or batteries?

Chapter 5

Biomass

Reptiles appreciate the stored heat of the sun when they bask on rocks, so there is little doubt the sun was the first energy source also employed by humankind for heat and light. Biomass was not far behind as humans discovered fire and used biomass, such as wood and animal waste, to stoke their blazes for heating, cooking, and illumination.

In earlier chapters, we noted how the sun was the source of all the renewable fuels addressed in this book except for geothermal energy. Biomass is no exception—the majority of it comes from plants that use photosynthesis to lock the sun's energy into their leaves, stems, and trunks. This carbon energy is then stored until released as CO when the plant matter decays or CO_2 and heat when it is burned.

Because humans can often control the time of burning, biomass has had a huge advantage over solar and wind power, which are intermittent and from which humans traditionally could benefit only when nature chose to punch its time clock. This availability of energy on demand made biomass a fuel of choice in the past. It is also attractive; practically any nation can stockpile a secure supply because it is available in some form anywhere plants grow.

According to data compiled for the National Renewable Energy Laboratory (NREL) 2009 Renewable Energy Data Book, biomass's 2009 contribution to overall U.S. electricity generation was about 54,000 million kWh. This figure is minimal in comparison to the approximately 1,800,000 million kWh generated by coal. However, biomass represents a significant proportion of the U.S. electricity generation attributed to non-hydropower renewables: 1.4 percent of the 3.6 percent total. *See* Figure 5.1.

While biomass may now represent more than a third of the renewable electricity pie, it is losing ground. Cumulative increases in nameplate capacity for biomass have been less than 2 percent in most recent years, in comparison to increases in solar and wind capacity that have averaged in the 30 to 50 percent range. *See* Figure 5.2

Fossilized biomass—In contrast to biomass materials, which start with plants in their unaltered state, fossil fuels such as coal represent a transformed version of plants altered over millions of years by chemical and thermal processes that concentrate and store the energy.

Figure 5.1: Renewable electricity as a percentage of total generation

	Hydro	Solar	Biomass	Wind	Geothermal	Renewables w/o Hydro	All Renewables
2000	7.2%	0.0%	1.6%	0.1%	0.4%	2.1%	9.4%
2001	5.8%	0.0%	1.3%	0.2%	0.4%	1.9%	7.7%
2002	6.9%	0.0%	1.4%	0.3%	0.4%	2.1%	8.9%
2003	7.1%	0.0%	1.4%	0.3%	0.4%	2.1%	9.2%
2004	6.8%	0.0%	1.3%	0.4%	0.4%	2.1%	8.9%
2005	6.7%	0.0%	1.3%	0.4%	0.4%	2.2%	8.8%
2006	7.1%	0.0%	1.3%	0.7%	0.4%	2.4%	9.5%
2007	6.0%	0.1%	1.3%	0.8%	0.4%	2.6%	8.5%
2008	6.2%	0.1%	1.3%	1.3%	0.4%	3.1%	9.3%
2009	6.9%	0.1%	1.4%	1.8%	0.4%	3.6%	10.5%

National Renewable Energy Lab., 2009 Renewable Energy Databook 29 (2010)

Growth of biomass resources for electricity generation may be relatively flat, but the same is not true for biomass developments in the transportation sector. According to the NREL 2009 data, U.S. ethanol production has grown dramatically, from 1.3 percent of the overall U.S. gasoline pool in 2000 to 7.8 percent in 2009. The United States led the world in ethanol production in 2009 with 10,750 of the world's total of 19,535 million

Figure 5.2: U.S. renewable generation by technology

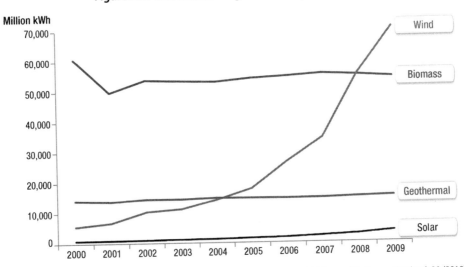

National Renewable Energy Lab., 2009 Renewable Energy Databook 28 (2010)

Figure 5.3: U.S. corn ethanol production and price trends

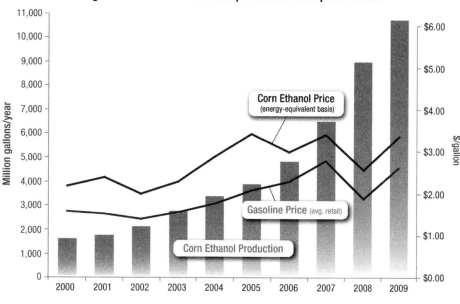

National Renewable Energy Lab., 2009 Renewable Energy Databook 101 (2010).

gallons, and in the same year, the United States was ranked third, behind Germany and France, for global biodiesel production with 545 of the 4,385 million gallons. *See* Figure 5.3. Reams of pages have addressed energy and transportation issues, so this chapter will instead provide a broad overview of biofuels in the context of biomass generally.

To study biomass, we must foray into the chemistry lab. The process of converting plant materials into fuel for energy can take many forms. We do not have space here to address in detail the various technologies for creating biomass such as gasification, hydroprocessing, or synthetic biology. Figure 5.4 provides a summary of some of the possibilities and resulting molecules such as ethanol, butanol, or hydrocarbons.

From a legal standpoint, we must first understand the sources of biomass, so this chapter will start with a discussion of biomass feedstocks. Special emphasis will be placed on the waste-to-fuel processes as these hold promise for simultaneously solving waste disposal and energy problems. This first section will also address emerging technologies that are producing second and third generation advanced biofuels. After laying the foundation, the last two sections of the chapter will explore a few of the significant drawbacks with biomass as a major renewable energy resource for the future: efficiency and environmental concerns both in growing it and in burning it.

A. Sources of Biomass Energy

Many resources fall under the purview of the term "biomass." Firewood is often the first that comes to mind. Yet using virgin wood for fuel is problematic in many ways. In the United States, shortages of virgin wood date back to the early 1800s when wood was our primary energy source. Similarly, deforestation remains a serious problem in developing countries that still rely on wood-type products for heating and cooking. Wood requires

Figure 5.4: Range of biofuels research

Feedstock	Processing	Energy product
Solid biomass/cellulose (virgin wood; woody crops such as willow & poplar; forest residues; pelletized residues; agricultural residues such as corn stover, straw, etc.; perennial grasses; switchgrass)	Direct burning in stoves OR converted to electricity using the same process as in coal-fired power plants (i.e., a Rankine heat to steam turbine to electricity cycle). Sometimes biomass is combined with coal in the same power plant.	Heat or electricity OR cogeneration of both heat and electricity.
Wet biomass (animal waste, municipal organic waste, etc.)	Converted to biogas.	Heat or electricity OR cogeneration of both.
Sugar & starch plants (corn, beets, etc.)	Sugars converted to ethanol.	Liquid fuels primarily for transportation.
Oil crops (rapeseed, sunflower, etc.)	Vegetable oil.	Liquid fuels primarily for transportation.
Algae	Several processes being developed.	Liquid fuels primarily for transportation.

Andy Aden,*The Current State of Technology for Cellulosic Ethanol*,
National Renewable Energy Lab. 3 (Feb. 5, 2009),
http://www1.eere.energy.gov/biomass/pdfs/aden_20090212.pdf.

vast swaths of land for growing and large acreages for storing and processing. Thus, it is only a sustainable resource if we consume it at a rate below that at which it can be replenished. Consequently, U.S. consumption of energy from domestic fuelwood has remained fairly steadily in the 1,000 megaton per year range since 1900. L.D. DANNY HARVEY, ENERGY AND THE NEW REALITY 2: CARBON-FREE ENERGY SUPPLY 174 (EarthScan 2010) (using figure 4.2 graphs).

The United States is turning away from wood as its primary biomass feedstock and instead moving to promising, faster-growing alternatives and recycled waste materials. Figure 5.5 lists other sources that fall within the same solid biomass category as virgin wood, as well as other biomass feedstocks in the wet biomass, sugar and starch plant, oil crops, and algae categories. The reading below discusses some of these categories in more detail.

1. Feedstocks

While the following excerpt focuses on feedstocks for biofuels, it also provides an introduction to other biomass feedstocks and some of the relevant chemical processes for converting these feedstocks to fuel. Notice how many of the feedstocks are cultivated products and how many are also sources of edible foods for humans.

Figure 5.5: Biofuels conversion processes

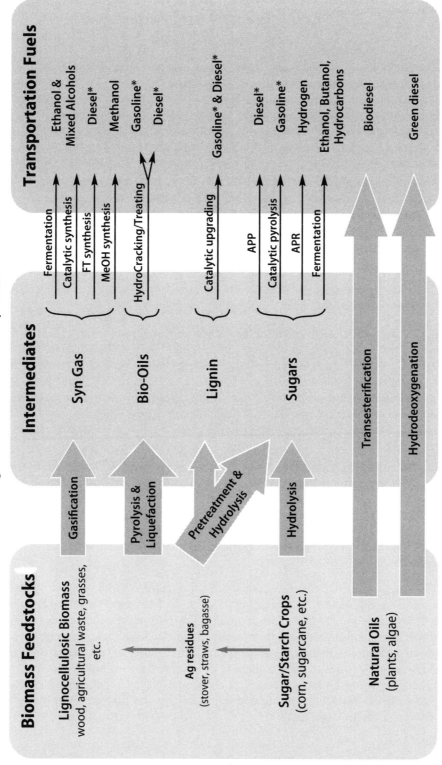

Created from data provided by John Ashworth of National Renewable Energy Lab.

Arnold W. Reitze, Jr., *Biofuels — Snake Oil for the Twenty-First Century*

87 Or. L. Rev. 1183 (2008)

The alcohols most commonly used in gasoline are ethanol ("CH_3CH_2OH") and methanol ("CH_3OH"). Ethanol can be made from any biomass feedstock, and is the same compound that is used in alcoholic beverages. Producing fuel alcohol is a four-step process. First, a carbohydrate (almost always corn in the United States) is reduced to a sugar solution. Next, it is fermented to ethanol and carbon dioxide. The ethanol then is removed by distillation to create a 95 percent alcohol solution. Finally, the water is removed. Because ethanol has about two-thirds the energy content of gasoline, a 10 percent ethanol blend results in a 2-3 percent decrease in mpg for the blend.

Methanol or wood alcohol can be made from wood, coal, biomass, municipal waste, or any other carbon-containing material. However, most U.S. methanol is produced from natural gas. The use of methanol results in reduced emissions of conventional pollutants, but its use has many negative effects....

* * *

Ethanol from Sugar and Grains

* * *

In the United States, 97 percent of the domestic ethanol production uses corn as the feedstock, and minor quantities of ethanol are produced from sorghum, cheese whey, and beverage waste. But corn-based ethanol with its high production costs, its need for high-quality farmland to produce corn, and the importance of corn for food, limits the potential expansion of ethanol production.... Whether food crops other than corn could be used for commercial ethanol production will depend on their cost of production, the cost of petroleum-based fuels, and, most importantly, the extent of government "carrots and sticks."

Ethanol can be produced from sugarcane, sugar beets, raw sugar, cane molasses, other molasses, wheat grain, sweet sorghum, Jerusalem artichokes, and other grains. A grain that has promise for ethanol production is hull-less barley. It can be grown in the winter, and if fertilizer is used efficiently the crop can "reduce erosion and nitrogen leaching from the field." There is not much experience in the United States using other food crops for ethanol so no cost data is available to use to accurately project costs.

* * *

The Energy Policy Act of 2005 established a sugarcane ethanol program within the EPA so that farmers and ethanol producers in Florida, Louisiana, Texas, and Hawaii can benefit from federal subsidies, and $ 36 million was authorized. But, the sugar industry, although heavily subsidized by taxpayers, was not granted any significant new subsidies to produce alcohol. The Energy Independence and Security Act of 2007 amended the CAA's Section 211(o)(1) to include sugar as an advanced biofuel, which could assist the sugar industry's quest for more subsidies.... The farm bill of 2008 continues the preexisting subsidies for the sugar industry, but does not offer any new programs aimed at biofuel production from the sugar industry.

Ethanol from Cellulose

Ethanol produced from cellulose or other nonfood inputs is a promising source of biofuels.... Advanced biofuels, such as perennial grasses, wood, and corn stover may be

Brazil's use of sugar for fuel — Brazil is second only to the United States in world ethanol production at almost seven billion gallons in 2010. While the United States relies primarily on corn for its biofuel feedstocks, Brazil's ethanol is made from sugar cane. Brazil first committed to ethanol during fuel shortages in World War II and renewed its vows with its 1975 "Pro-Alcohol" program after the OPEC petroleum export embargo. Although it suffered some slumps, Brazil has been enjoying increasing demand for its ethanol for more than a decade. Using Volkswagen's "flex fuel" motor, which in 2009 represented 94 percent of car sales in Brazil, motorists can burn almost any mix of ethanol and petroleum.

In 2005-2006, Brazil used over 50 percent of its sugarcane crop for ethanol. Sugarcane-based ethanol requires lower energy inputs than corn-based ethanol because corn must be turned into sugar before it can be distilled. According to the Brazilian Center for Sugarcane Technology, the EROEI (see section 5.B below) for cane to ethanol is 8.3 or better in comparison to 1.3 for corn. This also helps keep Brazilian ethanol more price competitive as Brazil can produce ethanol from sugarcane at about one-third of the estimated costs of producing ethanol from sugarcane in the United States.

used to produce fuel for transportation with less adverse social and environmental impacts than using food for fuel. But obtaining fuel from second generation feedstock is only in the early stages of development, and there is not yet commercially viable production. There were, however, fifty-five pilot plants and limited commercial facilities under construction in the United States in 2007.

* * *

Non-Food Ethanol

Ethanol can also be produced from trees, forest residues, and agricultural residues not specifically grown for food. This is an important source of ethanol feedstock because if we devoted all the U.S. corn and soybean harvest to ethanol and biodiesel production it would offset 12 percent of U.S. gasoline and 6 percent of U.S. diesel demand. Because of the energy input requirements, the net energy from ethanol and biodiesel would be about 2.4 percent and 2.9 percent of U.S. gasoline and diesel fuel. Nonfood inputs would allow marginal lands to be used for feedstock production that would not adversely affect food production. Moreover, cellulosic ethanol requires less pesticides and fertilizer than corn-based ethanol and offer the potential for a significant net energy balance. But, demand for cellulosic ethanol may result in adverse impacts on forests if they are cut to produce fuel and/or converted to plantations of fast-growing trees. If corn stover is used the benefits of using this material for soil conditioning and erosion control may be lost. Further, if abandoned lands, reclaimed mined land, or other lower value lands are utilized to produce cellulosic ethanol feedstock, the environment could benefit through improved soil conservation practices.

Lignocellulosic Ethanol

The Energy Policy Act of 2005, Section 932, identifies cellulosic material as lignocellulosic feedstock and lists barley grain, rapeseed, rice bran, rice hulls, rice straw, soybean matter, and sugarcane bagasse as being lignocellulosic. The 2005 act, section 941, amends the Biomass Research and Development Act of 2000 to expand biobased fuel research and development programs in an effort to overcome the "recalcitrance of cellulosic biomass." These programs are needed because using plant cellulose and extracting sugar to make ethanol is more difficult than getting sugar from grains....

Switchgrass and Woody Crops

Switchgrass (Panicum virgatum) or fast-growing woody crops such as hybrid willow and poplar are potential fuel sources. Switchgrass is a perennial Midwest and Southeast grass that can be harvested like hay once or twice a year, but has nearly three times the yield of hay. Switchgrass does not need water-supplied irrigation, and requires less fertilizer and pesticides than most crops. It requires no tillage, and its extensive root system reduces soil erosion and uses water efficiently. It grows up to ten feet in height, which provides habitat for wildlife. It can be used to produce a fuel with about three-quarters the energy of gasoline. Switchgrass, proponents claim, it [sic] can be harvested for fuel and still provide soil conservation benefits. Its use could be designed to complement the Conservation Security Program in the 2003 farm bill. But, environmentalists are concerned that soil and water conservation values and wildlife habitat will be undermined to encourage biofuel production.

* * *

Other Types of Cellulosic Ethanol

… Algae-based biofuel … has far more long-term potential [than cellulosic ethanol] as a fuel source, and can be processed to produce both ethanol and biodiesel. Current annual crop-based biofuel production is thirty gallons of fuel per acre using corn; sixty gallons per acre using soybeans; 150 gallons per acre from canola; 650 gallons per acre from palm; and 2000 to 5000 gallons per acre from algae. Moreover, algae needs 1 percent of the water of other crops used for ethanol production….

Biodiesel

Biodiesel is usually made from soybean oil, but it can be made from rapeseed oil (canola), palm kernel oil, sunflower seed oil, castor oil (i.e., mamona), groundnut oil, cotton seed oil, and coconut oil (copra). More recently, Jatropha has begun to be used as a biodiesel feedstock because it can be grown in tropical and semiarid regions. Germany is the world's largest producer of biodiesel, primarily from rapeseed (canola). It produced more than six times the U.S. production in 2005. In 2005 the United States produced seventy-five million gallons of biodiesel, which was about 0.02% of the 40.1 billion gallons of diesel fuel used for highway transport, but in 2008 biodiesel production was 700 million gallons. Feedstocks for biodiesel also are used for food, so there is a conflict as markets expand for biodiesel between food and fuel.

There are two types of biodiesel fuel which is a monoalkly ester of long chain fatty acids. The most common type is made from virgin vegetable oils. Soybean oil accounts for 90 percent of U.S. vegetable oil biodiesel production. The other type of biodiesel is made from nonvirgin vegetable oils or animal fats. To produce biodiesel from soybean oil, it is mixed with alcohol and a catalyst, such as caustic soda, and boiled at about 160 degrees F to create an ester. After boiling, the glycerin created by the process is allowed to settle, and it is then separated from the mixture. The excess alcohol and the catalyst is

---

Algae for Jet Fuel—The U.S. military, one of the world's largest consumers of gasoline, wants to decrease its dependence on foreign oil by substituting some algae-based fuel. The military needs a process that can produce inexpensive fuel on a large scale. Currently, algae-based fuels cost between $10 and $40 per gallon. AP, *Military Wants Algae Power*, Denver Post, July 5, 2009, at 10A.

> **Will manure-power be a staple of the future?** Some major corporations, such as Google and Yahoo, are locating new data centers in rural areas to take advantage of "one of the most abundant natural resources in America" — manure. If the price of technology to convert manure to energy can come down (the equipment is currently $5 million and costs $30,000 per year to operate), these corporations could see huge savings in their energy bills.

then removed, and the clear amber-colored biodiesel is ready to be used or mixed with conventional diesel fuel. The most common biodiesel blend is 20 percent biodiesel.

* * *

[*Animal Waste*—Ed.]

Another source of alternative energy is to use animal waste (i.e., manure) to produce methane, which is combined with animal fat or plant oil (often soybeans or corn) to produce biodiesel fuel. But, digesters, incinerators, and biodiesel plants are expensive to build and run. Because methane's Btu value is low in relation to the energy needed for its production, operating costs are high in relation to the value of the product. For this reason, anaerobic digestion of animal wastes is not considered an economically viable renewable energy source unless it is cost competitive with conventional waste management practices. Biodiesel production from manure, moreover, can be expected to have many adverse environmental and social impacts. The manure slurry created by biodiesel operations may exceed the volume of manure used in the process, and its disposal can create significant environmental problems. Moreover, these plants are likely to be an additional subsidy to industrial farms that can generate the volume of waste needed for efficient biodiesel production as well as producing significant water and air pollution. The Energy Independence and Security Act of 2007, Section 201 defines advanced biofuels to include biogas produced through the conversion of organic material from renewable biomass. This provision allows factory farms to provide biogas to ethanol plants and both operations would qualify for federal subsidies. The biodiesel program encourages the expansion of factory farms to the detriment of small farmers and the environment. One of the first major biodiesel plants in the United States is a facility servicing 500,000 pigs owned by Smithfield Foods in Utah.

The relatively insignificant contribution to the nation's diesel fuel supply made by biodiesel and the adverse impacts of using food for fuel should lead to caution when considering the desirability of federal subsidies and other incentives that distort the free market....

Notes and Questions

1. One of the criticisms of biofuels is the ethical concern raised when the United States squanders food stocks for human consumption as feedstocks for fuel at the same time that "[m]ore than 3.7 billion humans in the world are currently malnourished, so the need for grains and other foods is critical." David Pimentel & Tad Patzek, Editorial: *Green Plants, Fossil Fuels, and Now Biofuels*, BIOSCIENCE, Nov. 2006, at 875. When food is also used for fuel, food prices rise worldwide, making it difficult for poorer nations to feed the hungry. Do you agree there is an ethical obligation for richer countries to develop

non-food sources for biofuels? *See also* C. Ford Runge & Benjamin Senauer, *How Biofuels Could Starve the Poor*, 86 FOREIGN AFF. 41 (2007).

2. We saw that one of the draws of wind power is that it provides economic opportunities for poorer rural communities. Ethanol has been politically appealing for the same reason—it provides jobs in farming states. Iowa, which holds one of the first presidential caucuses, produces almost 40 percent of the nation's ethanol. Illinois and Michigan, other politically significant states, are also big ethanol producers. Historically, this has made it difficult to repeal subsidies. How big a role does, or should, politics play?

3. The Biomass Crop Assistance Program (BCAP) was created in the Food Conservation and Energy Act of 2008 (2008 Farm Bill) as a component of domestic agriculture, energy, and environmental strategy to reduce U.S. reliance on foreign oil and improve domestic energy security. BCAP was also intended to spur rural economic development by providing incentives for farmers, ranchers, and forest landowners to cultivate biomass crops for heat, power, bio-based products, and biofuels. In addition to biomass feedstocks, recent U.S. farm bills continued to subsidize biofuels, but emphasis has shifted from corn to advanced biofuels that do not use edible plants as a biofuel feedstock. Are such incentives to farmers sustainable policies? What do you think about Congress's role in subsidizing ethanol?

4. The Reitze article focused on biofuels so it only briefly addressed cellulose as a biomass feedstock. In addition to biofuels, the 2010 BCAP rules subsidized the following biomass feedstocks: forest waste, from thinning and post-disaster debris; non-edible crop residues, such as corn cobs and corn stover; and construction residues, such as wood chips and waste lumber. However, the BCAP regulations exclude animal waste, food waste, and yard waste from eligibility for matching payments. Should processing into fuel these products that otherwise would have to be processed in the waste stream warrant even greater support than BCAP provides?

2. Waste to Energy

The previous section primarily addressed the issue of incentivizing farmers and foresters to cultivate new renewable biomass feedstocks. However, there may be an even greater need to encourage recycling, converting biomass sources that would otherwise go to waste into energy. The following excerpt addresses one of the fastest growing potential sources for biomass energy—landfills.

Steven Ferrey, Symposium, *Smart Brownfield Redevelopment for the 21st Century: Converting Brownfield Environmental Negatives into Energy Positives*
34 B.C. ENVTL. AFF. L. REV. 417 (2007)

Abstract

There is a new paradigm for evaluating landfills. While landfills are contaminated repositories of hazardous wastes, they also are brownfields that can be redeveloped for renewable energy development. It is possible to view landfills through a new lens: As endowed areas of renewable energy potential that can be magnets for a host of renewable development incentives. Landfills also are critical resource areas for the control of greenhouse gases. Landfill materials decompose into methane, a greenhouse gas that is more than twenty times more potent—molecule for molecule—than carbon dioxide. This Article traces the molecular composition of waste in landfills, analyzing the chemical stew that brews in these repositories. Without doubt, landfills in America are brownfields. And many of these landfills leak and cause public health risks. This Article also analyzes the potential to utilize landfill gas for electricity production or as a thermal resource....

Introduction: from Environmental Negatives to Energy Positives

When discussing greenhouse gases, landfills are critical for several reasons. First, they constitute a large share of U.S. greenhouse gas emissions: as of 2000, the United States is responsible for approximately eleven percent of worldwide methane emissions. Approximately thirty percent of U.S. anthropogenic methane emissions, which is equivalent to 193.6 million metric tons, came from waste management in 2003. Landfills represent 92 percent of the 193.6 million metric tons of methane emissions, by far the single largest source. Approximately 5.2 million metric tons of the 178.1 million tons of landfill methane annually are captured as landfill gas (LFG); 2.6 million metric tons of this is used for productive energy, while 2.6 million metric tons of the recovered LFG are flared with no productive energy capture.

Second, the feedstock of LFG—municipal solid waste (MSW)—is the only increasing renewable resource. Total generation of MSW in the United States has increased more than fifty percent since 1980 to a level of 236.2 million tons annually. The per capita MSW generation rate is 4.45 pounds per person per day. This increase is not necessarily a positive attribute, but it is reality. MSW generation rates in European countries are significantly lower.

Third, landfills are the repository for the bulk of MSW. Approximately 56 percent of U.S. MSW goes to landfills as its final destination. Thirty years ago, in 1978, there were 20,000 operating landfills in the United States. The number steadily declined to approximately 1767 operating landfills in the United States in 2002. But declining absolute numbers of repositories belie the new larger mega-fills. While the number of landfills in the United States has been declining, their waste capacity has remained relatively constant. The currently available landfill capacity in the United States is estimated at 3.6 billion tons, which at current rates of disposal would provide twenty-eight years of additional disposal capacity.

Fourth, the bulk of MSW eventually degrades into methane molecules. About two thirds of the total MSW is organic matter that will degrade to release methane under anaerobic conditions. In 2002, landfills accounted for 6.9 million metric tons of methane emitted annually. These emissions can be captured and employed productively as a methane gas energy source, collected and flared for no productive purpose, or left alone to migrate into the environment as a potent greenhouse gas.

In addition to landfilling MSW and then capturing the methane produced as an energy fuel, the organic material can be directly combusted to release energy. Fourteen percent of MSW in the United States is incinerated; occasionally, incineration is coupled with a turbine to produce electricity. In 2002, there were 107 active waste-to-energy combustion facilities in operation in the United States. The most significant deployment of waste-to-energy combustion facilities to handle MSW is in New England, where 34 percent of the waste stream is handled in this manner. Waste-to-energy combustion of MSW in the United States generated 289 trillion British thermal units (BTU) of energy in 2001, representing approximately 0.3% of total U.S. electricity demand. There is a third destination alternative: 31 percent of the MSW waste stream in the United States is recycled or composted, almost a twofold increase from a decade earlier.

I. EVERY LANDFILL IS A BROWNFIELD

A. Why Municipalities May View Landfills as Hazardous Liabilities

1. The Presence of Household Hazardous Waste

All industrial nations are neck-deep in waste. What poses a potential liability for municipalities is that (1) so called "sanitary" MSW is actually hazardous, and (2) these landfills are leaking. All waste, whether liquid, gaseous, or solid, is characterized as solid waste. Some solid wastes are sanitary, although others are hazardous. Municipal garbage collected from households is designated as municipal solid waste and is appropriately disposed of in sanitary landfills.

Roughly 30 percent of the total waste collected in some communities are household wastes. Detailed surveys indicate that approximately 3 percent of MSW is recycled; the remainder is thrown out. In 1976, the United States officially produced and placed in landfills more than 360 million tons of solid waste created by municipal, commercial, and industrial sources. Congressional records from proceedings on the Resource Conservation and Recovery Act (RCRA) estimate that over 11 billion tons of waste are generated every year in the United States. In 1999, the United States produced approximately 545 million tons of solid waste, 374 million tons of which were placed in landfills. On a per capita basis, each American creates approximately 4.5 pounds of MSW each day. This amount varies depending on the "degree of urbanization," the season, the average income level, and level of economic activity. In 1984, 133 million tons of this waste were MSW; as of 2003 it had climbed to 236.2 million tons annually. In 2005, 245 million tons of this waste was MSW produced by U.S. industries, residents, businesses, and institutions. Nearly 131 million tons of the total MSW generated domestically was deposited in landfills.

* * *

II. WHAT GOES ON AT A LANDFILL STAYS AT A LANDFILL? LANDFILL GAS AS AN ENERGY RESOURCE

A. Overview of the Chemical Process

Americans annually dispose of millions of tons of waste in thousands of landfills across the country. Because waste is composed of a high percentage of organic materials, including paper, food scraps, and yard waste, over time, bacterial decomposition of organic material, the volatilization of certain wastes, and chemical reactions within the landfill create copious quantities of gas. This landfill gas is comprised primarily of carbon dioxide and forty to sixty percent methane, while containing smaller amounts of nonmethane organic compounds (NMOCs) and some other trace organic elements. For comparison, pipeline natural gas contains about 90 percent methane.

Landfill gas (LFG) constituents can pose health and safety problems. Methane in high concentrations can create an explosion hazard. LFG contains a variety of toxic gases and

Puppy Power? In Cambridge, MA, artist Matthew Mazzotta created the "Park Spark" project allowing owners to recycle their dog's poop into methane to power street lanterns. Other communities have initiated similar poop-to-power projects.

carcinogens that can have detrimental effects on the health of the surrounding community. Globally, methane and carbon dioxide released from landfills each are greenhouse gases contributing significantly to global warming. While both carbon dioxide and methane contribute to global warming, methane has twenty-one times the global warming potential of carbon dioxide.

B. Productive Energy Applications

1. Exploiting the Energy Potential of Landfill Gas Brownfields

After LFG is collected (in the vast majority of landfills in the United States and the world it is *not* collected) there are two disposal options. The first is an open flare system in which the gas is burned. The second option is using the gas for useful energy applications either in the gaseous form or as the fuel for electric production. Some states allow LFG projects to make direct sales of the methane to third-party customers without being regulated as a public utility. Small LFG-to-electricity projects are exempt from regulation in Connecticut, Florida, and Wisconsin.

The EPA encourages, but does not require or provide additional incentives for, the second option. Burning LFG converts methane into carbon dioxide, a gas less than five percent as damaging as methane in terms of global warming potential. EPA estimates that each megawatt of electricity generated from LFG has the same impact of planting 12,000 acres of forest, removing 8800 cars per year, or eliminating the need for 93,000 barrels of oil.

EPA maintains a database of more than 2300 landfills that are potential LFG-to-energy locations in the United States. EPA Landfill Methane Outreach Program (LMOP) tracks 395 operating LFG projects in the United States, and identifies more than 570 additional landfills as very good candidates because of their size and methane generation characteristics. These 570 candidate landfills have the potential of generating 695 million cubic feet (mcf) of LFG per day. The challenge is to get these developed amidst a variety of impediments. It is estimated that "each year ... 421 to 613 billion cubic feet of methane from landfills alone is wasted." That amount of methane could produce up to 4000 megawatts of electricity, enough to power three millions homes.

Additionally, there are one hundred landfills that are in the process of constructing LFG electricity projects. The 400 existing LFG projects generate about 9 billion kilowatt-hours (kW-h) of electricity annually, plus also produce approximately 200 mcf per day of LFG for direct thermal purposes. This is equivalent to planting nearly 19,000 acres of forest, saving 160 million barrels of oil, removing thirteen million vehicles from the road, or supplying the electricity and heating requirements of approximately one million homes.

2. Municipal Sewage Treatment Brownfields

The United States expends $25 billion every year to process and treat 33 billion gallons of wastewater. Many, but not all, cities and towns have sewage treatment facilities. Some of these also constitute brownfields because of contamination. They also consume significant quantities of electricity treating sewage. Treatment works are viewed by many municipalities exclusively as environmental negatives.

However, these facilities also can be adapted or redeveloped into environmental positives. The facilities can offer an energy generation or energy capture opportunity, rather than only being an environmental problem. There are several proven technologies to accomplish energy extraction from sewage.

One way to capture heat from raw sewage is to employ a heat pump to extract heat from the hot mixture and distribute the heat. Alternatively, an anaerobic digester collects the methane or "biogas" that bacteria convert consuming organic material anaerobically. The biogas produced is composed of about sixty percent methane, forty percent carbon dioxide, and approximately 0.2% to 0.4% hydrogen sulfide. The methane can be combusted for electricity or used thermally.

Such technologies are in use in the United States, and can be supported by renewable system benefit charges and trust funds, as evidenced by projects such as the Deer Island sewage treatment plant in Boston. This process also is finding application in developing countries in agricultural waste settings to create Carbon-Emission Reductions (CERs) pursuant to the clean development mechanism (CDM) of the Kyoto Protocol.

Other benefits of this energy extraction process include reducing the amount of waste remaining that has to be disposed of, and reducing the odor because volatile compounds have been removed. Sewage solids can be landfilled, burned, or recycled. When biomass—in this example sewage—is heated with little or no oxygen, it combusts, becoming a gas mixture of carbon monoxide and hydrogen known as syngas. This syngas then mixes with oxygen and burns more efficiently than the original solids in the waste stream, and can produce electric energy and/or heat. Plants that use biomass gasification have better efficiency of energy capture than plants that burn the waste solids, and also convert sludge to ash, which consumes less landfill space.

Sewage methane also can be used in advanced fuel cell technologies to produce direct current electricity for self-use or for wholesale export to the electric grid. For example, a New York sewage treatment plant employs a 200 kilowatt hydrogen fuel cell to supply enough electricity for sixty homes. Technology research is proceeding on a microbial fuel cell (MFC) which will not only create electricity from the sewage but also treat it.

In most states, all of the technologies that could be deployed at municipal brownfields that consist of either landfills or sewage treatment facilities would be eligible for subsidy pursuant to state renewable system benefit charges/trust funds schemes, or qualify to earn tradable Renewable Energy Credit pursuant to the state Renewable Portfolio Standard. They might also be eligible for Title II Clean Water Act grants for innovative systems.

C. Above Ground Energy Capture: Wind Power & Brownfields

While LFG-to-energy or direct thermal LFG applications might appear to be the logical first choice for use of landfill brownfields in many municipalities, it is not the only choice. Landfills are good sites for consideration of siting wind energy projects. Wind energy projects are possible where a landfill is too small, too old, or not sufficiently deep to allow LFG collection and beneficial use. However, a wind turbine can be placed at, or even on, a landfill that also is collecting LFG for beneficial purposes. A single wind turbine can not only lower a town's energy expenses, but can also result in green-energy certificates that the town can sell to utilities who need to comply with the state's Renewal Energy Portfolio Standard.

* * *

III. REGULATORY CONTROLS

There are significant regulatory requirements regarding landfill operation and management. However, far from discouraging productive energy use, these environmental regulatory requirements actually encourage the productive capture and use of landfill gas (LFG) at landfills. The federal Section 29 and Section 45 tax credits, the former of which added a subsidy of about one cent per kilowatt-hour (kW-h) to electricity generation from LFG, also provided incentives to make the capital investment at landfills to construct LFG projects.

A. Resource Conservation and Recovery Act (RCRA)

RCRA mandates that all large landfills operating after 1991 install a protective cap to prevent gas from escaping. Any landfill constructed or extended after October 1993 is required to install a protective lining around the sides and bottom of the landfill to prevent the lateral migration of LFG and groundwater contamination. RCRA requires that all municipal solid waste landfills have a methane gas concentration of less than "[twenty-five] percent of the lower explosive limit for methane." Methane gas is explosive between five and fifteen percent concentrations. Imminent hazards are deemed to occur where methane releases migrate to buildings or underground utility conducts at a concentration of ten percent of the lower explosive limit (LEL). RCRA also requires that the methane concentration at the facility's property boundary be less than the lower explosive limit for methane.

In order to know whether or not landfills are in compliance with these requirements, the owner/operators of the landfill must conduct a methane-monitoring program. Municipal solid waste landfill facilities must provide a report on their methane concentration levels quarterly. If the methane concentration levels exceed the limits, the owner/operator is required to initiate affirmative steps to correct the problem and must take the proper steps to ensure the health and safety of the people surrounding the landfill. A written record of the methane level, and the steps taken to protect human health, must be created within seven days of the detection. The state can order assessment and remedial action. The division of solid waste of the state environmental regulatory agency can require plan application and approval. Post-closure environmental monitoring is required.

B. Air Regulation for Landfills

1. New Source Performance Standards for Landfills

The New Source Performance Standards (NSPS) of the Clean Air Act (CAA) applies to any new landfill which began modification or construction after May 30, 1991. Under the NSPS, any landfill that has a design capacity in excess of 2.5 million cubic meters must monitor non-methane organic compound (NMOC) emission rates.... Separate rules apply to landfills that do not come under the NSPS.

... As there are increasingly fewer but larger landfills there are more landfills required by government regulation to capture and utilize or flare LFG. If the landfill expands from below the threshold to above it, the owner/operator must submit an amended design capacity report within ninety days of the increase in size so that it may now be treated as regulated under subpart WWW.

The owner/operator must calculate a NMOC emission rate and report it annually. If the rate exceeds fifty megagrams then a collection and control system will be required to be installed. The collection and control system must be designed in such a way to ensure capture of the gas generated by the landfill. Control of hazardous air pollutants is required.

Monitoring and testing is required as to gas pressure, flow, temperature, oxygen and nitrogen concentrations, and the operator must calibrate and maintain equipment. Information must be retained for at least five years. Individual permits establish units for emissions of NOx, CO, NMOC, PM, SO_2, VOC, and opacity (visible emissions)....

Under NSPS, semi-annual reports must be submitted regarding air limit excedances and gas bypass flow reports. Criminal liability can be imposed on the landfill owner if it commits a knowing violation of the CAA, or knowingly releases hazardous air pollutants, a violation punishable by a sentence of up to five years in prison. The CAA also authorizes significant civil penalties of up to $27,500 for each violation.

* * *

IV. REGULATORY INCENTIVES

A. Renewable Subsidies: System Benefits Charges and Renewable Trust Funds

The system benefits charge is a tax or surcharge mechanism for collecting funds from electric consumers, the proceeds of which then support a range of activities. In order to support demand-side management (DSM) or renewable resources, funds are collected through a non-bypassable system benefits charge to users of electric distribution services. The money raised from the system benefits charge is then used to "buy down" the cost of power produced from sustainable technologies on both the supply and demand side, so that they can compete with more conventional technologies. More than a dozen states have adopted these programs.

Between 1998 and 2012, approximately $3.5 billion will be collected by fourteen states with existing renewable energy funds. More than half the amount collected, at least $135 million per year, comes just from California. The funding level taxes range from $0.07 per megawatt-hour (MW-h) in Wisconsin up to almost $0.6 per MW-h in Massachusetts. Most only provide assistance to new projects, and not existing renewable projects.

The form of administration of renewable trust funds varies. Many states administer them through a state agency, while others use a quasi-public business development organization. Some funds are managed by independent third-party organizations, some by existing utilities, while two states allow large customers to self-direct the funds. For distribution, some states utilize an investment model, making loans and equity investments. Other states provide financial incentives for production or grants to stimulate supply-side development. Some other states use research and development grants, technical assistance, education, and demonstration projects.

* * *

B. Renewable Resource Portfolio Requirements

A resource portfolio requirement requires certain electricity sellers and/or buyers to maintain a predetermined percentage of designated clean resources in their wholesale supply mix. A number of variations of resource portfolios are possible, including a renewable resource portfolio requirement, a DSM portfolio requirement, and a fossil plant efficiency portfolio requirement.

Twenty states have adopted the renewable portfolio standard (RPS); two additional states have goals. The key to making the portfolio requirements work is to establish trading schemes for "portfolio obligations." Portfolio standards are flexible in that certain technologies can be included in the renewables definition, or certain subgroups of technologies can be targeted for inclusion at distinct levels. The standard allows market

Figure 5.6: Renewable Portfolio Standards map

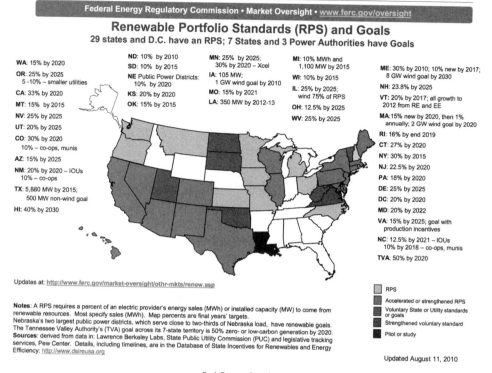

Renewable Power & Energy Efficiency Market: Renewable Portfolio Standards

Federal Energy Regulatory Commission • Market Oversight • www.ferc.gov/oversight

Renewable Portfolio Standards (RPS) and Goals
29 states and D.C. have an RPS; 7 States and 3 Power Authorities have Goals

WA: 15% by 2020

OR: 25% by 2025
5 -10% – smaller utilities

CA: 33% by 2020

MT: 15% by 2015

NV: 25% by 2025

UT: 20% by 2025

CO: 30% by 2020
10% – co-ops, munis

AZ: 15% by 2025

NM: 20% by 2020 – IOUs
10% – co-ops

TX: 5,880 MW by 2015;
500 MW non-wind goal

HI: 40% by 2030

ND: 10% by 2010

SD: 10% by 2015

NE Public Power Districts:
10% by 2020

KS: 20% by 2020

OK: 15% by 2015

MN: 25% by 2025;
30% by 2020 – Xcel

IA: 105 MW;
1 GW wind goal by 2010

MO: 15% by 2021

LA: 350 MW by 2012-13

MI: 10% MWh and
1,100 MW by 2015

WI: 10% by 2015

IL: 25% by 2025;
wind 75% of RPS

OH: 12.5% by 2025

WV: 25% by 2025

ME: 30% by 2010; 10% new by 2017;
8 GW wind goal by 2030

NH: 23.8% by 2025

VT: 20% by 2017; all growth to
2012 from RE and EE

MA:15% new by 2020, then 1%
annually; 2 GW wind goal by 2020

RI: 16% by end 2019

CT: 27% by 2020

NY: 30% by 2015

NJ: 22.5% by 2020

PA: 18% by 2020

DE: 25% by 2025

DC: 20% by 2020

MD: 20% by 2022

VA: 15% by 2025; goal with
production incentives

NC: 12.5% by 2021 – IOUs
10% by 2018 – co-ops, munis

TVA: 50% by 2020

Updates at: http://www.ferc.gov/market-oversight/othr-mkts/renew.asp

Notes: A RPS requires a percent of an electric provider's energy sales (MWh) or installed capacity (MW) to come from renewable resources. Most specify sales (MWh). Map percents are final years' targets. Nebraska's two largest public power districts, which serve close to two-thirds of Nebraska load, have renewable goals. The Tennessee Valley Authority's (TVA) goal across its 7-state territory is 50% zero- or low-carbon generation by 2020. **Sources**: derived from data in: Lawrence Berkeley Labs, State Public Utility Commission (PUC) and legislative tracking services, Pew Center. Details, including timelines, are in the Database of State Incentives for Renewables and Energy Efficiency: http://www.dsireusa.org

RPS
Accelerated or strengthened RPS
Voluntary State or Utility standards or goals
Strengthened voluntary standard
Pilot or study

Updated August 11, 2010

Fed. Energy Regulatory Comm'n, http://www.ferc.gov/market-oversight/
othr-mkts/renew/othr-rnw-rps.pdf (last visited Aug. 4, 2011).

competition to decide how best to achieve these standards. The standards become self-enforcing as a condition of retail sale licensure.

The renewable resource measures that states have incorporated into electricity restructuring and deregulation statutes vary. Some renewable energy measures create portfolio standards; others create trust funds to invest in the development and utilization of renewable resources. Some adopt both concurrently. How each defines an eligible renewable resource varies significantly.... Each defines differently what is an eligible renewable resource.... [Figure 5.6 is a map of the United States that summarizes each state's RPS goals—Ed.]

* * *

E. Financing

While there are many financing alternatives, landfill brownfields projects can take advantage of special bond financing and tax incentives.

1. Clean Renewable Energy Bonds

Under the federal Clean Renewable Energy Bond (CREB) program, electric cooperatives, public power systems and municipal utilities can issue or benefit from the issuance of clean renewable tax credit bonds (CREBs) to finance renewable energy projects as a less expensive alternative to traditional tax-exempt bonds.... The issuer of CREBs receives an allocation from the Secretary of the $800 million available for CREBs. Qualified issuers include:

- A clean renewable energy bond lender;
- A cooperative electric company; or
- A government body.

A mutual or cooperative electric company or governmental body can borrow CREB proceeds from the qualified issuer. An owner of a CREB is entitled to a tax credit, which is designed to be in lieu of or in substitution for any interest payments on the CREBs. Thus, it is interest-free borrowing. A CREB holder can deduct the amount of the tax credit from total income tax liability, with the proviso that the value of the tax credit is treated as taxable income.

Ninety-five percent or more of the CREB proceeds must be used for capital expenditures by qualified borrowers for qualified projects. "Qualified projects" includes any of the following producing electricity:

- Wind facilities
- Closed-loop or open loop biomass facilities
- Geothermal or solar energy facilities (solar energy facilities must be placed in service before January 1, 2006)
- Small irrigation power facilities

2. Tax Incentives

The original Section 29 tax credit was enacted in 1979 and extended repeatedly thereafter. It currently expires in 2007. [It was extended until the end of 2010, but not reauthorized after that—Ed.] For LFG projects, it requires the sale of a "qualified fuel" to an unrelated third party. This tax credit generates cash equivalent to about 75 percent of the capital cost of an LFG project during the first ten years of the project. It is a substantial incentive.

In its newer iteration, the old Section 29 tax credits, now redesignated as section 45(k) tax credits, provide credits for both direct thermal use of, and electric generation from, LFG. By comparison, the Section 45 tax credit applies only to electric generation. While the Section 29 credit benefits the owner of the LFG collection system, the Section 45 credit benefits the facility producing electricity from LFG. While less generous than the old Section 29 credit, it is worth about $350,000 annually for five years for a five megawatt LFG project.

The Section 45 credit was originally authorized for wind projects and later expanded to other technologies, and for wind projects is worth approximately 1.8 cents per kW-h for ten years of project operation. This credit is related to the amount of electricity produced and sold to a third party, while by contrast the section 29 credit is earned from the sale of a "qualified fuel" to a third party. These two tax credits cannot be taken simultaneously or "double dipped." These tax credits can be carried back one year or forward up to twenty years for federal income tax purposes.

Also available to some is the Renewable Energy Production Incentive (REPI) to subsidize local and state government owners of renewable energy projects, LFG projects, as well as non-profit electric cooperatives, during the first ten years of LFG project operation. Up to 40 percent of the REPI incentive program can be allocated to LFG projects; the remainder goes to other renewable power opportunities, including wind. These payments continue for a ten-year period and are worth approximately 1.5 cents per kW-h, adjusted for inflation

> **Project Financing** — As you might guess, renewable energy project financing is a complex and evolving art. For comprehensive coverage of this topic try Energy and Environmental Project Finance Law and Taxation: New Investment Techniques (Andrea Kramer & Peter Fusaro eds., Oxford Univ. Press 2010).

after 1993. There are ways to utilize and monetize the private tax credits even for municipal project owners, with careful legal guidance and proper project structuring and contractual relationships. This can also be done to monetize the REPI incentives, through the use of partnerships and LLCs.

V. THE LANDFILL BROWNFIELDS PARADIGM SHIFT

A. Economic Hierarchies

So where does this leave landfill brownfields seven years into the twenty-first century? From a cost-effective energy development perspective, LFG-to-energy development is the first option at a landfill. LFG has an energy content of about 550 BTU per cubic foot, or roughly half the energy density of pipeline quality gas. However, it is still capable at this energy density of running traditional electric-producing turbines or reciprocating engines.

EPA estimates the levelized generating costs of LFG-to-electricity technology as $45.67 per MW-h (4.57 cents per kW-h), which makes LFG electricity less expensive than either wind, geothermal, or solar photovoltaic resources, the other widely used renewable sources, and competitive with fossil fuel generated electricity. For every one million tons of MSW in a landfill, under anaerobic conditions, approximately 800 kilowatts of renewable electricity can be produced from the approximately 432,000 cubic feet per day of LFG creation. From the perspective of regulatory fit, LFG makes sense as a brownfield development strategy.

Approximately two thirds of the methane productively captured at landfills is utilized for electricity production, as opposed to direct thermal application. This methane could also be utilized in fuel cells or converted to methanol or ethanol. While the work horse of the LFG-to-electricity industry is reciprocating engines, there are approximately one dozen micro-turbines in operation at LFG facilities.

Landfills actively capturing and utilizing LFG for productive energy purposes tend to congregate in areas where there are extra incentives, such as renewable portfolio standards (RPSs). All states that have adopted an RPS system have elected, as a matter of state law, to recognize LFG as an eligible "renewable" fuel. Landfill gas is eligible as a "green" energy source in those thirty states that allow green power marketing and commands a price premium ranging from 0.5 cents to five cents per kW-h. In addition, LFG projects placed in service by 2006, under current legislation, unless and until it is extended or renewed, receive a Section 45 tax credit of one cent per kW-h during their first five years of operation. These tax credits historically have been extended in subsequent legislation. LFG projects are economically viable, environmentally positive, and add to the inventory of non-green-house-gas-producing energy sources.

B. Reconfiguring Landfill Land Uses

In an era of high world oil and gas prices, access to non-curtailable, reliable energy sources is a key advantage. With the rising cost of natural gas, landfills that can produce

The price of oil and natural gas has remained volatile over the last decades. Frequently, when those prices drop, fledgling alternative energy projects falter. For example, in November of 2010, Frontline BioEnergy suspended its operations. The company recently had several biomass and waste gasification projects that were technically successful but never were able to move beyond a conceptual engineering phase because of a lack of investment due to uncertainty about the regulation of carbon emissions and the low price of natural gas.

and deliver LFG and/or power become attractive sites for industries or commercial facilities that need reliable and/or low-cost natural gas or methane supply. Such facilities would normally not think about locating at a landfill, but for these rising energy prices.

The bulk of landfills in the country are municipally owned. Moreover, since the private sector has already developed LFG projects at many larger private sites, the majority of remaining best-candidate LFG-to-energy projects are at municipally owned landfills. This makes for many smaller landfill projects at sites owned by municipal entities—which traditionally have not collected LFG and are not familiar with LFG-to-electricity projects—prime candidates for application of this technology. Even a very small municipal project can yield net revenues to the municipality of $250,000 annually or more.

It is axiomatic to note that not all landfills are created equal. Some are better candidates than others for energy development. Like many things in life, this is a function of physical and longevity factors related to carbon-based molecules.

Key landfill development parameters include landfill size, years since closure, the type of waste accepted, and whether an LFG collection and control system is in place. Any landfills that accepted large quantities of ash, demolition, stump, sludge or soil offer less in terms of potential LFG generation. Larger sized landfills produce more LFG, since they have a larger waste mass. LFG production decreases annually after closure, therefore producing less LFG in older landfills. Any landfill with at least twenty to twenty-five acres or more and approaching approximately one million tons of MSW waste, closed in the past decade, offers potential.

Not all regulatory environments are created equal. The economic incentives for turning the organic content of waste into energy are greatest in Massachusetts and similarly disposed states for a variety of reasons: Massachusetts has the highest tipping fees for landfills in the United States, some of the highest electricity prices, and some of the highest natural gas prices. Of the approximately 701 inactive and closed landfills in Massachusetts, sixteen landfills have been developed with some type of LFG-to-energy project.

From a perspective of what energy source to develop first at a landfill, an LFG capture program should take priority for evaluation. Methane destruction has been a prime target of the campaign against global warming, because as a greenhouse gas methane is deemed to have twenty-one times the impact, molecule-by-molecule, compared to CO_2. When utilized productively, LFG is considered a carbon-neutral fuel, since its combustion releases carbon that was recently sequestered by the organic source materials before being placed in the landfill; those source materials, when degrading anaerobically, generate and release their methane content.

* * *

VI. BROWNFIELD LIABILITY FOR THE HAZARDOUS COCKTAIL

A. The CERCLA Scheme for Operator Liability

Operators of contaminated sites inherit legal liability, jointly and severally, for their cleanup under both federal Superfund and many state environmental statutes....

* * *

... [T]he liability uncertainties and impediments, which notably are a creation solely of law and not of any technical impediments, can swamp the energy revenue and environmental benefits perceived by third parties to develop energy projects at landfill sites. The joint and several liability risk of the Superfund statute often bludgeons the economic incentives and environmental benefits that many well-meaning parties and entrepreneurs would otherwise attempt to realize at these sites. It is the legal issues which create the most profound disincentives to landfills' productive reuse and development. These same issues have received by far the least attention, judging by the literature regarding landfills and EPA funding initiatives.

CONCLUSION

There are ways to bridge these chasms utilizing creative legal techniques. Before more of the tens of thousands of existing municipal landfills become too aged to support landfill gas-to-energy projects—and in the interest of methane containment and mitigation, renewable energy development goals, and energy efficiency—there should be greater effort redirected to providing the legal templates to insulate new energy project operators from preexisting legal liability at brownfields/landfills.... Creative legal reforms, not technical seminars, are the most urgently needed changes in this market. This can be a win-win situation for municipalities, the public, and state and federal enforcement agencies across the United States.

When covenants not to sue are in place, the critical joint and several liability risks associated with prior conditions are mitigated. This is one essential change to accomplish in order to make development possible at brownfields landfills. For private sector developers, these legal risks subsume the development opportunities.

A second essential factor is to cause municipal landfill owners to change their landfills paradigm. Most municipalities view their landfills as environmental negatives that must be hidden or ignored. They raise issues of contamination and liability. In fact, these landfills can be opportunities, at best, to capture landfill gas as an energy source, and at least to control landfill methane to mitigate global warming. There are opportunities both to utilize this methane, as well as to utilize the land area of existing landfills for wind or biomass facility siting.

The regulatory environment provides significant incentives for such renewable energy developments. Tax credits, tax-preferred Clean Renewable Energy Bonds, and renewable energy credits under state renewable portfolio standard laws in twenty-two states, as well as direct renewable trust fund subsidies in sixteen states, provide significant financial incentives for such renewable energy developments. Net metering also is available in forty states. Combined, these incentives should compel a much more vigorous development of energy generation potential at those brownfields that are existing landfills. Collectively, these factors create a new and different landfill paradigm, viewed through the lens of renewable energy potential, not just waste and contamination.

———————

Notes and Questions

1. In October of 2010, the EPA issued the RE-Powering America's Land Initiative Management Plan to provide incentives to site renewable energy on potentially contaminated land and mine sites. Action 3 in the RE-Powering management plan is to "Develop Technical Guidance on Siting Photovoltaic Solar on Closed Landfills." Embracing many of the concepts that Ferrey mentions in his article, the RE-Powering plan notes: "Closed landfills present a unique opportunity for siting solar energy; there are few reuses for closed landfills, closed landfills are located in all parts of the country, and landfills are often located close to roads and transmission lines." If you were working with the DOE and NREL to determine the legal considerations necessary to successfully site renewable energy on landfills, what are some of the issues you would raise?

2. What types of incentives might the EPA or states provide to encourage the use of brownfields for siting renewable energy projects?

3. Much has been written about Renewable Portfolio Standards (RPS), which are also sometimes called Renewable Energy Standards (RES). Since the time Ferrey published the above article, the number of states with mandates has increased to twenty-nine and an additional five have implemented voluntary standards. Also, the goals for several states have continued to increase. *See* Figure 5.6. The RPSs have been credited with the recent renaissance in renewable energy development in the last decades. The first requirement that utilities purchase renewable energy was created by the Iowa legislature in 1983. Colorado's Amendment 37, passed in 2004, was the first citizen-initiated renewable standard. Why do you think RPSs sprung up at the state level this way?

4. One problem with state-by-state RPSs is a lack of consistency in energy types that qualify as renewable under each separate standard. For example, waste-to-energy (WtE) or energy-from-waste (EfW) processes involve generation of electricity or heat from the combustion of non-hazardous wastes, often municipal solid wastes. A 2008 survey by the EPA showed 87 WtE plants across the United States generating approximately 2,500 MW. According to the Solid Waste Association of America, WtE plants emit less carbon dioxide than any other fossil fuel, including natural gas. Yet only 12.5 percent of all of the municipal solid waste in the United States was combusted in 2007 in comparison to over 60 percent in Japan and over 30 percent in many European countries. Could this discrepancy be explained by recognizing that municipal solid waste is not categorized as a renewable energy resource in many of the state RPSs?

5. In addition to variations in definitions for which sources qualify as renewable, the various state standards also lack consistency about the percentages of renewable energy that go into the mix and about the ability to trade bragging rights for renewable energy implementation through renewable energy credits or RECs. These and other issues have driven a call for a federal statute creating a national renewable-energy standard (RES). What do you think are the positives and negatives of a national standard? *See, e.g.*, James M. Van Nostrand & Anne Marie Hirschberger, *Implications of a Federal Renewable Portfolio Standard: Will It Supplement or Supplant Existing State Initiatives?*, 41 U. TOL. L. REV. 853 (Summer 2010); Benjamin K. Sovacool & Christopher Cooper, *Congress Got It Wrong: the Case for a National Renewable Portfolio Standard and Implications for Policy,* 3 ENVTL. & ENERGY L. & POL'Y J. 85 (Summer 2008); Mary Ann Ralls, *Congress Got It Right: There's No Need to Mandate Renewable Portfolio Standards,* 27 ENERGY L. J. 451 (2006).

Figure 5.7: Ethanol is the most mature biofuel technology

Technology Maturity (Low → High)	Key Drivers	Value Added
1 Ethanol	New market for grain and agriculture products. Large supply of lignocellulose.	High octane gasoline blend stock from carbohydrates.
1 Biodiesel	New market for excess oils, fats, and greases.	Petroleum compatible and biodegradable.
2 Green Diesel	Lower cost and higher product quality than FAME.	Utilize existing assets. High quality jet fuel or diesel.
2, 5 Butanol	New market for grain and agriculture products. Large supply of lignocellulose.	Better gasoline blending properties than ethanol.
2, 3 Syngas Liquids	Integration of biomass with Coal, Coke, Shale, or Heavy Oils.	High quality jet fuel or diesel. Reduced criteria for sequestration, and economy of scale (in combination with fossil).
2, 4 Bio-oil Derivative	Technical fit with woody biomass and liquid bio-crude.	Potential to integrate into existing large scale refinery and pipeline infrastructure.
2, 4 H2 from Biomass	Potential transportation fuel from any fuel/power source.	Ideal feed for fuel cells, and lowest tail pipe emissions.
6, 3 Diesel from Algae	Lg. source of biomass on non-arable land, and capture of CO_2.	High quality jet fuel or diesel yield per acre, with both off-shore and on-shore potential.
6, 2 Hydrocarbons from Carbohydrates	Better compatibility with petroleum products.	Potential for higher reaction rates than fermentation, and potential as H2 carrier.

Organizations Leading the R&D

1 Grain/Agriculture   2 Petroleum   3 Coal   4 Forestry   5 Chemical   6 Academia & Startups

Andy Aden, *The Current State of Technology for Cellulosic Ethanol*, National Renewable Energy Lab. 5 (Feb. 5, 2009), http://www1.eere.energy.gov/biomass/pdfs/aden_20090212.pdf.

3. History & Future Focus of Biofuels Research

Biofuels research by the U.S. government initially focused on the development of grain-based ethanol. Now that ethanol has gained market feasibility, research is expanding into second and third generation biofuels such as cellulosic ethanol and advanced biofuels from algae and plant oils. *See* Figure 5.7. The excerpt that follows provides some history and explains the U.S. government's efforts in more detail.

U.S. DEP'T OF ENERGY, ENERGY EFFICIENCY & RENEWABLE ENERGY
BIOMASS: MULTI-YEAR PROGRAM PLAN
1-1, 1-3 to 1-12 (Nov. 2010)

Growing concerns over national energy security and climate change have renewed the urgency for developing sustainable biofuels, bioproducts, and biopower. Biomass utilization for fuels, products, and power is recognized as a critical component in the nation's strategic plan to address our continued and growing dependence on imported oil. The U.S. dependence on imported oil exposes the country to critical disruptions in fuel supply, creates economic and social uncertainties for businesses and individuals, and impacts our national security.

Biomass is the only renewable resource that can supplant petroleum-based liquid transportation fuels in the near term. The United States has over a billion tons of sustainable biomass resources that can provide fuel for cars, trucks, and jets; make chemicals; and produce power to supply the grid, while creating new economic opportunities and jobs throughout the country in agriculture, manufacturing, and service sectors.

* * *

Expanded Program and Focus on Advanced Biofuels

While the overall mission of the Biomass Program is focused on developing advanced technologies for the production of fuels, products, and power from biomass, the Program's near-term goals are focused on the conversion of biomass into liquid transportation fuels. Historically, the Program's focus has been on RDD&D for ethanol production from lignocellulosic biomass. The driving factors behind the Program's historical focus on cellulosic ethanol are as follows:

i) Technology Readiness

- Over the last two decades, DOE-funded R&D has led to significant progress in the biochemical processes used to convert cellulosic biomass to ethanol. First generation technology for cellulosic ethanol production is now in the demonstration phase.

- DOE-funded R&D in this area has led to a well-developed body of work regarding the performance of ethanol as both a low-volume percentage (E10) gasoline blend in conventional vehicles and at higher blends (E85) in flexible-fuel vehicles (FFVs).

ii) Market Acceptance

- Starch-based ethanol is a well-established commodity fuel with wide market acceptance. Continued success and growth of the ethanol industry can help pave the way for the future introduction of cellulosic ethanol into the marketplace.

- FFV technology is commercially available from a number of U.S. automakers, and several have plans to significantly increase FFV production volumes and expand FFV marketing efforts in the coming years.

iii) Policy Factors

- Federal legislation predominantly focused on cellulosic ethanol production as a "second generation" biofuel to displace imported petroleum-based transportation fuels with domestic renewable fuels.

More recent national and DOE goals require the Program to expand its scope to include the development of other advanced biofuels that will contribute to the volumetric requirements of the Renewable Fuels Standard (RFS). This includes biofuels such as biobutanol, hydrocarbons from algae, and biomass-based hydrocarbon fuels (renewable gasoline, diesel, jet fuel).

Thus, while the Program's short-term objectives include demonstrating commercially viable cellulosic ethanol production, the investments the Program has made in technologies that can reduce the recalcitrance of lignocellulosic biomass will be leveraged toward the development of third generation advanced biofuels, bioproducts, and bioenergy.

Market Overview and Federal Role of the Program

Markets for biofuels, bioproducts, and bioenergy exist today both in the United States and around the world, yet the untapped potential is enormous. Industry growth is currently constrained by limited infrastructure, high production costs, competing energy technologies, and other market barriers. Market incentives and legislative mandates are helping to overcome some of these barriers.

Current and Potential Markets

Major end-use markets for biomass-derived products include transportation fuels, products, and power. Today, biomass is used as a feedstock in all three categories but the contribution is small compared to oil and other fossil-based products. Most bio-derived products are now produced in facilities dedicated to a single primary product, such as ethanol, biodiesel, plastics, paper, or power (corn wet mills are an exception). The primary feedstock sources for these facilities are conventional grains, plant oils, and wood.

To meet national goals for increased production of renewable fuels, products, and power from biomass, a more diverse feedstock resource base is required—one that includes biomass from agricultural and forest residues, and dedicated energy crops. Ultimately the industry is expected to move toward large biorefineries that produce a portfolio of biofuels and bioproducts, with integrated, onsite cogeneration of heat and power.

Transportation Fuels: America's transportation sector relies almost exclusively on refined petroleum products, accounting for over 70% of the oil used. Oil accounts for 94% of transportation fuel use, with biofuels, natural gas, and electricity accounting for the balance. Nearly 9 million barrels of oil are required every day to fuel the 247 million vehicles that constitute the U.S. light-duty transportation fleet.

Biomass is a direct, near-term alternative to oil for supplying liquid transportation fuels to the nation. In the United States, nearly all gasoline is now blended with ethanol up to 10% by volume, and cars produced since the late 1970s can run on E10. U.S. automakers have committed to increase their production of FFVs that can use E85 (blends of gasoline and ethanol up to 85%) to 50% of yearly production by 2012.

High world oil prices, supportive government policies, growing environmental and energy security concerns, and the availability of low-cost corn and plant oil feedstocks have provided favorable market conditions for biofuels in recent years. Ethanol, in particular, has been buoyed by the need to replace the octane and clean-burning properties of MTBE, which has been removed from gasoline because of groundwater contamination concerns.... [C]urrent domestic production of ethanol from grains has increased rapidly over the past five years, from under 4 billion gallons per year to nearly 13 billion gallons in 2010.

Over the last few years, commodity prices have fluctuated dramatically, creating market risks for biofuel producers and the supply chain. The national RFS legislated by EISA 2007 provides a reliable market for biofuels of 15.2 billion gallons by 2012. Blender's tax credits for ethanol and biodiesel have helped to ensure biofuels can compete with gasoline. Historically, when the blender's tax credit is subtracted from wholesale prices, biofuels are price competitive with petroleum fuels on a volumetric basis.

To successfully penetrate the target market, however, the minimum profitable cellulosic ethanol price must be cost competitive with corn ethanol and low enough to compete with gasoline. A minimum profitable ethanol selling price of $2.50/GGE (gallon gasoline equivalent) can compete on an energy-adjusted basis with gasoline derived from oil costing $75 to $80/barrel. Given the broad range of oil prices projected by the Energy Information Administration (EIA) in 2017 ($51 to $156/bbl), cellulosic technology may continue to require policy support and regulatory mandates.

Limited rail and truck capacity has complicated the delivery of ethanol, contributing to regional ethanol supply shortages and price spikes. Feedstock and product transportation costs remain problematic for the biofuel industry and have led many biofuel producers to locate near a dedicated feedstock supply or large demand center to minimize transportation costs.

Retail distribution also continues to be an issue. Although E10 is ubiquitous across the United States, a limited number of fueling stations for biodiesel and E85 exist. In 2009, less than 2% of fueling stations were equipped for dispensing these fuels. Some retail station owners are hesitant to offer higher percentage blends because the unique physical properties of the blends may require costly retrofits to storage and dispensing equipment. Independent station owners may also be uncomfortable with the market risk associated with novel biofuels and are reluctant to install new infrastructure.

Consumer attitudes about fuel prices and performance, biofuel-capable vehicles, and the environment also affect demand for biofuels. Consumers who are generally unfamiliar with biofuels have been hesitant to use them, even where they are available.

Products: Approximately 10% of U.S. crude oil imports are used to make chemicals and products such as plastics for industrial and consumer goods. Many products derived from petrochemicals could be replaced with biomass-derived materials. Less than 4% of U.S. chemical sales are biobased. Organic chemicals such as plastics, solvents, and alcohols represent the largest and most direct market for bioproducts. The market for specialty chemicals is much smaller but is projected to double in 15 years and offers opportunities for high-value bioproducts. These higher-value products could be used to increase the product slate and profitability of large integrated biorefineries. The price of bioproducts remains relatively high compared to petroleum-based products largely due to the high cost of converting biomass to chemicals and materials.

As the price of oil has increased, so has U.S. chemical manufacturers' interests in biomass-derived plastics and chemicals. Some traditional chemical companies are forming alliances with food processors and other firms to develop new chemical products that are derived from biomass, such as natural plastics, fibers, cosmetics, liquid detergents, and a natural replacement for petroleum-based antifreeze.

Biomass-derived products will also compete with existing starch-based bioproducts such as poly lactic acid. For biomass-derived products to compete, they must be cost competitive with these existing products and address commodity markets. New biomass-derived products will also have to compete globally and will, therefore, require efficient production processes and low production costs.

Power: Less than 2% of the oil consumed in the United States is used for power generation. Fossil fuels dominate U.S. power production and account for over 70% of generation, with coal comprising 48%, natural gas 21%, and oil 1%. The balance of power is provided by nuclear (20%) and renewable sources (9%) of which biopower accounts for 1%. New natural gas-fired, combined cycle plants are expected to increase the natural gas contribution, with coal-fired power maintaining a dominant role. Renewable energy, including biopower, is projected to have the largest increase in production capacity between 2008 and 2035.

Dedicated utility-scale biomass power applications are a potential route to further reducing our reliance on fossil fuels and improving the sustainability associated with power generation. Limits to the availability of a reliable, sustainable feedstock supply as well as competing demands for biofuels to meet EISA goals may constrain the feedstock volumes available for utilization in biopower applications and may also increase feedstock costs for both applications. A near-term opportunity for reducing GHG by increasing the use of biomass for power generation is to increase the deployment of cofiring applications for biomass and biomass-derived intermediates in existing power generating facilities.

State and Local Political Climate

States play a critical role in developing energy policies by regulating utility rates and the permitting of energy facilities. Over the last two decades, states have collectively implemented hundreds of policies promoting the adoption of renewable energy. To encourage alternatives to petroleum in the transportation sector, states offer financial incentives for producing alternative fuels, purchasing FFVs, and developing alternative fuels infrastructure. In some cases, states mandate the use of ethanol and/or biodiesel. Several states have also established renewable portfolio standards to promote the use of biomass in power generation.

Many states encourage biomass-based industries to stimulate local economic growth, particularly in rural communities that are facing challenges related to demographic changes, job creation, capital access, infrastructure, land use, and environment. Growth in the ethanol and biodiesel industry creates jobs through plant construction, operation, maintenance, and support. An ethanol facility producing 40 million gallons per year is estimated to expand the local economic base by $110.2 million each year through direct spending of $56 million and $1.2 million in increased state and local tax receipts. Several states have also recently begun to develop policies to reduce greenhouse gas emissions and are looking to biomass power and biofuels applications as a means to achieve targeted reductions.

International Political Climate

Oil is expected to remain the dominant energy source for transportation worldwide through 2030, with consumption expected to increase from 86.1 million barrels per day in 2008 to 110.6 million barrels per day in 2035. However, the use of renewable fuels is rising. Many nations are seeking to reduce petroleum imports, boost rural economies, and improve air quality through increased use of biomass. Some countries are pursuing biofuels as a means to reduce greenhouse gas (GHG) emissions. Brazil and the United States lead the world in production of biofuels for transportation, primarily ethanol, and several other countries have developed ethanol programs, including China, India, Canada, Thailand, Argentina, Australia, and Colombia.

As countries are developing policies to encourage bioenergy, many are also developing sustainability criteria for the bioenergy they produce and use within their countries. Both the United States and the European Union (EU) have focused on GHG reduction requirements for their fuel. The EU has also established a committee to coordinate the development of further biofuel sustainability criteria.

Several international groups, notably the Roundtable on Sustainable Biofuels, the International Organization for Standardization, and the Global Bioenergy Partnership, are in the process of developing criteria and standards for sustainability that could be utilized in evaluation of biofuel production and processing. These criteria will address environmental, social, and economic aspects of bioenergy production.

[Some sustainability problems with biofuels are addressed in the following section—Ed.]

The relationship between bioenergy, agriculture, and land use change has been the subject of increasing attention, particularly with regards to the conversion of old growth forests and native prairies into agriculture production. Policymakers, eager to address this issue, have encouraged scientists in the field of bioenergy to focus on researching the indirect impacts of bioenergy production in order to understand the magnitude of the linkage and to identify and protect any vulnerable areas valued for their role in preserving biodiversity and sequestering carbon.

In recent years, attention has focused on how the expanding production of bioenergy crops can influence international markets, potentially triggering price surges and price volatility for staple foods. Some governments have addressed this issue through discouraging the use of food-based feedstocks for bioenergy production. Recently, China halted construction of new corn-based ethanol plants and has worked to promote policies that encourage the production of biofuels from non-food feedstocks grown on marginal land. Many countries, particularly in the developing world, have identified ways by which bioenergy production can actually increase food security by generating employment, raising income in farming communities, and promoting rural development (Food and Agriculture Organization of the United Nations or UN FAO).

The principal technologies that compete with biomass today rely on continued use of fossil energy sources to produce transportation fuels, products, and power in conventional petroleum refineries, petrochemical plants, and power plants. In the future, as oil demand and prices continue to rise, non-traditional technologies will likely compete with biofuels in meeting some of the transportation fuel needs of the United States. Competing technologies include:

- **Hydrogen:** Hydrogen can be produced via water electrolysis, reforming renewable liquids or natural gas; coal gasification; or nuclear synthesis routes.

- **Oil Shale-Derived Fuels:** Oil shale is a rock formation that contains large concentrations of combustible organic matter called kerogen and can yield significant quantities of shale oil. Various methods of processing oil shale to remove the oil have been developed.

- **Tar Sands-Derived Fuels:** Tar sands (also called oil sands) contain bitumen or other highly viscous forms of petroleum, which is not recoverable by conventional means. The petroleum is obtained either as raw bitumen or as a synthetic crude oil. The United States has significant tar sands resources — about 58.1 billion barrels.

- **Coal-to-Liquids:** In terms of cost, coal-derived liquid fuels have traditionally been non-competitive with fuels derived from crude oil. As oil prices continue to rise, however, coal-derived transportation fuels may become competitive. It should be noted that conventional coal-to-liquid technologies can often be adapted to use biomass as a feedstock, both in standalone applications or blended with coal.

- **Electricity:** Electricity can be used to power electric vehicles (EVs). EVs store electricity in an energy storage device such as a battery or produce on-board power via a fuel cell, and power the vehicle's wheels via an electric motor. Plug-in hybrid electric vehicles (PHEVs) combine the benefits of pure electric vehicles and hybrid electric vehicles.

* * *

History of Public Efforts in Biomass RDD&D

Efforts in bioenergy were initiated by the National Science Foundation (NSF) and subsequently transferred to DOE in the late 1970s. Early projects focused on biofuels and biomass energy systems. In 2002, the Biomass Program was formed to consolidate the biofuels, bioproducts, and biopower research efforts across EERE into one comprehensive program. From the 1970s to the present, DOE has invested over $3.7 billion (including more than $900 million in ARRA funds) in a variety of RDD&D programs covering biofuels (particularly ethanol), biopower, feedstocks, municipal wastes, and a variety of biobased products.... While steady progress has been achieved in many technical areas,

Table 1

May 2009	Presidential Memorandum on Biofuels	Memorandum that, among other requirements, established a Biofuels Interagency Working Group to consider policy actions to accelerate and increase biofuels production, deployment, and use. The Group is co-chaired by the Secretaries of DOE and USDA and the Administrator of EPA.
February 2009	American Reinvestment and Recovery Act (ARRA)	• Provided funds for grants to accelerate commercialization of advanced biofuels R&D and pilot-, demonstration-, and commercial-scale integrated biorefinery projects. • Provided funds to other DOE programs for basic R&D, innovative research, tax credits, and other projects.
May 2008	The Food, Conservation, and Energy Act of 2008 (Farm Bill)	• Provided grants, loans, and loan guarantees for developing and building demonstration and commercial-scale biorefineries • Established a $1.01 per gallon producer tax credit for cellulosic biofuels • Established the Biomass Crop Assistance Program (BCAP) to support the production of biomass crops • Provided support for continuation of the Biomass R&D Initiative, the Biomass R&D Board, and the Technical Advisory Committee.
December 2007	Energy Independence and Security Act (EISA) of 2007	Supported the continued development and use of biofuels, including a significantly expanded Renewable Fuels Standard, requiring 36 bgy renewable fuels by 2022 with annual requirements for advanced biofuels, cellulosic biofuels and biobased diesel.
August 2005	Energy Policy Act of 2005 (EPAct)	Renewed and strengthened federal policies fostering ethanol production, including incentives for the production and purchase of biobased products; these diverse incentives range from authorization for demonstrations to tax credits and loan guarantees.

considerably more progress is required to make biomass utilization technology applications competitive in the marketplace.

Especially in recent years, several legislative, regulatory, and policy efforts have increased the focus on increasing and accelerating biomass-related RDD&D. These efforts are summarized [in Table 1].

Biomass Program Justification

Between 2008 and 2035, U.S. energy consumption is projected to rise by 14% while domestic energy production by 22%. Petroleum imports, which now serve more than 54% of U.S. energy needs, are projected to decline to 44% by 2035. Biofuels are projected to have the largest increase in meeting domestic consumption, growing from 3.5% to over 11% of liquid fuels. This decreased reliance on imported energy improves our national security, economic health, and future global competitiveness. In addition, the U.S. transportation sector is responsible for one-third of U.S. carbon dioxide (CO_2) emissions, the principal GHG contributing to global warming.

Combustion of biofuels and production of biopower also releases some CO_2, but that release is largely balanced by CO_2 uptake for the plants' growth. Depending upon how much fossil energy is used to grow and process the biomass feedstock, bioenergy can substantially reduce net GHG emissions. Biomass is the only renewable energy resource that can be converted to a liquid transportation fuel, and increased use of renewable fuels provides the best near-term option for reducing GHG emissions from the transportation sector.

The overarching federal role is to ensure the availability of a reliable, affordable, and environmentally sound domestic energy supply. Billions of dollars have been spent over the last century to construct the nation's energy infrastructure for fossil fuels. The production of alternative transportation fuels from new primary energy supplies like biomass is no small undertaking. The federal role is to invest in the high-risk, high-value biomass technology RDD&D that is critical to the nation's future but that industry would not pursue independently. States, associations, and industry will be key participants in deploying biomass technologies once risks have been sufficiently reduced by federal programs.

Notes and Questions

1. Notice that ethanol is the most mature biofuel technology. How close are some of the second and third generation biofuels to being market-ready? What potential legal issues might each technology face?

2. Biofuel production is highly dependent on federal tax policy. What should be the policy drivers? Supporting one type of technology? A particular feedstock? A particular type of product—*i.e.*, ethanol, diesel?

3. What other policy drivers could be considered? The Reitze excerpt above noted that a 10 percent ethanol blend results in a 2 to 3 percent decrease in miles per gallon. Also note the box below explaining "Energy Density." Should there be incentives for more energy-dense fuels? How about drivers that reward the development of less carbon-intense fuels or encourage less energy-intense processing facilities? What will be the differing impacts of each of these considerations?

B. Efficiency and Energy Return on Energy Invested (EROEI)

As Figure 5.5 illustrates, the end product of biomass production varies widely, with the most common products being direct heat, electricity, and biofuels. The technologies for converting the different sources of biomass to these end products also vary widely. In terms of direct heat, humankind originally turned to fossil fuels instead of biomass because fossil fuels can deliver up to twice as much energy by volume.

Energy Density—Although biomass may have been one of the first sources of energy captured for human energy needs, its limitations led to the use of fossil fuels. Per volume, wood only produces about half the energy contained in the more compact power punch of coal (at 28-31 gigajoules (GJ)/metric ton(t)) or oil (at 42 GJ/t). L.D. Danny Harvey, Energy and the New Reality 2: Carbon-free Energy Supply 175 (EarthScan 2010). When wood was our primary source of fuel in the United States in the 1850s, we had wood shortages for a U.S. population of fewer than thirty million. The current U.S. population is approximately 311 million, and our per capita energy use is much higher than in 1850.

Energy Return on Energy Invested (EROEI) (also called Energy Return on Investment or EROI) is another important parameter for energy development. It represents a ratio of the amount of energy from a particular source in relation to the amount of energy expended in producing that energy. For example, in the 1930s, the EROEI for domestic crude in the United States was 100:1. That figure had dropped to 20:1 by 2005 because of the increasing difficulty in finding and producing oil reserves. Cutler J. Cleveland, *Net Energy from the Extraction of Oil and Gas in the United States*, 30 ENERGY 769, 780-81 (2005). While the EROEI for most renewables is respectable—an average around 20:1 for wind power and 8:1 or better for solar PV—the EROEI for some biomass is below the break-even point of 1.0, meaning a net energy loss.

The conversion of biomass to electricity creates additional problems with EROEI. In contrast to the renewable technologies in previous chapters, the way biomass is converted to electricity is much less efficient. Distributed solar photovoltaics convert the sun's rays directly to electricity, which then can be consumed at the same location. The large bus-like box behind a wind turbine blade is the nacelle, which converts the kinetic energy of the turning blade directly into electricity to feed into the grid.

In contrast, the technology for a biomass-to-electricity facility is essentially the same as a coal-fired power plant. Physicists call it a Rankine steam cycle, named after William John Macquorn Rankine who first described the process of converting steam heat into work. First, chemical energy is converted to thermal energy as the biomass is burned. The thermal energy is used to heat water to steam. The steam then converts the thermal energy to mechanical energy as it turns turbines. These turbines finally convert the mechanical energy into electricity which can be surged to customers along transmission and distribution lines.

The Second Law of Thermodynamics informs us that useable energy is lost at each phase of this process as it is expended in the phase changes from one form of energy to another— *i.e.*, chemical to thermal. Consequently, according to research conducted by NREL scientists, coal-to-electricity plants may be reliable, but they are very inefficient—only about 33 percent of the energy of the coal conveyored into the boilers comes out as useable electric energy. PAMELA L. SPATH ET AL., NREL REPORT TP-570-25119, LIFE CYCLE ASSESSMENT OF COAL-FIRED POWER PRODUCTION (1999). *See also* PATRICIA NELSON LIMERICK ET AL., CENTER OF THE AMERICAN WEST, REPORT #4, WHAT EVERY WESTERNER SHOULD KNOW ABOUT ENERGY 2 (2003) (estimating that if the energy for mining and transporting the coal to the power plant and transmission losses are also included, "only 26 percent of the energy extracted from [coal in] the ground reaches your toaster."). See Figure 5.8 on the next page.

As Figure 5.8 shows, efficiencies for fossil-fuel generation can be improved by using combined heat and power (CHP) or cogeneration plants discussed further in the notes following subsection A of Chapter 7. In contrast to fossil-fuel generation, however, bio-mass-only plants deliver even a lower percentage of useable electric energy because steam-to-electricity turbines are more efficient at higher temperatures and most biomass burns at lower temperatures than fossil fuels.

The EROEI equation has been especially problematic in the context of biofuels. If one of the goals of biofuels is to reduce greenhouse gases (GHG), then it is relevant to factor in the amount of GHGs needed to produce that biofuel. Any fuel that requires more energy input to create in relation to the amount it provides is unsustainable. Critics have made the case that many biofuels have a negative return of energy input in comparison to the energy delivered by the fuel. The following two excerpts are discussions of that issue.

Figure 5.8: Conventional v. combined heat and power (CHP) generation

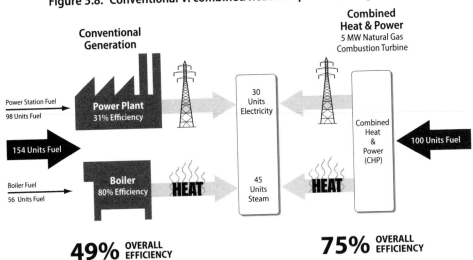

U.S. Envtl. Protection Agency Combined Heat and Power Partnership,
Introduction to C.H.P. Tech., 2 (2008), *available at*
http://www.epa.gov/chp/documents/catalog_chptech_intro.pdf

L. Leon Geyer, Phillip Chong, & Bill Hxue
Ethanol, Biomass, Biofuels and Energy: A Profile and Overview
12 DRAKE J. AGRIC. L. 61, 71-72 (2007)

Energy Balance of Ethanol Production

The core question surrounding corn-based ethanol is whether production consumes more energy then it creates. The controversy and uncertainty surrounding this question is in part a function of ethanol's complexity and "variations in data and assumptions used among different studies."

In June 2004, the U.S. Department of Agriculture updated its 2002 analysis of the issue and determined that the net energy balance of ethanol production was 1.67 to 1. (For every 100 BTUs [British Thermal Units] of energy used to make ethanol, 167 BTUs of ethanol is produced). In 2002, USDA had concluded that the ratio was 1.35 to 1.

Furthermore, a 2002 Michigan State University study "found that ethanol produced from corn provided 56 percent more energy than is consumed during production (1.56 to 1)."

Douglas Tiffany of the University of Minnesota attributed ethanol's energy balance to corn's "storage of solar energy in starch molecules." In addition to the positive energy balance, Tiffany reported that:

Corn-derived ethanol results in a six-fold displacement of liquid fuels. This means that every gallon of ethanol produced requires only one-sixth of a gallon of liquid fuels. This is due to the fact that corn production and ethanol processing utilize coal for electrical energy and natural gas for fertilizer production.

Despite the findings above by the USDA, Michigan State University, and Douglas Tiffany, there has been evidence to indicate a negative energy return for ethanol production using corn and liquid fuels from biomass energy. David Pimentel of Cornell University states that "ethanol production does not enhance energy security" due to the fact that energy needed to produce ethanol is greater than the energy output from its production. Pimentel's study supports his claim and concludes:

The total energy input to produce 1000 liters of ethanol is 8.7 million kilocalories.... However, 1000 liters of ethanol has an energy value of only 5.1 million kilocalories. Thus, there is a net energy loss of 3.6 million kilocalories per 1000 liters of ethanol produced. Put another way, about 70 percent more energy is required to produce 1000 liters of ethanol than the energy that actually is in the ethanol....

These findings were unanimously approved by many scientists....

The discrepancy that exists between the arguments for a positive and negative balance of energy can be found within the accounted inputs of ethanol production. Pimentel states that those who argue for a positive energy balance omit some energy inputs. Similarly, the United States Department of Agriculture states that those who argue for a negative energy balance "overestimate the amount of energy needed to grow corn and convert corn to ethanol." Ultimately, the consensus within the scientific community supports the findings of USDA.

Arnold W. Reitze, Jr., *Biofuels — Snake Oil for the Twenty-First Century*
87 OR. L. REV. 1183 (2008)

To convert these cellulosic biomass sources to ethanol involves significant pretreatment or mechanical separation before conversion, which increases the capital costs of these facilities. Steam is usually used to help break apart the glucose molecule. This is followed by enzymatic hydrolysis, which uses enzymes to break cellulose chains down to fermentable sugar. Then the lignin is separated from the mixture, which may be burned for power production. At this point the sugar is treated in the same way as corn-based alcohol production. Yeast is added and allowed to ferment. Then the alcohol is separated from the fermented mash, and a by-product called stillage is left. The ethanol is dehydrated to produce fuel-grade ethanol. In 2006 the industry was hoping to produce ethanol at sixty cents per gallon and sell it for two dollars per gallon at the pump, but such economic efficiency is unproven.

Four studies of cellulosic ethanol concerning the energy return on investment (r_E) [equivalent to EROEI — Ed.], which is the ratio of energy in ethanol compared to the nonrenewable energy required to make it, were surveyed and compared in a 2006 report. The Pimentel & Patzek study in 2005 found a negative energy return of 0.69, but a study published in 1993 found an r_E of 6.61 and two studies published in 2004 reported an r_E of 4.55 and 4.40. The low r_E reported in the Pimental & Patzek 2005 study may be attributable to their assumption that fossil fuel is used, but cellulosic production is expected to combust the lignin in the feedstock for the power needed for the ethanol conversion process. Thus, cellulosic ethanol could be produced using less nonrenewable energy than corn-based ethanol. A 2007 life-cycle study found that the energy requirements and adverse environmental impacts associated with chemicals used for production are low for switchgrass and hybrid poplar when compared to corn crops. If hybrid poplar or switchgrass are gasified to produce electricity, rather than being used to produce ethanol, the net energy obtained is doubled because the process of converting switchgrass to gas and then producing electricity is much more efficient that converting swithchgrass to ethanol. But this does not produce the liquid fuel needed by the transportation sector.

⋆ ⋆ ⋆

The Energy Independence and Security Act of 2007 expands federal support for advanced biofuel, which is defined broadly in an amended CAA Section 211(o)(1) to include ethanol from feedstocks other than corn and includes cellulosic ethanol, biodiesel, biogas (including landfill gas and sewage treatment gas), and other fuel derived from biomass including algae. Advanced biofuel must have lifecycle GHG emissions that are at least 50 percent less than baseline GHG emissions. Moreover, renewable biomass is defined in CAA section 211(o)(1)(I) to prevent existing forests from being cut to provide renewable biomass cultivation. In 2009 the 11.1 billion gallon renewable fuel requirement must be met using 0.6 billion gallons of advanced biofuel, and the amount of advanced biofuel required to be used increases until in 2022 when it is to comprise twenty-one billion gallons of the thirty-six-billion-gallon renewable fuel requirement.

Notes and Questions

1. Imposing a positive EROEI makes sense when we are considering the ongoing cultivation of new renewable feedstocks. But, consider biomass feedstocks that, if not recycled into energy, would decay and emit methane on their own, especially considering that the global warming impact of methane may be twenty times an equivalent unit of CO_2. Should EROEI calculi be as stringent when a biofuel source is accomplishing multiple goals?

2. Consider a biomass plant that uses multiple feedstocks—some renewable and some that are arguably waste. For example, the Biomass One plant in Figure 5.9 uses forest waste, which is covered by BCAP. But the removal of forest wastes, which are then burned as fuel in the plant, also serves the goal of fire suppression and the goal of reducing carbon emissions from decaying wastes in the forest. The Biomass One plant has been in operation since the early 1980s and works with four county landfills to processes local wood and yard waste as well as forest waste.

3. State incentives are a significant supplement to federal incentives for keeping some biomass plants going. For example, the State of Oregon has a Tax Credit for Production and Collection of Biomass for Use in Biofuel authorized under Or. Rev. Stat. § 315.141 (2011). In January of 2010, the administrative regulations for enforcing this tax credit created a requirement that a biomass facility must obtain a 40 percent thermal efficiency. Or. Admin. R. 330-170-0040(4) (2011). Co-generation facilities, that both generate electricity and use the waste heat for industrial processes, can achieve a 40 percent thermal efficiency, but most standalone biomass-to-electricity plants cannot, and the Biomass One plant averages below a 25 percent thermal efficiency. Should some adjustment be made for facilities that serve purposes beyond the generation of heat and electricity? What additional factors should be considered and how would you incorporate them into the tax credit calculation?

4. The EPA issued the following definition for renewable biomass in response to criticism about negative EROEI concerns. Notice that EPA employed a "lifecycle greenhouse gas

Figure 5.9: Biomass One plant in Oregon

Photo by K.K. DuVivier.

emission" metric instead of an EROEI ratio. How do the two metrics differ? What do you think are the advantages or disadvantages of each?

Regulation of Fuels
42 U.S.C. § 7545 (2010)

Not later than 1 year after December 19, 2007, [The EPA] Administrator shall revise the regulations under this paragraph to ensure that transportation fuel sold or introduced into commerce in the United States (except in noncontiguous States or territories), on an annual average basis, contains at least the applicable volume of renewable fuel, advanced biofuel, cellulosic biofuel, and biomass-based diesel, determined in accordance with subparagraph (B) and, in the case of any such renewable fuel produced from new facilities that commence construction after the date of enactment of this sentence, achieves at least a 20 percent reduction in lifecycle greenhouse gas emissions compared to baseline lifecycle greenhouse gas emissions.

––––––

C. Environmental Issues

While biomass has positives, it also has environmental negatives that must go into considerations about developing it as a green resource. The following excerpts address problems with air emissions, invasive species, and the challenges of quantifying the environmental and social impacts of biomass production.

1. Air Emissions

One of the greatest drawbacks of biomass is that it cannot generate heat or power without burning, and this burning emits noxious air pollutants and carbon dioxide.

The vast majority of the approximately forty-five exajoules of biomass the International Energy Agency estimates humans consume per year are used in the form of direct cooking and heating in lesser-developed countries. Biomass—in the form of wood waste, agricultural waste, and animal dung—supplied about 16 percent of the total energy for developing countries in 2005. Use of biomass in this way has created serious respiratory health hazards for people in these countries. Cooking with biomass is a problem because the efficiency is very low, 10-20 percent, and results in high emissions. It is estimated that nearly two million women and children die annually from breathing this toxic smoke.

In addition to the biomass used by lesser-developed countries, one fifth of the world's biomass consumption—approximately nine exajoules—goes to the developed world's use of biomass for heat, electricity, and biofuels. Modern scrubber technology can capture most air pollutants that a biomass-to-energy plant produces, but these biomass processes still have the drawback that they emit carbon dioxide. The following excerpt describes some of the irony of identifying biomass as a "green" resource.

Andrew P. Morriss, William T. Bogart, Andrew Dorchak, Roger E. Meiners, *Green Jobs Myths*

16 Mo. Envtl. L. & Pol'y Rev. 326, 350-51 (Spring 2009)

A different version of this problem can be seen in the way some analyses consider almost anything green if the technology does not use petroleum without considering the environmental impacts of the alternative's environmental impact. For example, the Mayors report [U.S. Conference Of Mayors, Oct. 2008, U.S. Metro Economies: *Current And Potential Green Jobs in the U.S. Economy* 9 (2008)—Ed.] touts biomass as a "group of technologies where additional investment and jobs will help to develop the nation's alternative energy infrastructure." Most of the green jobs literature extols the virtues of generating energy using "wood waste and other byproducts, including agricultural byproducts, ethanol, paper pellets, used railroad ties, sludge wood, solid byproducts, and old utility poles. Several waste products are also used in biomass, including landfill gas, digester gas, municipal solid waste, and methane."

Unfortunately, because biomass includes burning wood, "perhaps the oldest form of human energy production," [it is—Ed.] a means of energy production associated with smog, air pollution, and massive release of carbon. Yet biomass is included "because of the short time needed to regrow the energy source relative to fossil fuels." In other words, biomass counts as green because it is not petroleum, even though biomass causes environmental problems. Similarly, the Mayors report counts biodiesel and ethanol as green "because of their ability to reduce reliance on fossil fuels," overlooking arguments that growing corn or soy for ethanol or biodiesel requires agricultural practices that increase air and water pollution, bring marginal land into production reducing wildlife habitat, increase emissions of carbon dioxide and nitrous oxides, and increase the amount of nitrogen and pesticides in the environment.

Unintended Consequences of Trees? One of the efforts to deal with climate change is planting additional trees to sequester some of the carbon. An unintended consequence is that some of these trees may emit more volatile organic compounds (VOCs) than they absorb, thus contributing to high ozone levels. *See, e.g.,* http://www.fraqmd.org/Biogenics.htm; http://news.discovery.com/earth/trees-as-a-source-of-greenhouse-gases.html. Other unintended consequences of randomly planting trees, with the hope of sequestering carbon, is they may block solar access (*cf.* Chapter 2) or they may absorb significant water resources in areas where water is scarce. Thus, planting the wrong tree in the wrong place may have a detrimental, instead of a beneficial, result.

Notes and Questions

1. To address some of the problems with using biomass for cooking in developing countries, the United States contributed $50 million to the Global Alliance for Clean Cookstoves in September of 2010. These cookstoves can capture between 50 and 95 percent of harmful emissions. What do you think of this solution?

2. Emissions of carbon dioxide are part of the biomass burning process, but the process may be considered "green" if the amount of carbon dioxide emitted upon burning is equivalent to the amount taken up by the vegetative matter as it is growing. What issues do you see with attempting to put numbers on such a cycle to measure return of energy and to determine whether the result is no net carbon dioxide emissions?

3. The Climate Action Reserve (CAR) is a national offsets program. CAR helps quantify and verify greenhouse gas emissions by establishing regulatory-quality standards for various types of renewable energy. To facilitate renewable energy credit authentication and trading, CAR posts information on a publicly-accessible system. The California Climate Action Registry (California Registry), which was established by the State of California in 2001, helped establish CAR to expand the California Registry's mission throughout all of North America. How do you think CAR rates biomass offsets?

4. The excerpt above raised additional concerns about environmental degradation such as soil erosion, pesticides, the increased use of nitrogen fertilizer, as well as extensive water use to process raw plants into biofuels and the resulting noxious organic effluent. For a discussion of some of the environmental problems created by the use of water in biomass production, read NAT'L RESEARCH COUNCIL ET AL., WATER IMPLICATIONS OF BIOFUELS PRODUCTION IN THE UNITED STATES (Nat'l Acad. Press 2008). *See also* Jacqueline M. Wilkosz, *Thirsting for Change: How the Growth of the Biofuel Industry Can Stimulate Advancements in Water Law*, 2009 U. ILL. L. REV. 583 (2009) (arguing that biofuel development illustrates the inadequacy of our riparian rights system of water law because the reasonable-use rule favors intensive water users and lacks the ability to prospectively protect water rights).

2. Invasive Species

In seeking alternative feedstocks that grow quickly and robustly into biomass resources, we risk introducing plant species that will create environmental havoc, a story which unfortunately has played out with past efforts of humankind to tinker with nature.

Karen Ray, *Are Biofuel Crops the Next Kudzu?*
17 San Joaquin. Agric. L. Rev. 247 (2007-2008)

How Biofuel Crops May Add to the Invasive Species Problem

The United States is on the fast track to planting large-scale biofuel crops in order to meet the President's energy goal. There is a tremendous political and social pressure on the United States to push for the production of biofuels from cellulose plants. However, before there is any large-scale planting of biofuel crops, federal and state legislation should address potential ramifications if any of the biofuel crops escape cultivation and become invasive.

Overview of the Invasive Species Problem: Invasive Weeds and Plants

It is estimated that 25,000 nonnative plant species have been introduced to the United States, mainly for commercial or ornamental purposes. Approximately 5,000 introduced plant species have escaped and are now established in surrounding natural ecosystems. However, only a limited number of those species spread and cause severe harm. Even though a small percentage of nonnative plants ever become invasive, even one of those species can do significant damage. Invasive plants and weeds can "cause significant changes to the ecosystems, upset the ecological balance, and cause economic harm" to agriculture and natural sectors. Furthermore, they can choke out native plant species, alter wildlife and fish habitat, impact human health, and increase fire threats. Invasive plants and weeds threaten biodiversity and are a major contributing factor in the population declines of almost one half of the nation's endangered species.

The extent of the nation's invasive plants and weeds problem is enormous. Invasive plants and weeds spread into 4,600 acres daily. Annually, invasive plants claim three million acres, an area which is roughly twice the size of Delaware. Natural areas are significantly affected by invasive plants and weeds." The spread of invasive weeds in these nonagricultural areas is said to resemble an explosion in slow motion, and weeds now cover an estimated 133 million acres in the United States." "This is not natural evolution; rather changes ramped up by increased global mobility" and these changes are "caused by human decisions." Economically, the United States is impacted significantly. The United States spends thirty-six billion dollars annually addressing invasive weeds. Nationwide, invasive species cause an estimated $137 billion dollars of environmental damage per year.

An Objective of Executive Order 13,112 is to Identify Invasive Species Pathways

In February 3, 1999 President Clinton signed Executive Order 13,112. The intent of the Executive Order is to protect the United States from harm caused by invasive species. It creates an Invasive Species Council made up of the heads of several departments and agencies including the Secretary of State, the Secretary of Defense, the Secretary of the Interior, the Secretary of Agriculture, the Secretary of Transportation, and the Administrator of the Environmental Protection Agency. This council, as directed by the Executive Order, developed a national plan which includes addressing invasive weeds and plants. One of the goals of the management plan is to "include a review of existing and prospective ap-proaches and authorities for preventing the introduction and spread of invasive species, including those for identifying pathways by which invasive species are introduced and for minimizing the risk of introductions via those pathways...." When this order was signed in 1999, it appears cellulose biofuel crops were not considered yet as a source for renewable energy. This assertion is supported by the fact of the lack of legislation promoting cellulose biofuels at that time. This brings up the concept that agriculture, specifically large-scale planting of biofuel crops, could be an additional pathway for invasive species.

Intentional Introduction of Invasive Species: Agriculture as a Pathway

The majority of introduced plant species in the United States were introduced intentionally. Some of these plant species which have escaped cultivation and caused significant harm to the United States have stemmed from the ornamental and agricultural sector. However, not all intentional introductions are harmful. For example, many of the major crops growing in the United States today are nonnative and are also noninvasive. These crops, such as cotton, corn, and rice do not escape cultivation, and serve their intended purpose. Livestock and ornamental plants are also examples of intentional introductions of nonnative species that have proven to be very beneficial to the United States. Over 4,000 introduced plant species that were introduced for food crops do not display harmful or invasive characteristics. Nevertheless, a small percentage of these introduced plants for cultivation, such as for food, spices, and medicinal uses have escaped and invaded natural areas. Of the 300 nonindigenous weeds prevalent in the western United States, at least eight of those weeds have been cultivated as crops and twenty-eight have escaped horticulture and invaded other areas. Although the number is low in comparison to how many plant species do not escape cultivation, the harm caused by just one plant species that has become invasive can be insurmountable. Since grasses and other nonnative plant species are being considered for cultivation as biofuel crops, it is important to identify the potential legal, economic, and environmental ramifications if the cultivation of biofuel crops goes awry. Recognizing that agriculture can be a pathway for invasive species is consistent with the intent of Executive Order 13,112.

Historical Intentional Introductions of Plant Species to Solve Ecological Problems

Before there is large-scale planting of biofuel crops, it is necessary to consider past purposeful introductions of nonnative species to comprehend the gravity of the problem if these crops do escape and invade natural areas. Historically, the United States has accepted claims about the supposed benefits of nonnative plant species to solve ecological problems without solid substantiation. After the catastrophic consequences from the introduction of the wrong plant species in the past, legislatures and the public alike probably would not have rushed to introduce these species.

History of Kudzu

Kudzu (Pueraria lobata) was introduced into the United States in 1876 at a Philadelphia exhibition from Japan. In the 1910s, it was used as a forage crop. Approximately ten years later, the Georgia Railroad took interest in kudzu and distributed free kudzu plants to farmers in order for them to grow it as hay. In 1930, the government became the major player in promoting kudzu. During the Great Depression, massive soil erosion on southern farmlands seriously threatened the region's agricultural sector. The federal government launched a campaign to plant kudzu, through the Soil Erosion Service and later through the Soil Conservation Service, during the 1930s and 1940s as a solution for soil erosion. By 1950, the federal government had distributed eighty four million kudzu seedlings to southern landowners and offered them eight dollars per acre as an incentive to plant kudzu on their land. In 1934, there was an estimated 10,000 acres planted with kudzu and by 1946, acreage increased to almost three million acres. Even though farmers became concerned about kudzu's invasiveness, the federal government did not remove it from the list of permissible cover plants until the 1950s under the Agricultural Conservation program. Nearly twenty years later, the USDA identified kudzu as a "common weed." In

1997, almost a century after its introduction, kudzu was listed as a "noxious weed" under the Federal Noxious Weed Law. By that time, kudzu had invaded seven million acres in natural areas and it continues to spread to over 120,000 acres annually. In the United States, it is estimated that kudzu causes over $100 million dollars of damage per year, and if factoring in the nation's lost productivity in forests, the costs increases to over $500 million dollars per year.

Kudzu continues to alter the landscape of the United States' agricultural lands as well as the nation's natural areas. It stifles agriculture production as a result of its rapid growth and its ability to climb over plants and trees, smothering them by heavy shading. It also harms forest areas by inhibiting the process of tree renewal, which in turn prevents new growth of native trees. Kudzu out-competes native plants and ultimately disrupts wildlife habitats by diminishing vital resources and food.

Kudzu also harms the nation's power and transportation sectors. It causes significant problems to the rail system due to the slick pulp that is produced when the vines get on the track, which leads to derailments. Ironically, the railroad was one of its first promoters. In addition, kudzu overtakes power poles by weaving into the hot wire thus producing power outages. Countless manpower is devoted annually to clear the vines from these power poles. Kudzu is a cautionary example of how the government can promote a plant species to solve an ecological problem without establishing adequate safeguards.

Will History Repeat Itself?

There are several similarities between the federal government's involvement in the distribution of kudzu and its proposed large-scale planting of biofuel crops today. First, in the 1930s and 1940s, the massive soil erosion during the Great Depression is comparable to today's ecological problem of global warming and the decreasing supply of fossil fuels. Second, massive promotional campaign by the government to promote kudzu as a solution to soil erosion is strikingly similar to the increase in biofuel-related legislation to promote biofuel crops as a solution for the decreasing supply of fossil fuels and global warming. Last, the governmental economic incentives given to farmers to encourage them to plant kudzu on their land is analogous to the governmental economic incentives in the form of tax, grants, and loans offered to potential growers of biofuel crops. There are many lessons to be learned from this infamous example. Most critically, the long lag time before legislative response can make an enormous difference ecologically.

Kudzu as an Example of the Law of Unintended Consequences

Kudzu is an example of the law of unintended consequences which can be defined as "actions of people—and especially of government—always have effects that are unanticipated or 'unintended.'" The government was unaware of the consequences when it offered financial incentives to grow kudzu. When it realized the invasiveness of the vine, it responded slowly by not declaring it a "noxious weed" until a century after its introduction. The government today has chosen to ignore potential unintended consequences of biofuel crops and put them on the fast track for large-scale introduction without any adequate safeguards. By actively promoting large-scale biofuel crop introductions as a solution for the energy crisis, the government may cause the unintended result of contributing to the invasive species problem by introducing invasive species.

Examples of Other Government-Sponsored Intentional Plant Introductions that have Gone Awry

Not only have the federal, state, and local governments sponsored numerous nonnative plant introductions to provide solutions for ecological problems, but they have also sponsored planting projects such as for parks and trees as well as to provide shelter and wind barriers. These nonnative plant species have been selected for these projects because they possess qualities such as pollution tolerance and hardiness; however, these qualities allow the nonnative species to become invasive and spread by out-competing native species. An example of a nonnative species that has spread beyond its original purpose is the multiflora rose (Rosa multiflora Thunb). This is a thorny perennial shrub native to Asia. It was introduced to the United States in 1866 as a rootstalk for ornamental roses. In the 1930s, like kudzu, it was promoted by the U.S. Soil Conservation Service for erosion control and as a "living fence" to confine livestock. Soon after, several state conservation departments distributed rooted cuttings to landowners for free to be used as wildlife cover. Since multiflora rose is "tenacious" and has "unstoppable growth" which eventually crowds out native species, it is now classified a "noxious weed" in many states.

Not only has the government been slow to respond to species introductions that have gone awry, such as kudzu and multiflora rose, but sometimes it has chosen to turn a blind eye. An example of this is how it responded to the Elaeagnus species, a native to Asia. The United States Army Corps of Engineers recommended Elaeagnus for site restoration as late as the 1990s. This is surprising "because it was already known that the efficient dispersal of fruits by birds and the plants' continuous resprouting ability had enabled Elaeagnus to become one of the most numerous invaders in the United States." If some biofuel crop species unfortunately do escape cultivation and invade natural areas, hopefully the United States learn from the lessons of the past and respond more quickly.

Notes and Questions

1. This excerpt emphasizes consideration of unintended consequences, which in the context of renewable energy sources, are most likely to come from biomass. One of the feedstocks under consideration is genetically modified (GM) poplar trees. In 1999, the WWF released a report called *GM Technology in the Forest Sector* that warned, "the risk of genetic pollution [from GM trees] is high." Similarly, the Forest Stewardship Council bans research on genetically modified trees. What are some ways to avoid damage to neighboring vegetation by the inadvertent spread of non-native trees?

2. A company named Joule Unlimited has invented a genetically-engineered organism that secretes diesel fuel or ethanol wherever it finds sunlight, water, and carbon dioxide. Litigation has resulted from the inadvertent spread of genetically modified crops into adjacent fields. What might be some of the consequences of introducing Joule Unlimited's genetically-engineered organism into the natural environment?

3. Quantifying Environmental Problems

The excerpt below should help drive decisions about subsidizing biomass by proposing methods for governments to better quantify some of the additional environmental costs.

Jody M. Endres, *Clearing the Air: The Meta-Standard Approach to Ensuring Biofuels Environmental and Social Stability*
28 VA. ENVTL. L.J. 73, 74–105 (2010)

Biofuels are championed as an environmental savior, an agent of rural economic prosperity, and the pathway to energy independence, and they indeed have delivered on some of these promises. Until the recent collapse of the corn-based ethanol industry, biofuels provided a market for farmers' surpluses and jobs in rural areas of the Midwestern U.S. Since the mid-1970s, ethanol has played a primary role in Brazil's move towards energy independence. Many countries include biofuels as part of their strategy to reduce greenhouse gas emissions.

Recently, however, questions surrounding the environmental costs of biofuels have overshadowed their economic and national security achievements. Using food commodities as a feedstock for fuel can displace food production. Some speculate that the acceleration of ethanol production from grain crops not only caused the recent spike in world food prices, but also triggered destructive land use changes throughout the world. Production areas once dedicated to food crops, when displaced by fuel crop production, can move to virgin soils and cleared, native forests. Such cultivation not only releases copious amounts of soil carbon into the atmosphere, exacerbating already critical atmospheric levels of greenhouse gasses, but also threatens areas of high-value biological diversity. Environmentally damaging agronomic practices associated with fuel crop production, along with the social and economic well-being of workers and communities in developing and underdeveloped countries, also are attracting closer scrutiny.

By the mid-2000s, governments, companies, non-governmental organizations, and international bodies began to consider a definition of "sustainable" biofuel beyond merely carbon reduction to include the environmental and social effects of biomass production. "Sustainability," therefore, can be divided into "non-carbon" and "carbon." The distinction is important, as many standards delineate between the two.

The United Kingdom (U.K.) has taken concrete steps toward regulating carbon and non-carbon biofuel sustainability. Fuel suppliers in the U.K. are required to report whether the biofuels they sell in the U.K. meet both a carbon threshold and an environmental and social sustainability "meta-standard." A meta-standard sets basic requirements, and then allows the regulated community to use existing standards that have been benchmarked to the meta-standard to certify compliance. The European Union's (EU) Renewable Energy Directive (RED) and Fuel Quality Directive contain principles related to carbon reduction, biodiversity conservation, the prevention of soil, water, and air pollution, and social considerations such as labor and land rights. Although not yet implemented by the Commission, it is possible that biomass producers will be able to use existing standards to meet the RED requirements.

In the United States, regulators are just beginning to address sustainability standards, both carbon and non-carbon. The U.S. Energy Independence and Security Act (EISA) requires the Environmental Protection Agency (EPA) to develop regulations on biofuels' carbon footprint in relation to Renewable Fuel Standard (RFS) mandates, and to report on other broad-based sustainability issues. The EPA recently issued draft implementing regulations that seek comment on how certification schemes can assist in verifying and tracking "sustainable biomass." The draft regulations also seek solutions to the water, air, and soil impacts associated with increased biofuels production. The draft regulations, however, do not adopt a meta-standard approach. Draft climate change legislation contains

---

Biochar or charcoal is one current method of sequestering carbon from biomass. *See e.g.,* Darrell A. Fruth & Joseph A. Ponzi, *Adjusting Carbon Management Policies to Encourage Renewable, Net-Negative Projects such as Biochar Sequestration,* 36 Wm. Mitchell L. Rev. 992 (2010).

an amendment to the RFS containing similar requirements. California has taken more aggressive steps to curb greenhouse gas emissions, and is likely to incorporate sustainability standards for feedstock into its low carbon fuel standard by the end of 2009.

* * *

Sustainability Reporting: The RTFO Meta-standard

The RTFO's meta-standard approach takes existing, voluntary agro-environmental and social certification schemes and "benchmarks" them against the "RTFO Biofuel Sustainability Meta-Standard." The meta-standard sets minimum general principles, supporting minimum criteria (e.g., elements, conditions or processes), and performance indicators that are periodically measured to determine whether an obligated party meets all, or most, of the criteria. Benchmarking determines whether an obligated party can use part or all of the existing standard to meet the meta-standard. If the existing standard meets all of the criteria, then it is considered as meeting the full RTFO meta-standard. If it meets most, but not all of the criteria, the existing standard is said to "qualify" for the meta-standard.

The RFA's concern currently is focused on sustainability associated with biofuel feedstock production. Limiting sustainability reporting to feedstock production makes sense because sustainability standards for agricultural and forestry practices are better developed and adaptable to biofuel feedstock. Fuels derived from residues are not included in the sustainability reporting obligation. Once the reporting period ends in 2011, the RFA will reconsider extending sustainability standards to feedstock processing and transportation.

The RTFO Biofuel Sustainability Meta-Standard contains five environmental principles: (1) carbon reduction, (2) biodiversity preservation, and (3)-(5) soil, water, and air protection. The sustainability meta-standard also contains two social principles: (6) workers' rights and (7) land rights and community relations. The RFA has developed a set of minimum and recommended criteria and indicators contained in an annex to the Technical Guidance. The RFA has compared the criteria, indicators, and audit and certification qualities of a broad range of existing standards to the RFTO Meta-Standard. The RFA did not benchmark cross-compliance measures required under the EU Common Agricultural Policy (CAP) because member states implement measures differently and do not inspect annually all farms. Many of the benchmarked standards, however, contain requirements that producers comply with EU cross-compliance measures.

* * *

Biofuels Sustainability Efforts in the U.S.

Like its failure to lead in reducing global warming during the Bush Administration, the U.S. has fallen behind, in relation to the U.K. and EU, in developing biofuels sustainability standards. Instead, California has taken the lead in GHG regulation, and in recognizing the importance of biomass sustainability for the entire energy sector. The Council on Sustainable Biomass Production (CSBP), a private stakeholder group, also is in the process of developing sustainability standards for biomass, although no public, draft standard has been issued.

Federal-Level Initiatives

The U.S. Government began considering biofuels sustainability in the mid-2000s. In 2005, the U.S. Energy Policy Act (2005 EPAct) created the Renewable Fuels Standard (RFS). The RFS requires that growing amounts of renewable fuels be blended into the U.S. fuel supply. The Act did not contain any requirements vis-a-vis GHG emissions of fuels. The

2005 EPAct also added references to sustainability objectives in biomass research programs, although "sustainability" is not defined in the Act.

The 2007 Energy Independence and Security Act (EISA) created a second RFS (RFS2), which requires blending mandates to be satisfied with biofuels that achieve certain minimum levels of GHG reduction. Renewable fuels under RFS2 also must be made from "renewable biomass." Fuels from renewable biomass only qualify under the RFS2 if the biomass is derived from land that complies with certain use restrictions. For example, biomass cannot be sourced from land cleared or cultivated prior to EISA's enactment. Use-restrictions also will depend on EPA's final definition of terms in EISA, including what qualifies as "agricultural land," "fallow," "nonforested," "ecologically sensitive forestland," and restrictions regarding the use of forest residue ("slash") consisting of certain "ecological communities." EISA also requires the EPA to report every three years on the environmental impacts of the RFS, including air, soil and water quality, and "resource conservation issues."

On May 26, 2009, the EPA issued a Notice of Proposed Rulemaking for EISA implementation (Notice). In addition to the much-anticipated direct and indirect land use calculations for the GHG emissions of various biofuels, the Notice solicits comments on how the EPA can verify and track that biofuels are made with "renewable biomass." The EPA notes that in the search for a solution, it reviewed third-party agricultural and forestry certification standards. It concluded, however, that these standards (e.g., RSPO, Basel Criteria, RSB, FSC/SAN) are more comprehensive and stringent than what is necessary under EISA. The EPA points out three issues with using existing or developing standards to police EISA's land use restrictions: the limited types of biomass certified, the small number of forest acres under certification, and the imperfect match between EISA and third party standards' definitions....

* * *

California

California has been a surrogate for U.S. leadership in regulatory attempts to combat climate change. California Assembly Bill (A.B.) 32, The Global Warming Solutions Act of 2006, spearheads a multi-faceted and comprehensive regulatory scheme that includes, among other strategies, a Low Carbon Fuel Standard (LCFS) and a sustainable biomass roadmap. The A.B. 32 Scoping Plan guides GHG strategy.

The final draft of the LCFS was issued in March 2009. Like the draft amended EU Fuel Quality Directive, conventional and renewable fuel producers and importers of fuel sold in California must meet a yearly lowered fuel carbon intensity standard, beginning in 2011. The LCFS establishes default values for various fuels, based on the production pathway of a feedstock to fuel, using the California-modified Greenhouse Gases, Regulated Emissions, and Energy in Transportation (GREET) model for calculating direct emissions. Carbon intensity is measured on a life-cycle basis, based on the unit of energy produced. In the draft regulation, the California Air Resources Board (CARB) adopts the Global Trade Analysis Project (GTAP) model to calculate carbon emissions resulting from indirect land use change for fuels made from Brazilian sugar cane or U.S. corn feedstock, and it has incorporated this value into default carbon intensities. The resulting increase in carbon intensity is significant—so significant that CARB states that it must include the values to achieve the reductions necessary. The LCFS compares the carbon intensity of alternative fuels to gasoline and diesel. Obligated parties may also calculate carbon intensity by modifying the GREET model, either by modifying the calculation process or providing an alternative pathway.

* * *

Concluding Thoughts: Reasons For and Against the Meta-Standard Approach in the U.S.

Although the federal government has recognized that sustainability is critical to the viability and credibility of biofuels, other countries and international efforts have taken the lead in developing sustainability standards. The U.K. already has a regulatory scheme in place that measures the carbon and non-carbon sustainability of biofuels. In a short time, the U.K.'s reporting requirement will transition to an affirmative obligation on suppliers to source sustainable fuels. New EU directives on renewable energy and fuel quality were finalized recently, and the EU Commission will begin working on a system similar to that of the U.K. that contains a carbon intensity reduction and perhaps a meta-standard for other sustainability issues. Several efforts are ongoing at the international level to devise sound bases for both carbon calculation and non-carbon sustainability criteria.

* * *

U.S. and California regulators must confront the future need for increased supplies of sustainably-sourced biomass (and biofuels) from developing and third-world countries with less developed agro-environmental regulation. While regulators cannot disadvantage domestic biomass producers by applying less-stringent standards to imported biomass, many existing standards provide for continuous improvements in developing and third-world countries that can benefit the environment long-term while gradually satisfying sustainability standards. Benchmarking existing standards allows for different processes for raising biomass in third-world countries while achieving similar results. Environmental meta-standards could serve as a temporary proxy for countries that have yet to develop carbon regulations, as they arguably result in carbon reduction through improved agricultural practices and avoidance of high-value conservation areas with stored carbon reserves. With the income from increased biomass and/or biofuels exports, developing nations have the opportunity to build regulatory capacity.

Many standard-setting bodies (the RFA in the U.K. and the RSB) already have developed principles, criteria, and indicators for a biofuels sustainability scheme that the U.S. could draw upon. Industry has proven (e.g., Greenergy, SEKAB) that an "in-house" sustainability scheme is possible and even desirable, although these standards involve cost and still may require third-party verification. The meta-standard approach allows for the development of industry-initiated standards, while ensuring that important national and state principles guide standards. A meta-standard approach works particularly well in countries with well-developed environmental, agricultural and labor regulations and capacity.

... The ACCS [Assured Combinable Crops Scheme—Ed.] and Genesis QA qualifying standards under the RTFO incorporate relevant U.K. laws and regulations that govern agro-environmental practices and labor practices, in addition to other criteria. This gives an advantage to domestic feedstock producers, as agricultural enterprises must already comply with these laws. Pushback is likely from the conventional agricultural sector to the extent that good agricultural practices are not required or verified. Conventional producers fear that sustainability verification for biofuels will spread to other forms of agricultural production. The environmental impacts of agriculture are real and well understood, and stem in large part from systemic failure to require, assess, and verify good practices or stewardship claims. Meta-standards establish and verify minimum good practices for soil, air and water quality, and typically strive for continuous improvement. To the extent conventional agriculture maintains that it already follows sustainable practices, then meta-standards present the opportunity to verify sustainability claims to the marketplace, thus adding value in the long-term.

* * *

The meta-standard approach is not without weaknesses. The U.K. has not seen great initial success in fulfillment of environmental and social standards, particularly for biomass sourced from third countries. Although agro-environmental and labor standards may be well-developed in importing nations such as the U.K., EU, and U.S., relying only on existing agro-environmental laws may merely reinforce the status quo. In developing nations, agro-environmental regulation may be underdeveloped. Without consideration for the realities on the ground in developing countries regarding capacity, even a meta-standard may not be achievable.

... On the other hand, criteria and indicators in some certification systems may be too broad and vague. Further refinement of criteria and indicators to meet local conditions becomes costly and difficult, particularly in developing nations without agricultural, economic, and environmental institutional capacity. Macro-level concerns, such as food security, are not within the competence of private certifiers, and instead are best gauged by national governments rather than a meta-standard. A certification scheme can, however, provide governments with valuable, actual information on food crop displacement.

By its very nature, the meta-standard approach does not apply easily to carbon intensities of a given biofuel. But benchmarked standards can utilize principles and criteria that encourage, or require, agronomic practices that reduce carbon emissions and increase sequestration. Further, criteria and indicators of benchmarked standards need continual reassessment as scientific knowledge grows. This holds particularly true in the area of carbon reduction and biodiversity. No meta-standard or its benchmarked standards can truly assure that its carbon or biodiversity criteria are met without further scientific research. For example, there needs to be "an explicit calculation of the ecological footprint caused by large-scale cultivation of a given biofuel energy crop...."

... Meta-standards can assist in these efforts by pushing existing standards to assess environmental conditions and to identify courses of action that reach desired results, which in turn provides much needed data as to the environmental effects of biofuel production. Data gathered through diverse certification systems also can provide policy makers with information on the possible environmental benefits of biofuel production, such as creating habitat. Neither the U.K.'s meta-standard, nor its benchmarked standards, explicitly considers feedstock species selection in relation to the environmental and biodiversity benefits they may provide. They also do not consider the relation of food crops as biofuels to food insecurity issues. Once these issues are better understood, meta-standards could add species preferences.

Although scientists may believe that "biofuel sustainability has environmental, economic, and social facets that all interconnect," fitting this interconnectedness into legal standards undoubtedly will be challenging. Important questions arise as to whether such standards can truly incorporate countries' differing notions of social and economic welfare into a legal regime that governments can monitor and enforce. Although an international standard might be optimal, given the diverging views on criteria and indicators, it is unlikely that biofuel exporting and importing countries will reach consensus any time soon. With that realization, and armed with the lessons learned in the U.K. and EU, the U.S. and California can devise sustainability schemes that are economically feasible for a developing biofuels industry and that achieve greater sustainability.

In 2009, Congressmen Waxman and Markey introduced the American Clean Energy and Security Act of 2009 (ACES). On May 21, 2009, the Energy and Commerce Committee approved the bill. The bill defines a "renewable energy resource," as it relates to biofuel, as one that derives exclusively from "renewable biomass." The bill contains definitions, for purposes of a combined efficiency and renewable electricity standard, of "renewable biomass" and "high conservation land." These definitions are different from those contained in EISA and the EPA's draft regulations. The bill recognizes this fact and adds an amendment to the Clean Air Act replacing EISA definitions with ACES definitions. Verification approaches, however, are not added or clarified in ACES. Therefore, it appears that sustainability schemes for biomass will continue to depend on the EPA rulemaking pursuant to the RFS, at least until ACES becomes law and the EPA develops implementing regulations.

Notes and Questions

1. There are seven factors currently included in the U.K.'s Renewable Transport Fuels Obligation (RTFO) Biofuel Sustainability Meta-Standard. Would you add any others? How do you think each standard should be measured—especially the last two based on social principles? Do you think the U.K.'s Assured Combinable Crops Scheme and the Genesis Quality Assurance under the RTFO are sufficient?

2. In the full article, Endres recognizes some of the weaknesses of the meta-standard approach. She notes, "The U.K. has not seen great initial success in fulfillment of environmental and social standards, particularly for biomass sourced from third countries." Endres, *supra*, at 118. Do you think relying on existing agro-environmental laws may prevent developing countries from improving? Can the certification scheme at least provide helpful information about food crop displacement?

3. What are the difficulties with using a meta-standard approach for carbon intensities? Can it encourage agronomic practices to reduce carbon emissions and increase sequestration?

4. Finally, Endres notes that "'biofuel sustainability has environmental, economic, and social facets that all interconnect,' [but] fitting this interconnectedness into legal standards undoubtedly will be challenging." Endres, *supra*, at 119. Do you agree? Do you have additional suggestions to make the meta-standard approach effective?

5. Several bills were introduced during the 111st Congress to amend Section 201(o)(1)(I) of the Clean Air Act, 42 U.S.C. §7545(o)(1)(I), to change the definition of "renewable biomass." None of them went beyond the committee stage. What do you think of the following proposed amendment?

S. 636

To amend the Clean Air Act to conform the definition of renewable biomass to the definition given the term in the Farm Security and Rural Investment Act of 2002.

Be it enacted by the Senate and House of Representatives of the United States of America in Congress assembled,

SECTION 1. DEFINITION OF RENEWABLE BIOMASS.

Section 211(o)(1) of the Clean Air Act (42 U.S.C. 7545(o)(1)) is amended by striking subparagraph (I) and inserting the following: (I) RENEWABLE BIOMASS—The term 'renewable biomass' means (i) materials, pre-commercial thinnings, or invasive species from National Forest System land and public lands (as defined in section 103 of the Federal Land Policy and Management Act of 1976 (43 U.S.C. 1702)) that (I) are byproducts of preventive treatments that are removed (aa) to reduce hazardous fuels; (bb) to reduce or contain disease or insect infestation; or (cc) to restore ecosystem health; (II) would not otherwise be used for higher-value products; and (III) are harvested in accordance with (aa) applicable law and land management plans; and (bb) the requirements for (AA) old-growth maintenance, restoration, and management direction of paragraphs (2), (3), and (4) of subsection (e) of section 102 of the Healthy Forests Restoration Act of 2003 (16 U.S.C. 6512); and (BB) large-tree retention of subsection (f) of that section; or (ii) any organic matter that is available on a renewable or recurring basis from non-Federal land or land belonging to an Indian or Indian tribe that is held in trust by the United States or subject to a restriction against alienation imposed by the United States, including (I) renewable plant material, including (aa) feed grains; (bb) other agricultural commodities; (cc) other plants and trees; and (dd) algae; and (II) waste material, including (aa) crop residue; (bb) other vegetative waste material (including wood waste and wood residues); (cc) animal waste and byproducts (including fats, oils, greases, and manure); and (dd) food waste and yardwaste.

Chapter 6

Geothermal

Geothermal hot springs have long been special places. Archaeologists have evidence that over 10,000 years ago, Paleo-Indians gathered to them as sacred sites. When modern settlers explored the North American continent, they also laid claims to these spring areas because of their warmth and reputed healing properties. Early efforts to harness geothermal energy focused on locations where underground steam emerged at the earth's surface in the form of geysers or hot springs. Because there are relatively few of these locations, some believed that the United States' geothermal resources had been exploited to their fullest.

More recently, however, U.S. Secretary of Energy Steven Chu stated that the potential for geothermal energy is "effectively unlimited." Research shows that geothermal resources are available throughout the entire United States wherever the earth's temperature exceeds 300° Fahrenheit (approximately 150° Celsius). *See* Figure 6.1. In addition, new technologies are being developed to convert this energy into useful heat or electricity.

Recognizing the vast potential for newer Enhanced Geothermal System (EGS) technologies to generate power from otherwise unproductive geothermal resources, the Department of Energy has increased the funding for geothermal research. Only $3.5 million was appropriated in FY 2003, but the DOE issued Funding Opportunity Announcements for $20 million in FY 2009 and an additional $80 million specifically targeting EGS projects from the American Reinvestment and Recovery Act.

A. History

Section 1 explains the geology of geothermal systems and the history of how humans have harnessed them. Section 2 provides more detail about one particular area, The Geysers in northern California, which is currently the world's largest producer of geothermal electricity.

1. Geology & Geothermal Systems

This initial section provides information explaining the geologic origins of geothermal resources as well as a brief history of how humans have appreciated their benefits.

Figure 6.1

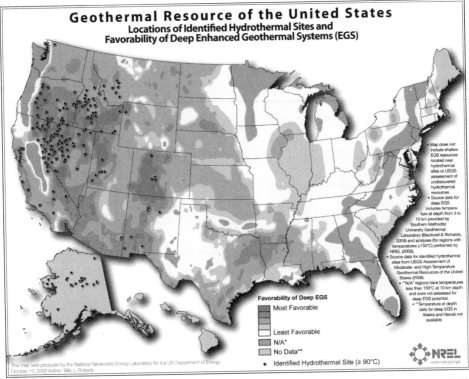

Billy J. Roberts, *Geothermal Resource of the United States,* National Renewable Energy Lab., (Oct. 13, 2009),
http://www.nrel.gov/gis/images/geothermal_resource2009-final.jpg

Wendell A. Duffield & John H. Sass, *Geothermal Energy: Clean Power from the Earth's Heat*

U.S. Dep't Of Interior, U.S. Geological Survey Circular 1249, at 1-7 (2003)

For centuries, people have enjoyed the benefits of geothermal energy available at hot springs, but it is only through technological advances made during the 20th century that we can tap this energy source in the subsurface and use it in a variety of ways, including the generation of electricity. Geothermal resources are simply exploitable concentrations of the Earth's natural heat (thermal energy). The Earth is a bountiful source of thermal energy, continuously producing heat at depth, primarily by the decay of naturally occurring radioactive isotopes—principally of uranium, thorium, and potassium—that occur in small amounts in all rocks. This heat then rises to and through the Earth's surface, where it escapes into the atmosphere. The amount of heat that flows annually from the Earth into the atmosphere is enormous—equivalent to ten times the annual energy consumption of the United States and more than that needed to power all nations of the world, if it could be fully harnessed.

Even if only 1 percent of the thermal energy contained within the uppermost 10 kilometers of our planet could be tapped, this amount would be 500 times that contained in all oil and gas resources of the world. How might we benefit from this vast amount of thermal energy beneath our feet? Where, by what means, and how much of the Earth's natural heat can be usefully harnessed? These are especially important questions to contemplate, because global population is expected to soon exceed seven billion and many scientists believe that the world's fossil-fuel resources may be substantially depleted within this century. Faced with such prospects, both the public and private sectors are working toward more fully utilizing the Earth's abundant thermal energy and other alternative energy resources.

A skeptic might question the wisdom of devoting much national effort to geothermal energy development, especially because many experts think that geothermal heat can contribute at most about 10 percent to the Nation's energy supply using current technologies. However, ongoing advances in exploration and heat-extraction technologies are improving our ability to use the resource and may substantially increase the geothermal contribution to the Nation's energy supply.

* * *

Geothermal in Ancient Times

Long before recorded history some ancient peoples must have been aware of geothermal features such as hissing steam vents, erupting geysers, boiling mud pots, and bubbling hot springs. One can only speculate on their reactions to such impressive natural phenomena, but some combination of fear, awe, and appreciation seems likely. By the time of recorded history, hot springs and other geothermal features were being used by people for food preparation and for bathing. The geothermally heated spas of the ancient Greeks, Romans, and Japanese have been imitated throughout history and today their modern counterparts attract many visitors for recreational and medical reasons.

Prehistoric and early historical uses of geothermal features were effectively limited to those found at the Earth's surface. With rare exception, such features produce water or steam with temperatures of less than 100°C (the boiling point of water at sea level); their relatively low temperatures restrict the variety of possible uses. Lack of knowledge and technical limitations prohibited any attempt to develop deeper, hotter geothermal energy. Still, many early civilizations benefited from the geothermal resources with which they were provided by nature.

Geothermal Now

With modern technology, drills can penetrate thousands of meters into the Earth in search of geothermal resources. Such drilling has resulted in the discovery of geothermal fluid as hot as 500°C, which can provide a resource of high-pressure steam to drive turbine generators at the Earth's surface. The traditional, ancient uses of geothermal water continue to have considerable scenic and recreational value, but the present-day capability to produce high-temperature fluid through drilled wells opens the door to diverse utilization of geothermal energy over a broad range of temperatures. Information gathered from measurements made during flow testing of geothermal wells can indicate how much power they can provide. A "typical" commercial geothermal well can power between 5 and 8 megawatts of electrical generation capacity (1 megawatt =1,000 kilowatts = 1 million watts).

Research and development have shown that the size and vigor of geothermal surface features are not necessarily representative of the entire subsurface system. For example, at The Geysers in northern California, where the surface geothermal features are relatively weak, drilling revealed the world's largest known reservoir of steam. Electricity generated from The Geysers geothermal field, which is fed into the regional power grid, is nearly enough to meet the energy demands of the nearby city of San Francisco. Similarly, in the Imperial Valley of southern California, rare and feeble hot springs belie the presence of many large subterranean reservoirs of hot water now partly harnessed to produce about 475 megawatts of electrical power. A major challenge for future exploration is to develop improved techniques that can help identify geothermal resources that have no surface expression whatsoever.

* * *

Geothermal Energy as a Natural Resource

Geothermal energy is present everywhere beneath the Earth's surface, although the highest temperature, and thus most desirable, resources are concentrated in regions of active or geologically young volcanoes....

* * *

Global Distribution

... Such "hot" zones generally are near the boundaries of the dozen or so slabs of rigid rock (called plates) that form the Earth's lithosphere, which is composed of the Earth's crust and the uppermost, solid part of the underlying denser, hotter layer (the mantle). According to the now widely accepted theory of plate tectonics, these large, rigid lithospheric plates move relative to one another, at average rates of several centimeters per year above hotter, mobile mantle material (the asthenosphere). High heat flow also is associated with the Earth's "hot spots" (also called melting anomalies or thermal plumes), whose origins are somehow related to the narrowly focused upward flow of extremely hot mantle material from very deep within the Earth. Hot spots can occur at plate boundaries (for example, beneath Iceland) or in plate interiors thousands of kilometers from the nearest boundary (for example, the Hawaiian hot spot in the middle of the Pacific Plate). Regions of stretched and fault- broken rocks (rift valleys) within plates, like those in East Africa and along the Rio Grande River in Colorado and New Mexico, also are favorable target areas for high concentrations of the Earth's heat at relatively shallow depths.

Zones of high heat flow near plate boundaries are also where most volcanic eruptions and earthquakes occur. The magma that feeds volcanoes originates in the mantle, and considerable heat accompanies the rising magma as it intrudes into volcanoes. Much of this intruding magma remains in the crust, beneath volcanoes, and constitutes an intense, high-temperature geothermal heat source for periods of thousands to millions of years, depending on the depth, volume, and frequency of intrusion. In addition, frequent earthquakes—produced as the tectonic plates grind against each other—fracture rocks, thus allowing water to circulate at depth and to transport heat toward the Earth's surface. Together, the rise of magma from depth and the circulation of hot water (hydrothermal convection) maintain the high heat flow that is prevalent along plate boundaries.

Accordingly, the plate-boundary zones and hot spot regions are prime target areas for the discovery and development of high-temperature hydrothermal-convection systems capable of producing steam that can drive turbines to generate electricity. Even though such zones constitute less than 10 percent of the Earth's surface, their potential to affect the world energy mix and related political and socioeconomic consequences is substantial, mainly because these zones include many developing nations.... [See the map of plate boundaries and the "ring of fire" in Figure 6.2—Ed.]

* * *

Figure 6.2: Map of tectonic plate boundaries

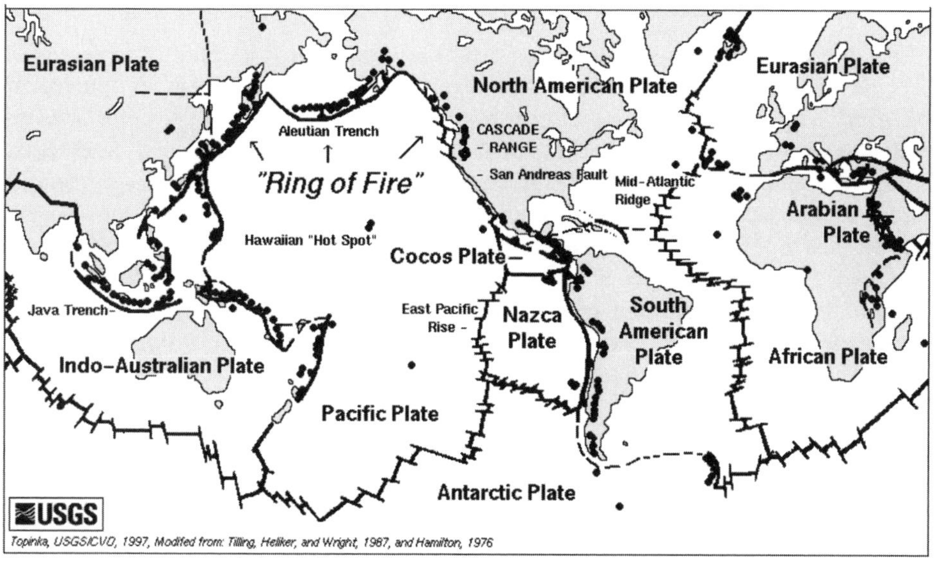

Active Volcanoes, Plate Tetonics, and the "Ring of Fire," U.S.G.S.,
http://vulcan.wr.usgs.gov/Imgs/Gif/PlateTectonics/Maps/
map_plate_tectonics_world_bw.gif (last visited Aug. 4, 2011)

In addition to a hot water-steam (hydrothermal) component, by far the majority of the thermal energy associated with a magma-volcano environment resides in the magma itself and in hot-but-dry rock around it. Using current technology, only the hydrothermal component can be exploited; meanwhile, studies are in progress to demonstrate the feasibility of possible future exploitation of the nonhydrothermal components of geothermal energy.

* * *

Comparison with Other Natural Resources

Geothermal resources are similar to many mineral and energy resources. A mineral deposit is generally evaluated in terms of the quality or purity (grade) of the ore and the amount of this ore (size or tonnage) that can be mined profitably. Such grade-and-size criteria also can be applied to the evaluation of geothermal energy potential. Grade would be roughly analogous to temperature, and size would correspond to the volume of heat-containing material that can be tapped. For mineral and geothermal deposits alike, concentrations of the natural resource should be significantly higher than average (the background level) for the Earth's crust and must be at depths accessible by present-day extraction technologies before commercial development is feasible.

However, geothermal resources differ in important ways from many other natural resources. For example, the exploitation of metallic minerals generally involves digging, crushing, and processing huge amounts of rock to recover a relatively small amount of a particular element. In contrast, geothermal energy is tapped by means of a liquid carrier-generally the water in the pores and fractures of rocks-that either naturally reaches the surface at hot springs, or can readily be brought to the surface through drilled wells. The extraction of geothermal energy is accomplished without the large-scale movement of rock involved in mining operations, such as construction of mine shafts and tunnels, open pits, and waste heaps.

Geothermal energy has another important advantage. It is usable over a very wide spectrum of temperature and volume, whereas the benefits of other natural resources can be reaped only if a deposit exceeds some minimum size and (or) grade for profitable exploitation or efficiency of operation. For example, at the low end of the spectrum, geothermal energy can help heat and cool a single residence. To do so requires only the burial of piping a few meters underground, where the temperature fluctuates little with the changing seasons. Then, by circulating water or some other fluid through this piping using a geothermal heat pump, thermal energy is extracted from the ground during the coldest times of the year and deposited in the ground during the hottest times. Together, the heat pump and the Earth's thermal energy form a small, effective, and commercially viable heating and cooling system. Heat pump systems are already in use at more than 350,000 buildings in the United States.

Toward the high end of the spectrum, a single large-volume, high temperature deposit of geothermal energy can be harnessed to generate electricity sufficient to serve a city of 1 million people or more. For example, at The Geysers in Northern California, fractures in rocks beneath a large area are filled with steam of about 240°C at depths that can easily be reached using present-day drilling technology. This steam is produced through wells, piped directly to conventional turbine generators, and used to generate electricity. With a generating capacity of about 1,000 megawatts electric, The Geysers is presently the largest group of geothermally powered electrical plants in the world. At current rates of per capita consumption in the United States, 1 megawatt is sufficient to supply a community with a population of 1,000.

Between these relatively extreme examples are geothermal resources that encompass a broad spectrum of grade (temperature) and tonnage (volume). The challenge, for governmental agencies and the private sector alike, is to assess the amount and distribution of these resources, to work toward new and inventive ways to use this form of energy, and to incorporate geothermal into an appropriate energy mix for the Nation and the world.

* * *

Notes and Questions

1. When Duffield and Sass wrote this article in 2003, hydrothermal steam was the only technically exploitable geothermal resource. Note how things have changed with recent developments addressed in Figure 6.3. Did the 2003 article anticipate some of these developments?

2. In the last portion of the excerpt, Duffield and Sass compare geothermal resources with other natural resources. What are some of the advantages of geothermal?

2. The Geysers

Because The Geysers is the world's largest geothermal generator, it provides a useful case study for how electricity is currently produced from geothermal resources.

Figure 6.3: Geothermal systems

Highest temperature: Magma environment, 650-1300° C (~1,200-2,350° F)

Magma is molten rock material in the Earth's crust. Magma is the "highest grade of geothermal ore" because it is the ultimate source of heat for all high-temperature geothermal environments in the crust, including both wet and dry systems.

Efforts have been made to tap the magma itself for energy. While tests in the laboratory and the field have shown it is technically feasible, the long-term economics are questionable because of uncertainties in the expectable life span of materials in contact with the magma and problems with accurately locating the magma body before development begins. Experiments were conducted at the Kilauea Iki Crater, Hawaii, in the 1980s, and at Sandia National Laboratories in New Mexico.

Hot Dry Rock (HDR) Systems

With decreasing temperature, the magma environment grades into hot solid rock that contains little or no available water.

Rocks at cost-effective drilling depths are hot enough to represent a huge inventory of potential geothermal energy. However, they are not naturally permeable enough to form a producing hydrothermal system. Research on developing ways to increase permeability through fracturing (fracing) may allow the exploitation of "dry rock" environments in the near future.

Enhanced Geothermal Systems (EGS)

Similar to the HDR environment, these areas are often shallower and adjacent to producing natural hydrothermal areas but have been considered uneconomic because of low permeability.

Only hydrothermal systems with sufficiently high temperatures and permeable, water-saturated rock have been commercially developed for generating electricity in the past. Research is currently underway to enhance uneconomic areas adjacent to natural hydrothermal areas by stimulating permeability through hydraulic fracturing, directional drilling to intersect favorably oriented fractures, and injecting groundwater or wastewater to replenish fluids and to reverse pressure declines.

Hydrothermal Electricity Systems

Distinguished from the categories above because not only hot, but also sufficiently porous and permeable to be saturated with fluids that mobilize the heat to generate electricity.

Approximately twenty geothermal fields in the United States generate electricity.
The three subcategories of hydrothermal electricity systems vary based on what turns the turbines:
  [a] steam (vapor dominated systems),
  [b] liquid hot water, or
  [c] a secondary fluid (using moderate-temperature water in a binary process).

a. Steam or vapor-dominated

When a potent heat source intersects with a restricted source of water, the pore spaces of rocks in a high-temperature hydrothermal system are saturated with steam, rather than liquid water, and only steam is produced through the wells and directly routed into turbine generators.

Vapor-dominated systems do not require the separation of steam from water, so the energy they contain is relatively simple and efficient to harness, making these systems the most desirable for electric power production. Vapor-dominated systems are rare compared with valuable, but less-simple-to-develop, hot-water systems. The largest vapor-dominated system developed in the world is at The Geysers in northern California.

Figure 6.3: Geothermal systems *continued*

b. Hot Water, 212-700°F

These systems are in porous and permeable rock naturally saturated with enough water to drive electric turbines. The water partly "flashes" into steam when it rises up production wells.

The hotter the hydrothermal fluids are, the more capable they are of producing steam and correspondingly generating electricity. To extract the most energy from the fluid, it sometimes can be "flashed" two or three times to drive additional turbine generators. Examples of hot-water systems are Coso and Imperial Valley in southern California.

c. Binary Systems, Below 212°F

Moderate-temperature hydrothermal systems are incapable of producing steam at high enough pressure to directly drive a turbine generator. They are, however, hot enough to produce a high-pressure vapor through heat transfer to a secondary working fluid.

A binary cycle generates power by transferring the heat from the geothermal fluid to another fluid whose boiling temperature is lower than that of water (for example, isobutene). Binary systems producing electricity include California plants at Mammoth Lakes, east of the Sierra Nevadas, and a plant in the Imperial Valley. By taking advantage of the more widespread distribution of moderate-temperature geothermal water, binary systems may contribute significantly to the overall generation of electricity from geothermal sources.

Direct Use Systems

>1,300 direct-use sites worldwide include Boise, Idaho; Klamath Falls, Oregon; Aquitaine Basin in SW France, and Iceland.

Some direct-use applications include heating for homes, businesses, greenhouses, fish farms, and dairies; industrial processing; domestic hot water heating; driveway and sidewalk snow melting; and recreational hot springs.

Low temperature Geothermal Heat Pumps (GHP)

GHP, also called Groundsource Heat Pumps or Geoexchange, employs normal ground temperatures in ordinary, relatively shallow surface rock and soil.

Not related to hotter-than-average magma systems, but instead based on solar isolation on the surface of the earth and retained because of the ground's thermal mass. Because this source of thermal energy is virtually everywhere, it represents a huge potential alternative energy source for heating and cooling individual buildings.

Wendell A. Duffield & John H. Sass *Geothermal Energy: Clean Power from the Earth's Heat*

U.S. Dep't Of Interior, U.S. Geological Survey Circular 1249, at 14 (2003)

The Geysers, a vapor-dominated hydrothermal system in northern California, has grown into the world's largest geothermal electrical development. At its peak in the late 1980s, about 2,100 megawatts of generating capacity were in operation. For comparison, 2,100 megawatts is roughly the equivalent of twice the electrical energy that can be generated by the turbines of GIen Canyon Dam, Arizona. Despite its name, there never were true geysers (periodically spouting hot springs) in the area; the surface features before drilling were restricted to weak stearn vents, warm springs, and mudpots, whose unimpressive character belied the huge resource below. Indeed, The Geysers is an unusual geothermal field in that its wells produce nearly pure steam, with no accompanying water.

It took decades for people to recognize the huge energy potential of The Geysers. The surface geothermal features were known to settlers in the region by the mid-1800s, but it was not until 1924 that the first production wells were drilled and a few kilowatts of electrical power were generated for use at a local resort. During the 1950s, wells were drilled as deep as 300 meters, and the main steam reservoir was thus discovered. At that time, however, few people had any idea that the steam reservoir could be developed to the extent that it was by the 1980s. Accordingly, development proceeded cautiously, from the first powerplant of 12 megawatts electric in 1960 to a total installed capacity of 82 megawatts by 1970. Major growth during the 1970s brought the electrical capacity to 943 megawatts by 1980, and even faster growth during the 1980s pushed capacity to over 2,000 megawatts by the end of the decade. Twenty-six individual powerplants had been constructed by 1990, ranging from 12 to 119 megawatts. More than 600 wells had been drilled by 1994, some as deep as 3.2 kilometers, and capital investment by then was more than $4 billion.

Located in mountainous, sparsely inhabited terrain approximately 120 kilometers north of San Francisco, the production area at The Geysers geothermal field is distributed over nearly 80 square kilometers and is surrounded by an area 10 times as large in which the amount of heat flowing upward through the Earth's crust is anomalously high. The Geysers is located southwest of and adjacent to the Clear Lake volcanic field, whose most recent volcanic eruptions occurred only a few thousand years ago. Accordingly, it is likely that The Geysers geothermal field is sustained by hot or molten rock at depths of 5 to 10 kilometers.

As a result of the rapid development at The Geysers during the 1980s and some subsequent but slower development, there has been a decline in the rate of steam production (and electrical generation) due to loss of pressure in production wells. Steam production peaked in 1988, and has declined since then.

Most of the geothermal energy of this system remains intact, stored in hot rocks that constitute the hydrothermal reservoir. A team of private industry and governmental agencies has devised a clever and effective solution to mitigate the decline of steam pressure in production wells and thereby extend the useful life of the resource. The solution also addresses how best to dispose of increasing volumes of wastewater from nearby communities. Simply put, the wastewater of treated sewage is injected underground through appropriately positioned wells. As it flows toward the intake zones of production wells, this wastewater is heated by contact with hot rocks. Production wells then tap the natural steam augmented by vaporized wastewater.

By 1997, a 50-kilometer-long pipeline began delivering about 30 million liters of wastewater a day for injection into the southern part of The Geysers geothermal field. This quickly resulted in the recovery of 75 megawatts of generating output that had been "lost" to the preinjection pressure decline.

This initial injection experiment is considered so successful that construction of a second pipeline is on schedule to deliver another 40 million liters a day by late 2003, to the central part of the field. Together, these two sources of "make-up" water will replace nearly all of the geothermal fluid being lost to electricity production. The injection program is expected to maintain total electrical output from The Geysers at about 1,000 megawatts for at least two more decades, and possibly much longer. The Geysers injection project shows how once-troublesome wastewater can produce electricity by one of the world's most Earth-friendly means. Industry, sanitation districts, the public, and the environment all win.

―――――――

Notes and Questions

1. The current average generating capacity of large coal-fired power plants in the United States is 500 to 1,000 MW. A large natural gas plant is usually in the 500 MW range, and Horse Hollow II wind farm in Texas, the nation's largest, has a generating capacity of 299 MW. (The current average U.S. terrestrial wind farm has a generating capacity of closer to 150 MW with actual output of 50 MW because of intermittency). How does The Geysers' generating capacity compare—at its peak in the late 1980s and currently?

2. The rapid development of oil and gas fields without appropriate controls can result in pressure drops preventing full recovery of the resource. Similarly, The Geysers suffered a decline in the rate of steam production and electricity generation because of drops in the pressure of the production wells. What are some of the ways the industry addressed this problem at The Geysers? How does this solution seem to create a win-win situation? Do you see any downsides or alternatives?

B. Legal Definitions of Geothermal Resources

Heat, or thermal energy, occurs in a wide range of geologic environments that sometimes transition from one to another without distinct boundaries. *See* Figure 6.3. The most common methods of classifying geothermal resources are based on the temperature of the resource and whether there is a transport mechanism—such as water in the liquid or gaseous state—bringing that heat energy to the surface.

How a geothermal resource is defined may impact how it is regulated. Section 1 will address the principal legal definitions for larger economic uses of geothermal resources. Section 2 discusses Geothermal Heat Pump or Geoexchange technologies that exploit the relatively constant temperatures of the earth just below the surface.

1. Hydrothermal & Direct-Use

Traditional geothermal resources were developed from above-average heats of the earth. There are five existing economic applications for the use of geothermal energy. The first four are used to produce electricity: (1) dry steam systems, (2) hot water systems, (3) hybrid geothermal brine systems, and (4) hot dry rock systems. The fifth current application is the use of low temperature geothermal waters for direct heating of buildings, domestic hot water, or spas. The Kochan & Grant excerpt in this section explains the legal definitions of each of these applications under federal law. Sometimes a state definition will vary from the federal definition. The Duffield & Sass excerpt here provides a bit more detail about developments at the lower end of the temperature spectrum.

Donald J. Kochan & Tiffany Grant, *In the Heat of the Law, It's Not Just Steam: Geothermal Resources and the Impacts on Thermophile Biodiversity*

13 Hastings W.-N.W. J. Envtl. L. & Pol'y 35, 40-45 (2007)

Tectonic Relationship of Earth, Plates, Magma, and Groundwater

Although the earth is primarily composed of solid rock, the extreme heat radiated from the core causes the inner earth to take on fluid-like characteristics; the material constantly shifts and rearranges itself. As mass moves away from the core and toward the surface, it begins to cool and takes on more rigid characteristics. At the surface, the cooled material forms into plates. These plates, in essence, float on top of the lower fluid layers of the earth. However, the plates do not stay in one place. They move around and collide into each other. Where two plates converge, one plate will move downwards in a process called subduction. Plates can also move apart and grow when magma comes to the surface to create new plate material; that process is termed divergence. Two plates can also slide past each other, and that movement is termed "transform boundaries." It is these processes that produce geothermal resources. A geothermal resource is made up of three parts: heat, water, and minerals. Magma, the molten rock that carries heat, is usually found deep below the earth's surface. However magma chambers close to the earth's surface can be found in areas of high tectonic activity, such as convergence zones along the Pacific Coast of California. When the magma chamber is sufficiently close to the earth's surface, the heat from the chamber heats the surrounding groundwater. The addition of high heat to water and rock causes chemical metamorphosis of minerals, some of which dissolve into the geothermal water. Thus, geothermal resources tend to have high mineral compositions. If there is a path for the groundwater to directly reach the earth's surface, the resulting flow of heated water is termed a hot spring, fumarole, or geyser, depending upon the specific characteristics of the flow.

Hot springs are the most common surface manifestations of geothermal activity and occur when there is a flow of water to the surface. They provide easy access to geothermal fluids and thus tend to be the most studied. A fumarole occurs when less water is present, and the vent emits gases and vapor instead of liquid water. Hot springs that undergo dry periods may become fumaroles. Hot springs and fumaroles can be important surface manifestations of geothermal fields, the term used to define areas of high heat concentrations and flow. Much of the categorization and local regulation of geothermal resources is based on the definition of various geothermal fields.

The composition and temperature of the reservoir determines to what economic application category the geothermal resource belongs: dry stream, hot water, hybrid geothermal brine, hot dry rock, and low temperature water. Resources may also be defined by physical states: liquid, vapor, or a mixture of liquid and vapor termed "dominant phase." Dominant phase resources are utilized in hybrid or "combined" systems.

Dry Steam Systems

When the resource emits predominantly vapor, the resource is defined as a dry steam geothermal resource. Dry stream resources are the most readily usable geothermal resource for generating electricity. Basically, a well is drilled to access the geothermal dry steam in a reservoir, which in turn passes through the drilled hole to the surface. Once at the surface, the steam expands and drives a steam turbine. Typically, after passing through a turbine, the steam is discharged to a condenser and mixed with cool water. The heated water is then pumped to a cooling tower where most of the condensate is evaporated. The water is then returned to the condenser. The excess water that does not evaporate is re-injected into the reservoir through re-injection wells.

Because it produces energy directly, dry steam is easily utilized. Also, due to its vaporous nature, it has a less corrosive effect on wells and machinery used in geothermal energy plants than other methods of production. However, the geological conditions that create dry steam fields are rare and only utilized in a few commercial fields globally.

One such field is Geysers Field in California. Located in Napa County, Geysers Field has been operating a dry steam turbine system since 1960....

Hot Water Systems

Most geothermal resources consist of liquids stored in reservoirs at very high temperatures and pressures. When the temperature of the water is higher than the boiling point, 212 [degrees] F, it is termed "hot water." It remains in liquid form only as long as it is subject to extreme underground pressure. Hot water resources can generate electricity through flash steam or binary processes.

Flash steam energy generation is used when water temperatures exceed 350 [degrees] F. The process takes advantage of the ability of highly-pressurized geothermal liquids to "flash" into steam as they reach the surface. In this system, the water is brought to the surface through a well and allowed to flash; the resulting steam drives a turbine. [Figure 6.4 illustrates how the three main systems convert hydrothermal heat into energy.—Ed.]

Hybrid Geothermal Brine Systems

In geothermal reservoirs with concentrated saline solutions, minerals from the surrounding rock leach into the solution. These solutions are termed "geothermal brines." Some of these brines contain a mixture of hot pressurized water and natural gas. Hybrid power systems are capable of generating electricity from both of the resources.

One plant that uses geothermal brines is the United States Department of Energy's Pleasant Bayou Hybrid Power System in Bayou, Texas, which started generating electricity in 1990. The Bayou plant first extracts natural gas from the geothermal brines. The natural gas is then burned in a gas engine to directly generate electricity. The exhaust heat from the gas engine is then combined with the geothermal brine heat to generate additional electricity.

Hot Dry Rock Systems

Hot dry rock technology is an artificial geothermal resource. Hot dry rocks are usually granitic, found at depths of 8,000 to 20,000 feet, and have high heat production. To exploit the heat generated at these locations, a well is drilled and water is injected into the rock at extremely high pressures to fracture the surrounding rock, thereby creating a reservoir. The injected water is subsequently heated as it flows through the rock. Secondary wells are then drilled to extract the heated water and generate electricity. The water cooled from electrical generation is then re-circulated and re-injected into the artificial geothermal reservoir. [Figure 6.5 illustrates a hot dry rock system—Ed.]

Low Temperature Water Systems

Low temperature geothermal water is any geothermal resource cooler than the boiling point of water. These waters are currently incapable of efficiently generating electricity.

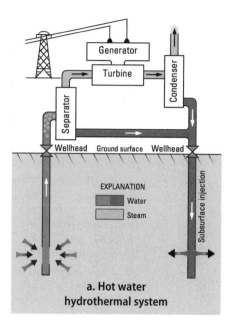

a. Hot water hydrothermal system

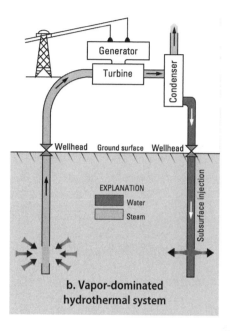

b. Vapor-dominated hydrothermal system

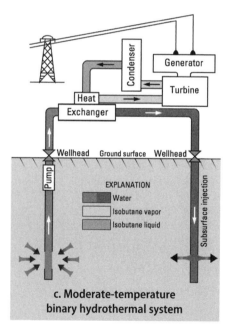

c. Moderate-temperature binary hydrothermal system

Figure 6.4: Three different hydrothermal systems

Diagrams showing how electricity is generated in (a) hot-water, (b) vapor-dominated, and (c) moderate-temperature hydrothermal systems. In a hot-water system, the part of the hydrothermal water that flashes to steam is separated and used to drive a turbine generator. In a vapor-dominated system, steam from the well is used directly to drive a turbine generator. A moderate-temperature binary system uses geothermal water to boil a second fluid (isobutane in this example), the vapor from which then drives a turbine generator. In all three systems, wastewater is injected back into the subsurface to help extend the useful life of the hydrothermal system.

Wendell A. Duffield & John H. Sass, U.S. Geological Survey, Geothermal Energy — Clean Power From the Earth's Heat 11 (2003), *available at* http://pubs.usgs.gov/circ/2004/c1249/c1249.pdf.

However, these resources can be used for direct heating applications. Communities have utilized direct heating to heat buildings in the United States since the 1800s. This resource can also be used in mineral spas and for heat in commercial applications, including greenhouses and food processing.

Figure 6.5: Hot dry rock system

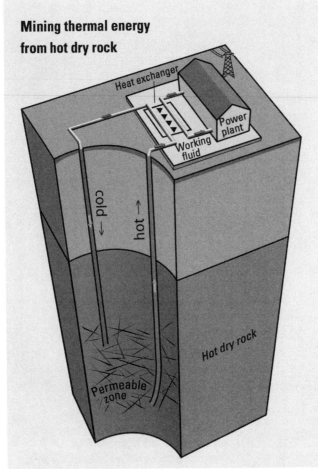

Mining thermal energy from hot dry rock

Two wells are drilled into the rock, and a permeable zone (pattern near base of wells) is then created by hydraulic fracturing. Cold water is then pumped down one well, and returns as hot water through the second well. At the surface, thermal energy is extracted in a heat exchanger and transferred to a working fluid, then the cooled water begins another circulation cycle.

Wendell A. Duffield & John H. Sass, U.S. Geological Survey, Geothermal Energy — Clean Power From the Earth's Heat 22 (2003), *available at* http://pubs.usgs.gov/circ/2004/c1249/c1249.pdf.

Wendell A. Duffield & John H. Sass, *Geothermal Energy: Clean Power from the Earth's Heat*

U.S. Dep't Of Interior, U.S. Geological Survey Circular 1249, at 16-17 (2003)

Electricity From Moderate-Temperature Hydrothermal Systems

Many hydrothermal systems contain water too cool to directly power steam-driven turbine generators, yet hot enough to boil another fluid whose vapor can drive a turbine. This method of power production—called a binary system—utilizes the combined properties of geothermal water and a second so-called "working fluid" in the energy-conversion cycle. A geothermal development employing binary technology with isobutane as the working fluid is currently in operation near Mammoth Lakes, east of the Sierra Nevada in central California.

Three binary-cycle powerplants near Mammoth Lakes account for a total net generating capacity of 37 megawatts electric. The plants are located within Long Valley Caldera, which contains numerous hot springs and other surface features of an active hydrothermal system. Wells, each about 200 meters deep, supply the binary powerplants with 170°C water. Heat exchangers transfer thermal energy from this water to the isobutane, which vaporizes and drives the turbine generators, is then condensed and revaporized to repeat the turbine-driving cycle.

The geothermal water for this development is kept liquid using pumps to maintain appropriate pressure and is injected through wells back into the subsurface reservoir once heat has been transferred to the isobutane. This injection avoids problems, such as chemical precipitation, often associated with boiling of geothermal water, and it minimizes decline in reservoir pressure while maximizing recovery of thermal energy stored in rocks of the reservoir. A flow rate of about 1,000 kilograms per second of geothermal water from the production wells is required for maximum powerplant output. In general, binary-cycle powerplants can produce electricity profitably from geothermal water at temperatures as low as 100°C.

The geothermal powerplants near Mammoth Lakes are built on both private and public lands. A portion of the revenue generated from these plants is returned through taxes and royalties to local, State, and Federal agencies to offset costs incurred in permitting and regulating their operation. The geothermal development is in an environmentally sensitive region that is a popular year-round resort. A program designed to monitor the effects of development on the local environment has been successfully implemented through a co-operative effort among the private developer, various regulatory agencies, and the U.S. Geological Survey.

* * *

Warm-Water Systems: Direct Use

Before the development of high-temperature drilling and well-completion technology, geothermal resources were limited to nonelectrical (that is, direct-use) applications. Thermal water too cool to produce electricity can still furnish energy for direct uses that range from heating swimming pools and spas, to heating soil for enhanced crop production at cool-climate latitudes, to heating buildings. The total capacity for direct use in 2000 amounted to about 600 megawatts thermal nationwide, substituting annually for the equivalent energy from 1.6 million barrels of petroleum. Worldwide, comparable figures are 11,300 megawatts thermal and 20.5 million barrels of petroleum. Low-temperature geothermal water is a relatively low-grade "fuel" that generally cannot be transported far without considerable thermal-energy loss, unless piping is extremely well-insulated and rate of flow through the piping is rapid. Yet, much of the world geothermal energy supply is consumed for direct-use applications. Warm-water systems—the most widely distributed of the hydrothermal systems—can locally compliment or supplant conventional energy sources. [Figure 6.6 illustrates the use of relatively low temperature hydrothermal waters by a spa. Note the deposits of minerals (geyserite or travertine) that precipitated out of the hot waters as they cooled—Ed.]

Extensive development of the warm-water systems, most commonly found in volcanic areas but also in a few non-volcanic areas, can significantly improve the energy balance of a nation. For example, the use of geothermal water for space heating and other direct-use applications in Iceland substantially benefits the economy of that nation. Similarly,

Figure 6.6: The Springs Resort and Spa, Pagosa Springs, Colorado

Photograph by K.K. DuVivier

people living in Klamath Falls, Oregon, and Boise, Idaho, have used geothermal water to heat homes and offices for nearly a century though on a smaller scale than in Iceland.

Great potential exists for additional direct use of geothermal energy in the Western United States. It might be advantageous for industry and municipalities to invest in installation (particularly retrofit) costs if energy prices stay at or near their current levels. To date, only 18 communities in the Western United States have geothermal district heating systems, whereas more than 270 communities have geothermal reservoirs suitable for the development of such systems.

———

Notes and Questions

1. Review the descriptions above and in Figure 6.3. Make sure you understand the distinctions because each difference can result in unique legal issues. For example, how do you think water disposal or contamination issues might vary between hot water systems and hot dry rock systems?

2. Look again at the technologies chart in Figure 6.3. The five applications discussed above are currently in use as energy resources. Research technologies are expanding the potential viability of geothermal reserves at both ends of the spectrum. New secondary fluids in a binary cycle generator may produce electricity from lower temperature waters, and hot dry rock experiments are making production from hotter, deeper, and drier

deposits more promising. Consider some of the additional issues raised by the systems included in the chart that are still in the development stage.

3. In 2004, Americans consumed .311 quadrillion Btus of electricity from geothermal sources and only .142 quadrillion from wind. Those numbers had flipped by 2008, with geothermal pretty steady at .314 quadrillion, but wind at .546 quadrillion. With so much potential, why do you think geothermal generation remains flat in comparison to other alternative energy sources?

2. Geothermal Heat Pumps

This section addresses Geothermal Heat Pump (GHP) technology, which varies from traditional geothermal energy in several respects. First, it is not related to hotter-than-average magma systems. Instead, it employs normal ground temperature reservoirs in ordinary, relatively shallow surface rock and soil. Second, GHPs do not provide just heat in winter but also cooling in the summer. Finally, because this source of thermal energy is virtually everywhere, in both dry and saturated environments, GHPs represent a huge potential alternative energy source for heating and cooling individual buildings.

Wendell A. Duffield & John H. Sass, *Geothermal Energy: Clean Power from the Earth's Heat*

U.S. DEP'T OF INTERIOR, U.S. GEOLOGICAL SURVEY CIRCULAR 1249, at 21 (2003)

Geothermal heat pumps can be used for heating and cooling buildings virtually anywhere. Though initial installation costs exceed those for conventional heating and cooling systems, monthly energy bills are always lower. Thus, within a few years, cumulative energy savings equal the extra up-front cost of installation. Thereafter, heating and cooling costs are less than those associated with conventional systems.

A heat pump is simply a machine that causes thermal energy to flow up temperature, that is, opposite the direction it would flow naturally without some intervention.... Thus, a heat pump is commonly used for space heating and cooling, when outside ambient air temperature is uncomfortably cold or hot, respectively. The cooling and heating functions require the input of "extra" work (usually electrical energy) in order to force heat to flow upstream, and the greater the "lift," or difference in temperature between the interior of a building and the outside, the more work is needed to accomplish the function. A geothermal heat pump increases the efficiency of the heating and cooling functions by substantially decreasing the thermal lift.

Because rocks and soils are good insulators, they respond little to wide daily temperature fluctuations and instead maintain a nearly constant temperature that reflects the mean temperature averaged over many years. Thus, at latitudes and elevations where most people live, the temperature of rocks and soil only a few meters beneath the surface typically stays within the range of 5 to 10°C.

For purposes of discussion, consider the functioning of a conventional air-source heat pump in a single-family residence, a system that exchanges thermal energy between air indoors and outdoors. Whereas such a heat pump must remove heat from cold outside air in the winter and deliver heat to hot outside air in the summer, a geothermal heat pump exchanges heat with a medium that remains at about 8°C throughout the year. As a result, the geothermal-based unit is almost always pumping heat over a temperature lift

Figure 6.7: Geothermal heat pumps

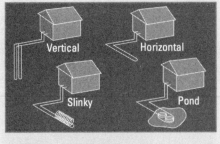

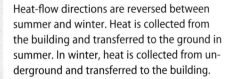

Configurations of heat exchange piping either underground or underwater for geothermal heat pumps.

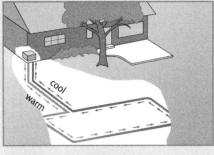

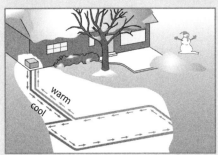

Heat-flow directions are reversed between summer and winter. Heat is collected from the building and transferred to the ground in summer. In winter, heat is collected from underground and transferred to the building.

Wendell A. Duffield & John H. Sass, U.S. Geological Survey, Geothermal Energy — Clean Power From the Earth's Heat 21 (2003), *available at* http://pubs.usgs.gov/ circ/2004/c1249/c1249.pdf.

much smaller than that for an air-source unit, leading to higher efficiency through less "extra" energy needed to accomplish the lift.

Some consumer resistance to geothermal heat pumps exists because initial purchase-and-installation cost is greater than that for an air-source system. The additional cost comes mostly from the need to bury piping through which fluid (water or antifreeze) is circulated to exchange heat with the ground or by drilling a shallow well to use ground water as the heat source/sink. Additional cost varies with the capacity and subsurface design of a given system. Experience to date indicates that the extra expense can be amortized in as little as 3 or 4 years for some systems. Other systems carry a longer pay-off period, but eventually all geothermal heat pumps provide savings that accrue as lower-than-normal utility bills. [Figure 6.7 shows various configurations for the heat pump piping and also explains the GHP process—Ed.]

Heat pumps provide significant energy savings, more than 75 percent as compared to electric baseboard heating and between 30 and 60 percent relative to other methods of heating and cooling. Many utilities, particularly in the Eastern United States, have subsidized the installation of geothermal heat pumps, also known as geoexchange systems, to help reduce peak demand for electric power. The lower electrical usage associated with the widespread use of geothermal heat pumps has allowed utilities to avoid or postpone construction of new power plants in areas where suitable land and transmission facilities are very difficult to acquire.

Worldwide, there are currently more than a half million geothermal heat pumps installed, for a total thermal output of over 7,000 megawatts. The United States accounts for most of these developments, with roughly 350,000 units whose combined output is about 5,000 megawatts.

Notes and Questions

1. In this chapter, I have used the term Geothermal Heat Pump or GHP because this is the terminology employed by Duffield & Sass and by the U.S. Department of Energy on its website. However, many engineers and scientists prefer to call this type of energy-delivery system a "ground-source heat pump" or "geoexchange." Why do you think this is so?

2. All of the renewable energy sources addressed in this book derive from the sun with the exception of geothermal, which instead taps the heat of decaying elements within the earth. While technically not renewable, geothermal energy is not likely to be depleted within our lifetime if we properly steward its use. The Geysers' story above is a cautionary tale about how the resource may be depleted if a geyser field is developed too quickly and pressures drop.

3. Geothermal Heat Pumps derive their heat from the relatively constant temperature of the earth right beneath its surface. While both the air and the substrata are warmed by the sun's rays, the subsurface temperature consistency is a result of the higher density of the substrata in comparison to the air above it. Thus, the thermal mass of the substrata allows it to better capture and retain the sun's energy. Do you see any legal significance in the fact that GHPs derive their energy from the sun instead of relying on an above-average geothermal magma source?

4. The plastic pipe used to collect heat or cooling for a GHP system can extend up to 500 feet below the surface. Could this collection field interfere physically with other subsurface development?

5. In *Rosette, Inc. v. United States*, 277 F.3d 1222 (10th Cir. 2002), the court held that a surface owner interfered with federal geothermal lease rights when it used geothermal hot water on the property for direct application in a greenhouse. How would you distinguish the *Rosette* case if you were a surface owner who wished to install a geothermal heat pump when the geothermal rights on your property were already leased to someone else? As you can see, the issue of ownership of the geothermal rights will be key to your answer. That issue is addressed in the following section.

C. Ownership Issues

Heat is the primary resource of value in geothermal production. However, this heat from the earth is often associated with other resources, such as water and natural gas, which can also serve as the vehicles for mobilizing the usable heat and moving it to the earth's surface. It is more complex to assert a property right over heat than to assert rights over tangible resources that can be possessed and stored. In addition, the association with other valuable resources, which may be claimed by others, also creates ownership conundrums. This section addresses some of the ownership issues that may make the development of geothermal resources problematic.

1. Background

The following two excerpts provide some context for the geothermal ownership cases in the second part of the section.

Kurt E. Seel, *Legal Barriers to Geothermal Development*
16 A.B.A. Sec. Env't, Energy, & Resources at 8-1 to 8-5 (2008)

Summary

Notwithstanding the concern over climate change and the demand for renewable energy proposed renewable energy developments encounter as much opposition from citizen and environmental groups as any other proposed developments, and often the objections are the same as for any other type of resource development. Although federal agencies are acting with relative haste in preparing programmatic geothermal leasing and pre-leasing environmental impact statements, in reality, the existing procedures for evaluating environmental impacts are slow, and legal challenges to environmental evaluations can delay development for many years, or preclude development altogether. In addition to navigating the traditional NEPA and "little NEPA" barriers, the new generation of geothermal developers will need to plan for an even greater scarcity of water resources; an increased protection of cultural resources and Native American sacred sites; and the potential impacts from possible geothermal induced seismicity. Furthermore, recent proposals to utilize existing oil and gas reservoirs and well bores, to produce or co-produce hydrocarbons and geothermal resources, will likely result in a new flurry of case law regarding owning and operating well bores; preventing economic waste; protecting correlative rights; and co-developing commingled resources....

* * *

What Is It And Who Owns It?

In general, the definition of "geothermal resources" varies from state to state, but in each case the resource usually consists of a small number of individual ownership interests, i.e., the bundle of sticks analogy. How to acquire, explore and most importantly, develop the heat resource will depend upon how the ownership bundle is defined in the locality where the resource is located.

For example, in Utah, "geothermal resources" are defined primarily as, "heat energy",[2] the ownership of which is derived "from an interest in land." (citation omitted). The Utah definition expressly excludes any ownership right to subsurface waters associated with the heat, i.e., the geothermal resource does not include any geothermal fluids."[3] Alternatively,

---

2. "'Geothermal resource' means (a) the natural heat of the earth, at temperatures greater than 120 degrees centigrade; and (b) the energy, in whatever form including pressure, present in, resulting from, created by, or which may be extracted from that natural heat, directly or indirectly through a material medium. Geothermal resources do not include geothermal fluids." 73-22-3(5) Utah Code Ann.

3. Geothermal fluids are deemed, in part, " ... a special kind of underground water resource, related to and potentially affecting other water resources of the state." 73-22-8 Utah Code Ann.

Tapping the Geothermal Potential of the Great Basin

The Great Basin contains the largest number of geothermal power plants in the United States, although most geothermal electrical production is at two sites elsewhere—The Geysers and Imperial Valley of California. Installed capacities of Great Basin plants range from 1 to 270 megawatts of electricity. Total installed capacity is 500 megawatts electric, about 17 percent of the national total. However, 500 megawatts electric is far less than the potential resource predicted (roughly 3,000 megawatts electric) by the U.S. Geological Survey (USGS) in the 1970s. Failure to reach this potential can be attributed in part to market forces and in part to overly optimistic assumptions about the character of Basin and Range geothermal reservoirs.

More than half of the geothermal electricity from the Great Basin is produced at Coso, in south-central California. The Coso area was long recognized as a potential geothermal resource, because it contains boiling mud pots and fumaroles within an area of many volcanoes whose little-eroded shapes indicate geologic youth. During the mid 1970s, a team of government and university scientists carried out a variety of field studies that suggested a potential geothermal resource of several hundred mega-watts electric. Subsequent drilling has resulted in about 270 megawatts electric on line by 2000; exploration and development continue.

A geologically similar, though smaller, resource has been developed at a young volcanic area called Roosevelt Hot Springs, Utah, but most geothermal resources of the Great Basin are not associated with volcanoes. Typically, resources occur where groundwater circulates deeply along the major fault zones that bound blocks of the highly extended crust of the region. This groundwater simply is heated as it circulates downward within the zones of fractured and therefore permeable rock.

Wendell A. Duffield & John H. Sass, *Geothermal Energy: Clean Power from the Earth's Heat*, U.S. Dep't Of Interior, U.S. Geological Survey Circular 1249, at 12 (2003).

the current federal definition of geothermal resources[4] focuses less on the heat energy characteristic, and more on the water and fluid resources associated with the heat energy. California, in a third attempt to put legal boundaries around the concept of geothermal resources, acknowledges that the primary stick in the geothermal bundle is the "natural heat of the earth, the energy, in whatever form, below the surface of the earth …" But then the California definition concedes that there are a variety of other "products", i.e., sticks, associated with "geothermal resources" including naturally heated fluids such as " … brines, associated gases, and steam, …" Although the legal nature of the individual sticks in the bundle can vary from state to state, as a general rule, geothermal resources usually do not include the rights to any oil, natural gas, nor other hydrocarbon substances. This is true even though geothermal resources may be closely associated, or commingled, with hydrocarbon bearing zones. (citation omitted). As explained later, the physical commingling of the legally separate heat energy, groundwater, and hydrocarbon ownership sticks could trigger some thorny legal issues for developers desiring to develop, or co-develop one or more resources. But, whether it's defined primarily as energy, or as steam,

---

4. The federal definition of geothermal resources is:
1. All products of geothermal processes, including indigenous steam, hot water, and hot brines;
2. Steam and other gases, hot water, and hot brines resulting from water, gas, or other fluids artificially introduced into geothermal formations;
3. Heat or other associated energy found in geothermal formations; and
4. Any byproducts. 43CFR 3200.1.

hot waters, brines, or as other "products," geothermal resources are unique. As a result, there is often little direct case law to guide the geothermal developer.

Surface Estate or Mineral Estate?

As indicated above, some states expressly define the heat energy aspect of geothermal resources to be part of the land, i.e., a real property interest. But what part of the land? Is it part of the surface estate, or the mineral estate, or some other real property interest e.g., a *profit a prendre*?

Early cases from California concluded that geothermal resources, including " ... all of the elements of a geothermal system — magma, porous rock strata, even water itself ..." are a "mineral" not a water resource, and therefore, owned by the mineral estate holder. United States v Union Oil Company, (9th Cir. 1977) 549 F.2d 1271, *cert denied* 934 U.S. 930, which states that "the surface owner [should] be entitled only to fresh waters that reasonably serve and give value to his surface ownership." As a result, the deep hot, non-potable geothermal resources are part of the mineral estate, not the surface estate, while relatively shallow, fresh groundwater resources are generally considered part of the surface estate. (citation omitted). Similarly, in split estates situations, e.g., federally reserved mineral rights, or when a mineral lease is silent as to geothermal rights, geothermal resources are generally considered part of the mineral rights. Although historically the mineral estate was the dominant estate and the holder possessed the right to access the surface estate for reasonable exploration and development of the mineral estate, recent state and federal laws have cut back on those rights. (citation omitted). At a minimum, one needs to acquire a geothermal mineral lease to obtain the right to explore and develop geothermal resources.

Who owns the transport mechanism?

Acquiring a geothermal mineral lease alone may not be enough to develop the resource. Energy in the subsurface isn't useful unless one can get that energy to the surface, which requires a heat transport mechanism. Usually, the transport mechanism is a heated brine or dry steam. Once the heated fluid is at the surface it can be used directly to provide a heat source, or provide power to a steam turbine to generate electricity. Dry steam from vapor dominated reservoirs can be used directly to drive a steam turbine, but more commonly the fluid is in the form of hot brines from water dominated reservoirs which can be converted to steam by flashing, or used to flash another liquid via a heat exchange mechanism (the "binary" method). (citation omitted). Another variant on the water dominated reservoir model is the geo-pressured reservoir, which generally consists of deep, hot, over-pressured brines containing dissolved methane. (citation omitted).

As described above, in states like Utah, the primary characteristic of a "geothermal resource" is the heat energy of the land; the hot, subsurface, geothermal fluids are the property of the state, deemed to be a "special kind of underground water resource", and are therefore subject to the primary jurisdiction of the Utah Division of Water Rights ("DWR"). (citation omitted). So in a state like Utah, unless the developer has another mechanism to retrieve their heat energy, they must understand and acquire the water rights associated with the subsurface heat in order to retrieve the heat energy.

Because of the similarities between developing geothermal resources and developing oil and gas resources, many states regulate both resources through a single agency, e.g., the California Department of Oil, Gas, and Geothermal Resources. Other states segregate geothermal from oil and gas conservation, even if that state regulates both resources similarly. For example, the Utah DWR regulates geothermal fluids, and the Utah Board of Oil, Gas, and Mining controls oil and gas development, yet both have similar jurisdictions and authorities over their respective resources, including the authority to order cooperative or unit operations, to prevent waste, protect correlative rights, possibly prevent drilling of unnecessary wells, and to maximize the ultimate economic recovery of the resources. (citation omitted). Which leads to the next group of questions—what if a geothermal developer doesn't own or control the entire geothermal reservoir? What if another developer is producing a fluid resource other than heat energy, e.g., hydrocarbons, which is commingled with the geothermal resource?

The Rule of Capture, Correlative Rights, Commingled Resources, and Multiple Resource Development

As a starting place, the Geothermal Steam Act of 1970, as updated by the Energy Policy Act of 2005 (Public Law 109-58, August 8, 2005) controls the exploration, leasing, and development of geothermal resources located on federal lands. At the state level, the laws governing geothermal exploration and development tend to follow the same state law concepts developed for state oil and gas programs. (citation omitted). As an example, in California, both oil and gas and the geothermal conservation laws are administered by the same agency, the Division of Oil, Gas & Geothermal Resources.

Federal courts deciding geothermal disputes often look to the state's oil and gas case law for precedent, oil and gas laws differ from state to state and can be surprisingly complex. In some states, such as Texas, oil and gas is considered an interest in the land itself while in others it is more akin to an exclusive hunting license, e.g., the exclusive right to drill for oil and gas and take possession of the hydrocarbons when they reach the surface. As noted above, California is a right-to-drill state, and when that right to drill is transferred from the mineral estate holder via a mineral lease, the result is usually something akin to a *profit a prendre*. (citation omitted). The differences between fluid mineral leases and how states treat concepts such as the rule of capture, correlative rights, pooling, drilling units, unitization, etc. are beyond the scope of this paper. But there are many well written papers which address these issues in both the oil and gas and the geothermal contexts. (citation omitted).

Multiple Resources Development

Even more difficult questions arise when multiple parties own separate resources which are commingled in a single reservoir, or when the development of one resource will necessarily impact a nearby fluid resource. For example, who owns the geothermal resources which are commingled with an oil and gas reservoir? Hot brines can be associated with the same subsurface reservoir in which hydrocarbons are present, and are controlled by the oil and gas lessees. As a result, the potential exists for there to be multiple lessees of the same general subsurface formation or reservoir. Some state laws expressly address concurrent multiple resource developments by either rule or order. (citation omitted). Other states' regulations provide less guidance. For example, who owns the methane dissolved in a geopressured geothermal reservoir? ...

———————

Wendell A. Duffield & John H. Sass, *Geothermal Energy: Clean Power from the Earth's Heat*

U.S. Dep't Of Interior, U.S. Geological Survey Circular 1249, at 17 (2003)

Geopressured Systems

A type of hydrothermal environment whose hot water is almost completely sealed from exchange with surrounding rocks is called a geopressured system. This type of system typically forms in a basin that is being rapidly filled with sediment, rather than in a volcanic area.

Geopressure refers to the hydrothermal water being at higher-than-normal pressue for its depth (that is, these systems are overpressured). Such excess pressure builds in the pore water of sedimentary rocks when the rate at which pore water is squeezed from these rocks cannot keep pace with the rate of accumulation of the overlying sediment. As a result, geopressured systems also contain some mechanical energy, stemming from the fluid overpressure, in addition to the thermal energy of the geothermal water. Moreover, these systems also contain potential for combustion energy, because considerable methane gas (otherwise known as natural gas) is commonly dissolved in the geothermal water. The bulk of the thermal energy of geopressured systems is accounted for by roughly equal contributions from the temperature of the water and the dissolved methane. During the 1970s and early 1980s, the Texas-Louisiana Gulf Coast served as a natural laboratory for offshore studies of geopressured systems sponsored by the U.S. Department of Energy (DOE). Presently, however, the economics of exploiting the geopressured environment are not favorable, and industry has shown no interest in following up the DOE studies.

Notes and Questions

1. The Department of Energy estimates that each barrel of oil pumped from the ground generates approximately ten barrels of hot brine, which is considered a waste product at oil and gas operations around the country. Now, companies in North Dakota, Wyoming, Colorado, Mississippi, Louisiana, and Texas are experimenting with coproducing geothermal energy by separating the water from the oil or gas and then using it in a binary system to heat a secondary fluid with a lower boiling point that will then flash to steam to run the turbines and generate on-site electricity for the well field. If you represented one of these oil and gas producers, what would you advise your client about how to secure ownership to the geothermal rights?

2. Suppose a company different from the oil and gas company that brought the hot water to the surface held title to the geothermal rights and decided to develop the geothermal resource. What additional complications might arise from a situation where you have two separate entities trying to maximize their own resource?

3. Geothermal fluids are frequently a soup containing a number of dissolved minerals. As the geothermal fluid is flashed into steam to run the turbines, the water cools and these dissolved minerals precipitate out. If they deposit on the turbine mechanisms, they can interfere with performance or cause corrosion. In the Imperial Valley, most of the "filtercake" wastes are silica and must be disposed. However, some dissolved mineral by-products are valuable, such as gold or lithium. If you represented a geothermal developer and sought a mineral lease from a private landowner for the development of "geothermal resources,"

how would you define that term to allow your company to develop these by-product minerals as well?

4. In drafting the geothermal lease in the previous exercise, consider the issue of royalties. If the primary purpose of the lease is to develop geothermal resources, the main royalties would come from the production of electricity. How would you phrase this portion of the lease? What about potential by-product minerals? Should royalties for these be addressed in the geothermal lease and if so what are some of the considerations?

2. Geothermal Ownership Cases

The following excerpts are from two key cases that have addressed ownership of geothermal rights. Consider how they clarify or complicate the status for purposes of geothermal development.

United States v. Union Oil Co.
549 F.2d 1271 (9th Cir. 1977)

Browning, J.

This is a quiet title action brought by the Attorney General of the United States pursuant to section 21(b) of the Geothermal Steam Act of 1970, 30 U.S.C. s 1020(b), to determine whether the mineral reservation in patents issued under the Stock-Raising Homestead Act of 1916, 43 U.S.C. s 291 et seq., reserved to the United States geothermal resources underlying the patented lands. The district court held that it did not. 369 F.Supp. 1289 (N.D.Cal.1973). We reverse.

Various elements cooperate to produce geothermal power accessible for use on the surface of the earth. Magma or molten rock from the core of the earth intrudes into the earth's crust. The magma heats porous rock containing water. The water in turn is heated to temperatures as high as 500 degrees Fahrenheit. As the heated water rises to the surface through a natural vent, or well, it flashes into steam.[1]

Geothermal steam is used to produce electricity by turning generators. In recommending passage of the Geothermal Steam Act of 1970, the Interior and Insular Affairs Committee of the House reported: "(G)eothermal power stands out as a potentially invaluable untapped natural resource. It becomes particularly attractive in this age of growing consciousness of environmental hazards and increasing awareness of the necessity to develop new resources to help meet the Nation's future energy requirements. The Nation's geothermal resources promise to be a relatively pollution-free source of energy, and their development should be encouraged." H.R.Rep. No. 91-1544, 91st Cong., 2d Sess., reprinted at 3 U.S.Code Cong. & Admin.News 5113, 5115 (1970).

Appellees are owners, or lessees of owners, of lands in an area known as "The Geysers" in Sonoma County, California. Beneath the lands are sources of geothermal steam. Appellees have developed or seek to develop wells to produce the steam for use in generating electricity. The lands were public lands, patented under the Stock-Raising Homestead Act. All patents issued under that Act are "subject to and contain a reservation to the United

---

1. Reich v. Commissioner, of Internal Revenue, 52 T.C. 700, 704-05 (1969), aff'd, 454 F.2d 1157 (9th Cir. 1972); H.R.Rep. No. 91-1544, 91st Cong., 2d Sess., reprinted at 3 U.S.Code Cong. & Admin.News 5113, 5114 (1970); Brooks, Legal Problems of the Geothermal Industry, 6 Nat.Resources J. 511, 514-15 (1966); Barnea, Geothermal Power, Scientific American, Jan. 1972, at 70, 74.

States of all the coal and other minerals in the lands so entered and patented, together with the right to prospect for, mine, and remove the same." Section 9 of the Act, 43 U.S.C. s 299. The patents involved in this case contain a reservation utilizing the words of the statute.[2] The question is whether the right to produce the geothermal steam passed to the patentees or was retained by the United States under this reservation.

There is no specific reference to geothermal steam and associated resources in the language of the Act or in its legislative history. The reason is evident. Although steam from underground sources was used to generate electricity at the Larderello Field in Italy as early as 1904, the commercial potential of this resource was not generally appreciated in this country for another half century. No geothermal power plants went into production in the United States until 1960. Congress was not aware of geothermal power when it enacted the Stock-Raising Homestead Act in 1916; it had no specific intention either to reserve geothermal resources or to pass title to them.

It does not necessarily follow that title to geothermal resources passes to homesteader-patentees under the Act. The Act reserves to the United States "all the coal and other minerals." All of the elements of a geothermal system magma, porous rock strata, even water itself[5] may be classified as "minerals." When Congress decided in 1970 to remove the issue from controversy as to future grants of public lands, it found it unnecessary to alter the language of existing statutory "mineral" reservations. It simply provided that such reservations "shall hereafter be deemed to embrace geothermal steam and associated geothermal resources." Geothermal Steam Act of 1970, 30 U.S.C. s 1024.[6] Thus, the words

---

2. The reservation reads:
    Excepting and reserving, however, to the United States all coal and other minerals in the
    lands so entered and patented, together with the right to prospect for, mine, and remove
    the same pursuant to the provisions and limitations of the Stock-Raising Homestead Act.
See 43 C.F.R. s 3814.2(a) (1976).
    5. Hathorn v. Natural Carbonic Gas Co., 194 N.Y. 326, 87 N.E. 504, 508 (1909); H.R.Rep. No. 91-1544, supra note 1, at 5126-27 (letters from Dep't of Interior); A. Ricketts, American Mining Law 64, 70 (4th ed. 1943); Webster's Third Int'l Dictionary 1437 (1961); 13 The New Int'l Encyclopedia 537 (Gilman, Peck, & Colby ed. 1913); 10 The Americana (1907-08) (unpaginated article on mineralogy includes water as mineral). See Kuntz, The Law Relating to Oil & Gas in Wyoming, 3 Wyo.L.J. 107, 109 (1949).
    Moreover, geothermal steam has been held to be a "gas." Reich v. Commissioner of Internal Revenue, 52 T.C. 700, 710-11 (1969), aff'd, 454 F.2d 1157 (9th Cir. 1972). See Geothermal Exploration in the First Quarter Century 185, 187 (Geothermal Resources Council 1973) (letter from George R. Wickham, Ass't Comm'r, Dep't of Interior, July 8, 1924 natural gas is a mineral within purview of mining laws).
    No one contends that water cannot be classified as mineral. Appellees argue only that the water should not be included in the term "minerals" in this statutory setting. This is basically a question of legislative intent, dealt with in detail later in the text. To the extent that the argument rests on the meaning of the word itself, however, the government is entitled to have the ambiguity resolved in its favor under "the established rule that land grants are construed favorably to the Government, that nothing passes except what is conveyed in clear language, and that if there are doubts they are resolved for the Government, not against it." United States v. Union Pac. R.R., 353 U.S. 112, 116, 77 S.Ct. 685, 687, 1 L.Ed.2d 693 (1957). See Caldwell v. United States, 250 U.S. 14, 20, 39 S.Ct. 397, 63 L.Ed. 816 (1919); Southern Idaho Conf. Ass'n of Seventh Day Adventists v. United States, 418 F.2d 411, 415 n.8 (9th Cir. 1969).
    Appellees argue that the term "minerals" is to be given the meaning it had in the mining industry at the time the Act was adopted, and that this understanding excluded water. This is a minority rule, United States v. Isbell Constr. Co., 78 Interior Dec. 385, 390-91 (1971), even as applied to permit conveyances. 1 American Law of Mining s 3.26, at 551-53 (1976).
    6. Members of the Subcommittee on Mines and Mining of the House Committee on Interior and Insular Affairs went to some lengths to make it clear that whether the term "minerals" as used in prior legislation included geothermal resources was a question for the courts, on which the official position

of the mineral reservation in the Stock-Raising Homestead Act clearly are capable of bearing a meaning that encompasses geothermal resources.

The substantial question is whether it would further Congress's purposes to interpret the words as carrying this meaning. The Act's background, language, and legislative history offer convincing evidence that Congress's general purpose was to transfer to private ownership tracts of semi-arid public land capable of being developed by homesteaders into self-sufficient agricultural units engaged in stock raising and forage farming, but to retain subsurface resources, particularly mineral fuels, in public ownership for conservation and subsequent orderly disposition in the public interest. The agricultural purpose indicates the nature of the grant Congress intended to provide homesteaders via the Act; the purpose of retaining government control over mineral fuel resources indicates the nature of reservations to the United States Congress intended to include in such grants. The dual purposes of the Act would best be served by interpreting the statutory reservation to include geothermal resources.[7]

Events preceding the enactment of the Stock-Raising Homestead Act contribute to an understanding of the intended scope of the Act's mineral reservation. Prior to 1909, public lands were disposed of as either wholly mineral or wholly nonmineral in character. United States v. Sweet, 245 U.S. 563, 567-68, 571, 38 S.Ct. 193, 62 L.Ed. 473 (1918). This practice led to inefficiencies and abuses. In 1906 and again in 1907, President Theodore Roosevelt pointed out that some public lands were useful for both agriculture and production of subsurface fuels, and that these two uses could best be served by separate disposition of the right to utilize the same land for each purpose. The President called the attention of Congress "to the importance of conserving the supplies of mineral fuels still belonging to the Government." 41 Cong.Rec. 2806 (1907). To that end, the President recommended "enactment of such legislation as would provide for title to and development of the surface land as separate and distinct from the right to the underlying mineral fuels in regions where these may occur, and the disposal of these mineral fuels under a leasing system on conditions which would inure to the benefit of the public as a whole." Id.[8]

---

of the 89th Congress was one of neutrality. See Hearings on H.R. 7334 et al. on Disposition of Geothermal Steam, 89th Cong., 2d Sess., ser. 89-35, pt. II, at 295-96 (1966). The point made here, however, is that in fact Congress thought the term sufficiently broad to encompass such resources.

7. The Stock-Raising Homestead Act "define(s) the estates to be granted in terms of the intended use ... The reservation of minerals to the United States should therefore be construed by considering the purposes both of the grant and of the reservation in terms of the use intended." 1 American Law of Mining s 3.26, at 552 (1976). Accord, United States v. Isbell Constr. Co., 78 Interior Dec. 385, 390 (1971). See also United States v. Union Pac. R.R., 353 U.S. 112, 77 S.Ct. 685, 1 L.Ed.2d 693 (1957); Caldwell v. United States, 250 U.S. 14, 21, 39 S.Ct. 397, 63 L.Ed. 816 (1919).

A similar approach has been taken in construing grants and reservations in deeds between private parties involving minerals. See, e. g., Northern Natural Gas Co. v. Grounds, 441 F.2d 704, 714 (10th Cir. 1971); Acker v. Guinn, 464 S.W.2d 348, 352 (Tex.1971). The "general intent (of the parties) should be arrived at, not by defining and re-defining the terms used, but by considering the *purposes* of the grant or reservation in terms of manner of enjoyment intended in the ensuing interests." Kuntz, The Law Relating to Oil & Gas in Wyoming, 3 Wyo.L.J. 107, 112 (1949) (emphasis in original).

8. The President said:

> If this Government sells its remaining fuel lands they pass out of its future control. If it now leases them we retain control, and a future Congress will be at liberty to decide whether it will continue or change this policy. Meanwhile, the Government can inaugurate a system which will encourage the separate and independent development of the surface lands for agricultural purposes and the extraction of the mineral fuels in such manner as will best meet the needs of the people and best facilitate the development of manufacturing industries.

41 Cong.Rec. 2806 (1907).

In 1909 the Secretary of the Interior returned to the same theme, arguing that "inducements for much of the crime and fraud, both constructive and actual, committed under the present system can be prevented by separating the right to mine from the title to the soil. The surface would thereby be open to entry under other laws according to its character and subject to the right to extract the coal. The object to be attained in any such legislation is to conserve the coal deposits as a public utility and to prevent monopoly or extortion in their disposition." 1909 Dep't Interior Ann.Rep. pt. I, at 7 (emphasis omitted). The Secretary made the same suggestion with respect to "oil and gas fields in the public domain." Id.

In the same year "Congress deviated from its established policy of disposing of public lands under the nonmineral land laws only if they were classified as nonmineral in character and enacted the first of several statutes providing for the sale of lands with the reservation to the United States of certain specified minerals. These statutes were soon followed by statutes providing for the sale of lands with the reservation to the United States of all minerals...." 1 American Law of Mining § 3.23, at 532 (1976).

The first of these statutes "separating the surface right from the right to the underlying minerals" was the Act of March 3, 1909 (35 Stat. 844), 30 U.S.C. § 81, followed shortly by the Acts of June 22, 1910 (36 Stat. 583), 30 U.S.C. §§ 83 et seq., April 30, 1912 (37 Stat. 105), 30 U.S.C. § 90, and August 24, 1912 (37 Stat. 496). See The Classification of the Public Lands, 537 U.S.Geological Survey Bull. 45, Department of Interior (1913). In the latter report, the Geological Survey pointed out that where lands were valuable for two uses, both uses could be served by "a separation of estates." The report urged adoption of legislation embodying "the extension of the principle of the separation of estates," plus the leasing of natural resources, as means of protecting such resources without delaying agricultural development.[10]

Appellees argue that the executive department statement preceding the enactment of the Stock-Raising Homestead Act dealt primarily with coal deposits. But the concern of the statements was with the conservation of underground energy sources, as the President's references to "fuel lands" and "mineral fuels" illustrate.

10. The report states (45-47):

The carrying out of the withdrawal policy for protecting the mineral and water resources of the public domain is in many cases rendered difficult and embarrassing by the agricultural value of the land withdrawn.... (S)ome of the best farming lands in the West are underlain by coal or phosphate, and some are so situated as to be of strategic importance in power development. Any hindrance to bona fide home building or other agricultural development of the public domain is indeed unfortunate, but in order to protect the public's natural resources withdrawals resulting in such hindrance have been necessary. For certain lands the situation has been relieved by the passage of acts separating the surface right from the right to the underlying minerals....

In carrying out its function of classifying the public lands and in making its fund of information available in the administration of the existing land laws the Geological Survey has become acutely cognizant of the need for certain new legislation. The laws desired are primarily of two types and embody two fundamental necessities first, the extension of the principle of the separation of estates, and second, the application of the leasing principle to the disposition of natural resources.

As has already been pointed out, the public lands can not be divided into classes each of which is valuable for one purpose only. Instead, the same tract of land may be valuable for two or more resources. In one tract for example, agricultural land that is underlain by coal both resources may be utilized at the same time without interfering with each other. In another tract for example, agricultural land within a reservoir site the land may be valuable for one resource only until it is utilized for another. In the first case the problem is so to frame the laws that no resource will be forced to await the development of the other. In the second case the problem is to permit the use of the land for one purpose pending its use

In 1914, within a year of this appeal, Congress began consideration of a forerunner of the Stock-Raising Homestead Act. The bill was referred to the Department of Interior for comment, revised by the Department, and reintroduced. H.R.Rep. No. 626, 63d Cong., 2d Sess., reprinted at 52 Cong.Rec. 3986-90 (1915). It was enacted into law the following year.

This background supports the conclusion, confirmed by the language of the Stock-Raising Homestead Act, the Committee reports, and the floor debate, that when Congress imposed a mineral reservation upon the Act's land grants, it meant to implement the principle urged by the Department of Interior and retain governmental control of subsurface fuel sources, appropriate for purposes other than stock raising or forage farming.[11]

We turn to the statutory language. The title of the Act "The Stock-Raising Homestead Act" reflects the nature of the intended grant. The Act applies only to areas designated by the Secretary of Interior as "stock-raising lands"; that is, "lands the surface of which is, in his opinion, chiefly valuable for grazing and raising forage crops, do not contain merchantable timber, are not susceptible of irrigation from any known source of water supply, and are of such character that six hundred and forty acres are reasonably required for the support of a family...." 43 U.S.C. § 292. The entryman is required to make improvements to increase the value of the entry "for stock-raising purposes." Id. § 293. On the other hand, "all entries made and patents issued" under the Act must "contain a reservation to the United States of all the coal and other minerals in the lands," and such deposits "shall be subject to disposal by the United States in accordance with the provisions of the coal and mineral land laws." Id. § 299. The subsurface estate is dominant; the interest of the homesteader is subject to the right of the owner of reserved mineral deposits to "reenter and occupy so much of the surface" as reasonably necessary to remove the minerals, on payment of damages to crops or improvements. Id.

The same themes are explicit in the reports of the House and Senate committees. The purpose of the Act is to restore the grazing capacity and hence the meat-producing capacity of semi-arid lands of the west and to furnish homes for the people, while preserving to the United States underlying mineral deposits for conservation and disposition under laws appropriate to that purpose. The report of the House Committee reproduces a letter from the Department of Interior endorsing the bill. The Department notes that "all mineral(s) within the lands are reserved to the United States." H.R.Rep. No. 35, 64th Cong.,

---

for another without losing public control of the development of the second. In both cases the answer is found in a separation of estates. The extension of this principle, now applied to coal, to withdrawn and classified minerals and to the uses of water resources would permit the retention of the mineral deposits and power and reservoir sites in public ownership pending appropriate legislation by Congress without in any way retarding agricultural development. Bills have already been introduced applying this principle to oil in other States than Utah and to phosphate in the State of Idaho. It is to be hoped that such bills will be passed and approved, or, better still, that a comprehensive act providing for the separation of the various estates will be introduced and enacted.

11. The court in Skeen v. Lynch, 48 F.2d 1044, 1046 (10th Cir. 1931) stated:
The legislative history of the Stock-Raising Homestead Act when it was reported for passage including the discussion that followed relevant to this subject leave us no room to doubt that it was the purpose of Congress in the use of the phrase "all coal and other minerals" to segregate the two estates, the surface for stockraising and agricultural purposes from the mineral estate, and to grant the former to entrymen and to reserve all of the latter to the United States. Although the Supreme Court of New Mexico specifically rejected the Skeen analysis in State ex rel. State Highway Comm'n v. Trujillo, 82 N.M. 694, 487 P.2d 122, 125 (1971), it did so in reliance upon the absence of an express provision in the Act, especially rejecting an invitation to examine the legislative history.

1st Sess. 5 (1916). The Department continues, "To issue unconditional patents for these comparatively large entries under the homestead laws might withdraw immense areas from prospecting and mineral development, and without such a reservation the disposition of these lands in the mineral country under agricultural laws would be of doubtful advisability." Id. Moreover, "(t)he farmer-stockman is not seeking and does not desire the minerals, his experience and efforts being in the line of stock raising and farming, which operations can be carried on without being materially interfered with by the reservation of minerals and the prospecting for and removal of same from the land." Id. This language is quoted with approval in S.Rep. No. 348, 64th Cong., 1st Sess. 2 (1916).

Commenting upon the mineral reservation, the House report states:

> It appeared to your committee that many hundreds of thousands of acres of the lands of the character designated under this bill contain coal and other minerals, the surface of which is valuable for stock-raising purposes. The purpose of (the provision reserving minerals) is to limit the operation of this bill strictly to the surface of the lands described and to reserve to the United States the ownership and right to dispose of all minerals underlying the surface thereof....

H.R.Rep. No. 35, supra, at 18.

The floor debate is revealing. The bill drew opposition because of the large acreage to be given each patentee. See, e. g., 52 Cong.Rec. 1808-09 (1915) (remarks of Rep. Stafford). In response, supporters emphasized the limited purpose and character of the grant. They pointed out that because the public lands involved were semi-arid, an area of 640 acres was required to support the homesteader and his family by raising livestock. E. g., id. at 1807, 1811-12 (remarks of Reps. Fergusson, Martin and Lenroot). They also pointed out that the grant was limited to the surface estate,[12] and they emphasized in the strongest terms that all minerals were retained by the United States.

For example, asked whether the reservation would include oil, Congressman Ferris, manager of the bill, responded, "It would. We believe it would cover every kind of mineral. All kinds of minerals are reserved ... (The bill) merely gives the settler who is possessed of any pluck an opportunity to go out and take 640 acres and make a home there." 53 Cong.Rec. 1171 (1916). It was pointed out that oil was not, technically, a "mineral." Congressman Ferris replied, "if the gentleman thinks there is any doubt about it we will put it in, because not a single gentleman from the West who has been urging this legislation wants anybody to be allowed to homestead mineral land." Id. During the closing debate on the Conference report, reference was twice made to the Department of Interior communication quoted above including the assertion that without a broad mineral reservation the grant would be unjustifiable, and the representation that "the farmer-stockman is not seeking and does not desire the minerals, his experience and efforts being in the line of stock raising and farming, which operations can be carried on without being materially interfered with by the reservation of minerals and the prospecting for and removal of same from the land." 54 Cong.Rec. 682, 684 (1916).

There is little in the debates to comfort appellees. Appellees cite a discussion between Congressmen Mondell and Ferris, in which Mondell objected to Ferris's describing certain laws as "surface-entry laws, for they are not." Congressman Mondell continued, "They convey fee titles. They give the owner much more than the surface, they give him all except

---

12. Representative Burke, explaining the earlier and, for our purposes, identical version of the Act (see 53 Cong.Rec. 1170 (1916)), stated that "Section 2 of the bill ... limits the entry to the surface and provides that the land must be chiefly valuable for grazing and raising forage crops...." 52 Cong.Rec. 1809 (1915).

the body of the reserved mineral." 53 Cong.Rec. 1233-34 (1916).[13] Representative Mondell was not referring to the Stock-Raising Homestead Act at all, but to three earlier statutes that reserved only particularly named substances, and not minerals generally.[14] Representative Mondell opposed the Stock-Raising Homestead Act's general mineral reservation for the very reason that it restricted the patentee's estate more than the earlier statutes, and to an extent Representative Mondell thought undesirable. Congressman Mondell remarked that the general reservation contained in the Act as adopted rested on "the monarchical theory" which, he asserted, "is to reserve all minerals to the crown, upon the theory that the mere subject is not entitled to anything except the soil that he stirs." 51 Cong.Rec. 10494 (1914). Although Representative Mondell eventually voted for the Act, he continued to protest the scope of the mineral reservation. His closing comment is worthy of notice. It confirms the view that the mineral reservation in the Stock-Raising Homestead Act was novel in its breadth. It also reveals that this broad reservation of subsurface resources was included at the insistence of the Department of Interior because of the large surface acreage granted under the Act:

> ... [T]he fact should be emphasized that the bill establishes a new method and theory with regard to minerals in the land legislation in our country. It reverts back to the ancient doctrine of the ownership of the mineral by the king or the crown and reserves specifically everything that is mineral in all the land entered. It was, it was claimed, necessary to accept a provision of that kind in order to secure the larger acreage. The Interior Department insisted upon it, and many supported that view. My own opinion is that that policy is not wise and that in the long run it will be found to be infinitely more harmful than beneficial or useful or helpful to anyone, either the individual or the public generally. When one takes into consideration the wide range of substances classed as mineral, the actual ownership under a complete mineral reservation becomes a doubtful question.

54 Cong.Rec. 687 (1916).

Appellees argue that references in the Congressional Record to homesteaders' drilling wells and developing springs indicate that Congress intended title to underground water to pass to patentees under the Act. These references are not to the development of geothermal resources. As we have seen, commercial development of such resources was not contemplated in this country when the Stock-Raising Homestead Act was passed. Moreover, in context, the references are to the development of a source of fresh water for the use of livestock, not to the tapping of underground sources of energy for use in generating electricity.[18]

---

13. Appellees also observe that the proviso to the mineral reservation in the Act originally stated that "patents issued for the coal or other mineral deposits herein reserved shall contain appropriate notations declaring them to be subject to the provisions of this act with reference to the disposition, occupancy, and use of the surface of the land," (italics added) and that the italicized phrase was stricken in the House. 53 Cong.Rec. 1233 (1916). The change was made by committee amendment, adopted without explanation or discussion. Even considered alone, its effect is unclear. It may have been thought, for example, that the stricken phrase might be construed to render the broad mineral reservation of the Act inapplicable to patents for a particular mineral, thus inadvertently broadening the mineral grant.

14. Act of Mar. 3, 1909, 35 Stat. 844, 30 U.S.C. s 81 (coal); Act of June 22, 1910, 36 Stat. 583, 30 U.S.C. ss 83 et seq. (coal); Act of July 17, 1914, 38 Stat. 509, 30 U.S.C. ss 121 et seq. (phosphate, nitrate, potash, oil, gas, or asphaltic minerals).

18. "A fair and reasonable (ruling) would hold the surface owner to be entitled only to fresh waters that reasonably serve and give value to his surface ownership. Salt water and geothermal steam and brines should be held the property of the mineral owner who owns such substances as oil, gas and

This review of the legislative history demonstrates that the purposes of the Act were to provide homesteaders with a portion of the public domain sufficient to enable them to support their families by raising livestock, and to reserve unrelated subsurface resources, particularly energy sources, for separate disposition. This is not to say that patentees under the Act were granted no more than a permit to graze livestock, as under the Taylor-Grazing Act, 43 U.S.C. §§ 315 et seq. To the contrary, a patentee under the Stock-Raising Homestead Act receives title to all rights in the land not reserved. It does mean, however, that the mineral reservation is to be read broadly in light of the agricultural purpose of the grant itself, and in light of Congress's equally clear purpose to retain subsurface resources, particularly sources of energy, for separate disposition and development in the public interest. Geothermal resources contribute nothing to the capacity of the surface estate to sustain livestock. They are depletable subsurface reservoirs of energy, akin to deposits of coal and oil, which it was the particular objective of the reservation clause to retain in public ownership. The purposes of the Act will be served by including geothermal resources in the statute's reservation of "all the coal and other minerals." Since the words employed are broad enough to encompass this result, the Act should be so interpreted.

Appellees assert that the Department of Interior has expressed the opinion that the mineral reservation in the Act does not include geothermal resources, and that this administrative interpretation is entitled to deference under Udall v. Tallman, 380 U.S. 1, 16, 85 S.Ct. 792, 13 L.Ed.2d 616 (1965), and similar authority. The documents upon which appellees rely do not reflect a contemporaneous construction by administrators who participated in drafting the Act to which courts give great weight in interpreting statutes.[19]

---

coal, since the functions and values are more closely related. Geothermal steam is a source of energy just as fossil fuels such as oil, gas and coal are sources of energy." Olpin, The Law of Geothermal Resources, 14 Rocky Mountain Mineral Law Institute 123, 140-41 (1968). See Reich v. Commissioner of Internal Revenue, 52 T.C. 700 (1969), aff'd, 454 F.2d 1157 (9th Cir. 1972); Allen, Legal and Policy Aspects of Geothermal Resources Development, 8 Water Resources Bull. 250, 253-54 (1972).

19. Zuber v. Allen, 396 U.S. 168, 193, 90 S.Ct. 314, 24 L.Ed.2d 345 (1969); Power Reactor Dev. Co. v. International Union of Electrical, Radio & Machine Workers, 367 U.S. 396, 408, 81 S.Ct. 1529, 6 L.Ed.2d 924 (1961); United States v. American Trucking Ass'ns, 310 U.S. 534, 549, 60 S.Ct. 1059, 84 L.Ed. 1345 (1940).

Appellees rely upon three letters by officials of the Department of Interior stating that "geothermal steam" is not a "mineral" within the meaning of the mining laws or the mineral reservation. Two of the letters, both dated Dec. 16, 1965, are responses by Edward Weinberg, Deputy Solicitor, to letters of inquiry from interested citizens. They are reproduced in an appendix to the district court's opinion, 369 F.Supp. at 1300-02, and as part of H.R.Rep. No. 91-1544, supra note 1, at 5126-28. The third letter was written by the Associate Solicitor for Public Lands to counsel for appellee Magma Power Company on Feb. 16, 1966, and apparently has not been published.

The letters do not reflect an agency view contemporaneous with the passage of the Act they were written a half century after the statute was adopted. Appellees also rely upon a Department of Interior memorandum from Edward Fischer, Acting Solicitor, to the Director of Bureau of Land Management, stating that geothermal steam is not a "mineral material" for the purposes of the Mineral Act of 1947, 30 U.S.C. s 601. Dep't Interior Mem. M-36625, Aug. 18, 1961. But this view is contrary to that expressed by Solicitor Stevens only seven months earlier in a letter to appellee Magma Power Company dated Jan. 19, 1961. Brooks, supra note 1, at 524 & n.56; Note, Acquisition of Geothermal Rights, 1 Idaho L.Rev. 49, 56 & n.44 (1964). This inconsistency, see Hearings on H.R. 7334 et al. before the Subcomm. on Mines & Mining of the House Comm. on Interior and Insular Affairs, 89th Cong., 2d Sess., ser. 89-35, pt. II, at 194-95 (1966) (statement of Emmet Wolter) is another factor indicating that we should not accord deference to the administrative construction. See Udall v. Tallman, 380 U.S. 1, 17, 85 S.Ct. 792, 13 L.Ed.2d 616 (1965).

Moreover, the expressions of opinion relied upon by appellees are weakly reasoned. They rest entirely upon the premise that geothermal resources are simply water. Water, the argument then proceeds, ordinarily is not included in mineral reservations by the courts, or treated as a mineral in

Nor is this a case in which Congress has approved an administrative interpretation, explicitly or implicitly. On the contrary, Congress noted the Department of Interior's interpretation, observed that a contrary view had been expressed, concluded that "the opinion of the Department is not a conclusive determination of the legal question...," and provided for "an early judicial determination of this question (upon which the committee takes no position)." H.R.Rep. No. 91-1544, 91st Cong., 2d Sess., reprinted at 3 U.S.Code Cong. & Admin.News 5113, 5119 (1970).

Appellees contend that enactment of the Underground Water Reclamation Act of 1919, 43 U.S.C. ss 351 et seq., three years after passage of the Stock-Raising Homestead Act, indicates that Congress did not consider subsurface water to be a "mineral." We disagree; indeed the more reasonable implication seems to us to be to the contrary.[21]

The district court granted appellees' motion to dismiss for failure to state a claim upon which relief could be granted. 369 F.Supp. at 1299. The State of California, as amicus, suggests that questions of fact are presented as to the nature of geothermal resources. We are persuaded that the facts necessary to decision are not disputed. The appeal presents only a question of law as to the proper construction of the statute, which we have answered.

Whether the United States is estopped from interfering with the rights of private lessees without compensating them for any losses they may sustain will be open on remand.

Reversed and remanded.

Notes and Questions

1. What did the *Union Oil* Court conclude? Is a geothermal resource a mineral?

2. What about the water associated with the geothermal resource? Who owns that?

---

public land laws. But all of the court decisions relied upon in the communications concern fresh water brought to the surface by means of a well. See Mack Oil Co. v. Laurence, 389 P.2d 955 (Okl.1964); Fleming Foundation v. Texaco, 337 S.W.2d 846 (Tex.Civ.App.1960). See Estate of Genevra O'Brien, 8 Oil & Gas 845 (N.D.Tex.1957) (charge of the court). And if geothermal resources are indeed "water," the later enactment of the Geothermal Steam Act has undercut the statement that "water" is not treated as a mineral in public land laws. But the principle deficiency in the documents relied upon by appellees is this: the sole question is the meaning of the statute; the answer therefore turns entirely upon the intent of Congress, and the documents do not mention that subject at all.

21. The Underground-Water Reclamation Act authorizes the issuance of permits to explore for underground water on not to exceed 2,560 acres of public lands in Nevada (s 351). The Act provides that if a permittee discovers and makes available for use a supply of underground water in sufficient quantity "to produce at a profit agricultural crops other than native grasses upon not less than twenty acres of land," he will be entitled to a patent on 640 acres of the public land embraced in his permit (s 355). The Act further provides for reservation of "all the coal and other valuable minerals in the lands" patented (s 359). Appellees argue that the term "minerals" in the latter provision must not include underground water, for if it did the reservation would deprive the patentee of the very water he had discovered.

But again, the obvious distinction is between underground water suitable for agricultural purposes and geothermal resources. The purpose of the Underground-Water Reclamation Act is fully realized and all of its provisions made fully effective if the term "minerals" is read to exclude the former but include the latter. As noted in the text, the significance of the Underground-Water Reclamation Act may be the opposite of that suggested by appellees when the statute is considered in conjunction with the Geothermal Steam Act of 1970, for the latter statute was adopted on the premise that existing legislation, presumably including the Underground-Water Reclamation Act of 1919, did not authorize the Department of Interior to dispose of geothermal resources in public lands. See, e. g., H.R.Rep. No. 91-1544, supra note 1, at 5115.

3. What is the significance of the reservation that the court was interpreting? Might the outcome be different if the wording were different?

4. What does this holding mean for surface owners?

5. Now compare the holding of the California State case below with that of the 9th Circuit in *Union Oil*.

Geothermal Kinetics v. Union Oil Co.
75 Cal. App. 3d 56 (Cal. Ct. App. 1977)

Scott, J., with Feinberg, J., and Draper, J., concurring.

Appellants' Petition for a hearing by the California Supreme Court was denied January 26, 1978. Mosk, J., was of the opinion that the petition should be granted.

The issue presented here is whether geothermal resources belong to the owner of the mineral estate or the owner of the surface estate. We conclude that the general grant of minerals in, on or under the property includes a grant of geothermal resources, including steam therefrom.

The owners of the surface estate, Union Oil Company of California, Magma Power Company, Thermal Power Company, and George and Hazel Curry, appeal from a judgment quieting title to the geothermal steam and power and geothermal resources in Geothermal Kinetics, Inc., the owner of the mineral estate. The subject of this action is a geothermal resource existing beneath the surface of approximately 408 acres of property located in an area of Sonoma County known as "The Geysers."

Geothermal Kinetics derives its title from a 1951 deed wherein the owners of the property conveyed to Geothermal Kinetics' predecessor in interest "all minerals in, on or under" the property. George and Hazel Curry succeeded to the surface estate and in 1963 leased to Magma and Thermal (who subsequently assigned a portion of their lease to Union Oil) the right to "drill for, produce, extract, remove and sell steam and steam power and extractable minerals from, and utilize, process, convert and otherwise treat such steam and steam power upon, said land, and to extract any extractable minerals."[1] At the time of execution of the lease, the Currys, the surface fee holders, apparently believed they owned the mineral rights. Geothermal Kinetics, however, has the only valid mineral lease. Therefore, appellants rely solely on their interest in the surface estate for the right to the geothermal resources. In 1973, Geothermal Kinetics, as holder of the leasehold of the mineral estate, drilled a geothermal well on the property at a cost of approximately $400,000.

I. Appellants' primary contention is that geothermal energy is not a mineral; they argue that the resource is not steam, rocks or the underground reservoir but the heat transported to the surface by means of steam. A mineral, appellants claim, must have physical substance and heat is merely a property of a physical substance. In support of this contention, appellants cite several definitions of "mineral" containing reference to "substance." Appellants

---

1. There is no contention here that appellants derived their title from the U.S. Government; therefore, the holding of United States v. Union Oil Co. of California (9th Cir. 1977) 549 F.2d 1271, cert. den. 434 U.S. 930, 98 S.Ct. 418, 54 L.Ed.2d 29, wherein the U.S. Government was deemed to retain the right to geothermal resources by virtue of its reserving mineral rights to the patented property, is not dispositive of the present appeal.

then reason that because they own everything in the property except for "mineral" substances, they own the geothermal resources, citing Civil Code section 829 which provides: "The owner of land in fee has the right to the surface and to everything permanently situated beneath or above it."

Respondent contends that since the parties did not specify particular minerals that were intended to be within the scope of the grant nor include any limitations on it, the grant conveyed the broadest possible estate. It urges that the "grant is to be interpreted in favor of the grantee." (Civ.Code, s 1069.) Respondent urges that we not adopt a mechanistic approach based upon textbook definitions of the term mineral; instead we should adopt a "functional" approach which focuses upon the purposes and expectations generally attendant to mineral estates and surface estates. Since normally the owner of the mineral estate seeks to extract valuable resources from the earth, whereas the surface owner generally desires to utilize land and such resources as are necessary for his enjoyment of the land, the geothermal resources should follow the mineral estate. We agree with respondent's contention.

II. Geothermal resources have been used commercially for several centuries, including their use to generate electricity in the early 1900s. In the United States, exploration and utilization of such resources has occurred generally in the western part of the nation, particularly in California. Commercial development of The Geysers area near Santa Rosa began in 1955 with the successful drilling of four wells. In 1960, Pacific Gas & Electric Company opened an electrical generating plant at The Geysers using the geothermal steam to power the generating turbines. Geothermal steam from respondent's well is piped to the P.G.& E. plant located about a mile away.

Geothermal energy is a naturally occurring phenomenon whose origin is the heat of the interior of the earth. The geothermal resources of The Geysers is apparently due to a layer of molten or semi-molten rock, called "magma," which has risen from the interior of the earth to a depth of 20,000 to 30,000 feet. Above this mass of magma, which constitutes the basic heat source for the area, are protuberances of magma called "plugs" or "stocks," which may rise within 10,000 to 15,000 feet of the surface of the earth. This intrusion of hot magma expels [sic] gases and liquids which combine with ancient water trapped in the surrounding sediment to form a geothermal fluid or brine. This fluid converts to steam which circulates in a sedimentary formation and transports mineral and heat from the magma toward the surface. Convection currents cause water to rise and cool, forming a mineral shell of silica and calcium carbonate which seals off the magma intrusion from the surface. This shell is approximately 1000 feet thick in the area of respondent's well. Immediately below this silicacarbonate seal is circulating geothermal steam and other gases; below these gases is boiling brine.

The seal over the steam reservoir permits only a small amount of ground water to penetrate. The amount of this ground water is insignificant compared to the volume of geothermal steam and brine; its penetration of the seal does not serve to materially deplete the general supply of ground water available for surface use. Hence, the ground water system and the geothermal steam reservoir are separate and distinct. Some geothermal steam escapes from the reservoir to the earth's surface through cracks in the silicacarbonate seal.

At The Geysers wells drilled through the silicacarbonate seal bring geothermal steam to the surface. Respondent's well is approximately 7,200 feet deep. The extracted hot steam, which contains minerals, powers steam turbines to produce commercially valuable electric power. The minerals in the condensed steam are generally toxic, requiring the reinjection of this water back below the silicacarbonate seal. Purification of the condensed steam so as to render it safe for agricultural or domestic purposes is not economically feasible. Geothermal resources are not necessary or useful to surface owners, other than as a source of electricity. The utilization of geothermal resources does not substantially destroy the surface of the land. The production of the energy from geothermal energy is analogous to the production of energy from such other minerals as coal, oil and natural gas in that substances containing or capable of producing heat are removed from beneath the earth. In fact, the wells used for the extraction of the steam are similar to oil and gas wells.

III. In the construction of a grant or reservation of an interest in real property, a court seeks to determine the intent of the parties, giving effect to a particular intent over a general intent. (Civ. Code, ss 1066, 1636; Code Civ. Proc., s 1859.) In the present case, the 1951 grant of mineral rights makes no specific mention of geothermal resources; hence, the general intent of the parties must be ascertained. In the absence of an expressed specific intent, several courts have sought to determine the general intent of the parties in construing the word "mineral" in a deed, rather than resort to attempts at rigid definition. (See United States v. Union Oil Co. of California (9th Cir. 1977) 549 F.2d 1271, 1274, fn. 7; Northern Natural Gas Co. v. Grounds (10th Cir. 1971) 441 F.2d 704, 714, cert. den. (1971) 404 U.S. 951, 92 S.Ct. 268, 30 L.Ed.2d 267; Acker v. Guinn (Tex.1971) 464 S.W.2d 348, 352.)

Initially, we observe that "as a general rule a grant or reservation of all minerals includes all minerals found on the premises whether or not known to exist." (Renshaw v. Happy Valley Water Co. (1952) 114 Cal.App.2d 521, 526, 250 P.2d 612, 615.) Thus, the fact that the presence of geothermal resources may not have been known to one or both parties to the 1951 conveyance is of no consequence.

Generally, the parties to a conveyance of a mineral estate expect that the enjoyment of this interest will not involve destruction of the surface. (See Bambauer v. Menjoulet (1963) 214 Cal.App.2d 871, 872-873, 29 Cal.Rptr. 874; but see Yuba Inv. Co. v. Yuba Consol. Gold Fields (1920) 184 Cal. 469, 194 P. 19; Trklja v. Keys (1942) 49 Cal.App.2d 211, 212, 121 P.2d 54.) In Acker v. Guinn (Tex.1971) 464 S.W.2d 348, 351, the deed of "oil, gas and other minerals in and under" the property did not convey an interest in the iron ore. The court observed that the parties to a mineral lease or deed usually think of the mineral estate as including valuable substances that are removed from the ground by means of wells or mine shafts, but "a grant ... of minerals ... should not be construed to include a substance that must be removed by methods that will, in effect, consume or deplete the surface estate." (at p. 352.)

Here, the trial court found that the exploitation of geothermal resources does not substantially destroy the surface of the property. Wells for the extraction of the energy of geothermal steam are similar to those wells used in drilling for oil. Appellant Union Oil Company apparently considered the development of geothermal resources to be a natural extension of their oil and gas drilling operations. The court found that the production of energy from geothermal resources is analogous to the production of energy from such other mineral resources as coal, oil and natural gas in that materials containing energy are extracted from the earth and transported to facilities where this energy is transformed into electrical energy.[2] The fact that extracted coal, oil and natural gas contain chemical

---

2. The first California legislation, in 1965, enacting a statutory scheme for the regulation of geothermal resources was made a part of Division Three of the Public Resources Code (s 3700 et seq.), which is entitled "Oil and Gas." The Geothermal Resources Act of 1967, relating to the leasing of public lands for the extraction of geothermal resources, is also located in the Public Resources Code in

energy while geothermal resources contain thermal energy is not significant; uranium ore is not denied the status of a mineral because it contains nuclear energy instead of chemical energy.

The parties to the 1951 grant had a general intention to convey those commercially valuable, underground, physical resources of the property. They expected that the enjoyment of this interest would not destroy the surface estate and would involve resources distinct from the surface soil. In the absence of any expressed specific intent to the contrary, the scope of the mineral estate, as indicated by the parties' general intentions and expectations, includes the geothermal resources underlying the property.

In United States v. Union Oil Co. of California, supra, 549 F.2d 1271, the court, dealing with other property in The Geysers area, interprets mineral reservations of "all the coal and other minerals" in patents issued under the Stock-Raising Homestead Act of 1916 to include geothermal resources underneath the patented land (at p. 1273). Although the basis for the holding is partly the Congressional intent to retain government control over energy resources, the court stated that "the words of the mineral reservation in the Stock-Raising Homestead Act clearly are capable of bearing a meaning that encompasses geothermal resources" (at p. 1274). The court further noted that "all of the elements of a geothermal system magma, porous rock strata, even water itself may be classified as 'minerals' " (at p. 1273). Also, in Reich v. Commissioner of Internal Revenue (1969) 52 T.C. 700, affd. (9th Cir. 1972) 454 F.2d 1157, wherein the Tax Court concluded that the geothermal steam at The Geysers was a gas for purposes of the oil and gas depletion allowance in the Internal Revenue Code, the court rejected the contention that heat, not gas, was being produced at The Geysers.

The cases cited by appellants involving the ownership of geologic formations, are readily distinguishable. Emeny v. United States (1969) 412 F.2d 1319, 188 Ct.Cl. 1024, holds that the owner of oil and gas leases did not have a right to use an underground geologic structure on the leased property to store helium gas produced elsewhere; the case deals only with the ownership of a geologic formation having value as a storage facility, and not an extractable commercially valuable resource. Contrary to appellants' suggestion, Edwards v. Sims (1929) 232 Ky. 791, 24 S.W.2d 619 is silent as to the ownership of underground geologic structures where the mineral and surface estates are severed. Edwards states that the owner of property is entitled to the free and unfettered control of his land above, upon and below the surface "unless there has been a division of the estate" (24 S.W.2d at p. 620).

Several courts have held that the grant or reservation of a mineral estate does not include rights to surface or subsurface water. (See Fleming Foundation v. Texaco (Tex.App.1960) 337 S.W.2d 846; Mack Oil Co. v. Laurence (Okl.1964) 389 P.2d 955.) However, such cases concern water that is part of the normal ground water system. As the trial court found, the water and steam components of geothermal resources are part of a separate water system cut off from these surface and subsurface waters by a thick mineral cap. Only insignificant amounts of ground water enter the geothermal water system. Unlike the surface and subsurface waters, the origin of geothermal water is not rainfall, but water present at the time of the formation of the geologic structure. Because rainfall does not replenish geothermal water, it is a depletable deposit. (See Reich v. Commissioner of Internal Revenue (1969) 52 T.C. 700, affd. (9th Cir. 1972) 454 F.2d 1157).

Division Six dealing with "Oil and Gas and Mineral Leases." It can be inferred from the placement of these statutes that the Legislature viewed geothermal resources as a mineral.

Not only is there a sound geologic basis for distinguishing between the usual ground water system and geothermal waters, but the rationale for recognizing the rights of the surface estate to these ground waters is largely inapplicable to geothermal waters. (See Bjorge, The Development of Geothermal Resources and the 1970 Geothermal Steam Act Law in Search of Definition (1974) 46 U.Colo.L.Rev. 1, 22-23; United States v. Union Oil Co. of California, supra, 549 F.2d at p. 1280, fn. 21; Olpin, The Law of Geothermal Resources (1968) 14 Rocky Mt. Min. L. Inst. 123, 140-141.) Several of the cases cited by appellants in support of the proposition that the surface estate includes rights to surface and subsurface waters, refer to the necessity of this water for the enjoyment of the surface estate. (See Mack Oil Co. v. Laurence, supra, 389 P.2d at p. 961; Vogel v. Cobb (Okl.1943) 141 P.2d 276, 280.) In the present case, the extraction of geothermal water for a domestic water source is impractical; the cost of respondent's well was approximately $400,000. In addition, geothermal water contains toxic minerals making it unfit for surface, agricultural or domestic use. Purification is not economically feasible. The water is so toxic that the Water Quality Control Board requires its reinjection deep into the earth. The analysis leading to the conclusion that geothermal resources are part of the mineral estate also leads to the conclusion that geothermal water is a mineral and thus, not part of the waters included in the surface estate. Recognition of rights of the owner of the surface estate to geothermal water would mean that resources consisting of hot rock without any fluid system belong to the mineral estate while fluid geothermal systems, like that in the present case, would be subject to a divided ownership with the surface estate owner having an interest in the water, and the mineral estate owner having an interest in any commercially valuable dissolved minerals. The difficulties of determining the type of system or systems on a particular property, as well as the confusion and complexity attendant to such an approach, are clear.

Examining both the broad purpose of the 1951 conveyance of the mineral estate and the expected manner of enjoyment of this property interest, it appears that the rights to the geothermal resources are part of the grant. A principal purpose of this conveyance was to transfer those underground physical resources which have commercial value and are not necessary for the enjoyment of the surface estate. (See Western Development Co. v. Nell (1955) 4 Utah 2d 112, 288 P.2d 452, 455.) The trial court correctly determined that the mineral grant herein conveyed to respondent the right to the geothermal resources located in, on or under the property in question.

Judgment is affirmed.

———————

Notes and Questions

1. Look at the language of grant in this case and compare it to the language of reservation in the federal deed in *Union Oil*. How does this change the analysis of the *Geothermal Kinetics* court in contrast to the reasoning of the *Union Oil* court?

2. The *Geothermal Kinetics* case involved a lease between private owners instead of one with the federal government. In such agreements, the rules of construction focus on the intent of the parties. But which party's intent is at issue? Who benefits from this focus? Is there an alternate mode of construction that is fairer?

3. Consider how the rules used here for interpreting mineral leases might apply in the context of wind or solar rights.

4. What is the result of the *Geothermal Kinetics* court's reasoning? Who owns the geothermal resource? Do you think that is the correct result?

5. How does the water resource relate? Is the water part of the geothermal right or does it need to be regulated separately under a state's groundwater regulation regime?

6. Who owns abandoned bore holes? May someone come back and use them to develop geothermal resources? How is this issue similar to or different from the question of who owns an abandoned mine or the pore space of a depleted oil or gas field that now may have value for carbon sequestration purposes?

D. Acquisition Process

This section briefly describes the process for acquiring the right to develop geothermal resources.

Donald J. Kochan & Tiffany Grant, *In the Heat of the Law, It's Not Just Steam: Geothermal Resources and the Impacts on Thermophile Biodiversity*
13 Hastings W.-N.W. J. Envtl. L. & Pol'y 35, 56-61 (2007)

Compliance with California geothermal law can be broken down into the following steps: acquisition of the right to develop a geothermal resource, meeting environmental requirements, and fulfilling development and extraction requirements. Acquisition of rights to develop a geothermal resource depends on whether the geothermal resource is federal, state, or private property. It should be noted that where surface and mineral estates are split, it is necessary to acquire rights to the geothermal resource itself and to "siting" rights, the right to construct a geothermal energy plant on the land above the resource. Siting rights are particularly important on federal leases for geothermal resources, where the United States has reserved mineral rights to land received under land grants. This section addresses the acquisition of rights to develop a geothermal resource on federal lands, including applicable case law, followed by federal environmental and development requirements. It then addresses the development of geothermal resources on California's state and private lands, through state environmental and development requirements.

Federal Geothermal Acquisition and Siting Rights

The majority of geothermal resources in the United States are located on federal land in the West. Prior to 1970, geothermal resource development was limited primarily to private lands, because federal government agencies including the Department of the Interior ("DOI") and National Forest Service under the Department of Agriculture were reluctant to dispose of geothermal resources on lands within their jurisdiction without congressional direction. To reduce this restriction on geothermal resource development, President Nixon approved the Geothermal Steam Act.

The Geothermal Steam Act ("Act") of 1970 is the basis of all federal geothermal jurisprudence. With two exceptions, the Act is the only means of acquiring rights to develop geothermal resources on federal lands. According to legislative history, the purpose of the Act was to "permit exploration and development of geothermal steam and associated geothermal resources underlying certain public domain land." The Act gave the Secretary of the Interior the power to issue leases to U.S. citizens for geothermal steam development and utilization on public lands, including national forests, as well as on lands conveyed from the U.S. to private entities. The U.S. may reserve geothermal steam and associated resources on land it conveys. The Act sets forth guidelines for leasing and royalties and states that a lessee is "entitled to use so much of the surface of the land covered by his geothermal lease to be necessary for the production, utilization, and conservation of geothermal resources." Certain federal lands and Indian lands are exempt from the Act.

The Act provides an exclusion clause for the development of geothermal resources within national parks when a significant thermal feature will suffer significant adverse effects. Section 1026 of the Act designates the monitoring and determination of adverse effects of proposed development within national parks, which are subject to notice and public comment. Specifically, the Act provides that the Secretary "shall determine on the basis of scientific evidence if exploration, development or utilization of the lands subject to the lease application is reasonably likely to result in a significant adverse effect on a significant thermal feature within a unit of the National Park System." For projects that the Secretary determines are "reasonably likely to result in a significant adverse effect on a significant thermal feature within a unit of the National Park System, the Secretary shall not issue such lease."

Outside the Act, offshore geothermal resources may be developed under the Outer Continental Shelf Lands Act, and the Department of Defense ("DOD") may develop geothermal resources on lands under its control. For example, a geothermal plant was developed at the China Lake Naval Weapons Center in Coso, California. The Coso plant utilizes a steam system which, at 311 [degrees] F (155 [degrees] C), generates energy to drive six turbine engines. Although the Act provides that lessees shall use all reasonable precautions to prevent the waste of steam and associated resources, the Act does not require the preservation of geothermal resources. The Act does provide that the Secretary shall prescribe rules for the development and conservation of geothermal and other natural resources, but does not mention conservation of geothermal organisms.

The primary question that arose from the Geothermal Steam Act was how to determine which lands conveyed by the U.S. are "subject to a [mineral] reservation of the geothermal steam and associated resources." The Ninth Circuit case United States v. Union Oil Company held that the U.S. reserved geothermal resources minerals on lands acquired under the Stock-Raising Homestead Act of 1916 ("SRHA"). As part of the effort to "civilize" the West, the SRHA transferred public lands to private ownership under patents subject to a reservation by the U.S. "of all the coal and other minerals." The SRHA did not directly address the reservation of geothermal resources or express an intent to reserve them, because Congress "was not aware of geothermal power" when it enacted the SRHA.

In Union Oil, landowners in the Geysers Field argued that the term "minerals" should be given the "meaning it had in the mining industry at the time the [SRHA] was adopted," and that geothermal resources should not be considered a "mineral" under the SRHA. The court instead looked at whether it "would further Congress's purposes to interpret" geothermal resources as minerals, and finding that it did, held that the mineral reservation under the SRHA included geothermal resources.

It should be noted that "nothing prevents a contrary result in a case involving private rights arising in another state," or under a statute other than the SRHA. In Bedroc Limited, LLC v. United States, the Supreme Court distinguished a mineral reservation under the Pittman Act from a mineral reservation under the SRHA. The Supreme Court had construed the SRHA to include a mineral reservation of gravel where the SRHA reserved "all the coal and other minerals." In Bedroc Limited, however, the Pittman Act reserved "all the coal and other valuable minerals." The Supreme Court noted that at the time the Pittman

Act was enacted, gravel was not a valuable mineral and therefore was not reserved to the United States. Likewise, a state or another federal land grant may treat geothermal resources differently than the SRHA does. For example, a court may find that geothermal resources do not fall under the Pittman Act either, because geothermal resources were not regarded as valuable at the time of enactment. The right to develop federal geothermal resources under the Geothermal Steam Act is obtained through a lease from the Bureau of Land Management ("BLM") as authorized by the DOI. After rights to develop the resource are acquired, rights to construct a geothermal energy plant must be obtained. The Geothermal Steam Act provides that a geothermal lessee "shall be entitled to use so much of the surface of the land as may be found by the Secretary [of the Interior] for the production and conservation of geothermal resources."

Lessees under the Act do not need consent of a private landowner in order to build. In Occidental Geothermal, Inc. v. Simmons, the holder of a DOI geothermal resources lease filed suit against two landowners who held surface rights where the U.S. had reserved mineral rights under the SRHA. Occidental sought, "among other forms of relief," declaration of its right to build and operate a geothermal plant without consent of the surface owners. The court held that power plant siting rights in lands under the SRHA were reserved to the U.S. and that the Geothermal Steam Act authorized such leases. The court noted that removal of geothermal resources is inextricably connected to their utilization, and to hold that geothermal lessees own the rights to geothermal resources and "yet do not have the right to exploit those resources without the consent of the owners of surface interests would reduce the holding of Union Oil to an empty theoretical exercise."

Federal Environmental and Development Regulations

After acquiring rights to develop geothermal resources and siting rights, geothermal energy developers begin the actual development of the geothermal resource. According to the Department of the Interior, the "development and production of geothermal resources involves six phases: exploration, test drilling, production testing, field development, power plant and power line construction, and full-scale operations."

Because the lease of federal geothermal resources requires the discretionary approval of a federal agency, geothermal resource development on federal land is subject to the National Environmental Policy Act ("NEPA"). NEPA was enacted to "ensure that all federal agencies consider the environmental impact of their actions" through the development of an environmental impact statement ("EIS"). A question arises as to which stage of geothermal resource development triggers NEPA compliance and the drafting of an EIS.

The Ninth Circuit addressed this question in Sierra Club v. Hathaway. The Sierra Club brought suit to prevent the Secretary of the Interior from executing leases to develop geothermal resources in the Alvord Desert Geothermal Area of southeastern Oregon, based on the DOI's failure to draft an EIS. The court, in holding for the DOI, noted that the lease in question was only in the exploration stage and that the DOI had conducted a programmatic EIS for leasing under the Geothermal Steam Act. The programmatic EIS concluded that exploration practices in the first or "casual use" stage do not significantly affect the environment, as those practices "do not ordinarily lead to any appreciable disturbance or damage to lands, resources, and improvements."

The court recognized that to undertake exploration other than casual use, the lessee must submit a detailed plan of operations to the United States Geologic Survey ("USGS") with proposed measures for "protection of the environment, including but not limited to, the prevention or control of (1) fires, (2) soil erosion, (3) pollution of the surface and groundwater, (4) damage to fish and wildlife or other natural resources, (5) air and noise

Development Barriers

The National Environmental Policy Act ("NEPA"), and the related state laws referred to as "little" NEPAs, e.g., California Environmental Quality Act, are procedural statutes which require some level of environmental evaluation for exploration or development of projects. The level of environmental review varies according to the range of the possible environmental impacts from the proposed project. The environmental evaluation performed pursuant to NEPA and little NEPA statutes often results in the greatest barrier to development, especially if evaluation is challenged in court. Treatises have been written on the law of NEPA and little NEPAs and it is beyond the scope of this paper to summarize in a few pages such a vast area of the law. In brief, proposed renewable energy developments encounter just as much opposition from citizen and environmental groups as any other proposed developments, and often the environmental issues are the same as for any other type of development. The Bureau of Land Management has recently released its draft "Programmatic Environmental Impact Statement" for leasing lands located in the western United States, Alaska and Hawaii for geothermal exploration and development. See "Programmatic Environmental Impact Statement for Geothermal Leasing in the Western United States," Vol. 1, May 2008, Table 2-7, page 2-35. This three volume document presents an excellent overview of the procedures for leasing public lands for geothermal exploration, the potential environmental impacts arising from geothermal exploration and development, and an introduction to geothermal energy in general. [You can read the PEIS at: http://www.blm.gov/wo/st/en/prog/energy/geothermal/geothermal_ nationwide.html — Ed.]

Kurt E. Seel, Legal Barriers to Geothermal Development,
16 A.B.A. Sec. Env't, Energy, & Resources at 8-5 (2008)

pollution, and (6) hazards to public health and safety during lease activities." Thus, geothermal energy developers are able to postpone the NEPA process until a development plan is prepared. It should be noted that although NEPA requires an EIS, it does not require that even significant environmental impacts be mitigated or avoided....

Notes and Questions

1. Based on the excerpt above, can you prepare a flow chart of the issues you should consider to acquire and develop a geothermal property?

2. Note the excerpt also discusses compliance with state agency requirements on lands that are not federally owned or regulated. If you plan to develop geothermal resources on state or private lands, what additional or alternative steps must you follow?

3. What is the role of NEPA in this process? Review the Development Barriers box above for additional considerations.

E. Environmental Concerns

Geothermal power plants provide a reliable source of baseload electricity. Once they are on line, they shut down only about 5 percent of the time. According to Duffield & Sass, geothermal power plants are also relatively clean, emitting up to 86 percent fewer kilograms of CO_2 per megawatt hour than the combustion of bituminous coal. In addition,

Duffield & Sass state that geothermal power plants emit fewer sulfurous gases than coal-fired plants. Despite these air emission advantages, geothermal development has environmental downsides, some of which are described in the following three excerpts.

Donald J. Kochan & Tiffany Grant, *In the Heat of the Law, It's Not Just Steam: Geothermal Resources and the Impacts on Thermophile Biodiversity*

13 HASTINGS W.-N.W. J. ENVTL. L. & POL'Y 35, 49-54 (2007)

This section discusses some of the environmental impacts of geothermal resource use. Defining geothermal energy production as "green" may be misleading. Although apparently less environmentally damaging than fossil fuel and nuclear energy production, geothermal energy does have adverse environmental impacts. Potential impacts include noise and air pollution associated with the construction of geothermal plants, subsequent thermal and mineral stream pollution from plant discharge, loss of thermophile biodiversity and geologic record, unknown groundwater depletion and subsidence, and tectonic effects of injection and re-injection of water into geothermal reservoirs. In addition, plant construction may conflict with the Endangered Species Act.

Noise, Air Pollution, and Endangered Species Habitat

As noted above, the need to harness geothermal energy close to the source requires the construction of energy plants in remote areas. The geothermal plant itself can add to noise pollution, which can be particularly harmful to the habitat of an endangered or threatened species. Many species, including birds, frogs, and mammals, rely on calls and sounds to mate, interact, feed their young, and warn of predators.

Although they generate significantly less pollution than a fossil fuel plant, geothermal plants can also harm air quality. Geothermal plants release hydrogen sulfide and carbon dioxide into the atmosphere. However, the U.S. Department of Energy reports that up to 99.9 percent of the hydrogen sulfide could be eliminated through technological processes.

Thermal and Mineral Stream Pollution and Geothermal Brine

Thermal stream pollution occurs when heated water is added to a surface stream, increasing its temperature. Fish and other aquatic or riparian species are intolerant to significant changes in temperature, which can result in population die-offs. Even minimal temperature changes can interfere with breeding by changing the temperature of eggs. Temperature pollution can be reduced using cooling towers. However, the use of local water by cooling towers can cause a "'severe strain on the available water resources.'"

Mineral stream pollution occurs when water with a high or different mineral concentration is added to a surface stream, increasing its mineral content. The addition of geothermal brines to streams can also add to environmental degradation. As discussed previously, minerals from surrounding rock leach into the solution, adding dissolved metals to the water. Discharge of these minerals into streams changes the chemical and mineral composition of the stream, which affects instream species.

Thermal Biodiversity and the Geologic Record

Geothermal resources contain extreme temperatures and concentrated mineral compositions, making them toxic for most prokaryotic species. However, certain species have adapted to live in these toxic ecosystems—hyperthermophiles and extremophiles, as well as thermophiles. "Ecological studies have shown that water-containing terrestrial, subterranean and submarine high-temperature environments harbor a great diversity of hyperthermophilic prokaryotes, growing fastest at temperatures of 80 degrees Celsius or above." Geothermal resources remain one of the last sources of uncategorized biodiversity on earth. "Microbial processes in these environments are of critical importance to the biosphere, and the noncultured bacteria residing there are a valuable resource for novel genomic information." Norman Pace, a molecular biologist at the University of Colorado, has noted:

> It has become clear over the past few decades that substantial microbial diversity occurs at very high temperatures. Hyperthermophilic organisms … promise a wealth of unknown biochemistry and biotechnological potential and challenge our comprehension of biomolecular structure. Nonetheless, relatively little is known about the diversity of life at high temperatures because of a traditional problem in microbial ecology: the inability to cultivate naturally occurring organisms.

Some argue that hyperthermophiles are the most primitive organism existing today. Hyperthermophiles are prokaryotes that require temperatures above 80C to 110C to live, functioning best at temperatures around 100C. Some species are able to "grow at up to 113 [degrees] C and, therefore, [represent] the upper temperature border of life." Cultures of some species are able to survive autoclaving, making them valuable for scientific research, especially microbial research.

Bioprospecting the species can yield pharmaceutical and industrial products and applications. The "enzyme Taq polymerase … was discovered through research on a thermally adapted microbe known as Thermus aquaticus" from sampling a hot spring in Yellowstone National Park. Both the enzyme and a resulting technique used in DNA identification called the PCR process were subsequently sold for $ 300 million in 1991 and reportedly generate annual revenues around $ 100 million per year.

Thermophile diversity is essential to such research. Diversity is threatened by human use of hot springs and fumaroles, and arguably bioprospecting of those resources. Loss of diversity may also occur on a subterranean level, due to depletion of groundwater sources, and injection and re-injection of non-heated water. As noted earlier, thermophiles are sensitive to changes in temperature and pressure. Injection of non-heated water into a geothermal reservoir can kill species if the temperature change is great enough to cool surrounding geothermal fluid, even briefly, to a temperature below tolerable levels.

In addition to adding to the advancement of science, geothermal resources may also add to the earth's geologic record by recording present fossil records. Hydrothermal springs have high rates of "microbial productivity, which often coexist with high rates of mineral precipitation, a situation generally regarded as highly favorable for microbial fossilization." Microbial fossilization, biosignatures, and their applicability to scientific innovation are beyond the scope of this article, but it should be noted that microfossilization of thermophiles and their biosignatures add to our scientific knowledge base, aid our understanding of evolution, and have the potential to drastically enhance medical and scientific research.

Unknown Replenishment Rates, Undefined Aquifers, and Unknown Chemical and Biological Compositions of Geothermal Resources

Many unknown conditions affect the sustainability and potential environmental impacts of geothermal resources. Research on replenishment rates and aquifer definition is primarily conducted in association with a proposed or currently utilized geothermal application.

The results are likely to be site or aquifer specific. The lack of scientific research and understanding of geothermal ecosystems may threaten thermophile biodiversity.

Geothermal resources have developed over tens of thousands of years, and their isolation created countless site-specific species. Separate species thrive as their environments vary in temperature and mineral composition. Species differentiate between geothermal fields and even between hot springs. Furthermore, a single pool may contain multiple species. Researchers have found different species at separate areas along the temperature gradient of a hot spring pool.

In addition, little is known about the subsurface mineral composition and biodiversity of geothermal resources. Research into the mineral, chemical and biological composition of subsurface geothermal resources is costly. In addition, it is difficult to maintain the integrity of the subsurface sample in collection. A report by the California Division of Oil and Gas notes:

It is possible to completely characterize the chemical composition of a bottled sample in the laboratory, but this sample may be much different from the geothermal fluid from which it was obtained. Certain chemical species will oxidize on exposure to air, and the partitioning of species between the gas and the liquid phases will change in response to changes in physical conditions.

All of these unknown factors add to the possibility of irreparable harm to the thermophile biodiversity of geothermal resources from the depletion and degradation of geothermal resources.

Kurt E. Seel, *Legal Barriers to Geothermal Development*
16 A.B.A. Sec. Env't, Energy, & Resources at 8-8 (2008)

Solid and Hazardous Wastes

Certain drilling fluids, produced waters and other wastes associated with the exploration, development, or production of oil, gas and geothermal energy ("E&P wastes") are exempt (or partially exempt) from federal and state hazardous waste laws. See e.g., 40 CFR 261.4(b)(5). But guidance regarding the scope of the federal exemption for geothermal wastes, and the applicability of the various state laws is often complex. For example, in California geothermal wastes are subject to the overlapping jurisdiction of the Department of Toxic Substances Control; the Regional Water Quality Control Board; and the Division of Oil, Gas and Geothermal Resources. Some of these agencies have authority delegated from the U.S. Environmental Protection Agency to implement selected federal programs, e.g., RCRA and geothermal injection wells, while at the same time the agencies have entered into Memoranda of Understanding regarding primary responsibility for addressing discharges of wastes, including geothermal wastes, at geothermal fields. Therefore, any evaluation of a geothermal prospect should also consider the costs associated with compliance with laws pertaining to E&P wastes.

Notes and Questions

1. What are some of the noise problems created by the development of geothermal energy? What about air pollution?

2. What is the difference between thermal stream pollution and mineral stream pollution?

3. What are some of the unknown conditions that can affect the sustainability and potential environmental impacts of geothermal resources?

4. The Seel excerpt makes the point that many drilling fluids associated with the exploration, development, or production of geothermal energy (and oil and gas) are exempt or partially exempt from federal and state hazardous waste laws. What is the impact of this exemption?

5. The focus of the Kochan & Grant article, from which many of the excerpts in this chapter derive, is the protection of thermophiles. The article makes the argument that the lack of scientific knowledge and understanding of thermophiles, and their geothermal environments, adds to the potential for the depletion and loss of thermophile biodiversity by current geothermal uses designed to exploit the resource for energy, recreation, or other uses. Review the additional discussion of thermophiles in the excerpt below.

Donald J. Kochan & Tiffany Grant, *In the Heat of the Law, It's Not Just Steam: Geothermal Resources and the Impacts on Thermophile Biodiversity*
13 HASTINGS W.-N.W. J. ENVTL. L. & POL'Y 35, 36-37 (2007)

... Because of their ancient origins, thermophiles may hold keys to some of the earth's most unfathomable mysteries, including how molecules evolved and how life formed on the planet. As an article in the New York Times reports, "Many biologists believe the earth's first organisms arose in the deep sea along volcanic gashes and that microbes known as thermophiles, which thrive today in such hot regions [where earthquakes or volcanic activity create fissures that bring heat to the surface through water or otherwise], are their direct descendants." Because of their ancient origins, thermophiles may hold keys to some of the earth's most unfathomable mysteries, including how molecules evolved and how life formed on the planet.

* * *

Current geothermal laws and regulations are directed toward the development, exploitation, and depletion of California's geothermal resources. This is exemplified by the recent zoning restriction in the Desert Hot Springs community of Southern California, which prohibits the development of land in the "hot water district" unless the landowner agrees to extract the underlying geothermal resource for commercial purposes. Such zoning forces landowners who would otherwise refrain from extracting the resource, to submit to geothermal pumping and extraction simply to develop their property. [Yet few laws exist to conserve or preserve geothermal and thermophile biodiversity—Ed.]

* * *

This article has outlined the basic mechanisms by which geothermal resources and thermophile biodiversity function. From understanding the complex web of plate tectonics, groundwater, magmatic heating, and thermophile biodiversity, we acknowledge that these incredibly specialized species of thermophiles are important to our understanding of biological life on the earth. Keeping in mind that geothermal resources are finite resources and not a renewable source of energy, the exploitation of geothermal resources without concern for thermophile biodiversity is tantamount to allowing at least some thermophile extinction or dissolution.

Many unknowns exist concerning exploitation of geothermal resources, including unknown replenishment rates, undefined aquifers, and lack of knowledge about thermophiles and how they should be categorized. In addition, exploitation of a geothermal resource for energy production can deplete groundwater, cause land to subside, and increase the risk of earthquakes.

This article has noted the importance of the preservation of thermophile biodiversity for future medical and scientific research. Though invisible to the naked eye, thermophiles may well have value independent of, and perhaps superior to, energy production.

There is certainly much more than steam or water in geothermal resources. In the heat of the law—and the heat of the desire to develop energy resources—we need to examine and consider what might be lost in the exploitation and extraction of these age-old resources, and strengthen the laws to protect them.

Notes and Questions

1. Do you think that geothermal power can be developed without compromising thermophiles? What, if any, accommodations can be made?

2. The acronym "TANSTAAFL" for "There Ain't No Such Thing As A Free Lunch" has been used to describe the tradeoffs we make for the lifestyles we have chosen. George (Rock) Pring, et al., *The Impact of Energy on Health, Environment, and Sustainable Development: The TANSTAAFL Problem*, in BEYOND THE CARBON ECONOMY: ENERGY LAW IN TRANSITION (Donald N. Zillman ed., Oxford Univ. Press 2008). Is lack of thermophile preservation a tradeoff that we must be willing to make in exchange for sustainable energy?

F. Additional Considerations

This section addresses three additional areas of consideration in the development of geothermal resources outside of traditional environmental impacts: induced seismicity, conflicts with sacred lands, and other competing interests.

1. Induced Seismicity

The injection of materials and removal of groundwater in the process of developing geothermal resources can trigger earthquakes that might not otherwise occur. The following excerpt describes this problem briefly.

Donald J. Kochan & Tiffany Grant, *In the Heat of the Law, It's Not Just Steam: Geothermal Resources and the Impacts on Thermophile Biodiversity*
13 HASTINGS W.-N.W. J. ENVTL. L. & POL'Y 35, 54-56 (2007)

Groundwater is the water contained in interstitial areas between rock compositions on the earth's surface. The use of hot or low temperature water systems inevitably requires the depletion of associated groundwater.

Groundwater carries a "considerable portion of the earth's ground load," the weight of the ground surface. "Land subsidence is 'a gradual settling or sudden sinking of the earth's surface owing to subsurface movement of earth materials.' Though several different earth processes can cause subsidence, more than 80 percent of the subsidence in the United States is related to the withdrawal of groundwater." Subsidence results in "damage to roads, buildings, and other structures," in addition to damaging the environment and habitat in the area of subsidence. A Geothermal Literature Assessment by Gawell & Bates notes:

> Because geothermal operations take place in areas that are very tectonically active, it is often difficult to distinguish between geothermal-induced and naturally occurring events. However, geothermal energy production has been shown to at times result in land subsidence. This occurs when the withdrawal of a fluid from an underground reservoir results in a reduction of pressure, thereby causing subsidence. However, subsidence can generally be defined as any slow ground movement, whether it is horizontal movement or vertical movement. Such subsidence occurs not only in geothermal fields, but in petroleum reservoirs as well. The most serious problems have occurred outside of the U.S. and may be the result of different approaches to re-injection technology. Weakening of underground support is suspected as being the cause of massive subsidence at Wairakei geothermal field in New Zealand—the largest subsidence ever recorded[,] which is generally thought to be human-induced. It is also suspected as being responsible for the large landslide at the Zunil geothermal field in Guatemala. Re-injection has been shown to help reduce the effects of subsidence.

As noted, a remedy to subsidence is to re-inject water back into the reservoir. However, little is known about the geologic and tectonic effects of re-injection into hot water systems and injection of water into dry rock systems. Utilization of geothermal resources has also been noted to add to the induced seismicity of geothermal fields:

> Geothermal resources are almost always found in places that are very tectonically active, which means that these areas will be subject to a great deal of geological activity even in the absence of field development. Therefore, seismic activity in geothermal regions raises questions about whether the calamity was due to natural causes or was man-made. The literature appears to indicate that geothermal operations can indeed cause some seismic activity, but the earthquakes that are generated are extremely small and weak, and usually require sensitive instrumentation to be detected at all, even directly above the epicenter. These microearthquakes appear to be associated with the subsurface pressure changes caused by production and injection operations.

Although unclear and relatively unstudied, there appears to be a significant question of how utilization of geothermal resources affects tectonic activity.

———————

Notes and Questions

1. Must we assume the risk of seismicity to develop geothermal resources? Often magma activities are already in seismically active areas. How can a link be made directly to a geothermal plant as the cause?

2. If increased seismicity could be directly linked to a geothermal plant, do you think that insurance or compensation for those injured in an earthquake might appropriately be factored into the cost of development?

3. Are geologists and geophysicists who induce earthquakes criminals? Geologist Markus Häring (aka Haering), one of the founders of Geopower Basel—AG, a Swiss partnership, designed an enhanced geothermal project to supply power to the city of Basel, Switzerland. The company's drilling and injection of pressurized water into rocks approximately three miles directly under Basel induced about thirty earthquakes that caused approximately $9 million in damage to buildings in the region. Geopower Basel reimbursed Basel residents for their property damages. However, criminal charges, including murder charges, were filed against Häring personally. Although the murder charges were dropped, Häring potentially faced a sentence of five years in prison if convicted of intentional property damage. Ultimately, in December of 2009, the Basel criminal court found no evidence that Häring either (1) acted carelessly or (2) intentionally caused earthquakes or property damage. If geothermal projects in the United States caused similar seismicity, what types of criminal charges might be brought?

4. Regardless of the potential for criminal charges, financial liability alone may be enough to dampen enthusiasm for similar enhanced geothermal projects in the United States. Shortly after the shutdown of Geopower Basel's project, one of the big U.S. projects, AltaRock Energy near The Geysers, was abandoned. Starting with the Price-Anderson Act in 1957, the U.S. government has long encouraged the development of nuclear power generation by capping the liability of private companies that operate the power plants and providing federal insurance to the nuclear industry. Should the federal government consider similar support for enhanced geothermal power?

2. Conflict with Cultural Resources or Native American Sacred Sites

For centuries, hot springs have been gathering places and sacred sites for humankind. Any potential developer should be aware of the responsibilities not to violate any laws established to protect cultural treasures.

Kurt E. Seel, *Legal Barriers to Geothermal Development*
16 A.B.A. Sec. Env't, Energy, & Resources at 8-6 to 8-7 (2008)

Even public lands being managed for multiple-use can be difficult to explore or develop if they are considered cultural resources, used by Native Americans for cultural or spiritual purposes, or are subject to the National Historic Properties Act. For example, pursuant the National Historic Preservation Act, the federal government has declared some large areas of United States Forest Service Lands to be eligible for the National Register of Historic Places as Traditional Cultural Property ("TCP") and as Traditional Cultural Places Districts ("TCDs"). The land is particularly eligible where it has been used by Native Americans for cultural or spiritual purposes. As a result, geothermal and development of those areas, even where valid mineral leases have been issued, may be difficult or impossible.

Recently, the BLM identified the Medicine Lake/Glass Mountain area of northern California as having the potential to generate 480 MW by 2015. See "Programmatic Environmental Impact Statement for Geothermal Leasing in the Western United States," Vol. 1, May 2008, Table 2-7, page 2-35. Earlier this year the U.S Forest Service initially rejected an application to conduct non-impacting, geophysical data gathering activity over a 50 square mile area of multi-use, national forest lands located in the Medicine Lake area. Normally non-impacting geophysical data gathering activities of this type are categorically exempt from intensive NEPA evaluation, and can be conducted even if no mineral leases have been issued (citation omitted)....

However, in the above Medicine Lake situation, the area was designated a TCD, and as a result the U.S. Forest Service concluded that: "[A]lthough no "physical" effect has been identified, the tribes have indicated that such a survey would be "disrespectful" to the overall spiritual nature and sacred quality of the MLHTCD [Medicine Lake Highland Traditional Cultural Places District], an effect which could not be mitigated." The U.S. Forest Service letter then stated that its denial " ... is not subject to administrative appeal ..." nor is it a proposed action under NEPA. In summary, the US Forest Service concluded that collecting geophysical data from anywhere within the fifty square mile area of the public forest lands, without leaving any physical effect, was so invasive and disrespectful of the sacred nature of the lands to deny any such exploratory activity. The Forest Service subsequently reversed it position and is currently leaving it to the BLM to conduct an Environmental Assessment, and make a decision, regarding the proposed exploration activity. Developers need to understand all regulatory restrictions, and the cultural aspects, of the lands they intend to explore and develop.

Tribal members have also successfully challenged, and continue to challenge, the validity of federal geothermal mineral leases and proposed geothermal projects, including two proposed 49.5 MW plants in the Glass Mountain/Medicine Lake area. (citation omitted).

Notes and Questions

1. Is commercial development of geothermal energy irreconcilably incompatible with preservation of sites that are sacred to Native groups?

2. Who should have authority to regulate these areas and make decisions about their development?

3. What principles might be evoked to set priorities?

3. Competing Interests

This section continues with a theme we have seen addressed throughout this book: how do we balance green energy development against other environmental concerns?

Donald J. Kochan & Tiffany Grant, *In the Heat of the Law, It's Not Just Steam: Geothermal Resources and the Impacts on Thermophile Biodiversity*

13 HASTINGS W.-N.W. J. ENVTL. L. & POL'Y 35, 45-49 (2007)

Geothermal resources have different competing interests, depending upon the classification of the resources as dry steam, hot water, hybrid, and hot rock, or alternatively, low temperature water. This section first addresses competing interests and environmental opposition to the energy-producing geothermal resources. Although geothermal energy production is thought of as "green energy," construction of geothermal plants may nonetheless be met with environmental opposition, much like the debate over whether windmill farms harm birds. Second, this section concludes with how uses of low temperature water may cause conflict.

Dry Steam, Hybrid, Dry Rock, and Hot Water Energy Generation

The growing need for efficient renewable energy in California and elsewhere has led to research into green energy sources such as wind, solar, and geothermal energy production. This section addresses the economic limitations of geothermal resources and the environmental opposition to their use.

The expense of harnessing and transferring geothermal energy adds to the scarcity of geothermal resources, to make the industry "high-risk and capital intensive." The risk means larger energy companies with sufficient capital dominate the industry. Companies have to build the plants, transfer the power once it is generated, and mitigate the effects of high mineral content and toxicity on equipment and operating staff.

Geothermal resource utilization requires that the geothermal resource be used "at or very near the site of its production." Geothermal resources generate electricity because of their high temperatures. Thus, any transport of water or vapor over long distances would result in significant cooling of the resource and a loss of potential electrical generation. Due to the need to harness the heat energy at the nearest possible surface location, geothermal plants are usually constructed in remote areas.

Once power is generated, it must be transferred through power lines to users. Power transfer from remote areas requires the "acquisition of power line easements and the installation of transmissions lines." Such limitations on remote locations significantly affect the cost effectiveness of geothermal power plants. The remote location of the power plant also contributes to the environmental concerns.

The mineral content of the water or steam also contributes to the expense of energy generation. For example, California's Salton Sea Geothermal Field "produces high temperature, high salinity water. Maximum temperatures range from 220 [degrees] C to 360 [degrees] C ... and the concentration of dissolved solids is up to 35% by weight." The combination of high temperatures and high concentrations of dissolved solids causes severe erosion of the carbon steel well casing "used to bring the resource to the surface." Erosion of the well casing and related equipment can result in blowouts. Therefore, geothermal well equipment must meet blowout prevention standards, adding to the costs.

Geothermal power plants are usually proposed in rural areas that for the most part remain in a natural state, and therefore they often trigger environmental opposition. Northern California groups have argued that geothermal plants will "further carve up forests" that local Native American tribes consider to "be part of their sacred lands." One pro-industry article claims that environmental opposition to geothermal plants in California is "no less hostile than [the] attitude toward all other forms of man-made power. After the installation of hundreds of 'alternative' energy plants in the state ... the greens have begun to reject one renewable power technology after another." Some of the reasons for the strong environmental opposition are noted in the environmental concerns section of this article.

Low Temperature Uses

Low temperature geothermal water is used for many different applications including direct heating, mineral spas, and religious ceremonies and customs. For naturally-occurring hot springs, conflict arises between religious users, local users, and environmental protection interests. Private property rights and tourism concerns have clashed in areas with geothermal fields capable of providing geothermal resources for mineral spas.

While energy producers in the U.S. rarely heat buildings with low temperature resources, the method has the potential to be a cost effective local heat source. Low temperature systems developed in Idaho in the 1800s are still in use today, heating private homes and government buildings, including the state capital building. Globally, direct heat is often used in geothermal regions; for example, Iceland warms 90 percent of the buildings in its capital with direct heating. Direct heating can also be used in food processing applications and in heating commercial greenhouses.

In California, naturally occurring hot springs provide public enjoyment and income for private spa enterprises. Conflict may arise between local users who have developed the resource by constructing bathing pools and groups exercising religious or cultural customs. Many eastern cultures believe that the mineral water aids in longevity and health. To comply with health regulations, owners of hot springs used for human bathing must regularly drain and clean the pools. This can result in runoff of human waste, cleaning compounds, and thermal pollution to nearby streams, thereby affecting the local ecosystem health and biodiversity. In addition, the cleaning and scrubbing of the hot spring necessarily results in the loss of thermophile species growth. Some areas, such as Yellowstone National Park, completely forbid human bathing or soaking in natural hot springs.

Conflict also arises in areas where low temperature geothermal resources are extracted from wells for use in mineral spas. For example, the City of Desert Hot Springs in southern California passed a zoning ordinance that illustrates these conflicts. The city is located directly above the Desert Hot Springs Geothermal Field, which has supported the development of a significant mineral spa tourist industry. In order to further profit from the tourism and with "strong lobbying support from local spa owners," the city enacted a hot water spa district in its zoning code that strictly limits development. The zoning "essentially puts a moratorium on new home construction within the newly designated zones." The zoning change has met opposition from local landowners who purchased land within the district with the intention of building single-family homes. Larger developers are required to include mineral spa hotels with any new development. The zoning may harm geothermal resources because it effectively requires the development, depletion, and ultimate exhaustion of the resources.

Notes and Questions

1. Although geothermal is considered a green energy alternative, what additional impacts does the excerpt above identify?

2. What are the positives and negatives of developing geothermal power plants in rural areas?

3. What do you see as some of the potential problems with also generating electricity from low temperature resources that currently are used as mineral spas?

4. In some of the previous notes, we considered the TANSTAAFL Problem. If the negative impacts of geothermal energy cannot be mitigated, how do they compare to the current environmental consequences of our dependence on fossil fuels, which must be extracted from the earth, are depleteable, and are contributing to climate-changing emissions of CO_2?

Chapter 7

Energy Efficiency

The five preceding chapters of this book catalogued supply-side sources of renewable fuels. However, a significant part of the total energy equation is reducing primary energy requirements, and this chapter will address what Jim Rogers, the Chairman and CEO of Duke Energy, calls the "fifth fuel"—energy efficiency.

Growth in peak electricity demands has exceeded transmission growth by almost 25 percent every year since 1982. Yet, expenditures by electric utilities on research and development are some of the lowest of any U.S. industry. According to a 2008 study by the Brookings Institution, energy sector R&D was approximately 0.3 percent, well behind R&D investment of 2 percent by the health care sector, 2.4 percent by agriculture, and 10 percent by both the information technologies and the pharmaceutical industries. James Duderstadt et. al., Brookings Institution, Energy Discovery-Innovation Institutes: A Step Toward America's Energy Sustainability (2008), http://www.brookings.edu/reports/2009/0209_energy_innovation_muro.aspx?sc_lang=en.

One of the least expensive and most promising areas for research and development is energy efficiency. According to John Holdren, Director of the White House Office of Science Technology Policy, and Co-Chair of the President's Council of Advisors on Science and Technology, "[T]he cleanest, fastest, cheapest, safest, surest energy supply option continues to be increasing the efficiency of energy end use—more efficient cars, more efficient buildings, more efficient industrial processes, more efficient airplanes. We have gotten more new energy out of energy efficiency improvements in the last 35 years than we've gotten out of all supply side expansion put together in the United States. That's even without trying all that hard. For most of that period, we haven't had anything that you could call a really coherent set of energy policies supporting increasing energy efficiency. We need ... a more coherent set of policies." Elizabeth Kolbert, *Obama's Science Advisor Urges Leadership on Climate*, Yale Env't 360 (2009).

Although energy efficiency and conservation evoke sharply contrasting visuals—sleek, painless modernization for energy efficiency as opposed to conservation's deprivation and asceticism—they are actually closely intertwined. Consequently, we will use "energy efficiency" or "EE" collectively here to address both efficiency and conservation measures.

This chapter will explore EE technologies and approaches as well as the implications of utilities moving into a less centralized and more distributed model of electricity generation, including that suite of technological innovations we anthropomorphically call Smart Grid.

A. History

Energy efficiency is the renewable resource that shows the greatest promise for energy savings. One of the first strategies from a utility perspective is to improve the conversion of one form of energy to another in the electricity generation cycle.

To convert coal to electricity, a coal-fired power plant requires three phases: (1) the solid coal is combusted, converting the chemical energy of the coal to thermal energy in steam; (2) the thermal energy of the steam is converted to mechanical energy as the steam turns the turbine blades; and (3) the mechanical energy of the turbine converts into the desired end-product—electrical energy. Because energy is expended in each conversion, coal-fired power plants lose two-thirds or more of the energy of the coal in the process. *See* Figure 7.1, and for more discussion and citations to sources about the conversion of coal to electricity, go to Chapter 5.B.

This section addresses creating new efficiencies in the traditional electricity generation process as well as the origin of our current system of electricity delivery and how that structure has encouraged consumption rather than conservation values.

Sidney A. Shapiro & Joseph P. Tomain, *Rethinking Reform of Electricity Markets*
40 Wake Forest L. Rev. 497, 499-511 (2005)

Electricity and the Environment

The generation of electricity is one of the leading causes of environmental problems in this country. According to Environmental Protection Agency ("EPA") data, electric utilities are the biggest polluters in the United States, with emissions far exceeding those of other industries such as chemical manufacturing and refining.

* * *

The Traditional Model

The electricity industry—like other network industries, such as telecommunications and natural gas—has exhibited a discernable historic pattern. That pattern is the result of a combination of technological developments, economic theory, and supporting government regulations. Regardless of the particular network industry, these three elements have given rise to an industrial structure and a regulatory regime with remarkable persistence....

* * *

Brief History

The story of electricity is a particularly interesting one, involving colorful characters from the very inception of the industry right up until today's headlines. Imagine the classic American inventor Thomas Edison toiling at his workshop in Menlo Park, New Jersey, inventing the incandescent light bulb. While this picture of Edison is an accurate one, it is also a partial one. Edison was a remarkable inventor; he was also quite a business genius.

Figure 7.1: Efficiency flow charts

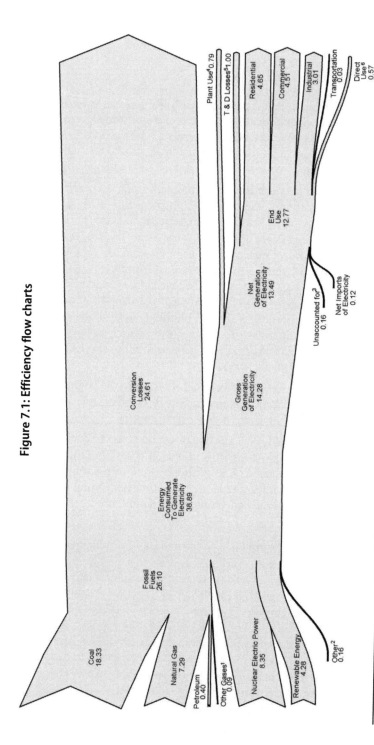

[1] Blast furnace gas, propane gas, and other manufactured and waste gases derived from fossil fuels.
[2] Batteries, chemicals, hydrogen, pitch, purchased steam, sulfur, miscellaneous technologies, and non-renewable waste (municipal solid waste from non-biogenic sources, and tire-derived fuels).
[3] Data collection frame differences and nonsampling error. Derived for the diagram by subtracting the "T & D Losses" estimate from "T & D Losses and Unaccounted for" derived from Table 8.1.
[4] Electric energy used in the operation of power plants.
[5] Transmission and distribution losses (electricity losses that occur between the point of generation and delivery to the customer) are estimated as 7 percent of gross generation.
[6] Use of electricity that is 1) self-generated, 2) produced by either the same entity that consumes the power or an affiliate, and 3) used in direct support of a service or industrial process located within the same facility or group of facilities that house the generating equipment. Direct use is exclusive of station use.

Notes: • Data are preliminary. • See Note, "Electrical System Energy Losses," at the end of Section 2. • Net generation of electricity includes pumped storage facility production minus energy used for pumping. • Values are derived from source data prior to rounding for publication. • Totals may not equal sum of components due to independent rounding.
Sources: Tables 8.1, 8.4a, 8.9, A6 (column 4), and U.S. Energy Information Administration, Form EIA-923, "Power Plant Operations Report."

Electricity Energy Flow 2009, U.S. Energy Info. Admin., http://www.eia.gov/totalenergy/data/annual/diagram5.cfm (last visited Aug. 4, 2011).

Once the incandescent light bulb was invented, it was necessary to illuminate that bulb with electricity....

While it is the case that Edison did not invent either the electric light or electricity, it was his particular genius to develop product distribution. Edison's business genius was the construction of the distribution system to deliver the electricity to light the lights....

* * *

The AC/DC battle is more than a story of rival technologies; it involves the development of the electricity industry. In short, at the turn of the nineteenth century, the electric industry was competitive and highly localized. Since electricity could be transmitted longer distances, however, the economics of competition changed dramatically.... Samuel Insull, who had also once worked for Thomas Edison[,] ... recognized that profits could be made in the electric industry once two fundamental costs were recouped: fixed costs and operating costs.

The electric industry, like other network industries, has high front-end capital costs. Significant investment must be made in plants and equipment before production can begin. These capital costs are particularly sensitive in the electric industry because it is difficult to store electricity in any significant quantity. You need only think of the battery on your laptop computer to realize how frequently it must be charged. Because end users for manufacturing purposes or home convenience need a reliable supply of electricity, it became necessary to build sufficient generation plants so that service could be delivered without interruption. Additionally, generation plants are expensive, costing millions of dollars at the beginning of the twentieth century and hundreds of millions of dollars today. Thus, there are high fixed costs. By comparison, there are comparatively low operating costs, including costs for fuel, labor and the like. Nevertheless, both of these costs need to be recouped in order to have a profitable firm.

Insull recognized that by charging users relatively higher prices at the beginning of a use period and then lowering prices with more consumption, he could capture both fixed and operating costs. Such a pricing scheme also induces consumption....

The point of this brief history is a simple one. The electric industry started competitively. However, technological change gave rise to a change in corporate form primarily through concentration, which led to manipulation. And with manipulation came cries for government regulation. It is at this point in the electricity story that economic theory takes the stage, for economic theory justified government regulation.

* * *

Again, we can turn to Insull for the central insight into how and why the government regulated the electric industry the way that it did. It is the case that a single producer of electricity can produce electricity at a lower cost than multiple producers. However, that single producer most likely will become a monopolist. The regulatory response to this exercise of monopoly power ... was to impose a government-sanctioned monopoly on that single provider through what has come to be known as the regulatory compact. The terms of the compact are fairly simple. An electric utility is given an exclusive franchise area and is obligated to provide service within that franchise area. The government, to counteract monopolistic pricing, is then given ratemaking authority over the electric utility. Ratemaking is a shorthand way of saying that the government controls utility prices and profits....

To encourage continued investment in utilities, regulators designed what we refer to as the traditional rate formula that allowed utilities to recover operating costs and a return on investment on all capital costs. Such ratemaking is a form of cost-plus pricing. Known as cost-of-service ("COS") ratemaking, the traditional formula functioned in such a way that as long as a public utility operated prudently and, for the most part, as long as customers received service, then a utility would stay in business. COS ratemaking had another feature which favored industrial growth and expansion: declining block rate design. Utility customers are charged for the amount of electricity that they consume and for the cost of providing the service. However … customers do not pay for exactly the electricity that they consume at the time that they consume it. Rather, they pay an average cost and do so in "blocks.". … Customers, under the declining block rate design, pay more for the first block of electricity that they consume and less for additional blocks. In this way, the utility has the opportunity to recover the more expensive capital costs in the beginning of the consumption period before going on to recover operating costs.

For an investor, COS ratemaking may look too good to be true. Although there was no guarantee that a profit would be made, rarely were profits lost. For many years the market reflected this low risk investment with low, although reliable, rates of return to utility investors, and every diversified portfolio contained some utility stocks or bonds.

* * *

… Historically, because of the need to provide electricity on demand and because of an obligation for reliable service, the industry would over-build and have its excess capacity accounted for in what is known as a reserve margin. In other words, a utility invested in an excess plant so that it could satisfy demand. That excess plant constituted the reserve margin. Historic reserve margins during the decades of the 1960s, 1970s, and 1980s ranged from 23.8 percent to 31.8 percent and fell to 19.9 percent during the 1990s. The question then became: Were the reserve margins too high, and could they be lowered? … The problem, however, was that utilities were granted government protected franchises, thus discouraging new entrants. The electric industry was surprised by the reaction to a piece of legislation known as the Public Utility Regulatory Policies Act ("PURPA"), which was part of President Jimmy Carter's National Energy Act legislation of 1978.

The National Energy Act was intended to be comprehensive and to respond to the energy crises that affected energy prices and the economy more generally during the 1970s. President Carter's legislation had several purposes, including responding to growing dependence on foreign oil, finding alternative sources of energy, and engaging in resource conservation and energy preservation measures. PURPA was intended to experiment with innovative ratemaking and rate designs.

PURPA was intended to move away from COS ratemaking and to try market-based rate strategies. As part of the rate regulation reform, PURPA encouraged new forms of electricity generation, including the promotion of small power producers and co-generators. … Small power producers, generators of eighty megawatts ("MW") and less, and co-generators were encouraged to enter the market and to connect to the local public utility with a guarantee that the local public utility would pay for the electricity generated by these two "qualifying facilities" at the utilities' own avoided cost. …

… As a result, electricity "deregulation" began in earnest. … What is being deregulated or what is attempting to be deregulated is the pricing of electricity at the wholesale and retail levels. To date, the restructuring is continuing at the wholesale level and has been largely discontinued at the retail level. Through a series of FERC rule-makings, the industry is indeed being restructured at the wholesale and at the interstate levels. …

Figure 7.2 illustrates electricity demand throughout an average hot summer day in most U.S. service areas. Note that the demand for electricity is low during the evening, when most people are sleeping and cooler night temperatures reduce air conditioning needs. On this

Figure 7.2: Electricity demand profile for a typical U.S. service area on a hot summer day

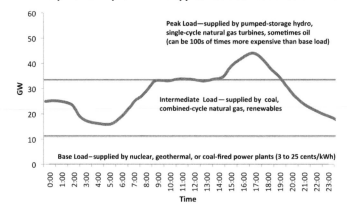

hypothetical profile, the maximum demand or "peak load" period spans five hours. Electricity demand is high during this time because air conditioners are turned up to ward off afternoon heat. This is also the time period when businesses are cranked up to full speed, restaurants are preparing dinners, and families returning home from work consume their highest loads of the day for activities such as cooking, washing, using the computer, or watching television.

Because there is such a wide swing in the demand throughout the day (between 15 GW and 45 GW on this chart), a utility will use different power sources to meet different loads. For example, a nuclear power plant, which is very difficult to ramp up or down, is used for the "base load" (below 10 GW on this chart). Coal-fired power is also used for base load. More variable sources can follow the intermediate load demands. "Peak load" is most problematic for utilities because to meet their reliability requirements, they must have sources of power available for the occasional times when the demand exceeds the intermediate load capacity. This peak load also needs to be available on very short notice to deliver power to their customers' maximum demands without causing any outages.

Sources that can be turned on quickly, such as natural gas or pumped-storage hydro, are often used to meet peak-load requirements. However, the cost of producing power during peak-load times can be many times more than the cost of producing base-load power because of the fuel sources or technologies used. Peak load is also costly because utilities will maintain "peaker plants" to meet demand that may only be needed a few hours a year. In a July 20, 2010 story in the Star-Ledger, reporter Abby Gruen noted that PSEG Power paid 13 million dollars a year to keep its Jersey City peaker plant operational. Yet, this low-capacity-high-cost plant had been used only five days that year.

Most utilities average the costs of peak power and base-load power on customer bills, so consumers are not aware of the actual cost of the power they use. Thus, there is no incentive to modify one's consumption during these costly peak-demand times. One way of "shaving the peak" is for a utility to invest in ways of cutting back on volatile customer demands through demand-side management as discussed in the next excerpt.

James W. Moeller, *Electric Demand-Side Management Under Federal Law*

13 Va. Envt'l. L.J. 57, 57-59 (1993)

The term "demand-side management" (DSM) is synonymous with electric power conservation. Thus, DSM embraces all measures associated with electric power conservation, from the use of energy-efficient light bulbs in the home to the introduction of energy-efficient processes in energy-intensive industries. DSM also reflects the principle that the nation's need for electric power should not be met solely through supply-side management, which includes the construction of additional electric power production capacity, but also through electric power conservation, which is in itself a source of electric power.

The environmental advantages of electric DSM relative to the construction and operation of electric power generation and transmission facilities are well documented. Unlike adding new coal and natural gas facilities, the use of DSM does not contribute to acid rain, global warming, or stratospheric ozone depletion. Unlike new nuclear power plants, using DSM does not produce radioactive waste.

* * *

The environmental and economic advantages associated with DSM have prompted electric utilities to invest heavily in DSM, resulting in significant decreases in residential, commercial and industrial use of electric power.... The DOE predicts that, with respect to growth in electric power demand between 1993 and 2010, "price-induced conservation, legislative action, and utility investments in demand-side management (DSM) programs are expected to increase efficiency in end-use electricity markets and, thereby, dampen that growth."

Notes and Questions

1. How does a declining block rate work? This rate structure was designed partially to lure large consumers away from purchasing their own individual generation equipment. Declining block rates encourage consumption; those who consume enough to qualify for a higher-use block benefit with a more favorable pricing structure. Also, the cost-of-service (COS) ratemaking model encouraged utilities to expand excess capacity by building additional power plants to meet peak and anticipate future increases in demand. Some states with scarce water reserves have the opposite pricing structure for water use—charging a base rate for average needs then charging a higher amount per unit for those exceeding the basic-use threshold. Many cell phone companies also charge customers a basic monthly rate but significantly more for minutes or texts over the monthly allocation. Do you think either of these models might work in the electric utility context? Do you have other suggestions about ratemaking models that will encourage conservation and also discourage larger consumers from abandoning the utility to construct their own power sources?

2. The utility industry bandies about the terms "demand-side management" and "avoided cost" for measures that reduce consumer demand and allow utilities to avoid the cost of building additional power plants to meet that demand. As a demand-side management initiative, many local utilities have programs to give away or to provide rebates on energy-efficient light bulbs. Can you think of other examples?

3. Sometimes a technology is caught between definitions. Active solar thermal and domestic hot water (DHW) systems, described in Figure 2.2 above, should qualify as demand-side management (DSM). They can meet up to 70 and 95 percent of a home's space and water heating needs, respectively, reducing consumer demand for power from the utility to achieve these same functions. But many utilities do not count them in the DSM programs because hot water and space heating are usually achieved with natural gas, and the utilities focus only on reducing electricity use, not total power demands. In addition, solar thermal does not fit nicely into some rebate programs because of its longer payback time. Finally, solar thermal systems do not currently fit within many states' definitions of

a renewable energy source for purposes of renewable energy credits (sometimes called renewable energy certificates (RECs) or tradable renewable certificates (TRCs)), which usually focus on electricity rather than total energy use.

4. PURPA required established utilities to allow small producers to connect to the utility grids. PURPA also encouraged efficiencies within the utilities themselves. Here are three technologies that promise significant efficiency savings:

(a) *Cogeneration or combined heat and power (CHP)* is the use of waste heat from electricity generation or from an industrial process for additional purposes such as curing lumber or generating more electricity. A common example is a combined cycle gas turbine (CCGT) plant: a gas turbine generates electricity, and heat in the exhaust is used to make steam, which in turn drives a steam turbine to generate additional electricity. For additional information about CHP and its efficiencies go to Figure 5.8 in Chapter 5 above.

(b) *Variable Speed Motors* allow industries to run their machinery at less than full speed. Industrial motors are the single largest consumers of electric power representing 65 percent of all electricity consumed by industry. By converting inflexible, full-speed-all-the-time motors with variable speed motors that run only as much as, and when, needed, industries can save up to 60 percent on electricity use. "All of that saved energy represents power that generators don't have to make and the transmission and distribution system doesn't have to deliver. Fuel costs can therefore be avoided while maintenance can be deferred. And it's megawatts that can be used to serve other purposes during peak periods—all without a single dollar invested in transmission." Bob Fesmire, *Is 'Efficiency' the New 'Reliability'?*, ARIZONA ENERGY, Apr. 16, 2007, *available at* http://www.arizonaenergy.org/News_07/News_Apr07/news_Apr07.htm

(c) *Distribution transformers* represent the final step in the electricity transmission to distribution process, cutting down the voltage of transmission lines to the level used by customers. In 2007, the Department of Energy proposed standards that would increase efficiency in these distribution transformers by a few percentage points, which would produce 8.5 billion kWh in savings each year. *Id.*

With such potential savings at the utility level, why do we also approach energy efficiency from the demand-side consumer perspective? Does the fact that the United States has the second highest per capita rate of electricity consumption—five times the world average—inform your answer from an energy justice standpoint? *See* Figure 7.3.

B. Efficiency Overview

The following excerpt addresses some of the policies for, and barriers to, encouraging energy efficiency among consumers.

Edward H. Comer, *Transforming the Role of Energy Efficiency*
23 NAT. RESOURCES. & ENV'T 34, 34-38 (Summer 2008)

Electric companies working with state regulators and policymakers are transforming the role of energy efficiency. The electric power industry has begun creating a new regulatory framework that treats investments in energy efficiency in essentially the same manner as those for generation, transmission, and distribution. This change in focus opens the door to greater benefits for consumers, the environment, and electric companies. The industry will continue to face a number of well-recognized barriers in improving the nation's

Figure 7.3: Annual per capita energy use worldwide

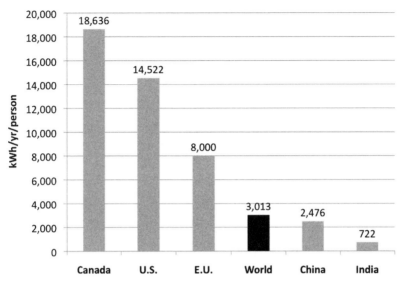

Data from 2007 U.N. Energy Stat. Y.B. 549, 554, 558, 562, U.N. Doc. ST/ESA/STAT/SER.J/51

energy efficiency. But today's twin challenges of reducing carbon dioxide (CO_2) and other greenhouse gas (GHG) emissions, while meeting the country's steadily rising demand for electricity, make the shift in focus essential. The key to this transformation is developing new regulatory and business models that turn energy efficiency into a sustainable business for electric companies.

The electric power industry began promoting energy efficiency after the oil embargoes in the 1970s. These early efforts consisted of distributing educational information on how electricity is used in the home and how consumers could make their home more energy efficient. During the 1980s, as energy prices started to rise, more electric companies expanded their efficiency efforts into formal programs that offered financial rebates and other incentives to motivate customers to become more energy efficient. These programs focused on getting customers to increase insulation levels in their home, replace their old appliances with energy-efficient ones, and participate in demand-management programs, such as allowing the utility to cycle their air conditioners on and off during hot summer afternoons in return for credits on their monthly bills.

Most electric companies scaled back their efforts to promote energy efficiency during the 1990s when states began deregulating the industry. Nevertheless, efforts to promote energy efficiency have made a difference. Going back to 1989, electric efficiency efforts have saved almost 860 billion kilowatt-hours, which is enough electric energy to power over 76 million homes for one year.

Energy efficiency is valuable to the electric system for several reasons. Efficiency programs help customers gain greater control over their energy bills and maintain the reliability of the power system for the benefit of all consumers. Additionally, the programs help utilities serve customers by helping ensure that there is enough electricity to meet demand.

* * *

Energy efficiency gives the industry and its customers a way to make immediate reductions in their carbon footprint. In 2005, the latest year for which data are available, the industry's energy-efficiency programs were responsible for approximately 13 percent, or more than 34 million metric tons, of carbon-equivalent GHG emissions being reduced.

Transforming the role of energy efficiency will also help the industry to keep pace with America's steadily growing demand for electricity. The government is predicting that the nation will be using 30 percent more electricity in 2030 than it does today, even after taking into account energy-efficiency improvements due to market-driven efficiency, including stricter building codes and appliance standards, and other efficiency standards mandated by the Energy Independence and Security Act of 2007.

Even with an aggressive approach to energy efficiency, the industry will not be able to eliminate the need for new plants or transmission lines. To keep up with the forecasted growth in demand, an estimated 224 gigawatts of incremental generation capacity will need to be added by 2030. The new and replacement power plants to meet this rising demand are expected to cost about $560 billion through 2030. Investment in transmission and distribution assets together will require nearly twice the investment in generation— $900 billion, under current trends and policies. However, a greater emphasis on energy efficiency, both within and outside of the electric power industry, has the potential to decrease this capacity need by 33 percent to 151 gigawatts.

Barriers Remain

A number of well-recognized practical barriers stand in the way of achieving energy-efficiency savings.

These barriers include:

- Market barriers—Homebuilders and commercial developers are not motivated to invest in energy efficiency for new buildings because they do not pay the energy bill.

- Transaction cost barrier—Consumers and small-business owners are reluctant to pay the higher cost of energy-efficient appliances and products when less efficient products cost less. Not understanding the true cost over the life of the appliance is one factor. Another factor is that consumers pay an average price for electricity—not the actual cost of electricity—which varies throughout the day.

- Customer barriers—Customers need more information on energy-saving opportunities as well as how energy-efficiency programs make investments easier. Both will be essential if we are to overcome the consumers' inherent reluctance to undertake the expense and trouble of making their homes more energy efficient. With an estimated 110 million existing households in the United States, this barrier represents a significant hurdle for enabling energy efficiency to do more in reducing the demand for energy.

The practical difficulties of improving the nation's energy efficiency are illustrated by the continued growth of electricity use in the country's residential sector. Retail sales of electricity to homes totaled 1.3 trillion kilowatt-hours (kWh) in 2003, and increased use of electricity accounts for 68 percent of the projected increase in residential delivered energy use between 2003 and 2025. Ultimately, efficiency improvements must be accepted and implemented by individuals and families. Ironically, even as the nation is developing more stringent appliance standards and labeling information, the demand for electricity to power appliances is projected to increase rapidly, particularly for home electronics. EIA projects electricity consumption to grow 3.5 percent annually for color TVs and computer equipment through 2025, to more than double the level of consumption in 2003. Both EPRI and Natural Resources Defense Council (NRDC) report that relatively new telecommunications appliances, including HD cable boxes, DVRs, and flat-screen TVs, are approaching the energy use levels of more traditional appliances. In part, our success at finding new uses for electric-powered technology seems to be steps ahead of our efficiency efforts. Continued growth of new housing in the South, where almost all new homes use central air-conditioning, is also expected to contribute to an increase in household electricity demand.

* * *

Electric companies maintain the key supporting infrastructures (e.g., rates, metering, billing), which are essential for the delivery, verification, and pricing of many efficiency services. And electric companies have more reasonable costs of capital than most consumers, which means they will be able to pass their savings on to customers.

* * *

New Regulatory Approaches

A key to unleashing utilities to promote energy efficiency is changing the industry's regulatory structure. This structure has historically rewarded utilities for building infrastructure (e.g., power plants, transmission lines, pipelines) and selling energy. Under traditional regulatory frameworks, efficiency programs that result in fewer sales make it more difficult for the utility to recover fixed costs, to create an adequate return for the investors' risk, and to keep Wall Street and shareholders confident.

States and electric companies are now working together to fix the old model. They are addressing three key issues. The first involves finding a way to recover the costs that arise from promoting energy efficiency. These costs include the cost to educate customers and the cost to provide efficiency programs, such as energy audits and rebates for efficient windows, insulation, and light bulbs.

* * *

The second issue the new business models need to address concerns the recovery of a company's fixed costs. A loss of revenues is likely to result if a company is successful in assisting its customers in using electricity more efficiently. The electric company, like every company, needs those revenues to pay its fixed costs. Lower revenues also make it harder for the company to earn an adequate return on its investment. Determining how to remedy an apparent fundamental disconnect between how companies receive revenues and how they achieve energy-efficiency goals has become a priority in some states.

A final issue to deal with is creating a financial incentive for the electric company to pursue energy-efficiency goals. Many states have adopted or are considering shareholder incentive mechanisms. Four incentive methods are beginning to emerge. These models are broadly being referred to as Shared Savings; Bonus Return on Equity (RoE); Save-A-Watt [or Virtual Power Plant—Ed.] ... ; and Energy Service Company (ESCO).

* * *

Energy Services Company. The final business model incentive is different in one essential way. Energy-efficiency services are sold directly to retail customers on a fee-for-service basis through a subsidiary of the electric company. General rates are not used to collect any part of the program costs incurred or shareholder incentives received by the ESCO, as they were in all three approaches discussed above. The ESCO family aligns the utility's and customer's incentives, and these models are less sensitive to the ebb and flow of regulatory interest in energy efficiency, which may create a more sustainable environment for growth in these services over time.

There are many different ways to structure ESCO contracts to recover costs and a competitive profit margin, which is what this efficiency business model is likely to support because this approach is open to competition. Since the cost recovery and shareholder incentives and payments are both within the contract, there is no regulatory treatment required. However, there are lost fixed revenues and recovering them (if the company chose to do so) would require regulatory approval.

* * *

Looking Forward

Transforming the role of energy efficiency holds much promise and potential. But it will not be easy. Besides actions by the electric-power industry, expanding the role of energy efficiency will involve electricity customers taking action, too. For state regulators and electric companies, it will also mean changing paradigms from a supply orientation to considering both supply and demand. And it will be a challenge to measure and verify the costs, benefits, and effectiveness of energy-efficiency programs.

Notes and Questions

1. What three barriers does Comer list as standing in the way of achieving energy efficiency (EE) savings? What three issues must be addressed to change the utility industry's regulatory structure to promote efficiency? Can you suggest some specific EE solutions?

2. Shared savings is one of the approaches Comer lists to provide financial incentives for electric companies to pursue EE goals. "Shared savings" allow a utility to earn a certain percentage of the total net benefits from an EE program. How do you think this approach would be administered? What do you think are the advantages? What is the benefit to the utility or regulators of a longer or shorter period for collecting the savings bonus?

3. A second type of financial incentive for electric companies is Bonus Return on Equity (Bonus RoE). This approach allows a utility to capitalize its EE investments and then earn a rate of return on them. While the utility's recovery of a rate of return would be similar to what it could recover on a power plant or transmission-line investment, this option allows the utility to earn an additional bonus on the equity portion of its capital structure. What do you see as some of the advantages or disadvantages of this approach?

4. What is an ESCO? How does it work? Why do you think its structure might be more or less successful than the other incentive models Comer describes?

5. The last of the four incentive models Comer lists is the virtual power plant model described in the following section.

C. Specific Solutions

This section of the chapter addresses five specific energy efficiency solutions: (1) virtual power plants; (2) green building codes; (3) distributed generation and Smart Grid, (4) energy efficiency technologies, and (5) behavioral changes. Considering the world's energy needs, we will, no doubt, need all of the above.

1. Virtual Power Plants

A virtual power plant is a way of combining some of the other energy efficiency solutions discussed in this subsection to substitute for building actual plants.

Carol Sue Tombari, Power of the People: America's New Electricity Choices
76-79 (2008)

During the 1990s, the municipally owned utility in Austin, Texas, "constructed" an energy efficiency power plant.[1] As it planned to meet the needs of its growing population and economy, Austin decided to put its money not into the bricks and boilers of a conventional power plant, but to invest instead in programs that resulted in energy efficiency throughout its utility service territory.

This is a good time to differentiate between energy conservation and energy efficiency. Conservation is a wonderful thing. It is rooted in human behavior, such as turning off lights, turning down the thermostat, disconnecting plug loads, etc. During California's natural gas crisis of 2000, energy conservation is what saved the system. Within twenty days, Californians voluntarily reduced their electricity consumption enough to avert major blackouts that otherwise would have occurred.

But let's face it: humans are unreliable. We often forget to turn off lights, unplug the computer and TV, lower the thermostat. Customarily a utility cannot rely on our behavior to reduce its load requirements in perfectly predictable fashion over a sustained length of time.

In contrast, a utility can rely on energy efficient technologies that, once installed, continue to work and reduce load. These would include high-efficiency lights and appliances, added insulation, passive solar design in new buildings — all of these and more.

This is what Austin did. It created programs to encourage investment in energy efficiency. This included energy efficient building codes that actually were enforced. (Many jurisdictions do not effectively enforce their codes, usually for budget reasons, as enforcement is labor intensive). Austin also underwrote rebates for high-efficiency equipment and appliances, and instituted other programs. The utility kept track of these investments and did the engineering calculations to ascertain how much electricity was offset when the equipment was installed. The monitoring and tracking processes were more complicated than I make them sound, but this was the gist of it.

After about a dozen years — which, coincidentally, is about how long it would take to acquire the permits and construct a coal-fired power plant — Austin had booked 550 MW of affordable and sustained energy efficiency on its system. Because it had "constructed" this virtual power plant, Austin took a 450 MW coal-fired power plant off its planning books. This was during a time when the local population almost doubled and the local economy grew by about 46 percent.

Part of Austin's economic growth may well have been related to its investment in energy efficiency.... [U]nlike central station power plants that are likely located somewhere else, energy efficiency requires local labor for retailing, distribution, installation, and other

---

1. [Austin actually called it a "conservation" power plant, as this effort commenced before the term *energy efficiency* became popular — Ed.]

functions. Money spent on those salaries is likely to be spent locally, and then respent again, creating a multiplier effect. The multiplier for energy efficiency in Osage, Iowa, was $2.23, in comparison to $1.66 for a central station power plant.

In addition, both Osage and Austin found that energy efficiency is cheaper than constructing new power plants. This enabled them to attract desirable new industries, lured by the promise of affordable electricity for business operations and enhanced quality of life for their employees.

How do we know energy efficiency is cheaper than new power plants? Regulated electric utilities across the country are required, and have been for almost 20 years, to create and carry out energy efficiency programs of one kind or another. One investor-owned utility, headquartered in the Midwest and serving multiple states, recently reported that its energy efficiency programs (including rebates) "produced" energy efficiency at a cost of less than 3 cents per kilowatt-hour.

Austin's virtual power plant was built exclusively from cost effective energy efficiency, and the energy savings achieved from any one measure were small. Yet, when aggregated across the utility's service area, the savings totaled the equivalent of a power plant. It didn't take any longer to accumulate these massive savings than it would have taken to construct the coal-fired plant they took off the books.

How many similar or larger plants could be built across the country? The Alliance to Save Energy, a non-profit organization dedicated to the advancement of energy efficiency, looked at this issue a couple of years ago. It examined a suite of four measures and policies and concluded that, in the aggregate, they would equate to 557 power plants. Those policies included increased appliance-efficiency standards, efficiency standards for commercial air conditioning, energy efficient design and construction of new buildings, and efficiency upgrades in existing building stock.

Austin "constructed" its virtual power plant before rooftop solar electric had become cost-effective or commercially available on any meaningful scale. Imagine what the savings of conventional electricity-generating fuels would be if distributed renewable energy technologies were added to the mix and a "green" virtual power plant were constructed.

Notes and Questions

1. What is a virtual power plant? What are some of the benefits of "constructing" a virtual EE power plant in comparison to investing that same money "into the bricks and boilers of a conventional power plant"?

2. The Comer article in the previous section discussed a similar virtual power plant model called Save-A-Watt by Duke Energy in North Carolina: "Duke's proposed model puts energy-efficiency investments on par with supply-side investments. It does so by compensating the electric company for meeting customer demand, whether by saving a watt or generating a watt. In the Duke Energy proposal, the company must spend at least 1 percent of its gross revenues on energy-efficiency measures. It is then allowed to recover 90 percent of the depreciation and operating costs it avoids by producing energy savings. Under the Save-A-Watt regulatory approach, the company does not specifically recover its energy-efficiency costs and lost sales through customer rates. A key feature of this creative model is that it spreads the costs out in time directly proportional to the assumed life of the energy-efficiency savings. Moreover, the company is only compensated for the savings it does achieve. The company waives the right to collect any costs incurred in im-

plementing the efficiency programs. As a result, this model gives incentive to the company to keep program costs low and saving results high." Edward H. Comer, *Transforming the Role of Energy Efficiency*, 23 Nat. Resources & Env't 34, 37 (Summer 2008). How does the Duke model compare with the Austin model?

3. One advantage Tombari notes with the virtual power plant model is the "local multiplier effect." Explain what this is and how it works.

4. Would you like to play a videogame that focuses on balancing energy demands and energy sources including EE/conservation? Go to "willyoujoinus.com" and play the "Energyville" game.

2. Green Building Codes

Carol Dollard, Energy Engineer at Colorado State University, proposes converting every kilowatt-hour your home uses into one pound of coal in your driveway because that is approximately how much coal a conventional power plant would need to produce 1 kWh. According to the EIA, the average home in the United States uses 901 kWh per month, so visualize that 901-pound pile of clinkers greeting you outside your window.

Energy efficiency measures for buildings could reduce that coal pile with attention to building design and added efficiency features, but no austerity or deprivation. As U.S. Energy Secretary Steven Chu stated in his 2009 address to Harvard's graduating class, "Energy efficiency is not just low hanging fruit; it is fruit lying on the ground." Chu then specifically mentioned that the United States could cut its carbon emissions by a third just by making our buildings more efficient. With the building efficiency programs adopted by the European Union and discussed here, we could cut our mountain of coal down by closer to two thirds.

Section a here explains the role of buildings in the EE equation; Section b addresses legal concerns with EE building codes.

a. In General — Buildings in the Energy Efficiency Equation

The following excerpt illustrates why building design is a critical component in our nation's energy use.

Carol Sue Tombari, Power of the People: America's New Electricity Choices
74-76, 79-84 (2008)

We can't talk about a new electricity model without specifically addressing electricity use in buildings. Seventy-one percent of the electricity consumed in this country (and 40 percent of primary energy) is in buildings. The 81 million buildings in the United States comprise the largest single energy-consuming sector in our nation's economy. This is the consequence of a building's orientation toward the sun, its architectural design, the tightness and insulation of the exterior shell, insulation level, quality of windows, general quality of construction, and—increasingly—the electricity appetite of equipment inside the building. With regard to how we use that equipment, especially heating, ventilation, and air conditioning (HVAC), we "operate" buildings.

Start at a building's beginning. First, orient it properly on the lot in order to take full advantage of the heating and natural day lighting properties of our solar system's power plant. Minimize the amount of window space on the east and west sides, because this is where the building soaks up unwanted solar heat gain and glare in the morning and afternoon. Developers and municipal planners should plot streets so that as many of the structures as possible can face south.

Next, design the buildings to maximize the sun's benefits. This is passive solar design. Size the roof overhangs so that the summer sun doesn't overheat interior spaces, especially in warm climates. Install windows with glazing to minimize glare and heat gain. Design your window space to optimize the sun's natural daylighting properties without paying the penalty of unwanted heat gain. Add the appropriate level of insulation for the climate zone. Make sure the building is well constructed and without leaks.

Top it with a cool roofto reduce solar heat gain. Until recently, that meant painting the roof white, to create an albedo effect and reflect heat just as arctic glaciers do. Aesthetically that worked okay for flat commercial roofs, but maybe not so much for homes. Now, however, one can purchase "cool roofs" in a variety of colors.

A building that is oriented toward the south and that includes the foregoing passive solar design features uses some 60 percent less energy than conventional structures. That percentage goes up even higher if you install energy efficient appliances and add rooftop solar water heating and electricity. These days, PV can be integrated into properly oriented and sloped roofs, or even the sides of buildings, as an alternative (albeit an expensive one) to conventional PV arrays.

Now we have what is called a net zero, or near zero, energy building. Futuristic? The future is now. Several production builders in California have done this already. Shea Homes, for example, offered near-zero energy homes in one-third of its upscale Scripps Highlands development. Production home construction is a market that can absorb the added cost of PV: builders can reap the cost savings of bulk power purchases and home buyers can put the added cost on their mortgage. Moreover, builders can contract with PV suppliers to provide turnkey installation services.

* * *

By offsetting the need for the electricity that a power plant would have had to generate, transmit, and distribute to these buildings, these homes have effectively created a demand-side resource of electricity. Imagine the impact of structures like this across a utility service area. Although it did not envision zero energy buildings, one utility in particular aggregated energy efficiency improvements across its service territory and benefited enormously by doing so.

* * *

Building-Based Power

Energy Efficiency

> *"Every watt not used is a watt that doesn't have to be produced, processed, or stored."*

—Richard Perez, Home Power magazine

Dubbed the "fifth fuel" by a utility executive, energy efficiency is about equipment, measures, and practices, not behavior. Examples include the following: CFLs, high-efficiency air conditioners, variable speed motors, optimum insulation levels, high-performance windows, and the like. These are not electricity-generating products, but in the aggregate they can offset the need for a significant amount of electricity, totaling hundreds of power plants.

Key to our twenty-first century electricity model is electric utilities' changing their corporate self-concept. They can provide energy services, not just electrons.

The Good

When you save energy, you save more than the amount saved at the point of use. You also save the energy that would have been used to generate the electricity. Some 25 percent of our domestic fossil fuel consumption occurs in the production and delivery of energy itself. You also save all the electricity that vaporizes during transmission and delivery. In this sense, energy efficiency is actually an energy resource—a demand-side resource.

A common example of the power of energy efficiency can be found in most homes in the United States: the common household incandescent light bulb. Ninety percent of the electrical energy that goes into the bulb is thrown off as waste heat. Only 10 percent actually lights the light. Until CFLs came into the marketplace in recent years, light bulbs were essentially the same technology as invented by Thomas Edison and his colleagues in the nineteenth century. Despite all the other technological innovations we've seen over the years, lightbulbs remained pretty much unchanged until recently.

If it's summertime or if you live in a warm climate, you probably pay to cool the interior of your home or business. Consequently, your light bulb is costing you even more: it's adding, perhaps significantly, to your cooling load and associated costs. CFLs, in contrast, have a higher price tag, but they use about three-quarters less electricity than incandescents, they last seven times longer, they produce a higher quality light, and they're better for the environment.

CFLs repay your initially higher cost through energy savings, probably within a year. This illuminates a really important concept: life cycle cost-benefit analysis. When making purchasing decisions about energy-using products, one should always factor in the cost of the energy needed to operate it for the anticipated lifetime of the equipment. The purchase price (first cost) of energy efficient equipment might be higher than standard models. But you'll save sufficient money in reduced energy bills to pay back the added purchase price. You have to decide if the payback comes fast enough to make it cost-effective by your standards.

* * *

You might not think that the energy savings from a lightbulb are very impressive. Even the energy savings from all the lightbulbs in your house might not seem to produce dramatic savings in any one utility bill. But how about over the course of a year? How about if those savings are multiplied across many homes all across the country? It all adds up, and it gets us back to the notion of virtual green power plants constructed all across the United States.

The really good news about energy efficiency, aside from its affordability, is its tremendous upside potential. This is the direct consequence of our wasteful energy habits to date. It is not unreasonable to expect, as a rule of thumb, at least 20 percent savings from energy efficiency retrofits alone. Imagine the 60 percent potential if we design and construct buildings for maximum potential energy efficiency.

The Bad

The many energy efficient products and practices represent different technologies and expertise, and they're not organized as one energy efficiency industry. This creates an incredibly fragmented "voice" for energy efficiency, exacerbated by the fragmented nature of the residential construction industry, in which workers are often hired on a day-to-

To calculate carbon use in your home, go to Xcelenergy.com/infosmart to access online tools such as the carbon calculator/home energy analyzer.

day basis. Quality control is difficult to achieve day in and day out, and is even more so if new measures and practices are introduced on-site.

In addition, builders and developers have the perception that energy efficiency costs more. This used to be true. Over the years, however, costs have come down. Today, added first costs for energy efficient design and construction amount to 2 percent or less — recouped within a few short years (sometimes months) through energy savings.

Nevertheless, perception tends to become reality, and builders and developers resist what they believe to be unnecessarily expensive building practices. In addition, it's difficult to change building practices, because so many different products, people, and actions are involved at each stage of construction. This requires diligent, sustained oversight of daily construction operations — very difficult to do in the real world.

Added first costs create a deterrent for home buyers. If given the choice between adding granite countertops or energy efficient windows and high-efficiency appliances to the mortgage, the home buyer usually chooses the granite countertops.

Why are consumers making economic decisions that clearly are not in their financial interest with regard to energy efficiency? Two reasons: (1) It's not clear to them that they're making financially suboptimal decisions, because they lack information about their choices and the consequences; and (2) energy efficiency improvements are not as sexy as, say, the aforementioned granite countertops or even rooftop solar, for that matter. There is nothing obviously cool — or even obvious, in most cases — about energy efficiency.

This suggests the need for greatly improved consumer education. It also suggests the need to redefine our notion of beauty in our homes and businesses.

The Balance

Whether you're trying to construct a virtual green power plant or making purchase decisions for your home or business, always consider energy efficiency first. First and foremost, this is because the low hanging fruit of the efficiency tree is customarily cheaper than any others. After an energy efficiency measure is installed, it continues to save money on energy bills. With the money saved, one can accumulate funds to purchase additional energy efficient equipment or save for a renewable energy investment such as PV.

Efficiency is also important to undertake first because it enables one to downsize other energy-using equipment and reap the resulting cost savings. For example, builders can install smaller HVAC [Heating, Ventilation, and Air Conditioning — Ed.] systems in energy efficient structures. They know that the building will retain the air the occupant has paid to condition (whether heated or cooled) if it has optimum levels of insulation in the walls, if the windows are energy efficient, and if leaks are sealed. A smaller HVAC system costs less, so this enables the builder to reduce the first costs accordingly. The homeowner also reaps the benefit of the reduced operating costs of a smaller HVAC unit.

Similarly, energy efficiency permits downsizing of renewable energy technologies. Even the most committed solar researchers or advocates always advise consumers to invest in efficiency first, in order to minimize the size and associated cost of a solar system.

Finally, energy efficiency makes investment in renewables such as PV more affordable. If one bundles efficiency with renewables and averages the costs, the higher-priced equipment becomes more affordable. Just like cost averaging stock market purchases.

Notes and Questions

1. In the context of CFL lighting, Tombari mentions "life cycle cost-benefit analysis." This concept is one of the hurdles to consumer adoption of EE technologies. Instead of focusing on the higher purchase price, consumers need to consider the energy and cost savings over the lifetime of the product. Go to the website below. Which energy improvements pay for themselves most quickly? What are some ways of overcoming the EE payback perception hurdle and educating the public about the benefits of different options? http://en.wikipedia.org/wiki/File:BloombergNewEnergyFinance2030USMACC.jpg

2. One problem Tombari mentions is that EE benefits are often invisible in comparison to other amenities in a home such as granite countertops. One possible approach is to require energy performance ratings at the time a home is sold and to include the EE performance score in the real estate listing to increase consumer awareness. The Residential Energy Services Network (RESNET) Home Energy Rating system (HERS) is the most universally recognized energy performance rating system in the United States. Current appraisal practices also may not reflect the value of EE features if there are no efficient homes in the vicinity that can serve as comparable sales. Can you suggest legislation that would address these issues? Do you think it should be enacted at the federal or state level?

3. What are some of the factors Tombari lists that might make builders hesitant to include EE products and practices in their construction? She specifically notes that "diligent, sustained oversight of daily construction operations" are required to ensure that the EE materials are properly installed and running. This is significant from a legal standpoint because some builders have been sued for failing to achieve green building certifications that might have been promised. *See, e.g.*, Mary Jane Augustine, *Project Owner Strategies for "Greening" Design and Construction Contracts*, 565 PLI/Real 121 (2009); Frederick R. Fucci, *Alternative Energy in Commercial Real Estate and Multi-Family Housing—Application of Distributed Resources—Practical and Legal Ramifications*, 565 PLI/Real 313 (2009); The Law of Green Buildings: Regulator and Legal Issues in Design, Construction, Operations and Financing (J. Cullen Howe & Michael B. Gerrard eds., A.B.A. 2010); Carolynne C. White, CLE Presentation, Legal and Regulatory Issues in Green Building (Apr. 27, 2010). If you were asked to draft an agreement for a contractor who planned to build a green home for a client, what recommendations would you make?

4. EE improvements are making great inroads in the area of government subsidized low-income housing. Almost 80 percent of New York City's greenhouse-gas emissions come from buildings, and 40 percent of those are caused by housing. To help improve New York's EE building performance, Mayor Bloomberg required all new affordable housing stock to meet Enterprise Community Partners criteria that increase construction costs by only 2 percent, which is paid back by lower running costs. Habitat for Humanity and HUD are also incorporating EE principles with newer construction. Why does it make especially good sense to require EE in the government subsidized low-income housing context?

5. Passive House (or Passivhaus) is perhaps the most energy efficient building certification program in the world (with efficiencies 70-80 percent better than the average U.S. home). Developed in the 1990s by German engineers, Passive House construction combines

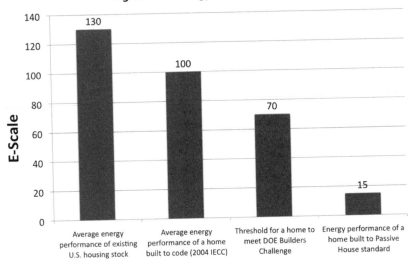

Figure 7.4: EnergySmart home scale

Data from Passive House Institute US (PHIUS); U.S. Dep't of Energy,
Energy Efficiency and Renewable Energy,
http://www1.eere.energy.gov/buildings/challenge/energysmart.html.

passive solar and superinsulation designs. It is popular in the European Union, especially in Germany, the Scandinavian countries, and Austria (where it accounts for at least 17 percent of new homes). As of 2010, more than 25,000 buildings had been constructed to the Passive House standard, including single-family homes, commercial buildings, low-income housing, and apartments.

Passive House Institute U.S. (PHIUS) was founded in 2007. Unlike most other certification standards such as LEED, Passive House is entirely performance-based, focusing on specific energy consumption of the building, rather than site energy. Figure 7.4 shows how much more efficient the Passive House standard is than most of the standards used in the United States. In 2008, the European Parliament passed a resolution that called on the European Commission to make Passive House construction a binding requirement for all new buildings in the EU. Eur. Parl. Res. No. P6_TA(2008)0033 of 31 Jan. 2008, ¶ 6. As you read the following excerpt, consider whether the United States should likewise adopt the Passive House standard.

b. Federalism and Other Legal Concerns

As we saw in Chapter 5, many state renewable portfolio standards have been the driving force behind the renewable energy renaissance in the last decade. However, the definition of what will be incentivized by a state under its particular standard varies dramatically from state to state. Energy efficiency is addressed in some RPSs, but not in others. As a result, energy efficiency is more often regulated at the local level in building codes.

Many early building codes evolved in reaction to crisis: "[T]he Great Chicago Fire of 1871 spawned the requirement that buildings in proximity to property lines be provided with parapets to mitigate fire spread, and the deadly 1980 fire at the MGM Grand Hotel and Casino in Las Vegas led to new code requirements for the sealing of exit system penetrations to prevent the passage of smoke." Yves Khawam, *Aligning Regulation with Sus-*

tainability, BUILDING SAFETY J. 62 (Dec. 2007). Others arose for social, environmental, or aesthetic reasons.

Historically, building codes have been creatures of local legislation, justified for the protection of the health and safety of a particular community. The following excerpt proposes a shift to national-level regulation for buildings and the legal concerns connected with such a change.

Shari Shapiro, *Who Should Regulate? Federalism and Conflict in Regulation of Green Buildings*
34 WM. & MARY ENVTL. L. & POL'Y REV. 257, 258-79 (2009)

History of Green Building Regulation in the United States

What Are Green Buildings?

Before examining the issues related to regulating green buildings, it is critical to define what constitutes a "green building." There can be no doubt that the built environment has an enormous environmental impact. "In the United States alone, buildings account for: 72 percent of electricity consumption, 39 percent of energy use, 38 percent of all carbon dioxide (CO_2) emissions, 40 percent of raw materials use, 30 percent of waste output (136 million tons annually), and 14 percent of potable water consumption."

Therefore, a "green building" is a structure that is designed, built, renovated, operated, or reused in an ecological and resource-efficient manner. Ideally, green buildings are designed and operated to meet certain objectives such as protecting occupant health, improving employee productivity, using energy, water, and other resources more efficiently, and reducing the overall impact on the environment.

Although there is some controversy about the actual achievement of these benefits, proponents of green buildings allege that they provide environmental benefits including energy consumption reduction; enhancement and protection of ecosystems and biodiversity; improved air and water quality; reduction of solid waste; and conservation of natural resources. In addition to the environmental benefits, green buildings allegedly "[r]educe operating costs; [e]nhance asset value and profits; [i]mprove employee productivity and satisfaction; [o]ptimize life-cycle economic performance ... [i]mprove air, thermal, and acoustic environments; [e]nhance occupant comfort and health; [m]inimize strain on local infrastructure; [and] [c]ontribute to overall quality of life."

Types of Green Building Regulations

According to an American Institute of Architects' study, state and local regulation and policies in support of green building have exploded since 2003. Since then, the number of counties with green building programs has risen from eight to thirty-nine, an increase of 387.5 percent. Simultaneously, the federal government has enacted tax incentives to encourage green building practices, regulated the energy efficiency of equipment, and encouraged green building through federal building programs. Most recently, the House of Representatives passed a national energy efficiency building code as a component of the [Waxman-Markey Act]....

There are four major types of green building regulations currently being utilized: 1) government construction regulations; 2) mandatory green building requirements; 3) financial incentives; and 4) non-financial incentives.

Government Construction Regulations

Where the state or other government entity acts as a private market participant, its freedom to regulate the terms of its purchases is very broad. Some government entities have passed regulations mandating that buildings built by the government entity must meet specific green standards. Others have extended such requirements to space that the government entity rents. Some government entities have even extended these requirements to buildings that receive funding from the government entity.

Mandatory Green Building Regulations

Some government entities have also enacted mandatory green building requirements that are much like traditional "command-and-control" environmental regulations, the Clean Water Act and the Clean Air Act being preeminent examples. Some regulations mandate specific green building practices or the achievement of a green building standard, such as LEED.

[The article describes provisions of the Waxman-Markey Act passed by the House of Representatives in June of 2009. The Act was never approved by the Senate. Section 201 of that Act would have created a national energy efficiency building code for residential and commercial buildings—Ed.]

* * *

Financial Incentives

Financial incentives are the third type of green building regulation. Some financial incentives take the form of direct grants from government entities. Others are structured as tax incentives or rebates. Yet others are rebates of the typical government-related costs of building, such as permit fees.

Non-Financial Incentives

The fourth type of green building legislation involves non-financial incentives. Non-financial incentives should be attractive to municipalities because they do not deplete public finances directly and should, therefore, be easier to pass in difficult financial times or with reluctant constituencies. There is some evidence that developers may value non-monetary incentives as much or more than monetary ones.

Examples of non-financial incentives for green buildings include increased floor- to-area ratios and expedited permitting processes.

In short, there has been active experimentation with green building regulation and incentives at every level of government. Given the enormous impact of buildings on both the environment and climate change, this is not surprising. Little attention, however, has been devoted to analyzing which approach is most effective, or which government entity is best equipped, from a legal and practical standpoint, to implement effective regulations.

Federalism Considerations in Green Building Regulation

There has always been conflict over the scope of the regulatory authority of the federal government versus that of the state governments. The Constitution established various mechanisms for determining the scope and extent of each level of governmental authority. With state governments, local governments, and now the federal government seeking to regulate green buildings, federalism conflicts were swift to arrive....

Federal Preemption

Article VI of the Constitution established the supremacy of federal laws over conflicting state laws....

———————

A case of direct preemption has already emerged in the green building field. On August 29, 2008, the Air Conditioning, Heating, and Refrigeration Institute and other HVAC and water heating equipment trade organizations, contractors, and distributors (collectively "A.H.R.I.") filed for an injunction in federal district court against the City of Albuquerque in order to stop components of the city's high performance building code from taking effect.

A.H.R.I. argued that the Energy Policy and Conservation Act of 1975 ("EPCA"), as amended by the National Appliance Energy Conservation Act of 1987 ("NAECA") and the Energy Policy Act of 1992 ("EPACT") preempted the building code's provisions related to energy efficiency of HVAC products. Together, these laws establish nationwide standards for the performance of HVAC equipment, and contain a preemption provision that "prohibits state regulation 'concerning' the energy efficiency, energy use, or water use of any covered product with limited exceptions."

The Albuquerque Energy Conservation Code ("Code") was part of the City's attempt to significantly reduce carbon dioxide and greenhouse gas emissions. The Code consisted of two volumes.... [Options for reducing energy use were included in each volume—Ed.]

* * *

A.H.R.I. argued that the Code's regulations would preclude them from "selling non-compliant HVAC and water heating products" in Albuquerque. Additionally, the Code would cause equipment costs to increase and, thereby, induce consumers to repair, rather than replace, their products. Furthermore, new home costs may have increased due to the increased equipment costs, and, thus, have impacted new home sales. Finally, A.R.H.I. argued that the Code would precipitate confusion with regards to the standards by which "manufacturers, distributors and contractors" were to abide.

On October 3, 2008, Chief District Court Judge Martha Vazquez not only granted the preliminary injunction, but laid out her opinion that the Albuquerque Code was indeed preempted. After analyzing the particular provisions, Judge Vazquez concluded that "[t]here is no doubt that Congress intended to preempt state regulation of the energy efficiency of certain building appliances in order to have uniform, express, national energy efficiency standards."....

Although the A.H.R.I. case presents an example of express federal preemption, the case could easily have been subject to an implied preemption analysis if the EPCA did not contain express preemption provisions—posing a harder case for Judge Vazquez. If the EPCA had simply regulated the energy efficiency of heating and air conditioning equipment, Judge Vazquez would have had to determine if Congress intended to dominate the field with its regulation. As federal regulation of the components of green buildings, like energy efficiency, water, and so forth, become more pervasive, the courts will doubtless be increasingly called on to make this type of determination.

State Preemption

In addition to federal preemption, another layer of intergovernmental conflict impacts green building regulation-state preemption. State preemption works like federal preemption, except that the regulatory authority of local governments is constrained by regulation taken at the state level.

A great example of the impact of state preemption on green building regulation comes from the Commonwealth of Pennsylvania. In 2004, Pennsylvania adopted a Uniform Construction Code ("UCC") as the common building code for all municipalities in Pennsylvania. The UCC, in itself, does not prevent local governments from passing green building regulations related to the building code as long as: the requirements are equal to or more stringent than the UCC; the local government secures approval from Pennsylvania's Department of Labor and Industry; and the local government provides appropriate public notice.

* * *

In Schuylkill Township v. Pennsylvania Builders Ass'n, the Commonwealth Court held that townships must prove that "conditions there were so different from the statewide norm that the uniform standards were not appropriate to use in the Township" in order to satisfy the "clear and convincing" standard for an exception to the UCC. This case is on appeal to the Pennsylvania Supreme Court to determine whether the Pennsylvania law implementing the UCC "requires a municipality to prove that there are unusual local circumstances or conditions atypical of other municipalities that would justify" an exception to the UCC. [The Supreme Court of Pennsylvania affirmed the heightened requirement to justify departure from the U.C.C. in *Schuylkill Twp. v. Pennsylvania Builders Ass'n*, 7 A.3d 249 (2010) — Ed.]

If the Supreme Court determines that atypicality is required, local governments would have a very difficult time passing green building standards which required building practices different from those in the UCC due to the difficulty of arguing that the benefits of green building are any different in one township than any other in Pennsylvania. The UCC would essentially have preempted the local governments from developing independent green building requirements.

Commerce Clause

In addition to the Supremacy Clause, the Commerce Clause also poses significant federalism concerns for green building regulation....

* * *

Most broadly, the current jurisprudential position has three basic tenets. First, where a state attempts to discriminate against interstate commerce, the law is per se unconstitutional. Second, where a state acts as a market participant-for example, by sourcing exclusively in-state materials for its own construction projects-the regulation is not restricted by the Commerce Clause. Finally, the remaining cases are judged under a balancing test that seeks to balance legitimate state interests with those of protecting interstate commerce.

Green building regulations can run afoul of the Commerce Clause very readily. For example, the A.H.R.I. plaintiffs specifically alleged that the Albuquerque green building regulations violated the Commerce Clause....

A hypothetical example could emerge with the sourcing of locally produced materials. Many of the green building rating systems include locally produced materials as a component of determining the "green-ness" of the building, as such materials require fewer resources to transport. If a green building regulation required in-state sourcing for private projects, it would likely be considered discriminatory against out-of-state market participants, and therefore in violation of the Commerce Clause.

When higher levels of government act to regulate, as in the case of the EPCA or the Pennsylvania UCC, lower levels of government can be constrained in their ability to regulate, and face stiff Constitutional challenges. With the many, often conflicting, priorities of the federal government, however, it is often desirable for states and localities to act. As with any federalist system, there is no perfect solution. Therefore, it is critical to analyze the pros and cons of regulation of green buildings at the federal, state, and local levels, and to determine which regulatory authority, or combination of authorities, will be able to regulate green buildings most effectively. . . .

Imperfect Solutions-Exclusive State or Federal Regulation of Green Buildings

State and Local Regulation of Green Buildings

Since the beginning of the administration of President George W. Bush, local and state governments have become increasingly responsible for environmental regulations. The cross-border effects of environmental damage have triggered federal environmental regulations. During the 1970s and 1980s, the federal government began to adopt environmental regulations in response to the increasing cross-border effects of toxic sites and pollution.

The lack of federal action on global warming has created a well-spring of creative legal experimentation in regulating green buildings, from creative incentive programs to strict state-wide green building codes. It has been a classic case study in states as "laboratories of democracy."

There are several benefits to state and local regulation of green buildings. The primary advantage is that, historically, building regulations have been a local concern. Building codes are developed at the state or local level, with huge variability from state to state and even within states. Part of the reason for this local control is the variability among local-ities-building regulations which apply in earthquake-prone areas, like California, would probably be inappropriate for flood prone regions of the Midwestern states. Another basis for local land use regulation is the intimacy of the regulation. Because people's homes, businesses and communities are directly impacted by building regulations, access to the levers of power which influence these regulations has been seen as critical. Finally, because of the history of local control of building regulation, every state and community with a building code has an enforcement mechanism already in place.

In addition to the history of land use regulation, there are potential regulatory advantages to local control of green building regulations. First, states and localities have already begun passing and implementing green building regulations. Implementing a federal green building standard might retard those programs already in place. Second, because coalitions should be easier and less expensive to develop at the state and local levels of government, it may be easier to pass green building regulations with more stringent environmental standards, thereby achieving greater reductions in natural resource consumption.

State and local regulation of green buildings is not without its drawbacks, however. A primary issue is simply lack of will to regulate; for example, as of the publication of this article, thirteen states have no statewide commercial building code, or have not updated their code within the last ten years, and eleven states have no statewide residential building code, or have not updated their code within the last ten years. Indeed, a patchwork of state or local regulations could lead to a "race to the bottom" where states and localities seeking additional development implement more lax green building regulations (or none at all). This is particularly troubling in an economic moment which has seen a virtual standstill in new residential and commercial development. [Figure 7.5 shows which states have adopted various versions of the IECC—Ed.]

Figure 7.5: Residential state energy codes status map

Online Code Env't & Advocacy Network (July 1, 2011),
http://bcap-ocean.org/sites/default/files/Residential%20Status%20Map.pdf
Courtesy of the Building Codes Assistance Project

Federal Regulation of Green Buildings

The inherent problems with state and local regulation of environmental concerns was essentially what led to the first wave of environmental regulations in the early 1970s. Federal regulation of environmental issues, including green building regulations, has some major advantages. First and foremost is national uniformity. Cross-border issues are eliminated if everyone must adhere to the same standards. States and localities which have, thus far, failed to regulate the built environment would be forced to regulate. Another potential benefit is cost reduction-if standards are nationally uniform, producers of green building materials need only design products to a single set of requirements. The federal government, however, has many priorities. Strict regulation of buildings may give way to other considerations. In addition, national interest groups have greater sway at the national level than at the state and local levels. For example, section 201 of the Waxman-Markey Act calls for the development and adoption by state and local governments of a national energy efficiency building code. Already, national organizations such as the National Association of Home Builders ("NAHB") and the International Council of Shopping Centers ("ICSC") have begun lobbying nationally against such a requirement. Due to conflicting priorities and strong interest groups, the regulations that are ultimately passed may be weaker than those implemented by individual states. Finally, because building regulation has historically been a state and local concern, the Federal government does not currently have an administration in place to implement national green building regulations.

Cooperative Federalism in Regulation of Green Buildings

* * *

[Under the Clean Air Act, if—Ed.] the EPA determines a SIP [State Implementation Plan] would not meet the NAAQS [National Ambient Air Quality Standards], the EPA must reject it. Should a state fail to repair the SIP's defect, federal highway funds may be withdrawn as a penalty. Additionally, a two-to-one offset for any new stationary sources may be imposed. If, after imposing penalties, the state still does not generate an acceptable SIP, the Federal Implementation Plan must be imposed on the state. On the other hand, "[o]nce EPA approves a SIP, federal agencies may not take, approve, or fund any activity that does not conform to the SIP."

A regulatory scheme similarly based on a Cooperative Federalism model could effectively balance the benefits and drawbacks of state versus federal green building regulation. First, the federal government would implement national green building metrics. These metrics should include at least energy efficiency, water conservation, materials and resource usage, site selection and indoor air quality considerations.

Once the national metrics have been established, states and localities should bear the responsibility of developing local codes to meet the national metrics, as well as implementing the system once in place. This structure would allow for the effective utilization of the code administration entities already in place at the state and local level, as well as allowing for variability to meet particular local needs and desires. As many states and localities are starved for funds, the federal government must provide resources for development and adoption of the new codes.

As with the Clean Air Act, if a state fails to develop or adopt a satisfactory plan, the federal government would have to provide an alternative code, or punish non-compliance by withholding funding. Administration for enforcement of the Cooperative Federalism system would be much less difficult and costly than implementing a nationwide code development and enforcement administration, however.

Notes and Questions

1. Look at Figure 7.6 and note all of the different organizations involved in EE code development, adoption, implementation, and enforcement. Can you see why there might be some confusion? Do you feel some groups are better situated than others to match regional needs? Might some groups have an incentive to be vested in the codes they create instead of searching for the "best" or most energy efficient code overall?

2. Based on the Shapiro article, what are some of the potential Constitutional challenges to a national energy efficiency code? Would a federal statute mandating state utilities to encourage energy efficiency encounter similar hurdles? Do you think national legislation is the best route? Why or why not? If so, what do you recommend be included in such bills?

3. California adopted the first state-wide energy building requirements in its building code in 1978. Today, over forty states use a version of the International Energy Conservation Code or better for residential buildings. Figure 7.5 shows which states have adopted the various versions of the IECC.

4. The American Recovery and Reinvestment Act of 2009 (aka the Stimulus) included several billion dollars in energy assistance grants for states. Section 410 of the Stimulus conditioned additional state energy grants on assurances by state governors of the following:

> (1) The applicable State regulatory authority will seek to implement, in appropriate proceedings for each electric and gas utility, with respect to which the State

Figure 7.6: Building Codes Assistance Project—code universe

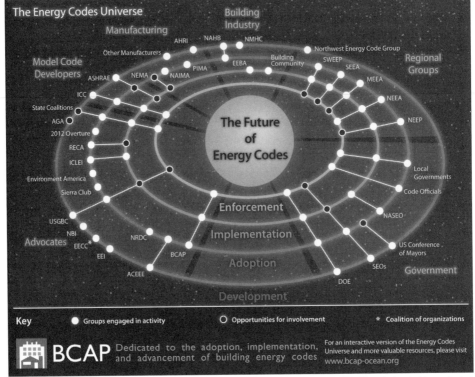

Image courtesy of Building Codes Assistance Project

regulatory authority has ratemaking authority, a general policy that ensures that utility financial incentives are aligned with helping their customers use energy more efficiently and that provide timely cost recovery and a timely earnings opportunity for utilities associated with cost-effective measurable and verifiable efficiency savings, in a way that sustains or enhances utility customers' incentives to use energy more efficiently.

(2) The State, or the applicable units of local government that have authority to adopt building codes, will implement the following:

(A) A building energy code (or codes) for residential buildings that meets or exceeds the most recently published International Energy Conservation Code, or achieves equivalent or greater energy savings.

(B) A building energy code (or codes) for commercial buildings throughout the State that meets or exceeds the ANSI/ASHRAE/IESNA Standard 90.1–2007, or achieves equivalent or greater energy savings.

(C) A plan for the jurisdiction achieving compliance with the building energy code or codes described in sub- paragraphs (A) and (B) within 8 years of the date of enactment of this Act in at least 90 percent of new and renovated residential and commercial building space. Such plan shall include active training and enforcement programs and measurement of the rate of compliance each year.

American Recovery and Reinvestment Act of 2009, Pub. L. No. 111-5, § 410, 123 Stat. 115, 146-47 (2009).

5. What do you think about the ARRA provisions in the section above? The 2009 IECC, referenced in 401(a)(2)(A), is not a very rigorous standard. The Energy Efficiency Codes Coalition (EECC) (created and housed by the Alliance to Save Energy) called for a "30 percent solution"—a proposal to boost new home energy efficiency by 30 percent. Although the 2009 IECC failed to adopt the 30 percent solution, the 2012 IECC now includes this goal. On its energycodes.gov website, the U.S. Department of Energy states, "This decision [to achieve the 30 percent solution—Ed.] represents the largest, one-step efficiency increase in the history of the national model energy code." As discussed at the end of the previous section, the EU has adopted the Passive House standard with 70-80 percent improved efficiencies for its buildings (in contrast to the upcoming 2012 IECC improvements of 30 percent). *See* Figure 7.4. Do you think the United States should likewise adopt the Passive House? If not, why do you think it is possible to have such a high standard in the EU and not in the United States?

3. Smart Grid

First, Section a here addresses how Smart Grid fits in with a distributed, as opposed to a centralized, generation structure. Then Section b explores criminal and commercial privacy concerns with Smart Grid.

a. Smart Grid & Distributed Generation

The three excerpts in this section provide context for understanding how distributed generation (DG) is distinguished from the more traditional centralized power plant model. They also introduce the Smart Grid concept and show how it relates to DG.

Carol Sue Tombari, POWER OF THE PEOPLE: AMERICA'S NEW ELECTRICITY CHOICES
67-73 (2008)

Twentieth-century technologies served us superbly well in the twentieth century, and they continue to perform well today. Nothing is perfect, and we had no choice but to live with the adverse consequences of coal, nuclear, and natural gas power.

Among the adverse consequences was the acid rain that created massive fish kills in New York's Adirondack lakes in the 1960s. Toxic rain was created by emissions from coal plants in the Midwest and pumped into the jet stream through tall stacks. Transporting the emissions via the jet stream to another region was an unforeseen and unintended consequence of regulatory compliance. This was an early warning of how much our planet is shrinking, figuratively, and an indicator that actions in one region can have unintended consequences in another.

Electric utilities do their best to manage and control the risks. By and large, they succeed on a grand scale that is unprecedented in engineering experience. In addition, twentieth-century electricity technologies enjoy what Hermann Scheer dubs the "home team advantage." If we gave any thought about our electricity system—though most of us don't—we probably would chose the path of least resistance and would stick with the devil we know rather than change the system.

But two things have changed. For one, some of the adverse consequences—such as accumulating greenhouse gases and resource depletion—are mounting. We as a society may be reaching our tipping point of tolerance for them. Second, we have new technology options that can be tools for change if we just reach out and use them.

It's understandable that many of us have not given much thought to what could go wrong, and it might seem that I dwell heavily on some of the drawbacks of twentieth-century technologies. However, given that the system is aging and stressed, it would be wise to be aware of potential problems and try to address them before they become crises....

Electricity 2000+: Power, Today and Tomorrow

> *"There are no silver bullets. There is only silver buckshot."*
> —Bill McKibben, author, educator, and environmentalist

Today we have technology options that we lacked yesterday. If we use them, we can start to mitigate the adverse consequences of yesterday's choices. More to the point, going forward we can reduce those consequences.

We cannot mothball our behemoth power plants. We thrive in an electricity-hungry economy, and we're not likely to put ourselves on that kind of austere diet. In fact, the hottest new entertainment and lifestyle gadgets, as well as commercial and industrial equipment, increase our collective appetite for electricity. Even if they are energy efficient, there are more of them and, in the aggregate, they use more electricity....

... Even the U.S. industrial sector has become increasingly electrified, in part for environmental compliance reasons, as electricity is cleaner than the direct combustion of fossil fuels.

All in all, U.S. electric utilities have served a 70 percent load growth in the past twenty-five years, without adding much base load generation or transmission capacity. The Edison Foundation, an arm of the electric utility industry association, predicts that demand for electricity will continue to rise in the foreseeable future, at rates ranging from 11 to 17 percent between 2006 and 2014.

Although we can't scrap our twentieth-century technologies, we can supplement them with the affordable alternatives that now are commercially available. Moreover, we can continue to add new technologies and products as they become available and affordable. Distributed energy technologies, such as rooftop PV or small wind, are to today's power plants as personal computers were to the mainframes of the 1980s.

Distributed Generation

To integrate renewables in the utility system will require a very different model of electricity generation, transmission, and distribution. Incredibly, and lucky for us, it will likely be cheaper than just adding more of the same-old to today's electricity grid. Simply building onto today's grid as it is currently designed is not the most efficient solution in any case.... [T]oday's grid is not, strictly speaking, one grid. It is, in fact, three regional grids that are not always connected where we need them to be, and not necessarily in sync with one another so that electricity can flow easily and where we want it to go.

Most important ... the grid was not designed to transmit electricity generated from renewable resources. Consequently, high voltage wires do not reach into remote sunny and windy regions. This creates stranded assets of sun and wind that can't be tapped for their electricity-generating potential. Upgrading the efficiency and adding enough high-power transmission capacity to meet our anticipated needs, if we simply build onto the current model, could cost billions for transmission additions alone. Given the myriad other pressures on our national wallet, this is not an investment that's likely to be made.

Picture, if you will, today's model: It's a large central station power plant—a nuke, a coal plant (with the associated coal piles and multiple railroad sidings), or natural gas. Transmission lines lead away from it and deliver electricity to distribution points. From there, distribution lines lead to your neighborhoods, homes, and businesses. In today's model, the electricity flows one way: from the power plant to you.

What if, however, you were to generate some of your own electricity? What if you put solar electric panels on your roof and actually fed electricity back into the line at sunny times when your system generated more electricity than you needed? If you own a commercial warehousing facility with a large flat roof, that urban roof could be a veritable field for solar electric arrays. It could become a mini power plant, generating electricity and feeding it back to the grid for consumption somewhere else in the system.

This is called distributed generation. The generating plants, whatever form they might take, are dispersed and located at or near the point of consumption. This increases transmission efficiency and would eliminate most of today's line losses in the distributed part of the system.

Now picture our utility model … but add these multiple mini power plants located at the point of consumption (i.e., your home or business). These power plants could be powered by solar, wind, or both. They could also be powered by biomass. Urban biomass supplies could include landfill gas, urban wood waste, suburban lawn trimmings. In rural areas, biomass could come from plant and animal wastes and perhaps even dedicated fuel supply crops such as switchgrass, drought resistant prairie grasses, or fast-growing poplar trees. Large facilities such as hospitals, schools, or industries could install fuel cells and create microgrids in building complexes or neighborhoods.

These mini power plants could take any form and would be distributed throughout the system. The end result is that they would take a "load" off the existing generation and transmission facilities. In the aggregate, they're likely to reduce the load enough that many (if not all) additional central station power plants would not need to be built. These distributed resources could also use local distribution lines and therefore could reduce the need to add expensive transmission capacity.

Maybe a lot of these distributed mini power plants would not be big enough to generate excess electricity. Maybe they would succeed mostly in reducing the need for electricity from the central station power plant. Maybe you would add energy efficiency to your personal mix, reducing the need for electricity even further. You might not have built a mini power plant, but you have succeeded in reducing—and, in some service areas, perhaps eliminating—the need for an added nuclear or coal plant.

In fact, one utility combined distributed renewable energy sources and aggressive energy efficiency to make up for the loss of its nuclear power plant in the 1990s. The Sacramento Municipal Utility District (SMUD) took a planned nuke off the planning books after the citizens directed it to do so in a referendum. Within five years of the referendum, SMUD installed 2 megawatts (MW) of PV on-site at the mothballed nuke, 5 MW of wind, 134 MW of geothermal electricity, and 4 MW of PV distributed on residential rooftops.

SMUD's rooftop PV program provides an interesting model for other utilities. The utility owned the PV arrays and used the homeowners' roofs, similar to leasing land for

Figure 7.7: Smart grid

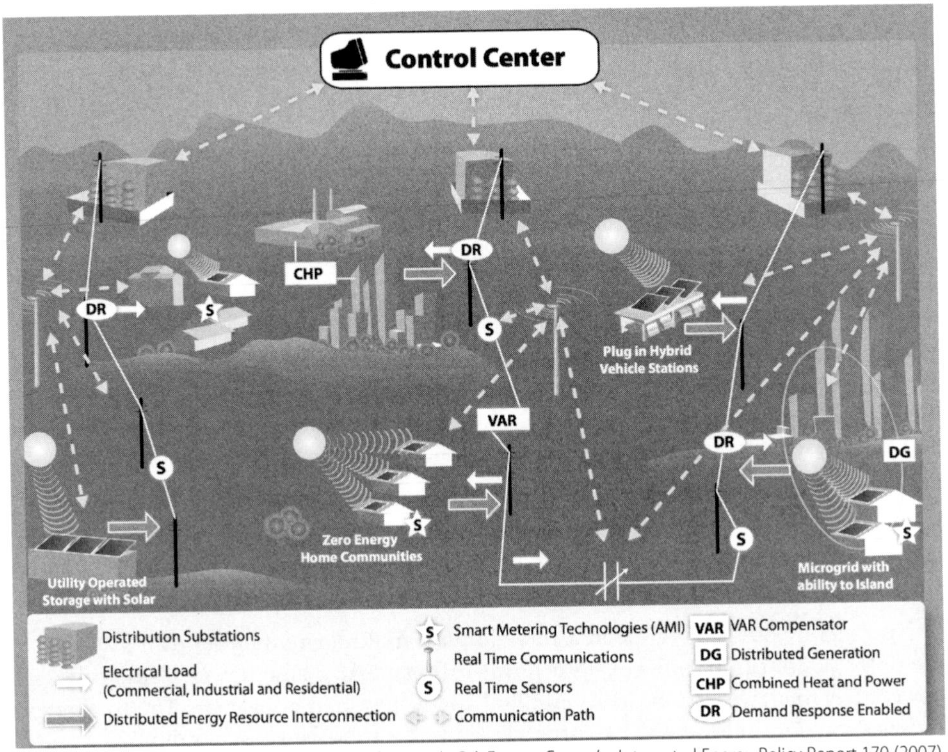

Susanne Garfield et al., Cal. Energy Comm'n, Integrated Energy Policy Report 170 (2007), *available at* http://www.energy.ca.gov/2007publications/CEC-100-2007-008/CEC-100-2007-008-cmf.pdf.

wind farms. Utilities, with the obligation to serve, are understandably nervous about turning over "the keys to the car" in terms of letting other generators onto the system, whether they are industrial cogenerators or dispersed small generators. SMUD retained ownership of the equipment and controlled operation of the system.

In addition, SMUD was able to reduce its peak load by 12 percent through energy efficiency. It planted 300,000 shade trees, reducing indoor cooling requirements as much as 40 percent; it helped customers purchase more than 42,000 superefficient refrigerators; and it provided rebates for cool roofs (rooftops with sun reflective coating). SMUD spent 8 percent of its gross revenues on energy efficiency and succeeded in holding rates constant for ten years. In contrast, rates would have skyrocketed 80 percent had the utility completed construction of the nuclear power plant.

Sidney A. Shapiro & Joseph P. Tomain, *Rethinking Reform of Electricity Markets*
40 WAKE FOREST L. REV. 497, 518-19 (2005)

DG and micropower are dependent upon significant technological improvements throughout electricity production, transmission, distribution, storage, and consumption.

Most simply, the scale of generation units is reduced significantly, and they are widely dispersed. "Smart energy" technologies are intended to reduce the size of power generation units, to be closer to the source of consumption, to utilize "Smart Grids" which will transmit power more efficiently, and to use "smart meters" which will provide consumers with more information about their consumption patterns and about their choice of providers.

Another term for DG is micropower, which also involves new technologies including microturbines, hydrogen fuels, solar cells, landfill gases, and the like. In this regard, micropower is touted as a clean energy alternative. According to the International Energy Agency, these technologies are increasing in importance....

Smart electricity policy is a return to the electricity future. When Edison flipped the switch at Pearl Street Station in New York City in 1882, the first electricity company went into operation and did so on a small scale. Technological advances enabled the effective nationalization of the electricity grid in the early part of the twentieth century. Today, we find ourselves contemplating a return to small scale because it promises economic efficiencies by removing producers from the grid, environmental benefits through greater energy efficiencies and increased use of renewable energy resources, and energy security advantages from terrorist attack, international supply disruptions, or catastrophic accidents.

The Smart Grid: An Introduction

Litos Strategic Commc'n, U.S. Dep't of Energy 6-15 (2008)

The Grid as it Stands: What's at Risk?

Even as demand has skyrocketed, there has been chronic underinvestment in getting energy where it needs to go through transmission and distribution, further limiting grid efficiency and reliability. While hundreds of thousands of high-voltage transmission lines course throughout the United States, only 668 additional miles of interstate transmission have been built since 2000. As a result, system constraints worsen at a time when outages and power quality issues are estimated to cost American business more than $100 billion on average each year.

In short, the grid is struggling to keep up.

Based on 20th century design requirements and having matured in an era when expanding the grid was the only option and visibility within the system was limited, the grid has historically had a single mission, i.e., keeping the lights on. As for other modern concerns ...

Energy Efficiency? A marginal consideration at best when energy was—as the saying went—"too cheap to meter."

Environmental impacts? Simply not a primary concern when the existing grid was designed.

Power System Fact: 41 percent more outages affected 50,000 or more consumers in the second half of the 1990s than in the first half of the decade. The "average" outage affected 15 percent more consumers from 1996 to 2000 than from 1991 to 1995 (409,854 versus 355,204).

Customer choice? What was that?

Today, the irony is profound: in a society where technology reigns supreme, America is relying on a centrally planned and controlled infrastructure created largely before the age of microprocessors that limits our flexibility and puts us at risk on several critical fronts:

Efficiency: If the grid were just 5 percent more efficient, the energy savings would equate to permanently eliminating the fuel and greenhouse gas emissions from 53 million cars. Consider this, too: if every American household replaced just one incandescent bulb (Edison's pride and joy) with a compact fluorescent bulb, the country would conserve enough energy to light 3 million homes and save more than $600 million annually. Clearly, there are terrific opportunities for improvement.

Reliability: There have been five massive blackouts over the past forty years, three of which have occurred in the past nine years. More blackouts and brownouts are occurring due to the slow response times of mechanical switches, a lack of automated analytics, and "poor visibility"—a "lack of situational awareness" on the part of grid operators. This issue of blackouts has far broader implications than simply waiting for the lights to come on. Imagine plant production stopped, perishable food spoiling, traffic lights dark, and credit card transactions rendered inoperable. Such are the effects of even a short regional blackout.

National Economy: The numbers are staggering and speak for themselves:

- A rolling blackout across Silicon Valley totaled $75 million in losses.

- In 2000, the one-hour outage that hit the Chicago Board of Trade resulted in $20 trillion in trades delayed.

- Sun Microsystems estimates that a blackout costs the company $1 million every minute.

- The Northeast blackout of 2003 resulted in $6 billion economic loss to the region.

Compounding the problem is an economy relentlessly grown digital. In the 1980s, electrical load from sensitive electronic equipment, such as chips (computerized systems, appliances and equipment) and automated manufacturing was limited. In the 1990s, chip share grew to roughly 10 percent. Today, load from chip technologies and automated manufacturing has risen to 40 percent, and the load is expected to increase to more than 60 percent by 2015.

Affordability: As rate caps come off in state after state, the cost of electricity has doubled or more in real terms. Less visible but just as harmful, the costs associated with an underperforming grid are borne by every citizen, yet these hundreds of billions of dollars are buried in the economy and largely unreported. Rising fuel costs—made more acute by utilities' expiring long-term coal contracts—are certain to raise their visibility.

Security: When the blackout of 2003 occurred—the largest in US history—those citizens not startled by being stuck in darkened, suffocating elevators turned their thoughts toward terrorism. And not without cause. The grid's centralized structure leaves us open to attack. In fact, the interdependencies of various grid components can bring about a domino effect—a cascading series of failures that could bring our nation's banking, communications, traffic, and security systems, among others, to a complete standstill.

Environment/Climate Change: From food safety to personal health, a compromised environment threatens us all. The United States accounts for only 4 percent of the world's population and produces 25 percent of its greenhouse gases. Half of our country's electricity is still produced by burning coal, a rich domestic resource but a major contributor to global warming. If we are to reduce our carbon footprint and stake a claim to global environmental leadership, clean, renewable sources of energy like solar, wind and geothermal must be integrated into the nation's grid. However, without appropriate enabling technologies linking them to the grid, their potential will not be fully realized.

Global Competitiveness: Germany is leading the world in the development and implementation of photo-voltaic solar power. Japan has similarly moved to the forefront of distribution automation through its use of advanced battery-storage technology. The European Union has an even more aggressive "Smart Grids" agenda, a major component of which has buildings functioning as power plants. Generally, however, these countries don't have a "legacy system" on the order of the grid to consider or grapple with.

* * *

The Smart Grid: What it is. What it isn't.

Part 1: What it is.

The electric industry is poised to make the transformation from a centralized, power-controlled network to one that is less centralized and more consumer-interactive. The move to a smarter grid promises to change the industry's entire business model and its relationship with all stakeholders, involving and affecting utilities, regulators, energy service providers, technology and automation vendors and all consumers of electric power.

A smarter grid makes this transformation possible by bringing the philosophies, concepts and technologies that enabled the internet to the utility and the electric grid. More importantly, it enables the industry's best ideas for grid modernization to achieve their full potential.

Concepts in action

Advanced Metering Infrastructure (AMI) is an approach to integrating consumers based upon the development of open standards. It provides consumers with the ability to use electricity more efficiently and provides utilities with the ability to detect problems on their systems and operate them more efficiently.

* * *

Adoption of the Smart Grid will enhance every facet of the electric delivery system, including generation, transmission, distribution and consumption. It will energize those utility initiatives that encourage consumers to modify patterns of electricity usage, including the timing and level of electricity demand. It will increase the possibilities of distributed generation, bringing generation closer to those it serves (think: solar panels on your roof rather than some distant power station). The shorter the distance from generation to consumption, the more efficient, economical and "green" it may be. It will empower consumers to become active participants in their energy choices to a degree never before possible. And it will offer a two-way visibility and control of energy usage.

An automated, widely distributed energy delivery network, the Smart Grid will be characterized by a two-way flow of electricity and information and will be capable of monitoring everything from power plants to customer preferences to individual appliances. It incorporates into the grid the benefits of distributed computing and communications to deliver real-time information and enable the near-instantaneous balance of supply and demand at the device level.

The problem with peak

While supply and demand is a bedrock concept in virtually all other industries, it is one with which the current grid struggles mightily because, as noted, electricity must be consumed the moment it is generated.

Without being able to ascertain demand precisely, at a given time, having the 'right' supply available to deal with every contingency is problematic at best. This is particularly true during episodes of peak demand, those times of greatest need for electricity during a particular period.

Imagine that it is a blisteringly hot summer afternoon. With countless commercial and residential air conditioners cycling up to maximum, demand for electricity is being driven substantially higher, to its "peak." Without a greater ability to anticipate, without knowing *precisely* when demand will peak or how high it will go, grid operators and utilities must bring generation assets called peaker plants online to ensure reliability and meet peak demand. Sometimes older and always difficult to site, peakers are expensive to operate—requiring fuel bought on the more volatile "spot" market. But old or not, additional peakers generate additional greenhouse gases, degrading the region's air quality. Compounding the inefficiency of this scenario is the fact that peaker plants are generation assets that typically sit idle for most of the year without generating revenue but must be paid for nevertheless. [Ed. See Figure 7.2]

In making real-time grid response a reality, a smarter grid makes it possible to reduce the high cost of meeting peak demand. It gives grid operators far greater visibility into the system at a finer "granularity," enabling them to control loads in a way that minimizes the need for traditional peak capacity. In addition to driving down costs, it may even eliminate the need to use existing peaker plants or build new ones—to save everyone money and give our planet a breather.

Notes and Questions

1. The Litos Strategic Communication excerpt above does not include the portion about what smart grid is not. How do you think Smart Grid and smart meters are distinguishable? Are wind turbines, plug-in hybrid electric vehicles, and solar arrays part of the Smart Grid? If not, how do they relate?

2. In most areas of the country, the only way a utility finds out about a power outage is when a customer calls to report it. Smart Grid efforts hope to transform the grid from this passive, one-way flow of power to an interactive network. Electronics in homes and throughout the system will maintain power for critical loads (such as for hospital operating rooms or data storage electronics) and retime or curtail demand for less critical needs (such as water heating, water pumping, clothes drying). The ultimate success of the Smart Grid depends on (1) the effectiveness of the devices and (2) the ability to attract and motivate large numbers of consumers to embrace the devices and the energy saving adjustments they can make. What are some of the attractions and impediments to Smart Grid deployment?

3. Reliability is one of the goals of the Smart Grid. The Energy Policy Act of 2005 required FERC to establish mandatory standards through electric reliability organizations such as the North American Electric Reliability Corporation (NERC). One measure of reliability is the Average System Availability Index (ASAI), which has been set at "four nines" or 99.99% available (representing only 52.56 minutes of down time out of the

entire 525,600 minutes in a 365 day year). Consider the cost per minute of blackouts discussed in the Litos excerpt in this section. Are you surprised then that Google has its own backup power for a reliability of "nine nines" or 99.99999%? Do high standards for reliability and consistency of frequencies make the integration of renewable energy sources and Smart Grid technologies more challenging? In 2011, FERC conducted a test of the nation's electric grid that would allow more frequency variation without corrections. Some argue that this variation in frequency may cause clocks to be erratic and other appliances to break.

4. The Tombari article mentioned that SMUD leased homeowners' rooftops for solar PV. Does this make the arguments stronger for protecting solar access on host properties as discussed in Chapter 2?

b. Smart Grid & Privacy

One of the biggest impediments currently being discussed with the implementation of Smart Grid technology is the potential abuse of personal privacy. The following two excerpts explore this issue — the first from a criminal law Fourth Amendment search perspective and the second from a commercial consumer protection viewpoint.

Kevin L. Doran, *Privacy and Smart Grid: When Progress and Privacy Collide*
41 U. Tol. L. Rev. 909, 909-20 (2010)

Introduction

The U.S. electric grid—a staggeringly complex network of interconnected electric systems—is poised to undergo a major physical, operational and conceptual transformation with far reaching implications for the privacy of individuals. By incorporating literally millions of new intelligent components into the electric grid that deploy advanced two-way communication networks with interoperable and open protocols, the "smart grid" heralds a fundamental change in the electricity paradigm that has prevailed for more than a century.

This marriage of twenty-first century information technology with nineteenth century electricity technology is designed to achieve a spectrum of interconnected objectives ranging from greater efficiency, reliability, and grid security to more affordable, environmentally benign power and enhanced global competitiveness. According to the usually rhetorically staid U.S. Department of Energy, smart grid is the *sine qua non* that "enables us to approach this matrix of complex issues all at once."

The essential innovation behind the smart grid is information—highly detailed electricity usage data communicated by and between the utility, the consumer, and in many instances, third-party vendors. However, while this information—and the extrapolations that can be made from it—is what enables the smart grid to be "smart," it is also what makes the smart grid so potentially invasive of individual privacy. Smart grid data is a double-edged sword. The sharper the blade in terms of informational granularity, the more it can be wielded to achieve both societal benefits such as grid reliability and energy efficiency *and* invasions of privacy.

… Given the extraordinary level of detail contained in smart grid data—data that contains highly personal, real time information about activities that occur within the putative privacy of the home—the Supreme Court's exclusion of data communicated to third-parties from Fourth Amendment protection presents particularly troubling privacy and public policy concerns.

The Supreme Court has endeavored to draw a "firm line at the entrance to the house"— a "bright" demarcation between the outside world and the sanctity of the home wherein governmental authorities must, with few exceptions, obtain a warrant prior to conducting a search of the premises. In *Katz v. United States*, the Court famously extended this principle by holding the Fourth Amendment "protects people, not places." To determine whether a "search" has taken place—which is the initial question in cases involving a potential violation of the Fourth Amendment—Justice Harlan penned a two-part test in his concurrence to *Katz*, a test which the Court subsequently adopted. According to Justice Harlan's test, to state a legitimate Fourth Amendment Claim the individual must have "exhibited an actual (subjective) expectation of privacy," and that expectation must have been "reasonable." In essence, the question of whether a search has occurred—thereby triggering Fourth Amendment protection—depends on whether the individual had a "reasonable expectation of privacy" in the place or object in question. Absent this reasonable expectation of privacy, no Fourth Amendment search has occurred and thus no warrant is required.

In a series of decisions made in the 1970s known as the "third-party doctrine" cases, the Court categorically affirmed the view that when an individual voluntarily turns over information to a third-party, that individual no longer possesses a legitimate expectation of privacy in that material. As the Court in *Smith v. Maryland* explained: "a person has no legitimate expectation of privacy in the information he voluntarily turns over to third parties." This is true even where an individual turns over information to a third-party with the understanding that the third-party will not share it with others. In *United States v. Miller* the Court held that such voluntarily communicated information would no longer receive Fourth Amendment protection:

> [T]he Fourth Amendment does not prohibit the obtaining of information revealed to a third party and conveyed by him to Government authorities, even if the information is revealed on the assumption that it will be used only for a limited purpose and the confidence placed in the third party will not be betrayed. The defendant assumes the risk, no matter how small, that the information will end up in the hands of the police.

* * *

Criticisms of the Court's Privacy Jurisprudence

... [T]hree criticisms of the Court's privacy jurisprudence ... are of particular relevance to the deployment of smart grid technologies: (1) conflation of privacy with absolute secrecy; (2) misapprehension of what is voluntary with respect to third-party communications; and (3) normative inadequacy, particularly in the face of technological advances and changes in societal expectations of privacy.

Privacy as Absolute Secrecy

Privacy in everyday life is not a binary, either-or, concept. Individuals regularly communicate information to banks, internet service providers, video rental companies, hospitals, and sundry other entities, while retaining some expectation that this information will not be passed on to other parties without their permission. As Justice Marshall wrote in his dissent in *Smith [v. Marlyand]*, "Privacy is not a discrete commodity, possessed absolutely or not at all. Those who disclose certain facts to a bank or phone company for a limited business purpose need not assume that this information will be released to other persons for other purposes."

In terms of our expectations, privacy is also an elastic, situational quality that depends on the particular informational object in question (such as a telephone number or bank record), as well as the facts and circumstances that surround the creation, transmission, and acquisition of that information.... [A]nswering the question of whether or not society at large considers an expectation of privacy to be reasonable is a difficult and necessarily speculative task.

The Court's prevailing privacy jurisprudence, however, eschews much of this nuance in favor of a much more abstract and simplified notion of privacy. Taken together, the Court's majority opinions in *Katz*, *Kyllo*, and the third-party doctrine cases, create a concept of privacy that is akin to [a] door that can only be fully closed or fully open — nothing in between. In the Court's view, the instant information is communicated to a third-party, it is the same as if were communicated to everyone.

* * *

What this view fails to recognize, however, is that there are increasingly few things in contemporary life that are not, in some fashion, communicated to third-parties. The smart grid is an exemplar of this fact. The assorted suite of technologies associated with smart grid have the potential to communicate details on virtually every activity that takes place within a home to third-parties. The Court has endeavored to draw a "bright line" of Fourth Amendment protection around the home. What happens, however, when the home it-self — and all the activities therein — are digitized and communicated in real time to third-parties? The bright line then becomes a meaningless symbol, a testament to a jurisprudence that has failed to adapt to a reality wherein the four walls of the home no longer demark the boundary between what is kept private and what is not.

One Shade of Voluntary

An essential element of the Court's approach to determining whether an individual possesses a reasonable expectation of privacy in an object involves the question of whether that object was voluntarily communicated to a third-party. It is by voluntarily communicating information to a third-party that an individual "assumes the risk" that the information will be handed over to other parties, including government agents. The underlying rationale here is that individuals have a choice; they can choose to communicate information to a third-party or they choose not to. And if they choose to do so, they assume the risk "that the information will be conveyed by that person to the Government."

The Court's reliance on the distinction between voluntary and involuntary communications is similar — and indeed conceptually flows from (or vice versa) — to its understanding of privacy as a binary, only-off or only-on quality. It is, as a strict truth statement, accurate to say that having a bank account, a telephone or cell phone, a computer with internet access, electricity, a credit card, car insurance (or for that matter, a car), are all voluntary choices — choices that need not have been made.... This perspective, however, fails to recognize that there are many shades of voluntary when it comes to the associations and amenities that pervade modern life. One would hardly equate, for instance, renting a catamaran for a weekend vacation as the same, in terms of the voluntariness of the action, as paying one's electricity bill. Yet the Court would construe both as the same. The voluntariness of a communicative act, in the Court's view, has only one value. It is either present, in which case the individual has assumed the risk, or it is not. A more realistic understanding of the voluntariness of communications to third-parties would embrace the fact that some things, such as basic utility services, are less voluntary as compared to other things, such as renting videos for recreation.

* * *

Normatively Unsatisfying

Privacy is always at odds with security. It is a necessary balancing act, weighing the need to ensure against abusive governmental invasions with the need to provide law enforcement with sufficient investigatory authority to solve crimes. Technology complicates this balancing effort. As surveillance technology advances, the ability of the police to use technology to intrude into private spheres also increases. The danger for abuse becomes more acute. Technology also enables criminals to obfuscate and refine their activities, requiring law enforcement to employ similar techniques to level the proverbial playing field.

The Court has been cognizant of [the] impact of technology on the balance between privacy and security. "It would be foolish," wrote Justice Scalia in the majority opinion for *Kyllo*, "to contend that the degree of privacy secured to citizens by the Fourth Amendment has been entirely unaffected by the advance of technology." In *Kyllo* the Court wrestled with the question of "what limits there are upon this power of technology to shrink the realm of guaranteed privacy."…. To guard against the ever encroaching nature of technology, the Court sought to jurisprudentially seal what happens within the four walls from technological intrusions. While technology might be able to physically penetrate into the sanctum of the home, it would not be allowed to legally do so without constituting a Fourth Amendment search.

The problem with this approach is that it focuses on a distinction that is increasingly less and less relevant in contemporary life—the distinction between the inside and the outside of the home. With technologies such as smart grid, the concept of technology reaching into the home no longer accurately captures the full picture. Technology is now reaching out of the home, producing detailed accounts of what takes place within its four walls, and delivering that information over to the hands of third-parties.

From a normative perspective, this is troubling. As the Court in *Kyllo* notes, "'At the very core' of the Fourth Amendment 'stands the right of a man to retreat into his own home and there be free from unreasonable governmental intrusion.'" Yet while the government may not be able to physically or technologically intrude on the home, through the third-party doctrine it can, when working in concert with smart grid technologies, do precisely the same thing.

_____

Elias L. Quinn, *Smart Metering & Privacy: Existing Law and Competing Policies*

Framing Document for Colo. PUC High Profile Dkt. No. 091-593EG
(Order C09-0878), iv-8, 28-34 (2009)

Advanced metering infrastructure (AMI or smart metering) is being installed throughout electric networks both in Colorado and across the country. From these smart meters, detailed information about consumer electricity usage will flow from residences and businesses to electric utilities. Instead of billing customers for their monthly draw, electric utilities will know what customers are using in half-hour, fifteen-minute, or even five-minute intervals. [Figure 7.8 shows some of the information that can be acquired using certain Smart Grid technologies—Ed.]

Proper management of this new information pool could support energy efficiency efforts and demand-side management (DSM) initiatives. However, insufficient oversight

Figure 7.8: Household electricity demand profile recorded on a one-minute time base

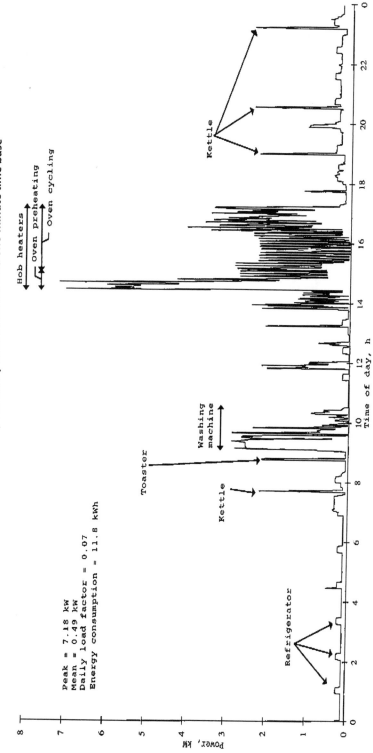

G. Wood & M. Newborough, *Dynamic Energy-consumption Indicators for Domestic Appliances: Environment, Behavior, and Design*, 35 ENERGY AND BUILDINGS 821, 822 (2003) (citing M. Newborough & P. Augood, *Demand-side Management Opportunities for the UK Domestic Sector*, IEE Proceedings of Generation Transmission and Distribution 146 (3) (1999) 283–293), as used in Elias L. Quinn, *Smart Metering and Privacy: Existing Law and Competing Policies*, Framing Document for Colo. PUC High Profile Dkt. No. 091-593EG (Order C09-0878) (2009), available at http://ssrn.com/abstract=1462285.

of this information could also lead to unprecedented invasions of consumer privacy. Many intricate details of household life can be gleaned from information obtained via advanced metering infrastructure.

A complicated network of risks and concerns bears on this issue. The more information gathered, the better supported DSM initiatives, efficiency investments, and conservation efforts. Yet such efforts are antithetical to traditional utility incentive structures, which tie returns to electricity sales. The use and sale of this information might play a role in reforming the business model of electric utilities; indeed, smart grid information is a potential revenue stream heretofore unexplored. As such, the formulation of privacy regulations should be seen, not only as consumer protection, but as incentive regulation.

However, information control regimes that centralize smart grid information disclosures by giving princip[al]control to the electric utility may work against innovation in service industries developing at the edge of the electric grid and provide new barriers to market entry. If privacy regulations make customer usage information [] too difficult or expensive to obtain, the regulatory regime could dampen the rampant growth and evolution of a promising new sector for economic development. The balance struck among these various factors will define any privacy concern related to smart grid information, which is ultimately founded on who has access with customer usage information, and what they can do with it.

* * *

The various interests converging on smart grid development are not strictly incompatible; a workable and perhaps even jointly beneficial compromise can be found. However, the integrated nature of these issues highlight that there is not a vision-neutral option before policy makers. Inaction on the construction of smart grid information controls favors some actors, while tailored regulation would likely favor others.

Additionally, three pressures urge that the privacy concerns be addressed earlier rather than later. First, the privacy concerns are real, and should be addressed proactively in order to protect consumers. Second and related, a salient privacy invasion—were it to happen and get press—could create significant opposition to smart grid deployment efforts. Third, information controls that govern which parties have access to smart grid information when, and what they can do with it, will be a critical part of the networking architecture and will inform—and constrain—viable business models for edge services.

* * *

... [C]omprehensive privacy protection requires a triptych of regulatory efforts:

[1] Regulations setting consent requirements for the disclosure of smart meter customer information to third parties;

[2] Requirements that both technological and procedural measures for the protection of customer data be in place as a prerequisite to gaining access to the data;

[3] Requirements that parties holding customer information inform those customers in the event their information is stolen or accessed by unauthorized individuals.

* * *

Electric Utility Information Bundling and Resale

The many things determinable from smart grid information analysis, and the many edge services and other ancillary parties that have a reason to seek it out, suggests that smart grid information has value. This gives rise to the question: could electric utilities turn the new information stream into a source of revenue?

* * *

These various issues are but flagged here for future consideration. The point for the purposes of this discussion is simply this: the systematic resale—and so disclosure—of individual electricity usage information is a real possibility. This possibility, while potentially having some economic and incentive benefits for the project of electricity provision reform, should be closely monitored if employed in order to protect consumers.

* * *

Conclusions

The information collected on a smart grid is a library of personal information, the mishandling of which could lead to the invasion of consumer privacy. However, the exchange of information lies at the very heart of the promise of the smart grid—both its environmental benefit, and as a growing home for investment and innovation. Several regulatory tools are available to policy-makers, which can be employed to strike any balance among the various privacy, environmental, and economic risks associated with information control restrictions. Regulations seeking to protect consumer privacy must be careful not to unnecessarily hinder the deployment of smart grid technologies and so plant an obstacle in the nation's path toward a new energy economy. Yet so too must they take care not to sacrifice consumer privacy amidst an atmosphere of enthusiasm for the project of electricity reform.

Notes and Questions

1. What are some of Doran's concerns with Smart Grid technology? How does he think the Supreme Court's privacy jurisprudence contributes to the problems?

2. Doran's article focuses on Smart Grid data revealing personal details about activities taking place within one's home. However, Smart Grid data does not necessarily stop in the home. In his article, Quinn wrote, "[E]lectric utilities may eventually collect usage information beyond the four walls of the home, *e.g.*, by tracking PHEV [Plug-in High Efficiency Vehicle] charges and battery-to-grid sales." What additional privacy considerations might such data raise?

3. Quinn is approaching the Smart Grid privacy question not from the perspective of preventing unwarranted criminal searches of that data, but instead from the utilities' perspective of how it can mine the customer information for marketing opportunities. What is the balance between privacy and commercial exploitation of the data available from Smart Grid that Quinn is trying to achieve?

4. What are the three methods of protecting privacy that Quinn recommends? Do you think these are sufficient? Would you recommend other alternatives?

4. Energy Efficient Technologies

Refrigerators are a U.S. success story for EE. In 1981 they used 1350 kWh per household per year. Federal government mandates in efficiency standards and advances in technology reduced that number by 60 percent to 425 kWh per year in 2001. Now, electronics have the dubious distinction of being the top seed for electricity consumption in most U.S. households, accounting for nearly 20 percent of household energy use. One high definition cable box averages 446 kWH per year, about 10 percent more than a 21-cubic-foot energy-efficient refrigerator. Plasma televisions are also some of the highest energy hogs in U.S. households, using as much as electricity as a refrigerator and costing consumers $200 a year.

In 2009, California passed the first mandate for electronic efficiency standards. The measure is expected to cut $1 billion per year off California's electricity consumption. Although federal efficiency standards forced refrigerators out of the top seed for electricity consumption, there is currently no national mandate for electronics.

The federal government does, however, provide some consumer guidance through the Energy Star Program. On May 1, 2010, the government rolled out energy star 4.0 for TVs and version 5.0 will become effective in May 2012. Manufacturers can still manufacture TVs that do not meet the standard, but they will not be able to earn the Energy Star logo if they fall below the minimums. Therefore, without government mandates for efficiency, consumers' only option is to vote with their pocketbooks by choosing higher Energy Star rated appliances and electronics. However, with cable boxes, customers have little choice and little individual bargaining power. Figure 7.9 shows the typical breakdown of U.S. residential electricity use.

The landscape is different for information technology companies, which are some of the biggest consumers of electricity. Google.org invested $45 million in breakthrough clean energy technologies in 2008. As part of its Renewable Energy Cheaper than Coal initiative (RE<C), Google created an internal engineering group solely dedicated to exploring clean energy. The goal behind RE<C is to develop electricity from renewable energy sources that is cheaper than that generated from coal and produce one gigawatt of renewable energy capacity in years, not decades. *Plug Into a Greener Grid: RE<C and RechargeIT*, Google.org (2010), http://www.google.org/rec.html.

In addition, in 2007, Google.org teamed with Intel and other industry partners to found the Climate Savers Computing Initiative. The group champions more efficient computing and is committed to cutting the energy consumed by computers in half by

Figure 7.9: U.S. residential electricity use

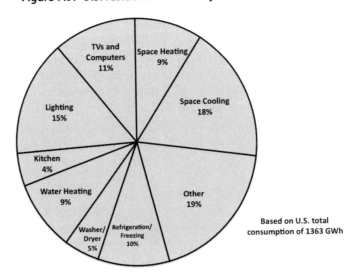

Based on U.S. total consumption of 1363 GWh

Data from U.S. Energy Info. Admin., http://www.eia.gov/tools/faqs/faq.cfm?id=96&t=3

2010. *Efficient Computing—Step 5: An Efficient and Clean Energy Future*, Google (2010), http://www.google.com/corporate/green/datacenters/step5.html

The following section discusses one of the largest consumers of retail electricity—data centers—and what IT companies are doing to improve efficiencies and address their unique energy needs.

William A. Tanenbaum, *Practical Steps to Contract for Energy-Efficient Data Centers and IT Operations*
981 PLI/Pat 247, 249-57 (2009)

Using energy-efficient IT and Green Outsourcing can provide four benefits: (i) improve computer technology; (ii) reduce IT costs (including maintenance and support); (iii) reduce energy consumption and associated costs in a company's IT ecosystem; and (iv) reduce the carbon footprint of corporate data centers and IT infrastructure, and thereby reduce the potential costs of buying carbon credits to offset a company's carbon footprint. There is also a potential fifth advantage: the ability to increase IT security by reducing manpower and thus conducting better background checks. There is a Homeland Security benefit to this for companies which are deemed to be part of the nation's "critical infrastructure" because they provide financial or other key infrastructure services.

ADAPTING GOVERNMENT-BASED TOOLS TO IMPLEMENT DATA CENTER ENERGY EFFICIENCY IN PRIVATE SECTOR AGREEMENTS

The thesis of this article is that two government-based programs should be adapted and used in private sector outsourcing and IT agreements. These programs and the proposal for how they can be adapted and used are set forth below.

The thesis of this article is that two government-based programs should be adapted and used in private sector outsourcing and IT agreements. These programs and the proposal for how they can be adapted and used are set forth below.

THE CONVERGENCE OF ENERGY EFFICIENT IT AND A COMPANY'S CARBON FOOTPRINT

A recent U.S. government study estimates that U.S. data centers alone use approximately 15% of the country's retail electricity. (A figure which the Wall Street Journal concludes equals the country of Argentina's energy usage.) A McKinsey analysis predicts that by 2020 the world's data centers will produce more Greenhouse gases than the airline industry.

Data centers and other parts of a company's IT infrastructure require electricity to power and cool the computer equipment. The amount of heat generated by even today's "off-the-shelf" servers makes these servers harder to cool than prior generations of equipment. This aligns the interests of the CIO and CFO in using new technology to reduce the costs of buying electricity necessary to cool data centers in particular, and IT equipment in general.

While energy efficient IT, or Green IT, can be justified on cost savings, the result of using Green IT is that the carbon footprint of the data center (and other corporate IT operations) is reduced. This, in turn, can reduce the carbon footprint of the company as a whole. And this, in turn, can reduce the costs a company could otherwise incur if it had to buy carbon credits to offset its overall carbon footprint. Thus, for many companies, "Greening" the data center—or making it more energy efficient—will reduce costs not only at the data center but elsewhere in the corporate budget. Put another way, the costs of re-engineering a data center to reduce energy consumption can result in lower data center operating costs as well as cost savings for the corporation's environmental sustainability program.

In addition, there is a facilities management and real estate expense aspect of energy efficient IT. For those companies which have or will soon run out of room at their data centers, and are interested in increasing data center capacity without incurring additional real estate costs, Green IT can provide a way to put more computer capacity in existing space, and thus avoid the expense of acquiring an additional facility.

Using energy efficient IT to reduce the costs of a company's overall Green compliance program is likely to become increasingly important if and when the U.S. adopts cap and trade legislation imposing restrictions on company emissions and energy use. Strategic using of outsourcing will provide an avenue for a company to reduce its energy use and emissions and, therefore, reduce the cost of compliance with new emissions laws.

NEW CONTRACT TECHNIQUES FOR ENERGY EFFICIENT IT

New contract techniques should be used to draft agreements to govern the adoption of energy efficient IT at data centers. These techniques are similar whether the agreement is one for outsourcing services or for the deployment of new IT for a company's internal IT operations.

Two basic contract requirements are needed. The first is a metric to measure the energy efficiency of the computer equipment. The second is a metric to measure the energy efficiency of the data center as whole. The baseline of the data center's energy efficiency needs to be established before energy efficient technology and methodology are deployed. The energy efficiency of the data center must be measured at intervals as the Green IT is adopted.

As a practical matter, in both outsourcing and internal technology agreements, a potential problem with the use of a standard in the agreement is that the vendor may resist the use of a standard if it believes that the standard favors another vendor or was developed by a standards body that is not neutral.

A solution to this is found in two government programs. The first is the EPEAT, which is a standard for certifying that computer equipment meets energy-efficiency and other standards. The second is the DC Pro software tool ("DC" stands for Data Center"). This is a software tool used to assess the energy efficiency as a whole. Together EPEAT and DC Pro can be used as metrics, methodology and software tools to implement Green IT in data centers and corporate IT infrastructures.

EPEAT AND ITS USES

EPEAT stands for "Electronic Product Environmental Assessment Tool," and it resulted from a decision made in 2006 by the Environmental Protection Agency ("EPA") to select the Green Electronics Council ("GEC") to manage a program to establish criteria to be used by government and private sector IT users to select computer equipment that meets energy-efficiency and environmental criteria and to certify that the equipment meets such standards. EPEAT standards govern: (i) energy efficiency; (ii) hazardous component materials; (iii) upgradability; (iv) recyclability; and (v) the length of the useful life of the product.

EPEAT has Bronze, Silver and Gold classifications, based on a set of environmental performance goals embodied in Institute of Electrical and Electronics Engineers ("IEEE")

Standard 1680. IEEE 1680 consists of 23 required criteria and 28 optional criteria in eight categories. To qualify for Bronze classification, the product must meet all the required criteria. To qualify for the Silver level, the product must meet the required criteria plus at least 14 (or one-half) of the optional criteria. To qualify for Gold status, the product must meet the required criteria plus at least 21 (or three-quarters) of the optional criteria. The GEC verifies that a product meets the Bronze, Silver or Gold requirements, and it maintains a website listing the qualifying products.

Federal agencies subject to Executive Order 13423 are required to use EPEAT products where there are EPEAT standards for the relevant computer equipment in at last 95% of their annual acquisition requirements. The "teeth" in the government contracting EPEAT requirements is that the use of EPEAT products in accordance with this standard directly impacts the rating the agency receives under the Office of Management and Budget's environmental scorecard. EPEAT is in the process of being integrated into the OMB's IT optimization plan.

The DC Pro software tool is part of the Department of Energy's ("DOE") "Save Energy Now" data center initiative. The Save Energy Now program is part of the DOE's Industrial Technologies Program which is designed to reduce energy consumption. The goal of the Save Energy Now program is to reduce energy usage at U.S. data centers by 10% by 2011. The Save Energy Now framework divides data center energy analysis into four areas: (i) IT, which includes virtualization technology and optimization of server use; (ii) power distribution and conversion, which measures the efficiency of the distribution of electrical power from the intake point at the data center to the servers located at the data center; (iii) cooling and air distribution, which analyzes cooling and the degree to which hot and cool air are mixed; and (iv) on-site power generation, which avoids power loss inherent in the transmission of electricity over the power grid.

The DC Pro software tool, which can be downloaded from an EPA website, is used to measure energy efficiency at data centers.

USING DC PRO AND EPEAT IN PRIVATE SECTOR AGREEMENTS

DC Pro and EPEAT can be used in private sector agreements in the following ways where they are entering into a contract to improve data center energy efficiency. Assume that part of the goal of the agreement is to improve energy efficiency by using energy efficient equipment, virtualization, using new network, wiring and cooling technology, adopting energy efficient methodologies, and re-engineering the data center to integrate these technologies and methodologies. In the case of servers, virtualization is adding a layer of virtualization software to servers to allow them to run multiple operating systems and/or software applications. Virtualization allows each server to run more application, and thus can reduce the number of servers needed at the data center. Virtualization can trigger two further issues in certain IT environments. It may require amendment to existing software licensing agreements that do not cover running software on virtualized servers. Also, if servers are heavily virtualized, the application software becomes a bottleneck in that the input and output spends of the software may slow down the optimal operation of the servers. In some cases, software may need to be modified to resolve this issue.

Improvements in wiring technologies, such as unified fabric, fiber channel over Ethernet and other technologies, can be used to make more energy efficient connections between data center servers and other equipment. These technologies also reduce manpower needs (for "racking and stacking," for example) and ongoing maintenance and support efforts.

CONTRACT ISSUES

The following issues need to be addressed to implement energy efficient technology and practices in outsourcing and technology services agreement.

First, the parties should determine a pre-contract baseline of data center energy efficiency by using DC Pro. Whether the customer conducts this assessment alone or in combination with the provider will depend, in part, on whether the assessment will be accepted as binding upon the provider when its performance and payments are measured by improvements over the baseline. DC Pro is fairly new, and therefore there is little empirical experience available on the cost, in time and dollars, required to run an assessment using this software in the private sector.

Second, the parties should determine which new equipment will be covered by an EPEAT standard.

Third, the parties should determine which of the covered products should be covered by Bronze, Silver and/or Gold classifications. They should then agree in the contract as to which EPEAT standards should be met, and at what cost. This will result from a cost-benefit analysis of performance vs. price.

Fourth, the overall budget should be determined which will include, among other items, costs for EPEAT equipment, virtualization of servers (including new servers, if necessary), adoption of new wiring technologies, deployment of cooling technologies, improving intra-data center power distribution, the other factors identified in the DOE's Save Energy Now methodology, other technologies and re-engineering agreed upon by the parties, and the professional and consulting services agreed upon. Moreover, the customer in an outsourcing IT transformational project may retain an outsourcing consulting or advisory firm, and this should be budgeted for as well.

Fifth, the contract should provide penalties or service credits if the provider fails to deploy the EPEAT certified agreement at the times and in the manner agreed to in the SOW or other governing contract document. This includes acceptance procedures. In this respect, the typical provisions of outsourcing contracts should be applied to both the provider's and customer's EPEAT obligations. This includes specifying the customer's responsibilities with respect to preparing the IT environment and discharging its agreed-upon obligations for installing and accepting EPEAT equipment.

Sixth, the parties need to agree on the extent to which the outsource provider or a third party equipment vendor is responsible for procuring (and installing) the EPEAT equipment.

Seventh, to the extent that EPEAT equipment will be installed in phases, these need to be agreed upon and documented in the agreement.

Eighth, EPEAT standards do presently not apply to all types of computer equipment, although standards are being developed for new classes of equipment. For example, servers are not yet covered because of the difficulty of developing a standard that balances performance vs. energy efficiency for servers of different "power." The agreement may require the addition of EPEAT equipment as standards are adopted for new categories of products.

Ninth, an alternative to specifying the use of specific EPEAT equipment, the parties may instead specify levels of data center efficiency that are to be met over time, and have the contract require the provider to meet those standards (or performance or service levels, in outsourcing parlance) by using whatever technology it chooses. For this purpose, technology includes not only the individual pieces of equipment but the manner in which they are integrated.

Tenth, the contract should impose an EPEAT compliance report requirement on the vendor so that customer has prompt information about any delays, whether caused by the provider or a third party vendor, in installing EPEAT equipment.

Eleventh, turning to the energy efficiency of the data center as a whole, the contract should specify the time intervals at which DC Pro should be re-run in order to measure whether the interim standards for data center efficiency have been met on time.

Twelfth, the DC Pro software is expected to be updated over time. The parties need to agree on which version of the software to use. More specifically, to the extent that new versions of DC Pro provide a way to measure new aspects of energy efficiency, the parties will need to agree on how to use future versions of the software where the functions of that software are not now known.

Thirteenth, the speed of technology implementation and data center re-engineering needs to be addressed in how the contract is structured. This author suggests that where the return on investment of new equipment or technology is two years or less, the requirements for deploying such technology should be made part of the base agreement obligations. To the extent that the return on investment will take longer, then they should be subject to a Statement of Work or Project Plan or the like. As noted above, the contract documents can either specify the equipment to be used, or they can establish energy efficiency standards and allow the provider to use whatever combination of software, hardware and other equipment will meet the standards.

Fourteenth, the parties need to adopt an outsourcing contract governance system and dispute resolution mechanism that addresses delays in deploying energy-compliance equipment and methodologies. Related to this is the fact that if the industry develops energy efficient technology faster than expected at the time of contract signing, the parties should consider how to modify contract obligations so that new technologies can be adopted sooner and so that the contractual requirements become a floor and not a ceiling.

Fifteenth, there are steps that the customer should take outside of the base Green IT transformation agreement. For example, if the customer is a company which has an online procurement system for its employees that cover laptops and other computer equipment, the company should consider modifying the procurement system so that only EPEAT certified equipment can be used when an EPEAT standard exists for the relevant product category.

PROPER DISPOSAL OF OLD DATA CENTER EQUIPMENT AND USING OUT-SOURCING TO ACCOMPLISH THIS

Re-engineering a data center to adopt EPEAT and other new technologies will often result in the need to dispose of the old equipment in the center. The disposal of such equipment is subject to the EPA and state regulations governing the disposal of hazardous waste. Cathode ray tubes ("CRT"), for example, often contain heavy metals and other components that are considered hazardous and cannot simply be buried in landfills. The hazardous waste laws distinguish between equipment which is discarded and equipment which is either donated or equipment or its component parts are recycled.

The short answer is that many companies will chose outsource the disposal, donation and recycling of old data center equipment to companies that specialize in the proper handling of such equipment. However, a prudent company will conduct due diligence on such outsource companies to protect itself against liability of their action and to ensure that the old equipment will be treated properly. The contracts with these outsource providers should impose the requirements and obligations on such providers that are appropriate to the equipment subject to disposal.

OBLIGATIONS TO PRESERVE DATA BUT REMOVE IT FROM DISPOSED SERVERS

Servers that are removed from data centers often contain sensitive data, such as personally identifiable information of corporate trade secrets. Such information should be removed from the servers and other storage media that disposed or recycled. Several technologies are available to accomplish this, ranging from purging the data from the disks in a computer science sense to physically destroying the disks themselves.

On the other hand, much of that information must be preserved by the company by transferring it to other storage media or devices. A range of data is subject to different statutory and regulatory requirements under U.S. law. These include personal heath data under HIPAA, personal financial data under Gramm-Leach-Bliley, information subject to SEC, stock and commodity exchange requirements, tax requirements and other regulations.

CONCLUSION

EPEAT requirements have been subject to government contracts. The provisions of these contracts should be adopted and expanded for use in private sector agreements. Similarly, the use of the DOE's Save Energy Now methodologies and its DC Pro software for assessment of the energy efficiency of data center should be used in private sector agreements. They should be combined with other provisions of private sector outsourcing and IT agreements to govern the development of performance standards and pricing of the deployment of energy efficient IT in corporate data centers and internal IT infrastructures and in outsourcing and technology services agreements.

_____

Notes and Questions

1. Have you noticed the glowing lights throughout your house on your computer, cellphone, microwave, or other appliances even when they are turned off? These lights are telltale clues of energy vampires. Many modern appliances draw power even when they are turned off and sometimes even when no light is glowing. Some estimate vampire energy costs U.S. consumers approximately $3 billion a year. What can you do to eliminate this consumption of energy you do not need?

2. What are some of the contract suggestions Tannenbaum makes?

3. Google is exploring generating electricity from cow manure and also has invested in a transmission line off the East Coast of the United States. Why do you think Google is so interested in these types of ventures?

5. Behavioral Changes

Although many of the measures addressed in this chapter integrate efficiency into design to minimize any conscious human effort, additional levels of efficiency and conservation can be achieved if we pay attention to our energy consumption. The following excerpt addresses the psychology of reducing energy use.

Hope M. Babcock, *Responsible Environmental Behavior, Energy Conservation, and Compact Fluorescent Bulbs: You Can Lead a Horse to Water but Can You Make it Drink?*
37 HOFSTRA L. REV. 943, 945-68 (2009)

Introduction

Despite professing to care about the environment and supporting environmental causes, individuals behave in environmentally irresponsible ways like driving when they can take public transportation, littering, or disposing of toxic materials in unsound ways....

* * *

... The agreed upon goal behind energy conservation is to reduce the country's reliance on fossil fuel-based energy production, thus reducing the emission of harmful airborne pollutants and greenhouse gases as well as the related environmental harms associated with coal production. One way to reduce residential energy consumption is to persuade individuals to switch to CFLs. Up to ninety percent of energy produced by incandescent bulbs is lost as heat; switching to CFLs is one way to prevent this energy loss. However, getting individuals to switch bulbs is not as easy as one might think because of various barriers that stand in the way of changing environmental behavior....

* * *

Barriers to Behaving in an Environmentally Responsible Way

Habits and self interest as well as the inconvenience and cost of the new behavior and the unavailability of alternatives are examples of common barriers that must be overcome before individuals will change their behavior. In addition, the persistence of the myth that only industry is responsible for environmental harm and the difficulty individuals have understanding how their seemingly minor actions ... can accumulate into more serious, widespread harm ... contribute to the resistance of individuals to changing their environmental behavior.

Individuals also employ cognitive heuristics (flawed problem solving techniques) that interfere with how they process information about environmental harms.... Often individuals resist changing their behavior because they disbelieve the reason for the behavior change or they question the legitimacy of the norm underlying the change. There are also social norms like the autonomy and reciprocity norms that get in the way of the environmental protection norm, the compliance with law norm, and the personal responsibility norm, norms which might otherwise encourage good environmental behavior.

There is one other barrier to good environmental behavior that particularly impedes compliance with the energy conservation norm...."The evidence suggests that esteem matters; many individuals care what others think of them." If external praise is not there, then it is less likely that an individual will feel proud of her good behavior and will engage in it.

* * *

Therefore, despite their obvious economic benefit to the individual and wider social benefit of reducing energy consumption, getting individuals to make the effort, spend the money, and adopt what could be seen as the less appealing option of swapping out their light bulbs for CFLs is not a frictionless endeavor. Nonetheless, it still may be easier to get individuals to replace a single incandescent light with a CFL than to turn off lights or to refrain from buying electronic appliances, which make our lives so much more convenient and pleasant, even though the social and individual benefits of the latter, in terms of reduced electricity use, may be greater.

* * *

Identifying and Assessing the Success of Different Approaches to Overcoming Behavioral and Structural Barriers to Using CFLs

[There are three—Ed.] possible approaches to getting people to reduce their energy consumption by switching to CFLs. Two of these approaches, smart meters and comparative billing, are already in use in some areas of the country. The third, personal incentives, has not been applied to reduce energy consumption, but has been used in a variety of other areas to get people to change their behavior....

Smart Meters

* * *

This means that to the extent that smart meters depend on the information they convey to reduce the amount of electricity individuals consume, they may not be able to achieve that goal by getting their customers to buy CFLs because information as a motivational tool is too problematic.

Comparative Consumer Information

Like the smart meter method, this approach also relies on information to persuade utility customers to decrease their use of electricity. However, the way in which the information is presented to the customer invokes additional motivators of personal action: the conformity norm, competition, and, to some extent, the use of shame.

* * *

While the conformity norm and competition may inspire individuals to improve their energy conservation performance when the information about their performance is positive or they win, the effectiveness of negative comparative information, or losing, to some extent depends on the individual feeling ashamed of her poor performance. However, shame is an extremely problematic motivator....

* * *

Personal Incentives

Personal incentives are a potential third way to encourage individuals to buy CFLs; the incentives can be economic or non-economic....

* * *

The negative attributes of economic incentives, and the motivational uncertainty of both economic and non-economic incentives, put into question their effectiveness as motivational tools to persuade individuals to consume less electricity, let alone their ability to overcome the reluctance of people to buy CFLs despite their individual and social benefits.

———————

Notes and Questions

1. Countries around the world are legislating to phase out of inefficient carbon filament incandescent light bulbs, which Thomas Edison invented in 1879. George W. Bush signed the U.S. version contained in the Energy Independence and Security Act of 2007 (Pub.L.

110-140), which among other things, required light bulbs in the United States to be 20 percent more efficient by 2012. Estimates are that the measure would have saved Americans over \$12.5 billion over the next nine years and would be equivalent in energy savings to thirty-three power stations. However, Tea Party Congressmen, casting it as government infringement on their freedom of choice, passed legislation on July 15, 2011, which effectively repealed the conversion to more efficient bulbs by prohibiting spending to enforce the new standard. Does Babcock's analysis seem in step with the brouhaha of resistance the incandescent bulb encountered?

2. In his address at the Harvard graduation in 2009, Secretary of Energy Chu stated, "In the coming decades, we will almost certainly face higher oil prices and be in a carbon-constrained economy. We have the opportunity to lead in development of a new, industrial revolution. The great hockey player, Wayne Gretzky, when asked how he positions himself on the ice, he replied, 'I skate to where the puck is going to be, not where it's been.' America should do the same." HARVARD GAZETTE, June 4, 2009. For a description of one person's efforts to skate where the puck is going to be by reducing his carbon dioxide output by 80 percent, *see* Peter Miller, *Energy Conservation: It Starts at Home*, NAT'L GEOGRAPHIC, March 2009, *available at* http://ngm.nationalgeographic.com/2009/03/energy-conservation/miller-text/1.

3. While much of the Smart Grid initiative is using electronics to make the grid more interactive, some systems will include feedback devices so individuals can monitor their energy use in real time and make adjustments. Will access to this information lead to reduced energy consumption or will "green fatigue" set in because the consumers are receiving so much information they eventually tune out and become paralyzed? Americans have embraced new technologies such as cellphones and social media. Do you see parallels with these technologies and behavioral changes to positively rein in our nation's energy binging?

4. What are the three approaches Babcock's article sets out for motivating people to reduce their energy consumption? Does she conclude that any of these are especially effective ways of ensuring positive human behavior to switch to energy saving CFLs?

The SMUD electric bills were created in response to research by Robert Cialdini, a professor at Arizona State University. Electric bills that tell you how much more or less electricity you used than your neighbors, and how much more or less it cost you, provide "social proof" or social norms for customers to compare themselves. (Some bills also included praise—such as smiley faces—and shame—such as frowney faces—to enhance the message, but research by the City of Fort Collins, Colorado, indicates this is less effective than the basic comparative data.) While SMUD's 2 percent reduction may seem minor, "[i]n energyspeak ... it's equivalent to taking 700 homes off the grid." Bonnie Tsui, *Greening with Envy: How Knowing your Neighbor's Energy Bill Can Help You to Cut Yours*, THE ATLANTIC 24 (July/Aug. 2009).

5. China set a five-year target to improve energy efficiency, energy consumption per unit of GDP, by 20 percent between 2005 and 2010. On January 7, 2011, Zhang Ping, chairman of the National Development and Reform Commission, announced that China had met its goal. John Holdren, Assistant to the United States President for Science and Technology, stated, "[First, this] is an extraordinary rate of improvement in energy efficiency. The second thing is that I would say the Chinese are already doing far more to try to contribute to the solution than they generally get credit for in the West. The Chinese have made enormous advances in energy end use efficiency in recent years. They are the world leaders, both in the pace of improvement in energy efficiency and the pace of deployment

of renewable energy technologies." Elizabeth Kolbert, *Obama's Science Advisor Urges Leadership on Climate*, YALE ENV'T 360 (2009). What do you think motivates the Chinese people? What lessons might the United States learn from other countries' efforts to reduce energy use and switch to more renewable energy sources?

Chapter 8

Renewables on Federal Lands

Throughout this book, we have explored the role of the federal government in the development of renewable energy sources. Obviously, that role is paramount in the context of renewables on federal lands. Historically, the United States flourished through generous giveaways of public lands for railroads, mines, farms, and ranches. At these times, exploitation of the U.S.'s vast public holdings was not only below market value or free, but also without consideration of the costs to the public common or to less powerful peoples who were negatively impacted.

Today, developments can be slowed or sometimes halted because federal statutes protect resources (such as wildlife, cultural, air, and water) and require the government to take a "hard look" at environmental impacts before acting. Finding an appropriate balance remains the challenge. As Justice Scalia noted, "'Multiple use management' is the deceptively simple term that describes the enormously complicated task of striking a balance among the many competing uses to which land can be put...." Norton v. Southern Utah Wilderness Alliance, 542 U.S. 55, 58 (2004).

Despite best efforts to be current, the material in this chapter changes with each new statute or regulation. Consequently, you should check government websites and current legislation to keep up-to-date. The goal here is to provide an overview of context and background so that you will better understand any recent changes.

A. Terrestrial Wind and Solar

The United States is one of the world leaders for terrestrial wind energy development vying closely with Germany and China. All six of the world's largest onshore wind farms are on private lands within the United States, and five of those six are in Texas. The Roscoe Wind Farm is currently the world's largest, and at 781.5 MW is more than twice the capacity of the world's largest offshore wind power facility.

Here is a good website for updates on renewable energy development on BLM lands: http://www .blm.gov/wo/st/en/prog/energy/renewable_energy.html.

As renewable energy sources began to be developed on federal lands around the start of the twenty-first century, agencies looked to their enabling statutes for how this development might be regulated. The Federal Land Management and Policy Act of 1976 (FLPMA) Title V "right-of-way" model appeared to be the best option for BLM-managed public lands. The excerpt below explains some of the shortcomings of this approach.

David J. Lazerwitz, *Renewable Energy Development on the Federal Public Lands: Catching Up with the New Land Rush*

55 Rocky Mt. Min. L. Inst. 13-1 (2009)

§ 13.01 Introduction

The federal public lands have long played a pivotal role in the development and expansion of our national agenda and aspirations. From the California Gold Rush of 1848 to the western Colorado oil shale boom of the late 1970s, these lands contain natural resources and environmental attributes central to the character of the American West. Today, a modern-day land rush promises to define a new era for the federal public lands, one that will require even greater vigilance to balance both the use of natural resources and protection of the environment. Unlike the resource booms that preceded it, however, this land rush focuses not on what is *in* the land but what is available *above* it—specifically, solar and wind resources, which are uniquely situated on the federal public lands and necessary to achieve national goals of energy independence and greenhouse gas emission reductions.

* * *

13.02 The Renewable Energy Land Rush Is Underway

(1) The Driving Forces for Solar and Wind Project Demand on the Federal Public Lands

The increased focus on developing renewable energy on the federal public lands results from a unique confluence of energy issues and characteristics inherent to the lands themselves. While the driving forces behind the surge in solar and wind project applications on the federal public lands are complex and varied, they can be grouped into three general areas: market forces, government intervention, and resource availability.

(a) The Confluence of Conventional and Renewable Energy Costs

The price gap between conventional fuels and renewable energy has narrowed substantially in recent years, leading to increased interest in solar and wind power projects. The period from 2003 to 2008 witnessed one of the largest increases in fossil fuel prices in our nation's history. During this time, the price of oil rose to a record level above $145.00 per barrel, and natural gas prices for electricity production peaked at record levels above $12.00 per million BTUs. Although more variable depending on the relevant geographic market, coal prices similarly reached record levels during this time period, striking $150 per short ton in northern Appalachia. While fossil fuel prices have retreated substantially since their highs, the potential for return to these levels and continued volatility in these markets remain concerns of consumers, utilities, and energy producers.

During the time period when conventional fuels reached their peak levels, renewable energy technology costs have substantially decreased. These technology efficiencies are largely attributed to the support for solar and wind technology in Europe where "feed-in tariffs" in Germany and Spain (which require utilities to purchase renewable power at above-market rates) drove an exponential growth in technology development and deployment. Between 2004 and 2009, thin-film photovoltaic solar panel manufacturers tripled the efficiency of the technology, reducing manufacturing costs from over $3.00 to $.98 per watt. Wind turbine manufacturers have gone a step further, claiming to reduce wind power production costs to $1.00 per watt installed. While there remains a significant price gap between fossil fuels and solar and wind on a per-kilowatt-hour cost basis, this gap is closing, and some predict "grid parity" as soon as 2012·····

(b) Government Intervention in the Market

In response to rising conventional fuel costs and increasing recognition of the adverse impacts from greenhouse gas emissions on air quality and climate change, public interest has driven greater government tax incentives to defray the cost of project development and mandates to develop renewable energy sources. On the incentive side, these programs include the federal production tax credit for wind power and investment tax credit for solar power, which provide tax incentives of up to 30%. Congress recently extended these programs in the American Recovery and Reinvestment Act of 2009, which went a step further by creating grants-in-lieu of tax credits and directing hundreds of millions of dollars to research, development, and loan programs.

Perhaps the government program with the greatest single impact on renewable energy demand is the establishment of state-based renewable portfolio standards (RPS), through which state governments require that regulated utilities generate certain percentages, or specified amounts, of renewable energy by specific deadlines. While some RPS were initiated as early as the 1990s, the vast majority were implemented after 2001. Moreover, since 2001, RPS legislation has become increasingly aggressive.... [See Chapter 5.2 and Figure 5.6 for more information about Renewable Portfolio or Renewable Energy Standards (RPS or RES) — Ed.]

(c) The Attraction of Public Lands for Solar and Wind Development

The siting of solar and wind projects sufficient to meet utility-scale power needs requires certain land and resource characteristics uniquely available on the federal public lands. While the land production capacity and grade requirements differ depending upon the relevant technology (even within the wind and solar fields themselves), as a general matter, utility-scale projects can range in size from a few megawatts (MW) to more than 1,000 MW. Such projects typically require large, open, and generally level, undeveloped tracts ranging in size from several thousand acres to more than 50,000 acres. They require access for interconnection to major transmission lines. Finally, and most importantly, for optimal efficiency, these projects need to be situated in areas with consistently high levels of sunshine and wind.

Each of the required characteristics is present in abundance on the federal public lands in the West, lands that remain largely undeveloped, crossed with major utility transmission lines, and recognized as containing the highest density of solar and wind resources in the United States. Recognizing this potential, in 2005 Congress mandated that the Secretary of the Department of the Interior (DOI) install 10,000 MW of non-hydropower renewable energy projects on the public lands by 2015. The Bureau of Land Management (BLM) itself estimates that it manages 30 million acres of public lands with solar potential, and another 20.6 million acres with wind potential. DOI Secretary Ken Salazar recently estimated that the public lands in the West could generate 206 gigawatts of wind energy and 2,900 gigawatts of solar energy — collectively about three times current national electricity generating capacity.

(2) The Resulting Impacts and Challenges for BLM's Administration of the Federal Public Lands

Driven by market and government forces and resource availability, the demand for siting large or utility-scale renewable energy projects on the federal public lands has skyrocketed in the past several years. While there was no installed solar generating capacity on BLM lands at the time of this writing, 223 solar project applications have been filed since 2005, covering more than 2.3 million acres of public land. At the same time, wind project applications have doubled, from 192 authorized projects in varying stages of development to more than 200 new applications with ever-increasing size and generating capacity. Nowhere is this demand felt more acutely than in California, where 156 solar and wind projects await approval for an area covering nearly 1.4 million acres, most of which are located in the Mojave Desert region.

The unprecedented rise in solar and wind project applications quickly outstripped BLM's capacity to process these applications, presenting a number of unique management and environmental challenges. On the management side, the sheer number of the proposed projects has overwhelmed staffing levels and BLM's ability to catalogue — let alone process — these applications through the myriad of Federal Land Policy and Management Act (FLPMA), National Environmental Policy Act (NEPA), Endangered Species Act (ESA), and other federal and state approvals. Moreover, these projects present new challenges to FLPMA's existing application process, which has historically focused on providing non-competitive "rights-of-way" (discussed in section 13.04[1]) for roads, pipelines, and transmission lines, and does not, on its face, address the types of applications filed with BLM today that may seek to develop thousands of acres of land.

Today's solar and wind project developers often seek to secure access to the optimal resource and transmission sites prior to securing the definitive utility power purchase agreement (PPA) often necessary to initiate and finance project development. Consequently, speculative applications have the potential to lock up large tracts of federal land for an indefinite period of time and, in many cases, could result in multiple applicants waiting in line for the opportunity to develop the same tract of land. BLM's existing rental fee structure — typically based on the fair market value of actual land involved — also appears ill-suited to address the valuable and long-term revenue stream to be derived from generating power from the public lands. [BLM has measured fair market value for renewable energy projects in terms of the capacity of the energy they can generate — Ed.]

The environmental challenges presented by large-scale solar and wind projects are significant. These challenges range from potential impacts to wildlife and habitat resulting from surface construction and development (and, in the case of wind turbines, from operation), to aesthetic impacts associated with installing rows of wind turbines or acres of solar panels, mirrors, or collectors. These potential impacts are compounded by relatively new and continually evolving technologies, with varying degrees of land-use impact (e.g., need for concrete footings or pads), resource needs (e.g., water for steam generation or cooling), and unpredictability regarding long-term effectiveness, operation, and time frame for decommissioning. Finally, unlike more traditional road or transmission line access, the significant size of these projects presents compatibility challenges for other uses of the federal public lands, including natural resource protection, recreational access, grazing, and mineral or oil and gas exploration.

The sudden rise in solar and wind project applications and unique challenges presented by these projects have forced BLM largely into a reactionary role. Given the earlier influx of wind project applications, in 2005 BLM completed a Programmatic Environmental

Impact Statement (PEIS) pursuant to NEPA for wind energy development in 11 western states, which resulted in amending 52 land use plans to address wind project development (discussed in section 13.03[3][a]). Faced with a backlog of 125 solar project applications in May 2008, BLM, along with the Department of Energy (DOE), issued a notice of intent to prepare a PEIS for solar project development and simultaneously instituted a two-year moratorium on filing new applications in Arizona, California, Colorado, New Mexico, Nevada, and Utah pending completion of its study. Within several weeks of BLM's decision, the agency rescinded the moratorium in response to widespread industry and public opposition. [The Draft EIS came out in December 2010 and 80,000 comments were received. On July 14, 2011, BLM indicated it would issue a supplemental PEIS — Ed.]

While the number of projects and the acreage covered presents a virtually unprecedented demand on federal agencies, the demand is unlikely to slacken. New government policies — including the economic stimulus push at the federal level, existing state and anticipated federal climate change legislation, and increasing state renewable portfolio standards — portend a continuing flood of project applications and mounting pressure to approve projects for development.

§ 13.03 The Current Regulatory Framework for Approving Solar and Wind Projects

(1) The Federal Land Policy and Management Act Right-of-Way Process

Pursuant to BLM's solar and wind development policies, BLM processes applications to site solar and wind projects on public lands pursuant to Title V of FLPMA and BLM's implementing regulations. FLPMA Title V governs the grants of rights-of-way "over, upon, under and through" the federal public lands and authorizes rights and privileges for a specified use of the land for a defined period of time and under terms and conditions imposed by the agency. FLPMA vests BLM with considerable — although not unfettered — discretion in approving or rejecting applications for rights-of-way. When BLM exercises its discretionary authority to reject a right-of-way application, however, it must provide a reasonable basis for its decision that is supported by the administrative record.

BLM's ultimate review of any proposed activity on public land centers around FLPMA's multiple use mandate and resource management plan (RMP) model. FLPMA directs BLM to conduct inventories and establish RMPs to manage tracts or areas of the public lands, taking into account, among other things, principles of multiple use and sustained yield; a systematic interdisciplinary approach to achieve integrated consideration of physical, biological, economic, and other sciences; and consideration of the present and potential uses of the public lands. As aptly described by the Supreme Court:

> "Multiple use management" is a deceptively simple term that describes the enormously complicated task of striking a balance among the many competing uses to which land can be put, "including, but not limited to, recreation, range, timber, minerals, watershed, wildlife and fish, and [uses serving] natural scenic, scientific and historical values."

Today, solar and wind energy only add to this complex balancing act.

BLM's implementing regulations describe those lands under the agency's jurisdiction that are available for right-of-way grants. The relevant areas include any lands under BLM's jurisdiction *except* where such lands are excluded from rights-of-way pursuant to statute, regulation, or public land order; where the lands are specifically segregated or withdrawn from right-of-way uses; or where the agency, in RMPs or in the analysis of an application, identifies areas that are inappropriate for right-of-way uses. Thus, unless the relevant public lands are excluded or withdrawn from right-of-way uses in one of the aforementioned ways, such lands are presumptively open for siting solar and wind projects, and BLM may process right-of-way applications pursuant to FLPMA. In this regard, BLM's RMP process is critical in identifying those areas that are either compatible or not compatible with solar or wind project development.

(2) BLM's Project-Specific Policies and Guidance

Beyond FLPMA and its implementing regulations, BLM has supplemented its procedures for processing and evaluating solar and wind project applications through separate internal guidance policies, known as Instructional Memoranda (IM). These policies are further expanded and clarified by additional agency guidance documents and orders and, particularly in the case of wind energy, identify development policies and best management practices (BMPs) (further discussed in § 13.04[3]). While the policies differ depending upon the applicable technology and contain specific details beyond the scope of this chapter, several elements common to both solar and wind project development merit attention and are outlined below.

(a) BLM's Promotion of Solar and Wind Energy Development

BLM's policies reaffirm the congressional directive contained in the Energy Policy Act of 2005 mandating the development of "at least" 10,000 MW of non-hydropower renewable energy projects on the federal public lands by 2015. In furtherance of this requirement, BLM's policies both "encourage" development of wind and solar energy and, in the case of solar, state that the agency's "general policy is to facilitate environmentally responsible commercial development of solar energy projects on public lands and to use solar energy systems on BLM facilities when feasible." BLM's 2007 Solar Policy further provides that BLM intends to identify right-of-way applications for solar energy projects "as a high priority Field Office workload" and to process those applications in a timely manner.

BLM's commitment to promote wind and solar development on the public lands received recent support from DOI Secretary Salazar through the establishment of a Departmental Task Force on Energy and Climate Change, which makes the development, production, and delivery of renewable energy one of DOI's "highest priorities." In an effort to begin to address the backlog of right-of-way applications and establish a more coordinated approach to process solar and wind project applications, Secretary Salazar also announced the opening of four new BLM Renewable Energy Coordination Offices.

(b) The Application and Plan of Development Process

BLM's processing of right-of-way applications proceeds through three principal levels of review under BLM's regulations, commencing with submitting an application (a Standard Form or SF-299 application), followed by filing a more detailed Plan of Development (POD), and concluding with environmental review pursuant to NEPA and related federal and state statutes. As an "action authorized, funded or carried out" by BLM, the issuance of a right-of-way grant also triggers compliance with consultation requirements under section 7 of the ESA—an issue particularly relevant for solar and wind development in light of many of the desert and alpine environments involved—and section 106 of National Historic Preservation Act (NHPA), as well as other applicable federal and state statutes.

Prior to submission of an SF-299 application, BLM's policies encourage the authorized officer to schedule a pre-application meeting with the prospective applicant. The purpose of this meeting is to facilitate preparation and processing of the application; to identify "potential issues and land use conflicts" that could impact the authorized officer's decision to grant or not grant the right-of-way authorization; and, if appropriate, to consider potential alternative site locations. The policies further require the submission and approval of a detailed POD for construction and operation, which BLM now requires prior to initiating its NEPA review process for project development and, in the case of solar projects, within 90 days following the filing of a right-of-way application.

(c) Determining Site Priority

Unlike competitive bidding processes that exist for oil and gas and geothermal leasing on the federal public lands, FLPMA's right-of-way grant process has historically addressed access issues (e.g., road rights-of-way) that are inherently noncompetitive. For both solar and wind project applications, BLM accepts and processes applications on a "first-come, first-served" basis. A competitive bidding process may, however, be initiated where an RMP specifically identifies an area for competitive leasing or, in the solar context, where other "public interest and technical" factors merit competitive leasing. Notably, until the issuance of a grant, it is generally recognized that the applicant does not possess a property interest in federal land and such application may be superseded by any number of authorized federal actions. [Recent rulemaking would give the BLM authority to withdraw areas from the mining law for up to four years if they are subject to a viable renewable energy application or are in prospective renewable areas — Ed.]

(d) Requiring Due Diligence

BLM must also deter land speculators from locking up tracts of public land that could impede actual development. This need, however, must be carefully balanced against the realities of utility-scale energy project development, which will typically require producers to first secure a PPA prior to proceeding with project financing and development. BLM has thus far addressed this issue in two ways. First, BLM's regulations require a detailed submission of applicant information, including the applicant's technical and financial capability to construct, operate, maintain, and terminate the project. Second, BLM requires that project construction commence within three years of the issuance of a right-of-way grant (for solar projects) and within two years of the issuance of a development authorization (for wind projects).

(e) Terms and Conditions on the Grant

FLPMA itself mandates that BLM impose terms and conditions to "minimize damage to scenic and esthetic values and fish and wildlife habitat and otherwise protect the environment … [and] require compliance with applicable air and water quality standards.…" The terms and conditions range from the requirement to comply with applicable federal and state laws and regulations to individual project-specific requirements. As discussed further below, BLM recently instituted BMPs for wind project development and at the time of this writing was in the process of doing the same for solar projects, which will result in a more uniform set of industry-wide terms and conditions.

As a general matter, the period or term of a right-of-way grant is not limited by FLPMA or BLM's implementing regulations. In recognition of the considerable project development costs and commitment, BLM's policies do envision the duration of solar and wind projects to extend through the useful life of the project technology, which for both solar and wind projects is generally 30 years and may be extended. The issuance of a grant is merely a possessory interest in the land and is by no means exclusive. BLM's policies make clear that the agency retains the authority to authorize other compatible uses within the scope of the right-of-way during the term of the grant. Moreover, BLM's regulations and policies require that project developers reclaim the relevant area following grant termination, and bonding to ensure compliance with the applicable terms and conditions in the grant.

(f) Expedited Review for Testing and Monitoring

While the issue has not yet been addressed for solar projects, in the wind context, BLM's policy provides for the expedited issuance of a three-year "site specific" or "project area" grant to permit preliminary site testing and monitoring. The site or area grant applicant need not file a POD and may be subject to use of a categorical exclusion from NEPA compliance depending on the scope of the proposed activity. Moreover, the issuance of an area grant temporarily precludes others from filing right-of-way applications for the designated area. In theory, this limited grant may not be renewed without the filing of a separate right-of-way development application and POD for project development, which places severe time constraints on wind developers to establish transmission access and execute a PPA in a relatively short period of time.

(g) The Interest Conveyed and Competing Uses

The issuance of a grant is merely a possessory interest in the land and is by no means exclusive. BLM's policies make clear that the agency retains ownership of subsurface and related resources and the right to authorize other compatible uses within the scope of the right-of-way, but the agency acknowledges that, at least in the solar context, other compatible uses are unlikely due to the intensive nature of both photovoltaic and concentrating solar technology. Moreover, the grant conveyed is subject to "valid existing rights," which recognizes the potential for preexisting and potentially conflicting mineral interests and oil and gas lease interests — an issue that should be addressed during the pre-application meeting and application process.

(h) Cost Sharing

The right-of-way grant process provides BLM authority to impose both cost sharing and rental fee requirements on project applicants, and these financial components continue to evolve as part of the regulatory process. As an initial matter, BLM requires applicants to enter into a cost sharing agreement with the agency through which the applicant submits a deposit and reimburses BLM for the agency's costs incurred in processing a right-of-way application (including NEPA compliance) and monitoring compliance after a grant has been issued.

(i) Rental Fees

Project proponents must reimburse the federal government for the "fair market value" of the relevant land. For the more typical linear rights-of-way, this fee is established based on a per-acre schedule tied to the land value for a particular use in the relevant geographic area or, in some instances is determined by appraisal. For wind power projects, BLM has established an annual rental fee of $4,155 per MW of the total installed project capacity, payable on an escalating basis. For solar power projects, BLM's existing policy provides that rental fees be determined based on the appraisal method for comparable lands in a similar stage of development. Reflecting a distinction in the extent of land use associated with solar and wind, however, BLM's solar policy provides that, "[s]ince the rental payment reflects the full use of the public land for solar facilities, similar to a lease for industrial purposes, there are no additional royalty payments for electric generation." While this policy makes sense in terms of recognizing compatible uses on wind versus solar sites, it does not reflect the vast distinctions among various solar technologies, some of which are more land intensive (e.g., thin film photovoltaic) while others are more invasive in

Figure 8.1: CSP solar resources on public lands

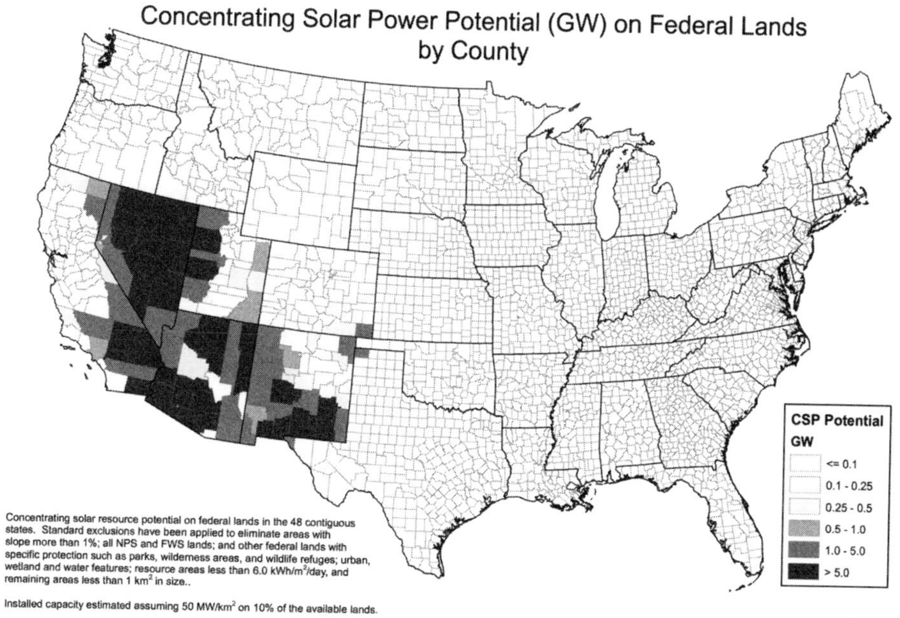

Concentrating Solar Power Potential (GW) on Federal Lands by County

Concentrating solar resource potential on federal lands in the 48 contiguous states. Standard exclusions have been applied to eliminate areas with slope more than 1%; all NPS and FWS lands; and other federal lands with specific protection such as parks, wilderness areas, and wildlife refuges; urban, wetland and water features; resource areas less than 6.0 kWh/m²/day, and remaining areas less than 1 km² in size..

Installed capacity estimated assuming 50 MW/km² on 10% of the available lands.

CSP Potential GW
<= 0.1
0.1 - 0.25
0.25 - 0.5
0.5 - 1.0
1.0 - 5.0
> 5.0

Prepared Statement, Director National Renewable Energy Lab. Dan Arvizu, (Mar. 17, 2009),
http://www.nrel.gov/gis/pdfs/arvizu/CSPMap.pdf

terms of grading and resource use (e.g., concentrating solar or solar thermal). This remains an issue to be resolved in conjunction with BLM's solar PEIS process.

(j) Incorporating Solar and Wind Projects into Land Use Planning

One of the central tenets of FLPMA is the land use or RMP planning process, which provides the framework for applying FLPMA's "multiple use and sustained yield" mandate within designated areas of public lands. The land use planning process considers both present and future uses of the public lands and requires designation and protection of areas of critical environmental concern. For projects implicating new and updated land use plans, BLM must identify and consider existing and potential areas for solar and wind energy development and the potential impacts on the local environment and community of making such lands available for development. Although BLM commenced this process by conducting a study of solar, wind, biomass, and geothermal sites in 2003, given the length of time necessary to update individual RMPs, few plans today include specific identification of solar and wind potential and related impacts and alternatives for planning purposes.

Where the relevant RMP does not already address solar and wind project development within the affected area (as is typically the case), BLM's policies make clear that issuance of right-of-way grants for wind and solar projects will require amendment of the relevant land use plan. BLM recognizes, however, that the RMP amendment and environmental analysis for the specific project proposal may be prepared and processed concurrently. To some degree, the Wind PEIS, discussed below, may have lessened this issue in the context of wind development, since that process resulted in the amendment of 52 individual RMPs to incorporate programmatic policies and BMPs and identify specific areas where development would not be allowed.

Notes and Questions

1. Instead of creating a *sui generis* mechanism for developing solar and wind on public lands, BLM worked within its existing FLPMA Title V right-of-way process. What are some of the problems Lazerwitz identifies with trying to fit the square peg of renewable development into the linear hole of FLPMA Title V? *See also* Rebecca Watson, *Renewable Power Projects on Federal Lands: Wind and Sun and the FLPMA Right-of-Way—Is It Working? in* Special Institute on Energy Development: Access, siting, permitting, and delivery on public lands, Paper No. 10 (Rocky Mtn. Min. L. Found. 2009).

2. How does the interest conveyed with a BLM right-of-way (or Forest Service Special Use Permit or SUP) vary from a mineral lease or a mining claim?

3. How does Lazerwitz indicate that the application process encourages speculation? What are the terms and conditions on the grant? BLM issued a 2011 Instruction Memorandum on due diligence to address these speculation concerns. What does the BLM now require? How is the federal oil and gas leasing process different?

4. How are rental fees determined under a FLPMA ROW model? Is the process different for solar than for wind? What about royalties? Because royalties usually represent a property interest in a mineral, how can BLM reserve a royalty in the wind or insolation?

5. Lazerwitz labels BLM's approach as reactive and notes, "The ultimate question is whether this reactive approach is sufficient, particularly in light of the challenges posed by the ever-increasing number and scale of utility-scale solar and wind development projects. Actual project proposals will very likely dwarf the relatively modest amount of acreage identified (or, in the case of solar, being identified) through the programmatic planning process. In the case of wind, for example, while the Wind PEIS identified some 160,000 acres of public lands for likely development in an 11-state region over 20 years, there are currently right-of-way applications for wind projects on over 957,000 acres of public lands in California alone. While increased technological efficiencies for both solar and wind may well lead to reduced project size and scale to achieve the same level of energy output (with less-intensive land and resource impacts), these efficiencies likely will further reduce the price differential with conventional fuels and allow the siting of projects in marginal locations not previously considered economically viable, thereby leading to even greater demand for public land access." Do you have alternative suggestions for how BLM can approach the situation proactively? BLM's current "Smart from the Start" approach attempts to be proactive. What do you think of this strategy?

6. Historically, multiple-use has been a key premise of public land use. How is that problematic for renewable energy development?

B. Offshore Wind and Wave Power

As of 2009, the United States had over 35,000 MW of installed electricity nameplate capacity of terrestrial (or on-land) wind power. However, as this book goes to press, the country had yet to install its first turbine in water—whether in the oceans or in the Great Lakes. Figure 8.2 shows the potential for such wind development in these areas.

Figure 8.2: U.S. offshore wind resources

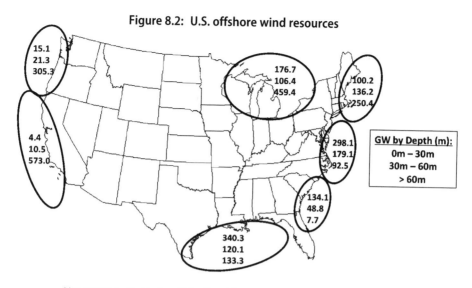

Map prepared with data from Walter Musial & Bonnie Ram, National Renewable Energy Lab. & Energetics, *Large-Scale Offshore Wind Power in the United States: Assessment of Opportunities and Barriers* 3 (2010), *available at* http://www.nrel.gov/wind/pdfs/40745.pdf

The technology exists—the European Union has over 2,000 MW of installed offshore wind capacity. Instead, a number of other factors are holding back offshore wind in the United States. Some of them are beyond our power to change including (a) weather—Unlike the North Sea, we have hurricanes off the East Coast of the United States; (b) topography—Because of plate tectonics, the West Coast of the United States drops off more steeply in its subduction zones, significantly diminishing the potential to develop any wind resources at depths of less than sixty meters; and (c) economics—Offshore wind currently is estimated to cost three times as much to produce as onshore wind at twenty-seven cents per kilowatt hour in comparison to the onshore average of nine cents per kilowatt hour in 2010. Others may be difficult to address, but they are human driven: policy; aesthetics; competing uses such as fishing, shipping, oil and gas drilling; federalism (the interrelationships between the federal and state governments); and legal hurdles.

A second promising source of renewable energy is offshore hydrokinetics—use of waves, tides, and other moving water to generate power. We will address first offshore wind and then ocean hydrokinetics in this section because many of the same federal and state management issues can apply to each.

1. Offshore Wind

This section will provide an overview of offshore wind. First comes some information about the federal process for obtaining an offshore wind right—which varies from the onshore wind process—from a different Interior bureau with different statutory authority than FLPMA. Second, we will explore some of the hurdles encountered for the very first offshore wind farm lease approved by the United States at Cape Wind off the coast of Massachusetts.

a. The Offshore Wind Leasing Process

The European Union holds bragging rights for offshore wind development with thirty-nine projects including the world's largest, the 300 MW Thanet wind farm almost due east of London in the North Sea. More than twenty years after Denmark installed the first offshore wind farm, the United States has yet to erect its first, even though the offshore wind potential for areas up to fifty nautical miles from U.S. shores is over 4,000 GW according to EERE and BOEMRE. U.S. Dep't of Energy, A National Offshore Wind Strategy: Creating an Offshore Wind Energy Industry in the United States 5 (Feb. 7, 2011). *See also* Figure 8.2.

Instead of using a FLPMA ROW, as a developer must do for onshore wind, an offshore developer acquires the wind rights through a lease, similar to how oil and gas rights are obtained offshore. While reading the following excerpt, consider how the process used for offshore wind varies from onshore and which appears to be a better model for wind development. Also pay attention to how the offshore wind leasing process compares to the offshore leasing process for oil and gas.

Joseph J. Kalo & Lisa C. Schiavinato, *Wind Over North Carolina Waters: The State's Preparedness to Address Offshore and Coastal Water-Based Wind Energy Projects*
87 N.C. L. Rev. 1819, 1821-44 (2009)

Introduction

As an alternative to traditional carbon-based energy sources, many look to the wind as a potential source of renewable energy. When wind energy developers, policy makers, and the public contemplate potential sites for wind energy facilities, many eyes turn toward coastal and offshore areas where the wind is "always blowing." A simplistic view of water-based wind energy facilities is: There is all that available open space out there. Wind resources are strong and consistent in coastal and offshore waters. The energy source is non-polluting. And, water-based wind energy projects have a significant advantage over land-based projects: the developer of a water-based project only has to deal with one landowner, either the State or, if the project is sited more than three miles from shore, the federal government. [Because Texas was its own nation before joining the United States, its sovereignty extends ten miles into the Gulf of Mexico — Ed.] Land-based projects may require dealing with a large number of individual landowners in order to acquire the acreage necessary for an economically viable wind energy project. Finally, looking to Europe, we see that it already has significant offshore wind energy generating facilities. In light of these facts, many ask why this country is slow to develop the wind energy potential of its coastal and ocean waters.

A careful examination of water-based wind energy, however, demonstrates that putting wind turbines and related equipment in coastal and ocean waters presents more, and different, technical and other difficulties than putting wind turbines and facilities on land. Furthermore, similar complexities also exist regarding the infrastructure that will be required to transmit the energy from the turbines to land and then to connect the produced energy to the power grid to be used in homes and businesses.

Unlike land-based wind energy facilities, placing an array of wind turbines in coastal and ocean waters is similar to putting a field of wind turbines in the middle of Yellowstone

National Park. Coastal and ocean waters are public waters held in trust for the people of the state and, similar to a park, are open to multiple uses, such as fishing, commercial and recreational boating, swimming, and other water activities. Therefore, siting wind energy facilities in coastal and ocean waters raises a number of important potential user conflicts which require careful analysis of what would be gained and what would be lost if a wind farm is sited in a particular location.

The recently promulgated federal regulations governing leasing of federal ocean waters and submerged lands for wind energy projects provide the necessary framework for developing their wind energy potential....

* * *

I. What's Happening with Water-Based Wind Energy

A. Developments Outside North Carolina

Presently, no operational wind facilities are in place in coastal or federal waters, but the wind-based energy industry hopes to begin constructing offshore facilities by 2010. The projects in the planning and permitting stage include the Cape Wind project, to be located off the coast of Cape Cod in Massachusetts, and the Bluewater Wind project, to be located approximately eleven miles off the coast of Delaware. The Cape Wind project in particular has engendered considerable local opposition from some quarters: fishermen, sailors, some environmentalists, boaters, and others. However, despite the opposition, Cape Wind seems to be on a path toward receiving the permits necessary to commence the project.

When completed, Cape Wind will consist of approximately 130 wind turbine generators capable of producing approximately 454 megawatts ("MW") of energy. The 3.6 MW wind turbine generators will be located approximately 0.3 to 0.5 miles apart and the total array spread over twenty-four square miles. Although the towers will extend only 257.5 feet above the water surface, each wind tower blade will reach 440 feet above the water. This wind facility will be located in federal waters in Nantucket Sound, sheltered on the north by Cape Cod, to the west by Martha's Vineyard, to the south by Nantucket Island, and to the east by the Great Sound Shoal. One reason Nantucket Sound was chosen as the location of this project is that it is relatively sheltered from significant Atlantic Ocean wave action and extreme storm waves. Its closest distance to shore will be 4.7 miles, and its furthest will be approximately 11 miles. This means a number of turbines will be visible from some points on the shores of Cape Cod and Martha's Vineyard but not from Nantucket. The cost of construction for this project is estimated to be as high as $2 billion.

In 2006, Bluewater Wind LLC ("Bluewater Wind") proposed a similar project, estimated to cost $1.6 billion, to be located in federal waters, at least eleven miles off the coast of Delaware in the Atlantic Ocean. At this distance, the turbines would be barely visible from the Delaware coastline. The future of this project is uncertain. The original project proposed the installation of more than 100 wind turbine generators capable of producing approximately 450 MW of electricity; however, the June 2008 power purchase agreement between Bluewater Wind and an onshore receiving utility company will only support the construction of fifty-five to seventy wind turbines. If Bluewater Wind decides to build more than seventy turbines, it will have to find another purchaser for the generated power. Another factor is the uncertain financial future of Bluewater Wind itself. In February 2009, Babcock & Brown, the Australia-based company that owns virtually all of Bluewater Wind, announced plans to liquidate its assets in order to satisfy creditor claims. This means that Bluewater Wind will need to find new financial backing for the Delaware project. [NRG Energy purchased Bluewater Wind in 2009 and has until September 2011 to seal a power purchase agreement with Delmarva Power and Light — Ed.]

A major difference between the Cape Wind project and the Bluewater Wind project is that the Bluewater Wind project is the first one proposed for open ocean waters; for that reason, it will confront significant location and construction challenges. Sea conditions in an ocean location may be one reason for the September 2007 official cancellation of a similar project proposed by the Long Island Power Authority ("LIPA") to be sited off the South Shore of Long Island. In 2003, when Cape Wind evaluated that site, the president of Cape Wind wrote a letter to LIPA stating "that the anticipated sea conditions in the Target Area pose unacceptable conditions. Both the significant wave and extreme storm wave are nearly three times that associated with current state-of-the-art offshore wind projects." The official LIPA reason for cancellation was the high cost of construction. The original projected cost in 2003 was $200 million but eventually ballooned to $811 million by the time LIPA decided to cancel the project.

Other states, such as New Jersey and Rhode Island, are also pursuing wind energy development off their coasts. New Jersey has adopted a renewable energy incentive program and an offshore wind rebate program for the installation of meteorological towers, in addition to awarding a $4 million grant to Garden State Offshore Energy for a 345.6 MW offshore wind facility tentatively to be located sixteen miles southeast of Atlantic City. In Rhode Island, interest in wind energy development in coastal and offshore waters will likely rise as the State seeks to achieve its renewable energy portfolio standard of sixteen percent by 2020. To help meet this goal, Governor Donald Carcieri announced in September 2008 that the company Deepwater Wind was selected to construct a wind energy project off Rhode Island's coast. The project will provide an estimated 1.3 million megawatt hours per year, which is approximately fifteen percent of the electricity used in the state.

* * *

Each of the projects described above is proposed for location in waters relatively near the shore because of technology and cost limitations. Current technology allows wind facilities to be located in waters deeper than twenty to thirty meters. In fact, existing technology would allow wind turbines to be sited in waters up to fifty meters in depth, but at the present time, it is prohibitively expensive to construct the foundations for and to locate facilities in water much deeper than twenty to thirty meters.... Until the cost of deeper water technology drops significantly, twenty to thirty meters is close to the economically feasible limit for offshore wind energy facilities....

B. Challenges to Siting Water-Based Wind Energy Facilities and their Relevance to North Carolina

Practical and economic factors make significant development of offshore wind energy difficult. First, at the present time, there is a limited supply of the necessary construction equipment. Second, construction costs, operational costs, and maintenance costs of offshore wind facilities could be double that of land-based wind facilities. Third, some turbine manufacturers are unsure of the durability of their equipment when placed in deep water. Fourth, even with current subsidies, the cost of generated offshore wind energy is not competitive with traditional onshore energy facilities. If oil prices continue to fall as they did in late 2008 and early 2009, the differential may be even greater. Finally, the chaos in the financial markets and money supply may make it more difficult to find financial backers for wind energy projects. State renewable energy portfolio standards, federal and state government subsidies and stimulus funds, and federal and state tax credits will continue to drive the interest in wind energy and could provide sufficient incentives to direct some capital into offshore projects. However, the lower costs of land-based wind energy may prove more attractive....

Potential user conflicts may also impact the development of nearshore and offshore wind energy facilities. Wind energy generating equipment and offshore and onshore support facilities and infrastructure may present a number of user conflicts. In areas heavily dependent upon coastal tourism and those with shorelines filled with very expensive vacation homes, the aesthetic impacts may be a significant concern. Commercial and recreational navigation and fishing, military airspace operations, marine mammal populations, seabird activity, the locations of beach quality sand and other non-living natural resources, and other water activities may also conflict with the siting of wind facilities in particular water locations. Difficult choices may have to be made between energy independence and other uses of coastal and ocean resources.

* * *

II. Wind Turbine Facilities in Federal Waters and the CZMA [Coastal Zone Management Act, 16 U.S.C. §§ 1451-66 (2011)—Ed.] Consistency Requirement

A. The Legal Framework Governing Offshore Wind Development in Federal Waters

Section 388 of the Energy Policy Act of 2005 grants authority to the Secretary of the Interior to issue leases and grant easements for alternative energy activities on the Outer Continental Shelf ("OCS"). Minerals Management Service ("MMS"), which also administers the OCS oil and gas leasing process, is the bureau within the Department of the Interior designated to develop the leasing program for OCS renewable energy activities. [Before the Macondo Well blowout from the Deepwater Horizon rig in the Gulf of Mexico in April of 2010, the MMS controlled offshore revenue generation as well as leasing and safety supervision. After the blowout, BLM reorganized the MMS to separate these two functions because there was some belief that the desire to generate revenue created a conflict of interest in MMS's safety inspection role. Currently, the MMS controls royalty payments, but the BOEMRE—Bureau of Ocean Energy Management, Regulation, and Enforcement—now controls leasing and safety. HENRY B. HOGUE, CONG. RESEARCH SERV., R41485, REORGANIZATION OF THE MINERALS MANAGEMENT SERVICE IN THE AFTERMATH OF THE DEEPWATER HORIZON OIL SPILL 1-14 (2010)–Ed.] On April 22, 2009, MMS issued its regulations.

The regulations contemplate the issuance of two different types of OCS alternative energy leases: commercial and limited. A commercial lease provides, subject to necessary approvals, the right to produce, sell, and deliver power on a commercial scale from an alternative energy source. Commercial leases grant a five-year term to conduct site assessment activities and a twenty-five year operations term. A commercial lease can be renewed, but there is no automatic right of renewal. MMS rejects the idea of an open-ended term, or automatic extensions and renewals, for alternative energy leases. Leases with such provisions are used for OCS oil and gas production, with continuation contingent upon drilling and production. However, in the context of evolving alternative energy technology, the concern is that an open-ended alternative energy lease could perpetuate inefficient and obsolete forms of alternative energy operations. MMS's judgment is that a fixed-term lease will promote and ensure diligent development and use of the most efficient alternative energy technology. MMS selected twenty-five years as a lease term because it matches the anticipated duration of power purchase agreements in which alternative energy lessees and onshore utilities are likely to enter. Limited leases are for periods of up to five years and grant access and operational rights for activities that support production of energy but do not directly result in the general production of electricity or energy for sale, distribution, or other commercial use. A company might seek such a lease to test energy-generating devices or collect data and other information.

B. Application of the CZMA Consistency Requirement to Wind Energy

Intended to provide large-scale, long-term commercial energy production, the issuance of an OCS alternative energy commercial lease and federal authorization of specific activities will be of significant concern to coastal states. For commercial leases, the regulations contemplate four stages: (1) lease issuance; (2) site assessment activities; (3) construction, operation, and conceptual decommission planning; and (4) actual decommissioning. For purposes of both the required National Environmental Policy Act ("NEPA") analysis and the Coastal Zone Management review, MMS has combined stages (1) and (2) to reduce the time needed to review competitive leases. This reduces the number of opportunities that an affected coastal state or states will have to voice any concerns and have them addressed. Any concerns about either the lease issuance or site assessment activities will have to be presented prior to the lease sale.

1. Lease Issuance

At the leasing stage, section 307(c)(1)(A) of the CZMA requires that any federal activity, including lease sales, that is reasonably likely to affect any land or water use or natural resource of a state's coastal zone must be consistent "to the maximum extent practicable" with that state's federally approved coastal zone management plan ("CMP"). Although a sale itself, which is nothing more than a paper transaction, would not directly affect any such land or water use or natural resource, it starts a chain of events that includes construction, maintenance, operation, and decommissioning, which could affect such land or water uses or natural resources. MMS must take into account such effects if they are reasonably foreseeable. If MMS determines the existence of such future effects, it must structure the terms of the lease in a manner that is consistent to the maximum extent practicable with the enforceable policies of the coastal state's CMP and submit a statement (consistency determination) to the State that the sale will be so conducted. If the State disagrees with MMS's consistency determination, then the State may file an objection. If an objection is filed, and MMS and the State continue to disagree, then the issue may have to be resolved through mediation or litigation in federal court.

What is important for coastal states, such as North Carolina, is having appropriate enforceable policies in their CMPs that would apply to wind energy projects. For a policy to be enforceable, it must be legally binding as opposed to advisory in nature. Thus, the application of the CZMA consistency provision to OCS alternative energy leases is directly related to how a coastal state treats similar projects proposed for state waters. For example, if a state (a) believes it is important to preserve existing sand resources for use in beach nourishment projects and, (b) to do that, prohibits placement of structures in areas where those resources exist, then it should promulgate a rule or rules prohibiting such activities in its own state waters. If the State enacts such a rule, then the rule would be applicable not only to projects in state waters but also, under the CZMA consistency provision, to projects in federal waters. In essence, the consistency provision directs a federal agency to treat a state's policies, which are legally binding as to activities within the State, as legally binding for the federal agency.

2. Site Assessment Activities

After the lease is issued, the next stage is for the lessee to submit a site assessment plan ("SAP"). The SAP describes the planned activities for site surveys, data gathering, and related facilities and operations. This plan must be approved by MMS before any site assessment activities begin. Under the process created by the MMS regulations, unless the SAP submitted by the holder of a commercially issued lease shows impacts different from those identified in the combined lease/site assessment NEPA document and CZMA consistency determination MMS prepared, the SAP would not be subject to a new NEPA/CZMA and other federal reviews. The process adopted by MMS raises a significant CZMA consistency issue. The degree of consistency required by the CZMA differs depending on whether the activity under review is a "federal agency activity" or an activity by "any person who submits … any plan for the exploration or development of, or production from, [OCS leased lands]." If it is a "federal agency activity," then it must "be carried out in a manner which is consistent to the maximum extent practicable with the enforceable policies" of a state's federally approved coastal zone management program. However, if the activity is one described in a federal OCS lands lessee's plan of exploration, development, or production, it must comply "with the enforceable policies of [the relevant state's] approved [coastal] management program and … be carried out in a manner consistent with such program." This means it must be completely consistent with the state's enforceable policies.

If MMS plans to incorporate both the lease issuance and site assessment activities into one consistency determination, it raises some significant issues with respect to the implementation of the mandates of the CZMA. If the activity is a federal agency activity and the State disagrees, then the applicable legal standards and the process for resolving the disagreement differ dramatically from the situation in which a State does not concur in the consistency determination of an OCS lessee. If the State objects to a federal agency's planned activity, it may mean mediation and a federal lawsuit. On the other hand, if the State does not concur in an OCS lessee's consistency certification, the planned activity cannot take place so long as the State objects, unless the Secretary of Commerce overrides the objection. MMS's decision to combine the lease issuance and site assessment activities consistency determination may expedite the review of commercial leases, but it presents serious questions about the administration of the CZMA consistency process and may conflict with a coastal state's right to object and block an OCS activity inconsistent with the state's enforceable policies.

3. Construction, Operation, and Conceptual Decommissioning Plan

After the site assessment is performed, the next stage is the submission of the Construction and Operations Plan ("COP"). The COP must cover all proposed activities and operations associated with the construction and operation of the alternative energy facility and demonstrate that the activities are safe, do not unreasonably interfere with other uses of the OCS, do not cause undue harm or damage, use the best available and safest technology, use the best management practices, and use properly trained personnel. MMS's review of the COP includes an assurance that the plan satisfies the requirements of NEPA and other applicable federal laws. At this time, a coastal state has another opportunity to address any inconsistencies between the proposed alternative energy operations and the state's enforceable policies under its federally approved coastal zone management plan. Any activities described in the COP affecting any land or water use of a natural resource in a state's coastal zone must be consistent with that state's enforceable policies.

One interesting aspect of the MMS regulations is the treatment of decommissioning. MMS considered postponing decommissioning regulations because there are no large-scale alternative energy facilities on the OCS as of yet, and it may be twenty to twenty-five years or more before any project yet to be built would be decommissioned. A lot could change between now and then. Nonetheless, MMS decided that decommissioning should be addressed so that lessees will know what would be required at the end of a project ahead of time. The COP would include a conceptual decommissioning plan. Although a coastal state has an opportunity to assert a consistency objection at the time the COP is presented, before the actual decommissioning takes place, the state should have another opportunity to raise any new consistency objections arising from new information or federally approved amendments to the state's CMP.

4. Actual Decommissioning

Minerals Management Service regulations state that a lessee must submit a decommissioning plan to MMS for approval before beginning actual decommissioning. It is only when the operator is actually ready to decommission the facility, files the decommissioning plan, and seeks MMS approval that the precise decommissioning details will be known. This decommissioning will take place many years after the approval of the COP and under potentially different ecological conditions and a changed legal environment, as new state coastal legislation or regulations are put into place or older statutes and regulations amended. The question is whether at that time another CZMA consistency review should be required.

The MMS regulations suggest such a review will only take place if the decommissioning plan in the submitted application results in "a significant change in the impacts previously identified," requires any additional authorizations, or "[p]ropose[s] activities not previously identified and evaluated." This implies that impacts associated with changed ecological conditions or a decommissioning activity not described in the COP will be subject to a consistency review at the time of the decommissioning application. However, the proposed rules do not discuss the relevance of any intervening changes in a state's enforceable policies in its CMP. On one hand, if the state had an opportunity to object to the decommissioning plan set forth in the COP but stated no objections at that time, then the federal lessee/operator should be able to rely upon its submitted COP. On the other hand, it seems that the consistency of an activity that was not intended to take place until some date long into the future should be based on compliance with the enforceable policies of the state's coastal management plan in existence at the time when decommissioning actually occurs. Because the Secretary of Commerce can override a state's consistency objection, the better path would be to require the decommissioning applicant to submit a consistency certification at that time, to allow the State to object if there are grounds, and, if no satisfactory resolution can be reached among the parties, to allow the applicant to appeal to the Secretary and seek an override of the state's objection.

C. Making the Most of the Consistency Review

The actual utility of the consistency review is dependent upon the State having the enforceable policies in its CMP to address the important ecological, environmental, and economic issues likely to be presented by locating alternative energy facilities in ocean waters.

––––––––

Notes and Questions

1. First let's start with an overview of some of the key statutes. The Submerged Lands Act, 43 U.S.C. §§ 1301-15 (passed in 1953), granted states title to the "lands beneath navigable waters" within the three miles adjacent to their coasts as well as the petroleum and mineral resources in those lands. At the same time, Congress passed the Outer

Continental Shelf Lands Act (OCSLA), which affirmed federal ownership of the resources located seaward of the three-mile boundary line. 43 U.S.C. §§ 1331-56 (2011). Then in 1972, Congress passed the Coastal Zones Management Act (CZMA), 16 U.S.C. §§ 1451-66 (2011). What does the CMZA require? What is a CMP and how does a consistency determination relate to one?

2. Note that the BOEMRE regulations grant a limited term for commercial wind leases. How long is that term? In contrast, oil and gas leases on the OCS have defined terms of ten years with the ability to get automatic extensions and renewals with production. Are alternative energy leases given similar terms? Explain the rationales for the discrepancies and determine whether you think they are justified.

3. List the four stages the regulations anticipate for large-scale, commercial energy production and the possible federal and state consistency issues the excerpt suggests might arise at some of these stages.

4. In chapter 4.D, we addressed some of the problems in the past when larger energy projects, such as federal dams, did not consider decommissioning at the early stages of the construction process. More modern statutory authority, including OCSLA and FLPMA, now requires decommissioning and reclamation with the posting of a bond before construction to ensure that reclamation takes places. The BOEMRE COP (Construction and Operations Plan) also requires a decommissioning strategy as part of the leasing process. At the time the COP is approved, coastal states have an opportunity to address any inconsistencies between the proposed operations and its CMP policies. Do you see any problems with this timing?

5. The 2005 EPAct also requires "coordination and consultation with the Governor of any State or the executive of any local government that may be affected by a lease, easement or right of way under [the Act]." EPAct 2005, Pub. L. No. 109-58, § 388(a) (codified at 43 U.S.C. § 1337). BOEMRE has looked to early local input to the planning process through regionally-constituted task forces as the primary mechanism for meeting the coordination and consultation requirement. 30 C.F.R. § 285.102(e). In addition, the federal and state governments have collaborated in Multi-purpose Marine Cadastral and Coastal and Marine Spatial Planning Plus maps. Do you think these two approaches are effective?

b. Cape Wind

No discussion of offshore wind in the United States would be complete without a little history about the Cape Wind project about five miles off the Massachusetts coast in Horseshoe Shoal, Nantucket Sound. The 420 MW project includes 130 400-foot-high wind turbines in a twenty-four square mile area. Although the project was first initiated in 2001, construction had not yet begun as this book went to press ten years later in the summer of 2011. Because the project is visible in parts of the affluent resort communities of Cape Cod and Martha's Vineyard, it faced a wide range of opponents from Senator Edward M. (Ted) Kennedy to two Wampanoag Indian tribes. Issues ranged from detrimental effects on property values, possible damage to birds, whales, fishing, aviation, and historic and cultural sites. The most recent lawsuit was filed by the Wampanoags on July 6, 2011, in the United States District Court for the District of D.C. The complaint alleges that permitting the Cape Wind project would cause irreversible alterations, significant adverse effects, and the destruction of historical, cultural, and spiritual tribal resources. Notably Greenpeace and Natural Resources Defense Council supported the wind farm development.

Figure 8.3: Proposed U.S. offshore wind projects and capacity showing projects with significant progress

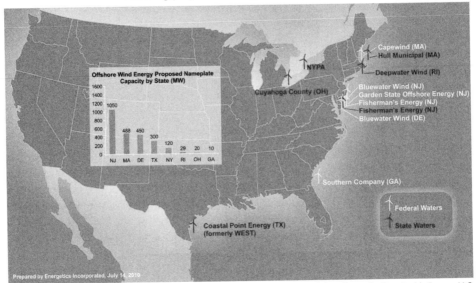

N. Carlisle, J. Elling, and T. Penney. Image provided courtesy of The Alliance for Sustainable Energy, LLC. Table 3-5 of NREL TP-500-40475, Walter Musial & Bonnie Ram, *Large Scale Offshore Wind Power in the United States: Assessment of Opportunities and Barriers* 29 (2010), *available at* http://www.nrel.gov/wind/pdfs/40745.pdf.

One of the primary opponents to the Cape Wind project is the Alliance to Protect Nantucket Sound. The Alliance has filed lawsuits to challenge the development process at almost every step, and the following case sets out some of the very first issues the group raised.

Alliance to Protect Nantucket Sound, Inc. v. U.S. Dep't of the Army

398 F.3d 105 (1st Cir. 2005)

Torruella, J.

On November 20, 2001, Cape Wind Associates, L.L.C. ("Cape Wind") submitted an application to the U.S. Army Corps of Engineers ("Corps") for a navigability permit under Section 10 of the Rivers and Harbors Act of 1899 ("Section 10"), 33 U.S.C. § 403,[1] to construct and operate an offshore data tower in an area of Nantucket Sound known as Horseshoe Shoals. Horseshoe Shoals is located on the Outer Continental Shelf ("OCS"), land subject to federal jurisdiction and control under the Outer Continental Shelf Lands Act ("OCSLA"), 43 U.S.C. § 1331.

The proposed tower was to consist of a platform and a fixed monopole approximately 170 feet high, supported by three steel piles driven into the ocean floor. Various instrumentation was to be attached to the data tower in order to gather data for use in determining the feasibility of locating a wind energy plant on Horseshoe Shoals. A separate permit ap-

---

1. Section 10 delegates authority to the Corps to issue permits for projects that impact on the navigability of United States waters. 33 U.S.C. § 403.

plication for the wind energy plant-a complex originally proposed to include 170 wind turbines with blade rotors rising 423 feet above mean sea level, occupying twenty-six square miles of Horseshoe Shoals-was submitted to the Corps in November 2001. That application is not at issue in the instant appeal, and we therefore will not engage in any analysis of the Corps's authority to permit construction of the wind energy plant.

... On August 19, the Corps issued a Section 10 permit authorizing Cape Wind to construct and maintain the data tower, subject to the imposition of sixteen special conditions.... The permit was accompanied by an Environmental Assessment ("EA") and Finding of No Significant Impact ("FONSI"), as required by the National Environmental Policy Act ("NEPA"), 42 U.S.C. §§ 4331-32.

Appellants subsequently filed an action against the Corps in the District of Massachusetts, arguing that (1) the Corps lacked authority to issue a Section 10 permit for the data tower; (2) the Corps acted arbitrarily and capriciously, in violation of the Administrative Procedure Act ("APA"), 5 U.S.C. § 706(2)(A), by granting Cape Wind's permit application in spite of Cape Wind's lack of property rights on the OCS; and (3) the Corps failed to comply with NEPA requirements for evaluating the data tower's environmental impacts. Upon the receipt of cross motions for summary judgment, the district court granted summary judgment in favor of the Corps and intervenor Cape Wind. We review that decision *de novo*, construing the evidence in the light most favorable to appellants. (citations omitted) We affirm the decision of the district court.

I. *Discussion*

A. Corps jurisdiction

The reach of the Corps's Section 10 permitting authority on the OCS turns on a question of statutory interpretation. Congress passed OCSLA in 1953 to assert federal jurisdiction over the OCS and to establish a regulatory framework for the extraction of minerals therefrom. *See* 43 U.S.C. § 1332; *see also Ten Taxpayer Citizens Group v. Cape Wind Assocs.*, 373 F.3d 183, 188 (1st Cir.2004) ("A major purpose of the OCSLA was to specify that federal law governs on the [OCS]....") (internal quotation marks omitted). Accordingly, OCSLA extended the Corps's Section 10 regulatory authority "to prevent obstruction to navigation in the navigable waters of the United States ... to "the artificial islands, installations, and other devices referred to in subsection (a) of this section." 43 U.S.C. § 1333(e) (2004)....

* * *

Congress made clear that "[t]he existing authority of the Corps ... applies to all artificial islands and fixed structures on the [OCS], whether or not they are erected for the purpose of exploring for, developing, removing, and transporting resources therefrom." Conference Report at 82. This express legislative intent is determinative of the scope of the Corps's authority. Accordingly, we hold that the Corps had jurisdiction to issue a Section 10 permit for Cape Wind's data tower.

B. Property interest

Appellants argue that the Corps failed to properly consider Cape Wind's lack of a property interest in the OCS land on which it sought to build the data tower when it granted the Section 10 permit.

* * *

... [A]ppellants argue that Cape Wind's affirmation that it possessed the requisite property interests was obviously false, as there exists no mechanism by which private entities can obtain a license to construct a data tower on the federally controlled OCS. The Corps's grant of a Section 10 permit on the basis of this false affirmation was therefore arbitrary and capricious, in violation of the Administrative Procedure Act, 5 U.S.C. § 706(2)(A).... Appellants' argument hinges on the veracity of Cape Wind's affirmation, which in turn depends, appellants argue, on whether authorization in addition to a Section 10 permit is necessary for construction of the data tower.[7] The first part of our opinion holds that a Section 10 permit is *necessary* for all structures on the OCS unless otherwise indicated by law, but does not determine whether such a permit is *sufficient* to authorize building on the federally controlled OCS.

Whether, and under what circumstances, additional authorization is necessary before a developer infringes on the federal government's rights in the OCS is a thorny issue, one that is unnecessary to delve into in the instant case. The data tower at issue here involves no real infringement on federal interests in the OCS lands. To start, the structure is temporary, of five years' duration, more than two of which have now passed. The tower is also not exclusive-it must accept data collection devices form the government and others, and it must give the data to the government. The tower is a single structure, and it provides valuable information that the Corps requires in order to evaluate the larger wind energy plant proposal. The Corps's public interest evaluation of the data tower resulted in a finding of "negligible impact" on property ownership and stated that collection of the data is in the public interest. Environmental Assessment at 4-5. It is inconceivable to us that permission to erect a single, temporary scientific device, like this, which gives the federal government information it requires, could be an infringement on any federal property ownership interest in the OCS.

Thus, the question of infringement of federal property interests is entirely hypothetical in this case. As a result, appellants' arguments based both on the arbitrary and capricious provision in the APA and the public interest standards discussed in *Alaska* are misplaced. We do not here evaluate whether congressional authorization is necessary for construction of Cape Wind's proposed wind energy plant, a structure vastly larger in scale, complexity, and duration, which is not at issue in the present action. Our analysis is limited to whether additional Congressional authorization is necessary for the data tower, which does not infringe on any federal property interest, and we conclude that it is not.

C. National Environmental Policy Act

The Council on Environmental Quality ("CEQ") is authorized to enact regulations to ensure federal agencies' compliance with NEPA. *See* 42 U.S.C. §§ 4342, 4344. Appellants argue that the Corps violated CEQ regulations by failing to circulate for public comment a draft EA and FONSI. We evaluate agency action to determine if it is "arbitrary, capricious,

---

7. Congress has established regulatory schemes for certain types of structures on the OCS. OCSLA itself sets up a system of oil and gas leases that require both a lease from the Secretary of the Interior as well as a Corps permit. *See* 43 U.S.C. § 1331 *et seq.* The Ocean Thermal Energy Conversion Act of 1980, 42 U.S.C. § 9101 *et seq.,* authorizes the creation of large thermal energy plants by requiring a license from the National Oceanic and Atmospheric Administration, while the Coast Guard is authorized to make rules ensuring safety of navigation. *Id.* §§ 9111, 9118. The Deepwater Ports Act of 1975, 33 U.S.C. § 1501 *et seq.,* requires a license from the Secretary of Transportation in order to authorize construction of deepwater ports. *Id.* § 1503. The National Fishing Enhancement Act of 1984, 33 U.S.C. § 2101 *et seq.,* in contrast, does not require approval for artificial reefs placed on the OCS beyond a Section 10 permit. *Id.* § 2104.

an abuse of discretion, or otherwise not in accordance with law." 5 U.S.C. § 706(2)(A), (C).

CEQ regulations require that an "agency shall involve … the public, to the extent practicable, in preparing [an EA]," … Appellees argue that the Corps met the requirement of involving the public "to the extent practicable" in preparing the EA by issuing public notice of Cape Wind's application, providing a comment period that they later extended to over five months, carrying out two public hearings, noting and responding to public comments in the EA, and conferring with federal and state environmental agencies. We agree. Nothing in the CEQ regulations requires circulation of a draft EA for public comment, except under certain "limited circumstances." 40 C.F.R. § 1501.4(e)(2).

Appellants argue that one of those circumstances applies to this case: A draft FONSI must be made available for public comment when "[t]he nature of the proposed action is one without precedent." *Id.* § 1501.4(e)(2)(ii). Appellants argue that the data tower proposal is "without precedent" because Nantucket Sound is a pristine, undeveloped area and because "there is no precedent for permitting a privately-owned structure for wind energy, or even related research, on OCS lands." The Corps, however, determined that "[t]here is precedent for this type of structure in Massachusetts's waters," in the form of a data tower in Martha's Vineyard. Environmental Assessment at 10. The district court agreed, relying on the Corps's findings that while "[t]here are no other similar structures or devices in Horseshoe Shoals," a data tower was permitted in state waters off Martha's Vineyard, and Cape Wind's data tower was "not inconsistent with other pile supported structures in the marine environment in Nantucket Sound." *Id.* at 2; *see Alliance,* 288 F.Supp.2d at 78–79.

We find that the Corps's determination that the data tower is not without precedent, on the basis of physically similar structures in nearby waters, was reasonable. We do not agree with appellants' argument that construction of structures like the data tower on the OCS without additional authorization from Congress is without precedent, but even if that were so, it would suggest only that issuance of the permit is *legally* unprecedented. The CEQ regulations, however, are designed to address environmental impact. Based on the Corps's findings about the existence of similar pile-driven structures in Martha's Vineyard and near the shore of Nantucket Sound, we can see nothing unprecedented about the way this data tower will impact the environment. Thus, we find that the Corps fully complied with its obligations under NEPA and CEQ regulations to engage with the public in preparing the EA and FONSI.

Notes and Questions

1. This lawsuit challenged the Army Corps of Engineers' granting of a Section 10 permit for simply installing a tower to collect data for possible construction of a wind farm. In addition to permits from the Corps, an offshore wind developer may have to obtain approvals from each of the agencies listed in Figure 8.4. And those are only the federal agencies; a permittee must also work with the state agencies discussed in the previous excerpt. Is the multiple permitting process more burdensome for offshore than onshore wind? Some federal efforts, such as DOI's Smart from the Start Initiative, intend to streamline the approval process. What do you think can be done?

2. Part of the challenge in this case was that Congress had not explicitly authorized use of the OCS for wind development. The court here upheld the Corps' conclusion that the impact of the data tower was negligible and that the collection of the data was in the

Figure 8.4: Key statutes and agencies involved in offshore wind permitting

Statute	Key Agencies
National Environmental Policy Act of 1969 (NEPA)	All federal agencies
Marine Mammal Protection Act of 1972	FWS; NOAA; NMFS
Magnuson-Stevens Fishery Conservation and Management Act	NOAA; NMFS
Marine Protection, Research, and Sanctuaries Act of 1972	EPA; USACE; NOAA
National Marine Sanctuaries Act	NOAA
Coastal Zone Management Act of 1972	NOAA Office of Ocean and Coastal Resource Management (OCRM)
National Historic Preservation Act of 1966	NPS; Advisory Council on Historic Preservation; State or Tribal Historic Preservation Officer
Federal Aviation Act of 1958	FAA
Federal Power Act	FERC; BOEMRE
Ports and Waterways Safety Act	USCG
Rivers and Harbors Act of 1899	USACE
Outer Continental Lands Act of 1953	DOI
Clean Water Act	EPA; USCG
Clean Air Act	EPA; BOEMRE

Data from U.S. Dep't of Energy, A National Offshore Wind Strategy: Creating an Offshore Wind Energy Industry in the United States 11-12 (2011), *available at* http://www1.eere.energy.gov/windandhydro/pdfs/national_offshore_wind_strategy.pdf

public interest. This court did not, however, opine whether development of a wind farm infringes on the federal government's rights in the OCS. How has Congress stepped in to avoid this consideration in future wind farm situations?

3. The Alliance argued that additional public input was required for the draft FONSI because there was no precedent for privately owned structures for wind energy on OCS lands. Do you agree that renewable projects without precedent should require additional NEPA input? If so, does this give renewables additional hurdles to development, disadvantaging their development in contrast to more established fossil fuels?

4. Another consideration for offshore development is maritime cabotage laws. "Offshore alternative energy projects involve the use of vessels that, when acting in U.S. waters, are regulated in a variety of ways. Several U.S. maritime trade laws, or 'cabotage' laws, restrict a number of offshore activities to qualified U.S.-flagged vessels. These laws are often referred to as the 'Jones Act'—even though that term genuinely applies to only one of the several relevant maritime cabotage laws. The principal maritime cabotage laws apply to

the transportation of merchandise and passengers, the towing of vessels, and dredging. Many commenters have assumed that these maritime cabotage laws apply to offshore alternative energy projects without either distinguishing the various laws or examining their jurisdictional bases." Constantine G. Papavizas & Gerald A. Morrissey III, *Does the Jones Act Apply to Offshore Alternative Energy Projects?*, 34 Tul. Mar. L.J. 377, 378 (2010). How do you think the Jones Act might impact offshore wind development in the United States?

5. Although Cape Wind may be the first offshore wind project to have received all governmental approvals, it may not be the first offshore wind farm *built* in the United States because of the ongoing litigation against it. Vying for first place are projects in Rhode Island; Atlantic City, New Jersey; and Galveston, Texas. *See* Figure 8.3. Coastal Point Energy, the developer of the project off the coast of Galveston, may have the advantage because it will not need to navigate through the shoals of federal regulatory agencies. When the Republic of Texas joined the United States in 1845, it retained jurisdiction over waters in the Gulf extending 10.3 miles from its coasts.

2. Ocean Hydrokinetics

Most of the hydropower discussed in Chapter 4 involved energy generated from the hydrostatic pressure of water moving from a higher elevation, usually from a reservoir, into a lower elevation point below a dam. A greater drop between elevations translates to a higher hydrostatic head with the potential for producing more power.

Hydrokinetics is a broad term used to label methods of generating power not through elevation drops but through harnessing moving water. Some hydrokinetic projects are

Figure 8.5: Four primary types of wave energy conversion

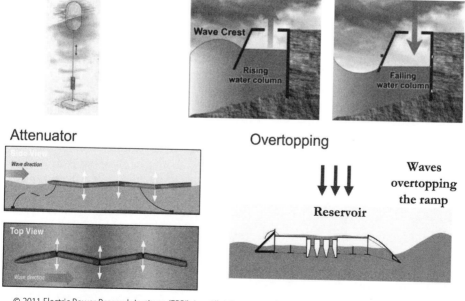

sited within inland rivers as discussed in Chapter 4, but the focus here will be offshore resources from ocean currents, tides, and waves.

Hydrokinetics is still in its infancy. There are dozens of experimental technologies, and no one has emerged as the best. Figure 8.5 shows four of the most common for capturing the energy of waves: point absorbers, oscillating water columns, attenuators, and overtopping devices. Figure 8.6 describes these and some other devices the Department of Energy is currently exploring.

Although little has been written about offshore hydrokinetics, the emerging technologies for harnessing the moving water of the ocean to generate power have significant potential for providing power to electricity-hungry areas along the coasts. According to Roger Bedard of the Electric Power Research Institute, hydrokinetics could generate 10 percent of the U.S.'s electricity demand, with 250-260 TWh/year for offshore wave energy potential and approximately 125 TWh/year for ocean current and tidal potential. Megan Higgins, *Is Marine Renewable Energy A Viable Industry in the United States? Lessons Learned from the 7th Marine Law Symposium*, 14 ROGER WILLIAMS U. L. REV. 562, 570, n. 33 & 34 (2009).

Figure 8.4 lists the federal statutes and related agencies involved in offshore wind power permitting. Unfortunately, development of hydrokinetic resources involves a similarly daunting process that additionally has been hampered by agency turf battles.

As we saw in Chapter 4 above, the Federal Power Act gave the Federal Energy Regulatory Commission (FERC) permitting authority for all water power generation facilities in the navigable waters of the United States. Because the statutory definition of navigable waters includes the Outer Continental Shelf, FERC logically interpreted its authority to include licensing for offshore energy projects.

However, another federal agency claimed overlapping authority for energy projects on the OCS. The Bureau of Ocean Energy Management, Regulation and Enforcement (BOEMRE) (formerly MMS) is an agency within the Department of Interior charged with overseeing the safe and environmentally responsible development of energy and mineral resources on the OCS. After the Deepwater Horizon explosion in 2010, we may be most familiar with BOEMRE's role in regulating offshore oil and gas development. However, BOEMRE also supervises offshore wind, as we saw above, and was explicitly given authority for regulating hydrokinetic projects under section 338 of the 2005 EPAct, which amended § 8(p)(1) of the OCSLA.

Despite Congress's effort to resolve the dispute between the two agencies, FERC continued to claim permitting authority for hydrokinetic projects based on a provision in the 2005 EPAct "stipulate[ing] that MMS authority does not supersede the existing authority of any other agency for renewable energy project permitting." In October of 2007, FERC announced its Hydrokinetic Pilot Project Licensing Process, and Figure 8.7 from FERC's website shows twenty-six tidal and eight wave preliminary permits it has issued. Moving forward under its separate authority, BOEMRE, also in late 2007, established an interim policy to issue limited leases authorizing "renewable energy resource assessment, data collection, and technology testing activities on the OCS."

Because of the confusion created by these overlapping regulatory regimes, DOI and FERC attempted to resolve the issue with a Memorandum of Understanding (MOU) signed in April of 2009. According to a BOEMRE press release, the 2009 MOU "clarifies their agencies' jurisdictional responsibilities for leasing and licensing renewable energy projects on the OCS. Under the agreement, the MMS has exclusive jurisdiction with regard to the production, transportation, or transmission of energy from non-hydrokinetic renewable energy projects, including wind and solar. FERC will have exclusive jurisdiction

Figure 8.6: Marine and hydrokinetic technologies

Dams & Current	
Horizontal Axis Turbine	Typically has two or three blades mounted on a horizontal shaft to form a rotor; the kinetic motion of the water current creates lift on the blades causing the rotor to turn, driving a mechanical generator.
Vertical Axis Turbine	The same mechanics of water current turning the rotor that drives the generator, but the rotor blades are mounted along a vertical instead of a horizontal shaft.
Oscillating Hydrofoil	Similar to an airplane wing but in water; control systems adjust their angle relative to the water stream, creating lift and drag forces that cause device oscillation; mechanical energy from this oscillation feeds into a power conversion system.

Wave	
Attenuator	Wave energy capture device with principal axis oriented parallel to the direction of the incoming wave that converts the energy due to the relative motion of the parts of the device as the wave passes along it.
Pitching/Surging/ Heaving/Sway (PSHS) device	Any of several devices that capture wave energy directly without a collector by using relative motion between a float/flap/membrane and a fixed reaction point; the float/flap/membrane oscillates along a given axis dependent on the device; mechanical energy is extracted from the relative motion of the body part relative to its fixed reference.
Oscillating Water Column	Partially submerged structure that encloses a column of air above a column of water; a collector funnels waves into the structure below the waterline, causing the water column to rise and fall; this alternately pressurizes and depressurizes the air column, pushing or pulling it through a turbine.
Overtopping Device	Partially submerged structure; a collector funnels waves over the top of the structure into a reservoir; water runs back out to the sea from this reservoir through a turbine.
Point Absorber	Wave energy capture device with principal dimension relatively small compared to the wave length and able to capture energy from a wave front greater than the physical dimension of the device.
Submerged Pressure Differential	Wave energy capture device, which can be considered a fully submerged point absorber; a pressure differential is induced within the device as the wave passes driving a fluid pump to create mechanical energy.
Oscillating Hydrofoil	Similar to an airplane wing but in water; yaw control systems adjust their angle relative to the water stream, creating lift and drag forces that cause device oscillation; mechanical energy from this oscillation feeds into a power conversion system.

Figure 8.6: Marine and hydrokinetic technologies *continued*

Ocean Thermal Energy Conversion	
Closed-cycle	These systems use fluid with a low-boiling point, such as ammonia, to rotate a turbine to generate electricity. Warm surface seawater is pumped through a heat exchanger where the low-boiling-point fluid is vaporized. The expanding vapor turns the turbo-generator. Cold deep-seawater—pumped through a second heat exchanger—condenses the vapor back into a liquid, which is then recycled through the system.
Open-cycle	These systems use the tropical oceans' warm surface water to make electricity. When warm seawater is placed in a low-pressure container, it boils. The expanding steam drives a low-pressure turbine attached to an electrical generator. The steam, which has left its salt behind in the low-pressure container, is almost pure fresh water. It is condensed back into a liquid by exposure to cold temperatures from deep-ocean water.
Hybrid	These systems combine the features of both the closed-cycle and open-cycle systems. In a hybrid system, warm seawater enters a vacuum chamber where it is flash-evaporated into steam, similar to the open-cycle evaporation process. The steam vaporizes a low-boiling-point fluid (in a closed-cycle loop) that drives a turbine to produce electricity.

Marine and Hydrokinetic Technology Glossary, U.S. Dep't of Energy, http://www1.eere.energy.gov/windandhydro/hydrokinetic/
techTutorial.aspx (last visited Aug. 5, 2011). http://www1.eere.energy.gov/windandhydro/hydrokinetic/techTutorial.aspx
For a database with the latest wind and wave technologies and status, follow this link: http://www1.eere.energy.gov/
windandhydro/hydrokinetic/listings.aspx?type=Tech

to issue licenses for the construction and operation of hydrokinetic projects, including wave and current, but companies will be required to first obtain a lease through MMS." Press Release, MMS, President Obama, Secretary Salazar Announce Framework for Renewable Energy Development on the U.S. Outer Continental Shelf (April 22, 2009), http://www.boemre.gov/ooc/press/2009/press0422.htm.

Remaining overlaps between FERC and BOEMRE's hydrokinetic-project jurisdiction may still be causing confusion, but they have not prevented some hydrokinetic projects from moving forward. In a May 23, 2011 press release, BOEMRE announced that it was preparing an Environmental Assessment for four proposed priority areas for testing ocean current technology and for collecting resource data off the shores of Florida. The comment period for the proposed lease of these areas to Florida Atlantic University closed on June 23, 2011.

Notes and Questions

1. FERC and BOEMRE formalized their Memorandum of Understanding (MOU) in April of 2009. Under the MOU, an applicant must first receive a lease, easement, or right-of-way from BOEMRE, and only then may the applicant receive a license to generate electric power, or an exemption from the licensing requirement, from FERC. The BOEMRE licensing process can take one to two years for noncompetitive leases and two to five years for competitive leases. The FERC licensing process can take another one to two years. If

Figure 8.7: Issued hydrokinetic preliminary permits

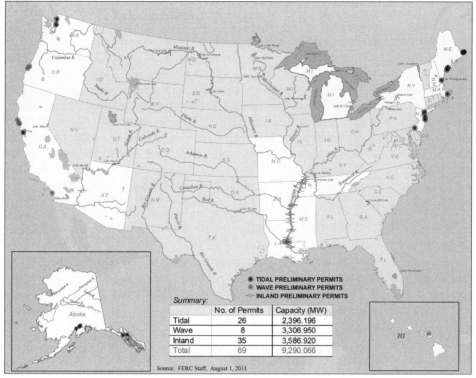

Fed. Energy Regulatory Comm'n, http://www.ferc.gov/industries/
hydropower/indus-act/hydrokinetics/issued-hydrokinetic-
permits-map.pdf (last visited Aug. 5, 2011).

the FERC process cannot be started until after the applicant has received the BOEMRE lease, the lead time for a project can stretch out to seven years or more. Do you think this might discourage investment?

2. FERC is liberal in granting preliminary permits. In only six months to a year, a developer may receive a free permit to study a project at a specific site for up to three years. Also, the FERC process guarantees a site to the first-to-file. In contrast, BOEMRE conducts an expensive competitive leasing process and currently has the power to lease land under a FERC preliminary permit to a different developer during the study period. How can these approaches be reconciled?

3. To read more about the 2009 MOU and the history of the dispute between BOEMRE and FERC, *see* Peter F. Chapman, *Offshore Renewable Energy Regulation: FERC and MMS Jurisdictional Dispute Over Hydrokinetic Regulation Resolved?*, 61 ADMIN. L. REV. 423, 425-38 (Spring 2009).

4. Do you have any suggestions for how the hydrokinetic development process might be streamlined? *See, e.g.*, Danielle Murray, *Dual Federal Regulations Slowing Public Wave Power Project Development*, 41 No. 6 ABA TRENDS 1 (July/August 2010).

C. Renewable Energy Development and NEPA

Although the National Environmental Policy Act of 1969 (NEPA) is often referenced as just a procedural statute, it has proved a powerful tool for curtailing development on federal lands because of the time and expense involved in preparing detailed environmental review documents such as Environmental Impact Statements (EISs). Large-scale renewable projects have impacts comparable to any other large-scale industrial project, and they should not receive an automatic pass on environmental requirements. However, renewable energy development promises environmental and economic benefits to the public, and the fact that renewable development is coming at a time of heightened environmental concern and public involvement has put renewable resource development at a disadvantage when compared to the development of minerals or fossil fuels on public lands. The following three excerpts explore the role of NEPA in renewable energy's future.

1. Programmatic Environmental Impact Statements for Terrestrial Wind and Solar

Pressure to develop renewable resources on public lands began before the 2005 EPAct in which Congress declared this to be a priority. Without Congressional guidance, federal agencies attempted to move quickly in efforts to address the demand. The Bureau of Land Management's response was to prepare a series of broad programmatic-level EISs to facilitate renewable energy project-specific NEPA review.

David J. Lazerwitz, *Renewable Energy Development on the Federal Public Lands: Catching Up with the New Land Rush*
55 ROCKY MT. MIN. L. INST. 13-1 (2009)

(3) BLM's Solar and Wind Energy Development Programs

In an effort to address the rising wave of solar and wind project applications and establish comprehensive program guidance for the development of solar and wind projects on the federal public lands, BLM initiated and, in the case of wind energy, adopted an energy development program and accompanying PEIS under NEPA. The Solar Energy Development PEIS (Solar PEIS), being jointly prepared with the DOE, was originally slated for issuance in draft in spring 2009 but at the time of this writing was anticipated in fall 2009. This process is particularly instructive in understanding both BLM's objectives and future directions for wind energy—and possibly solar—development on the federal public lands.

(a) BLM's Wind Energy Development Program and PEIS

In 2003, BLM embarked upon a process to develop a wind energy development program and, in conjunction with that process, prepared a PEIS pursuant to NEPA and a Programmatic Biological Assessment (PBA) pursuant to the ESA. The objectives of the program and PEIS were two-fold:

(1) to assess the environmental, social and economic impacts associated with wind energy development on BLM-administered land, and (2) evaluate a number of alternatives to address the question of whether the proposed action presents the best management approach for the BLM to adopt, in terms of mitigating potential impacts and facilitating wind energy development.

The geographic scope of the analysis covered all BLM-administered lands in 11 western states, excluding Alaska.

The Wind PEIS identified and addressed three potential program alternatives. These encompassed: (1) implementation of a wind energy development program on all BLM lands on which wind project development may be technically and economically viable under BLM's maximum potential development scenario (MPDS); (2) a more limited development program based on project development only in areas where projects currently exist or are in process of approval; and (3) a no-action alternative focused on the preexisting case-by-case project analysis. In its Record of Decision (ROD), BLM selected its MPDS alternative. From the 174.7 million acres of land BLM manages in the 11 western states analyzed, BLM's model predicted that 160,100 acres of land could be developed over the next 20 years, and BLM based its environmental impacts analysis and ultimate decision on this assumption. To put this in context, at the time BLM approved the ROD, there were three operating wind energy projects totaling 21,161 acres on BLM lands.

A critical component of BLM's PEIS and program adoption is the development and incorporation of specific programmatic policies and BMPs. BLM's programmatic policies provide, among other things, that the agency will not issue right-of-way grants for development in areas that are part of the National Landscape Conservation System (e.g., Wilderness Areas) and Areas of Critical Environmental Concern (ACEC), as well as areas where resource impacts cannot be mitigated or will conflict with existing or planned multiple use activities or land use plans. These policies further provide that, to the extent possible, wind energy policies shall be developed in a manner that will not prevent other land uses, such as mineral extraction, livestock grazing, and recreational use. For purposes of individual project NEPA compliance, BLM envisioned that some projects could proceed by tiered environmental assessment (EA) tied to the PEIS to the extent the PEIS addressed anticipated issues and concerns—an action that could streamline NEPA compliance in many instances but which remains far from certain in practice.

The BMPs, incorporated into BLM's 2008 Wind Policy, are applicable to all wind energy activities on BLM-administered public lands. As the agency explained, the BMPs "establish environmentally sound and economically feasible mechanisms to protect and enhance natural and cultural resources [and] identify the issues and concerns that need to be addressed by project-specific plans." BMP examples include use of existing roads where possible; monitoring environmental and species conditions during construction, operation, and decommissioning; configuration of equipment (such as wind turbines) to avoid landscape features known to attract raptors; integration of project features into the surrounding landscape; and development of storm water management plans.

In conjunction with issuing its ROD on the Wind PEIS, BLM amended 52 land use plans to incorporate the programmatic policies and BMPs. In some instances, these amendments also incorporated the identification of specific areas where wind energy development would be excluded, but these amendments were not comprehensive and did not include the designation of the specific areas approved for wind energy development. For those land use plans not amended—which include all of the plans for Arizona and California—BLM explained that those areas would be addressed in conjunction with ongoing or upcoming land use plan amendments.

(b) BLM's Solar Energy Development Program and PEIS

Confronted with a later and more significant surge in solar right-of-way applications, in May 2008 BLM and DOE announced the initiation of a process to develop a solar energy development program for utility-scale solar projects and conduct an accompanying

PEIS. The purpose of the proposed solar program closely tracks that of the wind energy development program. The program and PEIS aim to determine whether the agencies should develop and implement agency-specific programs that would establish environmental policies and mitigation strategies (e.g., BMPs) for solar development on BLM-administered land in six western states: Arizona, California, Colorado, New Mexico, Nevada, and Utah. DOE is providing technical support for BLM's analysis and independently evaluating the development of its own program of environmental policies and mitigation strategies to apply to projects supported by DOE on federal, state, and private lands.

Much like the wind energy development program, the agencies' proposed action is intended to identify BLM-administered land in the six-state study area upon which solar project development is likely to occur over the next 20 years, through a "reasonably foreseeable development" model. This analysis includes identifying those lands that may be environmentally suitable for solar energy development and, conversely, those areas to be excluded from such development. As a result, the proposed Solar PEIS scope excludes from consideration lands that BLM has previously identified as "environmental sensitive," including ACECs. BLM envisions amending the applicable land use plans to identify these areas and incorporate its environmental policies and mitigation strategies. Further, like the Wind PEIS process, the agencies anticipate that the Solar PEIS will facilitate, but not replace, project-specific environmental analysis through tiering to the PEIS.

Unlike the Wind PEIS, however, the agencies are considering whether the designation by BLM of additional electricity transmission corridors on BLM-administered lands is necessary to facilitate utility-scale solar energy development—a critical issue for project development that implicates transmission corridor studies occurring at both the regional and state level. In fact, on April 27, 2009, the agencies announced a postponement for issuance of the draft PEIS until fall 2009 to, among other things, await preliminary results of the Western Governors' Association's Western Renewable Energy Zone transmission study. On July 27, 2009, BLM and DOE extended the public comment period until Sept. 14, 2009.

On June 30, 2009, during preparation of the draft PEIS, BLM and DOE announced the location of 24 "solar energy study areas" on BLM-administered lands that the agencies would consider designating Solar Energy Zones (SEZs) as part of the Solar PEIS process. The agencies describe SEZs as "specific locations determined best suited for large-scale production of solar energy," but it is not yet clear whether or how BLM would administer SEZs for solar development, which could include competitive or noncompetitive procedures. The solar energy study areas each encompass at least 2,000 acres of land, are situated near access roads and transmission routes, have slopes of less than 5%, and exclude environmentally sensitive areas. In total, these areas encompass approximately 676,000 acres of BLM lands.

Interestingly, of the 220-plus solar project applications pending on BLM land, only 35 are for parcels situated within the solar energy study areas. Thus, the solar energy study areas being evaluated in the PEIS process—which could become SEZs—do not directly address the vast majority of pending solar project applications, many of which reflect executed PPAs with power purchasers, pending transmission interconnection requests, and the investment of significant time and resources in project-related studies. Despite the announcement of the solar energy study areas, BLM will continue to process solar applications filed prior to June 30, 2009, both within and outside of these study areas, and to accept new solar applications on lands outside of the study areas. These applications will be subject to BLM's existing application procedures, but to the extent a project is not approved until after issuance of the PEIS ROD, that project may be subject to mitigation requirements in the ROD. Applications filed after June 30, 2009, on lands within the solar energy study areas will, however, not be processed until after the ROD has been issued for the Solar PEIS.

§ 13.04 Towards a Comprehensive Regulatory Solution

The renewable energy land rush on the federal public lands is well under way and placing unprecedented pressures and demands on our existing public land management system. These challenges will continue to mount as new solar and wind project applications are filed, existing applications are processed, and public land managers struggle to balance mounting pressure to approve projects with multiple use and environmental responsibilities. Further complicating this balance will be the likely wave of litigation, both from public interest organizations opposing approved projects and project applicants challenging project denials and the imposition of terms and conditions. The important question that needs to be asked is whether we are doing enough to create a regulatory structure to address the multitude of issues presented by utility-scale solar and wind project development.

* * *

(2) The Unresolved Question of NEPA Compliance

Perhaps the most significant near-term issue facing solar and wind project development on the public lands is the uncertainty surrounding the NEPA process and timing. BLM's project application pipeline is already substantial, and that pipeline continues to grow with few large-scale projects moving into the "Notice of Intent" stage. Moreover, while BLM's programmatic processes and policies seek to streamline project-specific NEPA compliance through tiered EAs, it is not clear that this goal is being achieved for wind projects, or that it will be achieved for solar projects that will almost certainly require full Environmental Impact Statement compliance—a process that a recent report found averaged 3.4 years for all federal agencies.

One of the critical considerations is BLM's ability to tier to a PEIS without first amending, and conducting NEPA compliance for, the relevant RMP. This issue is not new as, in the coalbed methane context, BLM conducted resource-specific supplementation of RMPs, a process that may serve as a model for solar and wind project planning. Congress also possesses authority to exempt certain projects or classes of projects from, or otherwise streamline, the NEPA process. Although not specifically tied to renewable energy projects, Congress has required that projects funded under the American Recovery and Reinvestment Act of 2009 be completed on "an expeditious basis" and by "the shortest existing applicable process" under NEPA—a process which remains to be defined.

Notes and Questions

1. Which PEIS was initiated first—wind or solar? Why was this so?

2. How do the two PEISs vary? Was transmission a factor in each? Do you think it should be?

3. Things continue to move quickly in this area. In December of 2010, BLM approved several CSP and wind projects on federal lands to meet the ARRA goals. Environmental, financial, and litigation hurdles have prevented most of these projects from being completed, and DOI is working on changes to its EIS process. In a July 14, 2011 press release, the BLM announced the approval of four new projects on public lands and the launch of en-

vironmental reviews on three others. In addition, the press release stated that the DOI, in cooperation with the DOE, would prepare a targeted supplement to the Draft Programmatic Environmental Impact Statement for Solar Energy Development (Draft Solar PEIS) which was first released for public review in December of 2010 and which promoted the development of "solar energy zones" on public lands in six western states. Do these latest developments address some of Lazerwitz's concerns about NEPA issues facing solar and wind project development on public lands?

2. Leveling the Playing Field

The following excerpt explains some of the urgency for streamlining NEPA for renewable energy development. It also describes some of the expedited routes to NEPA compliance that Congress has provided as alternatives for nuclear power plants and fossil fuel production. Renewables enjoy no similar alternatives and still must jump through all of the NEPA hoops. According to a study by EMPS, Inc, the average time between a notice of intent and a final EIS in 2005-2006 was in the range of 500 to 2,000 days.

Irma S. Russell, *Streamlining NEPA to Combat Global Climate Change: Heresy or Necessity?*
39 ENVTL. L. 1049, 1056-63 (2009).

I. Introduction

* * *

The National Environmental Policy Act (NEPA) requires federal agencies to consider the environmental impacts of major projects they undertake. It added to each agency's mission the additional requirement of considering the effects on the environment of federal projects. To achieve its goal, NEPA mandates that "all agencies of the Federal Government ... utilize a systematic, interdisciplinary approach which will ensure the integrated use of the natural and social sciences and the environmental design arts in planning and in decisionmaking which may have an impact on man's environment." NEPA's policy seeks to foster conditions "under which man and nature can exist in productive harmony, and fulfill the social, economic, and other requirements of present and future generations of Americans." NEPA has made significant changes in the way federal agencies go about achieving their missions. Fulfilling the procedural requirements of NEPA takes time and money.

NEPA results in delays in virtually all major energy projects. It applies to projects requiring federal permits because permitting requirements make energy projects federal agency actions under NEPA. Thus, NEPA applies to traditional energy projects such as coal-fired utilities and, additionally, to energy projects aimed at supplying energy without the [greenhouse gases] (GHGs) associated with combustion, such as concentrated solar installations, wind farms, and wave technology....

* * *

III. The Significance of NEPA to Global Climate Change

The National Environmental Policy Act (NEPA) was passed by Congress in 1969 at the beginning of the most active legislative period for environmental protection. It has been called the "grandfather" of U.S. environmental law because it was the first major congressional act to insert environmental considerations into federal decision making.

Although the title of the Act is the National Environmental Policy Act, many people think of NEPA as the National Environmental Protection Act—an act for the protection of the environment. This is because, before NEPA, the charge for federal agencies did not include consideration of the effects on the environment of federal projects, and the effects of many projects undertaken or approved by federal agencies is enormous. Moreover, the Act's goal of considering environmental impacts has an implicit purpose of protecting the environment. NEPA states that:

> [I]t is the continuing policy of the Federal Government, in cooperation with State and local governments, and other concerned public and private organizations, to use all practicable means and measures, including financial and technical assistance, in a manner calculated to foster and promote the general welfare, to create and maintain conditions under which man and nature can exist in productive harmony, and fulfill the social, economic, and other requirements of present and future generations of Americans.

Additionally, NEPA "recognizes that each person should enjoy a healthful environment and that each person has a responsibility to contribute to the preservation and enhancement of the environment." The primary focus of litigation under NEPA is its requirement that federal agencies consider the environmental consequences of their projects. NEPA mandates that "all agencies of the Federal Government … utilize a systematic, interdisciplinary approach which will insure the integrated use of the natural and social sciences and the environmental design arts in planning and in decisionmaking which may have an impact on man's environment." Although courts have held that NEPA is primarily procedural and does not require that agencies choose the least environmentally harmful course of action, many people continue to see the Act as a protection of the environment from adverse federal actions. Among other things, NEPA requires that agencies prepare an impact statement for major federal actions significantly affecting the quality of the human environment. Hundreds of such federal actions occur each year in the form of permits or authorizations by federal agencies. Section 2 of the Act states that agencies must do the following:

> [I]nclude in every recommendation or report on proposals for legislation and other major Federal actions significantly affecting the quality of the human environment, a detailed statement by the responsible official on—
>
> (i) the environmental impact of the proposed action,
>
> (ii) any adverse environmental effects which cannot be avoided should the proposal be implemented,
>
> (iii) alternatives to the proposed action,
>
> (iv) the relationship between local short-term uses of man's environment and the maintenance and enhancement of long-term productivity, and
>
> (v) any irreversible and irretrievable commitments of resources which would be involved in the proposed action should it be implemented.

In addition to the statement required by section 2, now referred to as an environmental impact statement (EIS), NEPA imposes consultation requirements on federal agencies to contact and confer with other agencies having "special expertise with respect to any environmental impact" of a project and requires that agencies provide copies of the EIS and consider the input of federal, state, and local agencies. The Act also created the Council on Environmental Quality (CEQ), a new federal agency charged with overseeing the implementation of the Act and reporting to the President on the state of the environment and implementation of NEPA. The Council on Environmental Quality promulgated regulations to implement NEPA. The CEQ defines "major federal action" as "actions with effects that may be major and which are potentially subject to Federal control and responsibility." These include projects of different scopes and impact, ranging from general programs such as regional plans for forest management to specific construction projects of all kinds, and from roads to mineral sales and exploration for energy resources on public lands.

Clearly NEPA applies to alternative energy projects as well as traditional energy projects. The time-consuming processes of NEPA increase the costs of green projects. "Existing land use plans and planning efforts may be amended as necessary, with appropriate level of NEPA analysis and decision, to address this change in wind energy and [Areas of Critical Environmental Concern] policy...." The Bureau of Land Management (BLM) policy for the management of energy and minerals on public lands (part of multiple use mandate) states that "energy and mineral-related permit applications will be reviewed consistent with the requirements of NEPA and other environmental laws." The BLM Instruction Memorandum provides "guidance for the processing of right-of-way applications for wind energy projects on public land administered by the BLM." When an evaluation indicates that constructing a meteorological tower on adjacent nonfederal land could provide the ability to characterize wind patterns on public lands, the regulations require that a NEPA document be prepared "describing the Federal action as the issuance of a right-of-way grant with limited activities on the public land."

IV. Examples of Current Streamlining of NEPA in Energy Law

The fact that no NEPA streamlining applies to green energy such as wind installations should not lead us to assume that NEPA has retained its full force in relation to energy production generally. Many agencies have developed regulations that streamline the NEPA process, truncating or curtailing the application of NEPA. For example, the Oil Shale, Tar Sands, and Other Strategic Unconventional Fuels Act of 2005 (OSTSOSUFA) declares that "it is the policy of the United States that ... shale, tar sands, and other unconventional fuels are strategically important domestic resources that should be developed to reduce the growing dependence of the United States on politically and economically unstable sources of foreign oil imports." The declaration portion of the Act also states that the "development of oil shale, tar sands, and other strategic unconventional fuels, for research and commercial development, should be conducted in an environmentally sound manner, using practices that minimize impacts." Similarly, OSTSOSUFA notes the need for the "development of those strategic unconventional fuels" with an "emphasis on sustainability."

To implement its purposes, the Act empowers the Secretary of the Interior to make land available for leasing as necessary "to conduct research and development activities with respect to technologies for the recovery of liquid fuels from oil shale and tar sands resources on public lands." The Act applies to public lands within Colorado, Utah, and Wyoming. It instructs the Secretary of the Interior to prepare a programmatic EIS and establish a commercial leasing program for oil shale and tar sands in an expeditious manner—"not later than 18 months after August 8, 2005." The Act also instructs the Secretary of the Interior to issue final regulations on a fast time frame (within six months after completion of the programmatic EIS). In November of 2008, the BLM issued its final regulations opening state lands for commercial development of oil shale. The regulations allow for a single lease holder to develop up to 50,000 acres of public lands and 50,000 acres of acquired lands in each of Colorado, Utah, and Wyoming. This allows a leaseholder to develop up to 300,000 acres of land throughout the three states. Federal lands in the affected states are subject to leasing, with the exception of national parks, land within cities, and lands specifically excluded by either the Oil Shale Act or by BLM.

Before offering a lease, BLM must prepare the normal analysis mandated by NEPA. If the regulations required NEPA analysis on each tract of land, they would substantially maintain the NEPA process. The possibility of significant streamlining of NEPA exists, however, because the regulations provide for a cumulative review of tracts of land by means of a land use planning action by BLM. Such cumulative action can virtually dispense with the NEPA inquiry on significant portions of public lands and acquired lands. After an agency develops an EIS, NEPA normally requires a comment period before the agency may act on its proposal. This period is ninety days for a draft EIS, or thirty days for a final EIS. However, an agency may adopt a draft or final EIS in lieu of preparing a new one. As long as "the actions covered by the original environmental impact statement and the proposed action are substantially the same," the adopting agency can merely recirculate the EIS as a final statement. Even in cases that do not meet the criteria of "substantially the same" action, the adopting agency can "treat the statement as a draft and recirculate it." Together these regulations allow an agency to complete an EIS analysis on a relatively small section of land (perhaps a thousand acres), and then through adoption apply the results to tens or hundreds of thousands of additional acres. While agency regulations often contemplate the creation of a programmatic EIS, the possibility of an individual or project EIS under the umbrella of the programmatic EIS ensures full consideration of environmental values. Adjustment of the steps required by NEPA process in a way that dispenses with the project level analysis creates a real risk that decision makers ignore environmental values at a crucial stage of the process. When significant streamlining of the NEPA process occurs, the likely result is a reduction or loss of public input and scientific analysis relating to the affected lands.

* * *

The most dramatic streamlining of NEPA is found in the concept of categorical exclusions from the Act. The CEQ defines a categorical exclusion as a "category of actions which do not individually or cumulatively have a significant effect on the human environment and which have been found to have no such effect in procedures adopted by a Federal agency in implementation [of NEPA regulations]." The regulation also specifies the effect of finding that an action is within the definition of categorical exclusion. The regulation indicates that "therefore, neither an environmental assessment nor an environmental impact statement is required." Finally, the regulation also makes clear that finding an action falls within a categorical exclusion does not prohibit an agency from conducting an environmental assessment. The regulation states the following:

> An agency may decide in its procedures or otherwise, to prepare environmental assessments for the reasons stated in § 1508.9 even though it is not required to do so. Any procedures under this section shall provide for extraordinary circumstances in which a normally excluded action may have a significant environmental effect.

The use of categorical exclusions seems to undercut the original purposes of NEPA and provide a dramatic softening of the NEPA requirements, and categorical exclusions have been the subject of significant criticism. There is no doubt that the mandates of NEPA give way to a clear congressional mandate to restrict the NEPA process. Moreover, the mechanism of categorical exclusions was established early in the process of implementing NEPA through regulations.

Some energy sources have already received categorical exclusions under NEPA. For example, legislation charges both the Secretary of the Interior and the Secretary of Agriculture with managing public lands and National Forest System Lands by applying "a rebuttable presumption that the use of a categorical exclusion" under NEPA would apply "if the activity is conducted pursuant to the Mineral Leasing Act ... for the purpose of exploration or development of oil or gas." The activities classified as falling within the rebuttable presumption include the following:

(1) Individual surface disturbances of less than 5 acres so long as the total surface disturbance on the lease is not greater than 150 acres and site-specific analysis in a document prepared pursuant to NEPA has been previously completed.

(2) Drilling an oil or gas well at a location or well pad site at which drilling has occurred previously within 5 years prior to the date of spudding the well.

(3) Drilling an oil or gas well within a developed field for which an approved land use plan or any environmental document prepared pursuant to NEPA analyzed such drilling as a reasonably foreseeable activity, so long as such plan or document was approved within 5 years prior to the date of spudding the well.

(4) Placement of a pipeline in an approved right-of-way corridor, so long as the corridor was approved within 5 years prior to the date of placement of the pipeline.

(5) Maintenance of a minor activity, other than any construction or major renovation o[f] a building or facility.

The Nuclear Regulatory Commission has created numerous categorical exclusions by its regulations. Licensing nuclear power plants, and facilitating the operation of the plants, is aided by such exclusions. For example, NRC regulations indicate specifically that "[e]xcept in special circumstances ... an environmental assessment or an environmental impact statement is not required for any action within a category of actions included in the list of categorical exclusions." ...

* * *

The Indian Tribal Energy Development and Self-Determination Act of 2005 (ITEDSDA) (part of the Energy Policy Act of 2005) provides another example of an off-ramp from the standard NEPA process. The Bureau of Indian Affairs (BIA) rule authorizes tribes to assume authority for approving and managing leases, business agreements, and rights of way for energy resource development on tribal land. ITEDSDA makes NEPA inapplicable to energy agreements between tribal authorities and developers when a project meets the requirements of ITEDSDA. Under the terms of ITEDSDA, the Secretary of the Interior must apply the NEPA analysis when considering approval of a tribal resource energy agreement (TERA). Once the TERA has been approved, however, the tribe no longer needs Department of the Interior (DOI) approval for specific energy agreements entered pursuant to a TERA. The result of the removal of DOI approval is that NEPA no longer applies to the issuance of the specific energy agreements. This effect is clear from the statement in the regulations that the "scope of the Secretary's evaluation will be limited to the scope of the TERA." A TERA application can be fairly broad. Tribal authorities may even acquire control over activities normally administered by DOI by specifying the type of energy resource in the TERA application.

The environmental criteria for TERA approval include identification and evaluation of "all significant environmental effects," identification of "proposed mitigation measures," a process ensuring public input on the environmental effects, proper administrative support and technical capability, and tribal oversight of any third parties related to the

TERA. To the extent that a project meets the requirements of ITEDSDA, environmental considerations will be taken into account. It is not clear, however, that there is no effect of outsourcing environmental considerations from NEPA to ITEDSDA. One clear effect of this shift from NEPA to ITEDSDA is that the project no longer meets the category of a "federal action" of NEPA. Accordingly, judicial oversight provided for federal projects no longer applies. ITEDSDA also requires a quick response from the Secretary of the Interior in evaluating TERAs, and provides criteria under which the Secretary must approve the application. The Secretary is required to approve or deny TERA applications "[n]ot later than 270 days after the date on which the Secretary receives a tribal energy resource agreement." Even if the TERA is denied, the Secretary must, within ten days, notify the tribe of why the TERA was disapproved, identify what changes are required, and allow the tribe to resubmit the TERA. If the tribe does this, the Secretary must approve or deny the revised TERA within sixty days.

Achieving involvement of disadvantaged groups in developing policies is central to the purposes of NEPA as well as to a democratic approach. Participatory justice is an important value of environmental law and of environmental justice. It can play a major role in achieving a just regime of energy regulation. While the democratic values implicated in citizen participation in NEPA are of great significance, the climate change debate must not allow the process to produce calcification of the status quo of energy production or doom the move toward truly inexhaustible energy, which is required to further economic development while reducing the threat of global climate change. Clearly it will be difficult for tribal governments to pursue TERA effectively without technical and financial assistance from Congress. A failure of Congress to support the goal of tribal self-determination under the Act could result in support and control by private power companies and other entities, stripping the energy projects on tribal land of the NEPA process without effectively substituting the process envisioned by the ITEDSDA.

The government's action of incentivizing fossil fuel as a strategic fuel while devoting significant resources to green versions of energy seems to attempt to realize two possibly contradictory goals. NEPA presents a significant barrier to development of green energy, particularly in light of the streamlined process already in place for some fossil fuels and nuclear energy. The foregoing are merely examples of some of the ways that agencies streamline the NEPA process. The decision to streamline NEPA suggests that in these areas the purposes of NEPA give way to other needs. Whether Congress will make the move toward a green grid is unknown at this time. Given the shortcuts available in other energy sectors such as nuclear power and fossil fuel, green energy does not yet have a level playing field in the NEPA process.

Notes and Questions

1. What does the NEPA process require? Are renewable energy projects exempt from these steps? Should they be?

2. Define each of the following: EIS, PEIS, EA, and categorical exclusion.

3. Russell describes several examples of current NEPA streamlining in the context of energy law. Explain what NEPA streamlining the following uses receive: oil shale and tar sands development, oil and gas drilling in developed fields, nuclear power plant licensing.

3. Balancing of Equities

The following is an excerpt from a recent challenge to a large-scale wind farm in Nevada. Notice how the district court judge attempts to balance the equities of renewable energy development with NEPA requirements.

Western Watersheds Project v. Bureau of Land Management

No. 3:11–cv–00053–HDM–VPC, 2011 WL 1195803 (D. Nev. Mar. 28, 2011),
aff'd, No. 11–15799, 2011 WL 2784155 (9th Cir. July 15, 2011)

McKibben, J.

This case concerns approval of a wind energy facility in Spring Valley, Nevada. Plaintiffs are two environmental organizations—Western Watersheds Project and Center for Biological Diversity. Defendant is the Bureau of Land Management (BLM). Intervening defendant is Spring Valley Wind, LLC, the energy company developing the wind facility at issue.

Plaintiffs filed a motion for a temporary restraining order and/or preliminary injunction pursuant to Federal Rule of Civil Procedure 65 seeking to bar the BLM from issuing a Notice to Proceed or otherwise authorizing construction and site clearing for the Spring Valley Wind Energy Facility set to commence on March 28, 2011.

I. Factual Background

The Spring Valley Wind Energy Facility project is an industrial scale alternative energy project to be constructed in and around Spring Valley in east-central Nevada near Great Basin National Park. Approximately 430 acres is the total area estimated for use for the project (including short-term and long-term disturbance). This is approximately 5.6 percent of the total right of way. The project would advance United States' goal of providing renewable energy generation options to Nevada. It would generate enough energy to power 45,000 Nevada homes, up to $3 million in tax benefits to local school districts, and provide 225 jobs during the construction phase. The overall expected economic benefit for Nevada from the project is $45 million. Approval of the project makes it eligible for millions of dollars of federal financing under the American Recovery and Reinvestment Act, which requires that qualifying projects commence construction no later than September 30, 2011.

The project area is not untouched. The existing landscape has been modified through past and current human habitation, road development, ranching and mining activities, and transmission lines. Project construction would incorporate existing structures and include over 25 miles of new roads, between 66 and 75 lighted 400–foot tall wind turbines, two gravel pits, over nine miles of new fencing, a microwave tower, electrical lines, switchyard, and other facilities.

Project site clearing and construction is scheduled to begin the week of March 28, 2011. Erection of the wind turbines is scheduled for March 2012. The Spring Valley Wind Facility is expected to be commercially operational by June 2012.

Site clearing and construction for the project is set to begin March 28, 2011. This would impact native vegetation and wildlife, including the greater sage-grouse. There are 38 sage-grouse leks (mating grounds) in Spring Valley, three within a mile of the project site, but none in the project area. The project site itself is in low quality sagebrush habitat, the highest-quality habitat is located outside the project area, and the area already contains existing roads and transmission lines. In addition, to offset potential impacts, Spring Valley

Wind committed $500,000 (eligible for federally matched funding) to enhancing sagebrush habitat in the area.

The operation of the turbines beginning in 2012 would also impact local bat populations. The public land designated for the project is near a large seasonal bat cave in the Great Basin, the Rose Guano Cave. The Rose Guano Cave is located four miles from the Spring Valley Wind project site and is a seasonal roost site to over one million Brazilian free-tailed bats during their fall migration in August and September. The bats' migratory path takes them near the Spring Valley Wind Project site. The bats also travel up to 50 miles one-way at night to forage for insects, and may consume their body weight nightly.

Bats are vulnerable to mortality from operational wind turbines because wind turbines attract insects that the bats feed on and are perceived by the bats as potential migratory rest-stops or roosting sites. Bats are killed by contact with moving turbine blades and by "barotrauma." Barotrauma is a phenomenon that occurs when air pressure changes near spinning turbine blades. The change in air pressure causes the bats lungs to suddenly expand, bursting blood vessels. Ninety percent of bat fatalities near wind turbines may be attributed to barotrauma.

II. Procedural Background

In June 2009, the federal government announced plans for the BLM to "fast track" the approval process for renewable energy projects across the United States. "The fast track process is about focusing [BLM] staff and resources on the most promising renewable energy projects." (BLM Opp'n Ex. A) The Spring Valley Wind Facility was approved for a "fast track."

In December 2009 and July 2010, the BLM issued preliminary environmental assessments (EAs) for the project. The preliminary EAs concluded that the project would pose no significant environmental impacts.

In response to these documents, the BLM received over 67 public comment letters, containing almost 1,000 comments. Plaintiffs were among those who submitted written comments and met with the BLM over their concerns with the preliminary EAs. Several agencies and organizations, including the U.S. Fish and Wildlife Service, Nevada Department of Wildlife, National Parks Service, and Southern Nevada Water Authority, were also initially concerned about the preliminary EAs.

On October 15, 2010, the BLM approved the project through a Decision Record and Finding of No Significant Impact (FONSI) and issued a Final Environmental Assessment which addressed comments and concerns. As a result, it did not complete an environmental impact statement (EIS).

The final EA tiers to the BLM's 2005 Final Programmatic EIS on Wind Energy Development on BLM Administered Lands in the Western United States (Wind PEIS), a document that evaluates the consequences of wind energy development across BLM lands, and the 2007 Ely Resource Management Plan's Final EIS. The final EA also relies on a detailed Avian and Bat Protection Plan (ABPP) to mitigate project impacts on bats and birds. The ABPP mitigation measures include: (1) creation and utilization of a Technical Advisory Committee (TAC) to monitor bat and bird mortality and ensure the implementation of mitigation measures should the mortality rates reach BLM designated thresholds; (2) a radar detection system to monitor flight and migratory habits and potentially trigger turbine breaks and feathering during periods of high flight activity; (3) wind turbine operation curtailment and shut downs; and (4) a mitigation fund. The mitigation measures do not include the recommendation of orienting wind turbines parallel to bat and bird flight patterns because doing so would render the turbines useless based on area wind flow.

On October 22, 2010, the BLM issued two rights-of-way to Spring Valley Wind, LLC. One was for the wind generation facility and substation, and the other was for a switchyard, overhead electrical lines, fiber-optic cable, microwave tower, and associated facilities.

On November 13, 2010, the environmental plaintiffs filed an administrative appeal and petition for stay to the Interior Board of Land Appeals (IBLA). On January 11, 2011, those plaintiffs filed a notice of dismissal of their appeal.

On January 25, 2011, plaintiffs filed a complaint with this court alleging the BLM violated the National Environmental Policy Act (NEPA), 42 U.S.C. § 4321 *et seq.* On February 28, 2011, plaintiffs filed a motion for temporary restraining order and/or preliminary injunction, seeking to enjoin site clearing and construction of the project. Defendants BLM and Spring Valley Wind, LLC opposed the motion on March 15, 2011 in separate responses.

* * *

2. NEPA Requirements for an Environmental Impact Statement

NEPA requires federal agencies, like the BLM, to prepare an environmental impact statement EIS for all "major Federal actions significantly affecting the quality of the human environment." 42 U.S.C. § 4332(2)(c)(2006). This is to ensure that the agency "will have available, and will carefully consider, detailed information concerning significant environmental impacts; it also guarantees that the relevant information will be made available to the larger [public] audience." (citations omitted).

The requirement to prepare an EIS is triggered when a proposed project will "significantly affect" the environment. 42 U.S.C. § 4332(2)(C). An agency may prepare an EA "to decide whether the environmental impact of a proposed action is significant enough to warrant preparation of an EIS..An EA is a 'concise public document that briefly provide[s] sufficient evidence and analysis for determining whether to prepare an EIS or a finding of no significant impact' (FONSI)." (citations omitted). EAs may "tier" to other NEPA documents, but tiering does not eliminate the EIS requirement when a proposed project significantly affects the environment. 40 C.F.R. §§ 1502.20, 1508.28. If an agency decides not to prepare an EIS, it must provide a detailed statement of reasons explaining why the proposed project's impacts are insignificant. (citations omitted).

"An EIS must be prepared if 'substantial questions are raised as to whether a project ... may cause significant degradation of some human environmental factor.' " *Id.* (internal citations omitted). Plaintiffs need not show that significant effects *will* occur, it is enough to raise "substantial questions" whether a project *may* have a significant effect on the environment. *Id.* To determine if a project may have "significant" impacts, an agency must evaluate ten NEPA factors. 40 C.F.R. § 1508.27(b). The factors at issue in this case are: effects that are "highly uncertain or involve unique or unknown risks" or are "likely to be highly controversial"; "[u]nique characteristics of the geographic area such as proximity to historic or cultural resources, park lands, [ ] wetlands, [ ] or ecologically critical areas"; "[t]he degree to which the action ... may cause loss or destruction of significant scientific, cultural, or historic resources"; and the presence of cumulative impacts. *See* 40 C.F.R. §§ 1508.27(b)(3)-(5), (7)-(8). Just "one of these factors may be sufficient to require preparation of an EIS." (citations omitted).

An agency's decision to forego issuing an EIS may be justified by the adoption of mitigation measures to offset potential environmental impacts. (citations omitted). Further, if "significant measures are taken to 'mitigate the project's effects, they need not completely compensate for adverse environmental impacts.' " (citations omitted). The proposed mitigation measures must be "developed to a reasonable degree." *Id.* Mitigation measures with supporting analytical data are sufficient to support a finding of no significant impact. (citations omitted). "In evaluating the sufficiency of mitigation measures, [the court] consider[s] whether they constitute an adequate buffer against the negative impacts that may result from the authorized activity [, s]pecifically, ... examin[ing] whether the mitigation measures will render such impacts so minor as to not warrant an EIS." (citations omitted).

* * *

(c) EA Properly Tiered to Other Documents

"Tiering, or avoiding detailed discussion by referring to another document containing the required discussion, is expressly permitted" and encouraged under NEPA, so long as the tiered-to document has been subject to NEPA review. 40 C.F.R. § 1502.20. Tiered analyses are viewed as a whole to determine whether they address all the impacts. (citations omitted). A programmatic environmental impact statement (PEIS) may obviate the need for a site-specific impact statement. (citations omitted). However, new and significant issues that develop after an agency issues a PEIS should be evaluated in an EA. *Id.* Only where neither the general nor the site-specific documents address significant issues is environmental review rejected. (citations omitted).

The 2005 Wind PEIS contemplated site-specific tiering when it stated: "The level of environmental analysis to be required under NEPA for individual wind power projects will be determined at the [field office] level. For many projects, it may be determined that a tiered ... [EA] is appropriate in lieu of an EIS." (Wind PEIS A–2–A–8) The Wind PEIS analyzed the potential impacts of wind energy development on public lands; it specifically studied BLM lands in the western United States, and examined mitigation measures to reduce harmful impacts on natural, cultural, and socioeconomic resources.

Any new issues that developed after the Wind PEIS was published were addressed in detail in the final Spring Valley Wind EA. The EA specifically supplements the Wind PEIS with site-specific data on bats and sage-grouse on pages 52–53, 58–63, 96–98, 101–102, 105–111, 151–153, 165, 167. The EA considered barotrauma in bats, bat flight patterns and height, the Fish and Wildlife Service's decision to list sage-grouse as "warranted" for the endangered species list, and 2008–2010 telemetry data concerning active and inactive leks in the project area. (EA 97, 108–109, 58–59)

An EA need not consider all mitigation measures proposed in a PEIS. Measures should be evaluated objectively and on a site-specific basis before being implemented. (Wind PEIS 5–1) The BLM considered the mitigation measures proposed by the Wind PEIS and implemented the ones most suited for the project site. (EA 160–173) The Wind PEIS lists hundreds of potential mitigation measures. (*See e.g.* EA 161–171) It would not be possible to implement all the suggested measures. Notably, when the EA did not adopt a mitigation measure, it explained why. For example, the Wind PEIS suggests orienting turbines to bat and bird flight paths. The BLM considered this mitigation measure and determined it was infeasible at the project site because the turbines could not take advantage of the wind flow through Spring Valley oriented in that position. (EA 164) Tiering the EA to the Wind PEIS was proper.

(d) "Hard Look" and Cumulative Impacts

In determining whether an action requires an EIS, the agency must consider whether the action "is related to other actions with individually insignificant but cumulatively significant impacts." 40 C.F.R. § 1508.27(b)(7). The EA's discussion of cumulative impacts includes a detailed table that discusses past actions, present actions and future actions that may cumulatively impact the environment, including other impacts to the environment such as ranching and grazing and notes that adjustments may need to be made to maintain habitat quality of other species in the area, including utilizing existing fencing and vegetation treatment. (EA 148–151) It also tiers to the Wind PEIS and notes that "direct, indirect and cumulative impacts" are "quantified where possible" in its individual "discussions of impacts on each affected source." (EA 148) Impacts on bats and sage-grouse are addressed in more detail in other sections of the EA, as set forth in the discussions above. (EA 81–122, 96–98, 101–102, 108–110, app. F) By tiering to the Wind PEIS and incorporating new scientific data into its final decision, together with articulations of substantial mitigation measures, the court concludes that the BLM sufficiently considered the cumulative impacts of the project and took a "hard look" as required.

* * *

B. Irreparable Harm

Plaintiffs must show that irreparable injury is likely in the absence of an injunction.... Given the poor quality of sagebrush habitat within the project boundaries, the lack of sage-grouse use of the project area, the BLM's mitigation measures, and Spring Valley Wind's commitment to enhance existing habitat, it is unlikely the sage-grouse population will suffer irreparable harm if the court denies the plaintiffs' request for injunctive relief.

In addition, the initial stages of development of the project pose no threat to the bats. Any risk to the bat population arises from operational wind turbines. The wind turbines will not be operational until April 1, 2012. (SVW Opp'n Inlow Decl. ¶ 16) There is no risk of irreparable harm to the bats *before* a decision on the merits of this case is determined. (citations omitted).

Also, for the reasons set forth in detail above, the risks to the bats from operational wind turbines should be insignificant as well. As the studies considered and conducted by the BLM indicate, the Rose Guano Cave is a seasonal migratory stop-over for a large population of free-tailed bats, but the bats only use the cave for a limited period of time. (EA 61) Further, their foraging and migratory patterns tend to take them parallel to and away from the project site at high altitudes. *Id.* These habits when combined with the extensive mitigation measures proposed by the BLM — including but not limited to, radar detection and monitoring to break and feather turbine activity, phased turbine curtailment and shutdowns, a $500,000 wildlife fund, and a TAC to regularly monitor project impacts so that the project will not exceed the reasonable bat mortality threshold of 192 bats per year set forth in the EA — it is unlikely the bats will suffer irreparable harm. (EA 96, 98; EA app. F, at 15–31)

Accordingly, the court concludes that a denial of a preliminary injunction at this stage in the proceedings will not result in irreparable harm to either the sage-grouse or the free-tailed bats.

C. Balance of Equities

Delaying this project would harm federal renewable energy goals. The United States government has ordered developing renewable, alternate energy sources to reduce the country's dependence on foreign oil and address concerns over climate change. (BLM Opp'n Ex. E (Interior Orders 3285, 3289)) The Energy Policy Act of 2005 directs the Secretary of the Interior to approve renewable energy projects. Executive Order 13212 requires federal agencies to expedite renewable energy projects. (BLM Opp'n Ex. G)

The project is beneficial to Nevada's economic recovery. The project will generate enough energy to power 49,000 Nevada homes. (EA 5) Its property taxes will create over $1.65 million in tax revenue for the state. (EA 144) It will create 225 construction jobs, with employment preferences to Nevada residents and about $6 million in wages during the construction period. (BLM's Opp'n D'Aversa 2d Decl. ¶ 3; SVW Opp'n Hardie Decl. ¶ 16) It will create up to 12 permanent operation positions. *Id.* Wages over the life of the project would be about $15 million. *Id.* On the condition that the project is built, Spring Valley Wind has committed $750,000 in economic benefits to White Pine County over the next 20 years. (SVW Opp'n Hardie Decl. ¶ 16D)

The defendants assert that a preliminary injunction would result in the loss of the project. Spring Valley Wind will likely lose federal funding through the ARRA if it does not begin construction on the project by the end of September 2011. *Id.* ¶ 10. It would also threaten the project's eligibility for an investment tax credit grant. *Id.* Without these financial incentives, it is likely the project would not be built. *Id.* In addition, an injunction would hinder Spring Valley Wind's ability to honor its contracts with Nevada Energy. *Id.* ¶ 11. Under these contracts, Spring Valley Wind must obtain construction financing by June 30, 2011. *Id.* Finally, Spring Valley Wind has invested $11 million in the project thus far. *Id.* ¶ 15. It will commit an additional $12 million to ensure the project is operational by June 30, 2012. *Id.* Spring Valley Wind faces a financial loss of $23 million if the project is delayed. *Id.*

While the court recognizes that the denial of an injunction will result in the commencement of construction on the project, for the reasons set forth above, the court concludes that any disturbance of the sage-grouse and bat habitats will be minimal and will not significantly impact the environment as long as the mitigation measures set forth in the EA are complied with.

D. Public Interest

The public has a strong interest in the project. Congress has articulated the public policy that our nation should incorporate clean energy as a necessary part of America's future and it is essential to securing our nation's energy independence and decreasing green house emissions. (SVW Opp'n Hardie Decl. ¶¶ 9–10 (referencing ARRA of 2009 which amended Energy Policy Act of 2005)) It is also important to Nevada's economic and clean energy goals. The state's unemployment rate is 14.9 percent. (BLM's Opp'n D'Aversa 2d Decl. ¶ 3) The project would generate over 220 new jobs with priority to Nevada residents and over $20 million in wages. (BLM's Opp'n D'Aversa 2d Decl. ¶ 3; SVW Opp'n Hardie Decl. ¶ 16) Additionally, it would provide millions of dollars in property tax revenue. *Id.* Nevada is also committed to developing renewable energy sources. (SVW Opp'n, Ex. 3, Ex. C–1, Letter from Harry Reid to Mary D'Aversa ("I write to voice my support for the ... project[, which] ... represents an important milestone in developing Nevada's ... Clean energy resources.")) *See also* N.R.S. § 701A.220. The project, which has contracted with Nevada Energy will certainly help the state reach these goals.

While the public also has a strong interest in preserving the environment and protecting species like the free-tailed bats and greater sage-grouse, as noted above, that interest in this case at this stage in the proceedings is outweighed by the other interests articulated in this decision.

V. Conclusion

Having fully considered the administrative record and the arguments of the parties, and having weighed all relevant factors necessary for issuing a preliminary injunction — the likelihood of success on the merits, the likelihood of irreparable harm, the balance of equities, and the public interest — the court finds that the plaintiffs have failed to carry their burden of showing that a preliminary injunction should issue at this time. Plaintiffs' motion for a temporary restraining order/preliminary injunction (Docket No. 24) is DENIED.

IT IS SO ORDERED.

_____

Notes and Questions

1. Note that the court's decision is dated March 28, 2011. Is this date mentioned elsewhere in the order? What is its significance?

2. What relief did the plaintiffs in this case request? What is the court's decision here? Do you think the procedural posture may have made any difference in the outcome? Is the case over?

3. What does the court say about the tiered NEPA process used by the BLM? Does a PEIS automatically eliminate the need for a site-specific EIS? What is the role of an EA in supplementing a PEIS? What if individual actions are insignificant but cumulatively create significant impacts? What did the court decide about cumulative impacts here?

4. How did Judge McKibben balance the equities between environmental concerns and renewable energy development? Do you agree with his result? Why or why not?

5. The final section of Judge McKibben's order is labeled "public interest." What is the basis for finding a public interest in the development of renewable energy? Does it make any difference whether Congress expressly used this terminology? Consider our discussion in previous chapters about renewable energy's status as a public or private interest and how Congress might have the power to make it a priority in such balancing efforts by designating it as such.

D. Environmental Concerns

All of the previous chapters addressed some of the tensions between developing green energy resources and environmental concerns. The following excerpt addresses this tension, specifically in the context of large-scale solar projects.

Robert Glennon & Andrew M. Reeves,
Solar Energy's Cloudy Future
1 Ariz. J. Envtl L. & Pol'y 92, 94-122 (2010)

... [T]he solar power industry is uniquely positioned to help the United States reach its energy and economic goals and avoid the worst effects of climate change and global warming. But there are storm clouds on the horizon. As a matter of fundamental economics, solar power remains an emerging technology that is not competitive with fossil fuels. And while the *idea* of solar energy is appealing to every environmental organization, the *reality* of siting specific projects has turned out to be a contentious issue. The land mass required for utility-scale solar power installations is enormous. The plants are usually located far from urban areas, requiring upgrade or replacement of existing transmission lines — another contentious issue. And finally, there is the problem of water, which is intricately connected with the demand for energy.

In 2006, the U.S. Department of Energy predicted that the country's demand for energy will grow by 53 percent over the next twenty-five years. The National Energy Policy Development Group calculated in 2001 that the country will need 393,000 megawatts of new electrical power capacity by 2020. That amount of power would require that we build more than one power plant per week for the next twenty-five years. Yet, since 2007, Georgia, Idaho, Arizona, and Montana have denied permits for new power plants because there was not enough water to run them.

The United States' energy policy has almost totally ignored the water demands associated with various kinds of energy production. The energy industry consumes substantial quantities of water and the water industry, in turn, needs substantial quantities of energy. The roughly 60,000 water systems and 15,000 wastewater systems in the United States use about three percent of the nation's electricity to deliver and treat water and wastewater. And global climate change is expected to put strains not only on the availability of fresh water but also on the amount of energy generated by our hydroelectric facilities. Our thirst for energy to power our cell phones, light our homes, feed our Internet inquiries, and run our automobiles seems unlimited. But our water supply is not.

... The reallocation of both land and water from existing low-value farms to new solar production facilities offers a viable political, environmental, and economic alternative to siting projects on federal lands.

Ultimately, the article will attempt to show that — though the clouds on solar energy's horizon are dark and ominous — the future of solar power can be a bright one. It will take a major reorientation of federal incentives, an increasing commitment to the research and development of improved solar technologies, and a willingness of local citizens and environmental organizations to accept a significant number of solar projects on both private and public lands near their communities. If solar is ever to become more than a marginal force in this country's commitment to greener energy production, we must provide enhanced financial incentives to solar companies, utilities, and consumers; we must ensure that our environmental permitting system provides a deliberate, transparent process that does not erect endless and innumerable obstacles to actually siting renewable power projects; and we must recognize that solar energy has amazing potential to help us address climate change if, and only if, we address the money, land, and water issues associated with solar power.

II. SOLAR POWER TECHNOLOGIES, WATER CONSUMPTION, AND LAND FOOTPRINTS

It seems sensible to locate utility-scale solar facilities in the American Southwest, which obviously enjoys an abundance of sunshine. But some solar technologies use enormous quantities of water, a scarce resource in deserts. And all utility-scale solar projects, no matter their fundamental technologies, require large tracts of land.

A. Water Use for Various Solar Technologies

There are two basic kinds of solar power systems: photovoltaic cells (PV) and concentrating solar power (CSP). The first type, PV, converts solar radiation directly into electrical current. On the upside, photovoltaic systems require a minimal amount of water (essentially to wash periodically the solar panels and operating equipment) and can be built in stages—a major incentive for private companies requiring short-term profitability. Additionally, PV systems need not be built to utility scale. With continued improvement in "smart metering" and "smart grid" technology, private residents will continue to benefit from installing these solar panels while the owners of warehouses and urban commercial buildings may be able to install larger PV arrays on their rooftops to offset operational costs and create revenue by selling energy back to the grid. On the downside, however, PV systems present a major intermittency problem as PV cells are currently incapable of storing the energy they produce. Thus, when the sun is absent, either from uncooperative weather or darkness, PV cells are largely ineffectual.

The second type of utility-scale solar technology is CSP. A major advantage to CSP plants is their ability to address the intermittency problem that is such a liability for PV systems. Using thermal storage, hybridization with natural gas, or molten salts, CSP facilities can dispatch power to the grid even after the sun has set. CSP plants employ four different approaches: solar trough; linear Fresnel; power tower; and dish/engine. The first three use a steam cycle whereby an energy source is used to generate enough heat to boil water, to create exhaust steam, to spin a turbine that generates electricity. These three CSP technologies operate like coal, natural gas, or nuclear plants with one exception—the CSP technologies use the sun's heat instead of coal, nuclear fuel, or natural gas to boil water and begin the generation process. [Figure 8.8 shows the four primary CSP technologies: Linear Fresnel Reflectors, Parabolic Troughs, Heliostats (aiming to a power tower), and dish/sterling engines. Figure 8.9 explains briefly how these technologies work— Ed.]

All power plants involving a steam cycle use water to create steam. This water is highly purified and continuously recycled. The steam cycle begins when a heat source (here, concentrated sunlight) is applied, turning water into steam. The steam then turns the turbines, generating electricity. After leaving the turbines, the steam is passed through a condenser where it is cooled and condensed into liquid water. This liquid water is then returned back to the heat source to begin the steam cycle once again.

Because the water in the steam cycle is continuously recycled, the amount of water consumed by the steam cycle itself is quite small. On the other hand, substantial quantities of water are generally used in the cooling cycle. In most cooling cycles, cooling water is passed through the condenser where it picks up heat from the hot steam. The ultra-pure steam does not mix with the cooling water. Rather, as the hot steam comes into contact with cool tubes of cooling water inside the condenser, the heat from the steam is transferred to the cooling water. This heat transfer warms up the cooling water as it simultaneously cools and condenses the steam....

In an "open-loop" cooling system, cooling water is passed through the condenser only once before being returned to the environment. Large quantities of cooling water are removed from a river or other large body of water in an open-loop system. However, nearly all of that water is quickly returned, albeit at a higher temperature.

In a "closed-loop" cooling system, the cooling water is not returned to the environment but is recycled after passing through the condenser. Although the cooling water is recycled, significant quantities are lost with each turn of the cycle. This occurs for two reasons. First, before the cooling water can be reused, it must itself be cooled. In drier climates, this cooling generally occurs in large cooling towers, where a significant portion of the water is intentionally evaporated to chill the water.51 Much as the human body is cooled by sweat that evaporates from the skin, some of the cooling water must evaporate in order to cool the water that remains. A second reason why cooling water is lost in a closed-loop

Figure 8.8: The four primary CSP technologies

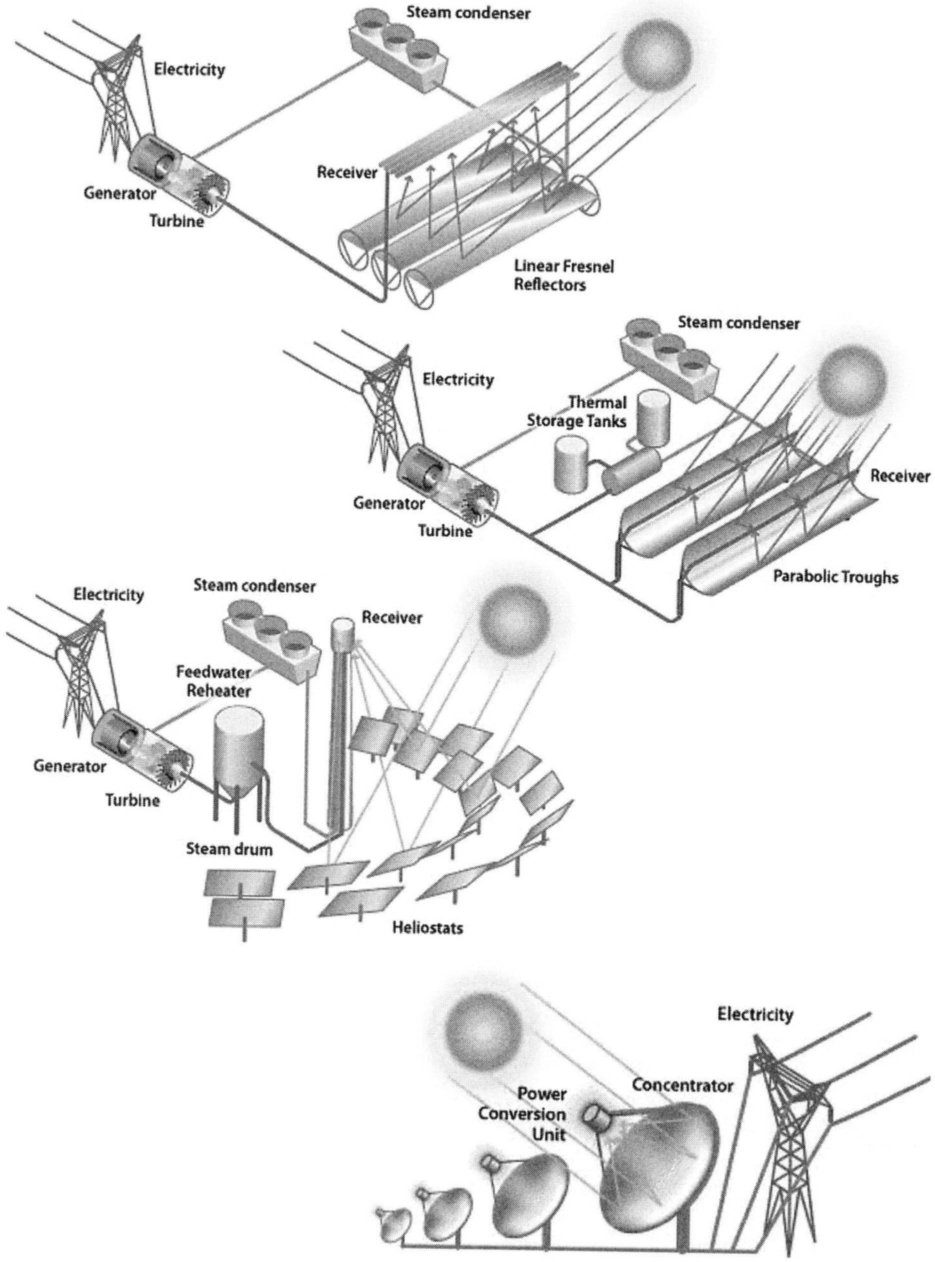

Concentrating Solar Power, National Renewable Energy Lab.,
http://www.eere.energy.gov/basics/renewable_energy/csp.html (last visted Aug. 5, 2011)

system has to do with the fact when water evaporates it leaves behind natural salts. Left unchecked, these salts would reach concentrations so high that they would damage the equipment. In order to prevent such a problem from occurring, a portion of the cooling

Figure 8.9: Concentrating solar thermal technologies

Linear Fresnel Reflectors	Large mirrors reflect and focus sunlight onto a linear receiver tube containing a fluid that is converted into superheated steam to spin a turbine producing electricity through a typical Rankine steam cycle.	In contrast to the direct conversion of the sun's energy to electricity, the CST process often requires as much water as conventional fossil fuel generating plants.
Parabolic Troughs	Parabolic mirrors focus sunlight onto pipes containing water, salt, or other heat exchange mediums.	The superheated fluids run through a heat exchanger to be stored or to generate steam for the Rankine steam cycle. This process also requires water.
Power tower	Liquid in a "power tower" reaches very high temperatures because of the concentrating effect of heliostats (flat, sun-tracking mirrors) surrounding it that simultaneously beam the sun's energy toward the tower.	The higher temperatures achievable from the concentration of the sun increase the efficiency of the steam cycle used to produce electricity. This process also requires significant amounts of water.
Dish/engine	Dish/ engine (also sometime called "Dish Stirling" because of the type of engine used) is one of the newest and most efficient methods of converting the sun's energy to electricity. Concave mirrors in a dish direct the sun's rays to a focal point holding the engine, which runs from the heat generated.	Dish/ engine is more efficient than either parabolic trough or power tower technologies. Although it has no storage capacity and it still requires large land areas, dish / engine does not use any water. Some complaints have arisen, however, about the noise from the Stirling engines.

(CSP stands for Concentrating Solar Power and can apply to both CST and Concentrating PV described in Figure 2.2 above.)

water must be discharged from the cooling cycle (called "blowdown") and replaced with fresh water.

A third cooling system is air or "dry-cooling" which does not use any cooling water. Here, steam cools by transferring its heat through the walls of the condenser directly to the surrounding air. The process is similar to a car's radiator which transfers heat to the air under the hood or (when the driver turns on the radiator) to the air in the passenger compartment of the vehicle. Although effective when ambient air temperatures are low (such as in the winter), air/dry-cooling is less efficient in the hot summer months— especially in desert areas where temperatures frequently exceed 120 degrees. One power plant using air/dry-cooling technology was found to produce five percent less energy over the course of a year, thereby increasing the electricity cost seven to nine percent over a water-cooled plant.

One problem associated with closed-loop, wet-cooled CSP plants is water consumption. Table 1 summarizes the "Water Use Intensity"—the number of gallons of water required at the power generation facility to produce one megawatt hour of electricity—for various power producing technologies.

Two paradoxes emerge from this data. First, if water scarcity resulting from consumption is a major concern, why not utilize open-loop technology as, across the board, it *consumes*

Table 1: Water use and consumption at power plants by various sources of electric power generation

| Plant Type | Cooling Process | Water Use Intensity (gal/MWh) | | Other Uses |
| | | Steam Condensing | | |
		Withdrawal	Consumption	
Fossil / Biomass	Open-loop	20,000–50,000	~300	~30
	Closed-loop	300–600	300–480	
	Air/Dry	0	0	
Nuclear	Open-loop	25,000–60,000	~400	~30
	Closed-loop	500–1,100	400–720	
	Air/Dry	0	0	
Natural Gas Combined Cycle	Open-loop	7,500–20,000	100	7–10
	Closed-loop	~230	~180	
	Air/Dry	0	0	
Coal Integrated Gasification Combined-Cycle	Closed-loop	~250	~200	130–140
Geothermal Steam	Closed-loop	~2,000	~1,400	Not Available
Concentrating Solar Power: Trough	Closed-loop	760–920	760–920	8
	Air/Dry	0	0	78
Concentrating Solar Power: Tower	Closed-loop	~750	~750	8
	Air/Dry	0	0	90

less water than closed-loop systems? In short, despite consuming less actual water, the other environmental hazards of open-loop systems are considerable. Because, in a standard open-loop system the power plant is located near a river or other large body of water, the water reintroduced into the source is returned at a much higher temperature than when it was originally extracted. This temperature differential can wreak havoc on the ecosystems connected to the water supply.

The second paradox involves closed-loop systems. If CSP plants use closed-loop thermal technologies similar to traditional coal, natural gas, and nuclear plants, why is it that (with the exception of geothermal steam) they consume, on average, 300 percent more water? The answer is that solar plants are less efficient at electricity production, and therefore require more water for steam production used in generating electricity through turbine electricity production. Thus, utilizing data from Table 1, it can be seen that, while a closed-loop nuclear and closed-loop CSP tower system may each withdraw 500 gallons of water to be used for energy-production, the nuclear plant—able to achieve steam production at a much more efficient rate—will likely *consume* about 350 gallons of the water while the CSP plant will consume all of it. Simultaneously, the 500 gallons will be used more efficiently in the nuclear plant and will be able to produce one-megawatt hour of electricity. With the CSP tower, however, the 500 gallons will likely only produce about two-thirds of a megawatt hour of electricity.

Apart from the possibility of utilizing air/dry-cooling technology, a hybrid system that has both wet-cooling and air/dry-cooling capabilities is possible. Though more expensive, hybrid systems are attractive because, when ambient air temperatures are lower, air/dry-cooling can effectively be utilized and, in the summer, when high temperatures make air/dry-cooling less effective, wet-cooling can be employed. Still, as a consequence of the added cost of maintaining a dual system, the overwhelming preference for utilities is wet-cooling.

In Arizona, for instance, the U.S. Bureau of Land Management has received thirty-two requests for solar plants to be located on federal land and twenty-eight of these plants intend to employ some form of CSP technology—many of which are likely to be wetcooled. With mounting pressure from environmental groups, politicians, and concerned citizens, however, it seems likely that some of these projects will change their plans to embrace dry- or hybrid-cooling technology. Further, in California, the California Energy Commission (CEC) is opposed to the use of fresh water for power plant cooling and the Nevada State Engineer in a 2002 ruling stated:

> [T]he State Engineer does not believe it is prudent to use substantial quantities of newly appropriated ground water for water-cooled power plants in one of the driest places in the nation, particularly with the uncertainty as to whatquantity of water is available from the resource, if any....

A fourth CSP technology has been developed by Stirling Energy Systems, which uses parabolic-shaped dish reflectors to focus sunlight on a generating unit that produces electricity directly without requiring cooling water. The first commercial-scale Stirling dish system, developed by Tessera Solar North for a project in Peoria, Arizona, came on-line in December 2009.[75] But the disadvantage is that the technology has thus far not allowed for thermal storage, which makes it of less use to utilities that need consistent, uninterruptable power. Still, as with PV systems and CSP technologies utilizing air/dry-cooling, the Stirling system requires significantly less water than wet-cooling systems and can be used if economic and performance penalties can be reduced or tolerated.

B. Land Use for Various Solar Technologies

In addition to water consumption issues, the land use impacts of solar energy are considerable. Sandia National Laboratories, a government-owned, contractor-operated (GOCO) facility run by the Lockheed Martin Corporation, has produced an estimate of the land requirements for various kinds of electrical power generation. According to their data, to produce 1,000 megawatts of power, a coal plant requires 640–1,280 acres of land, as does a nuclear plant; a natural gas combined-cycle plant requires at least 640 acres; but a concentrating solar thermal plant would require approximately 6,000 acres.[76] Wind power

---

76. It is important to note that these figures do not account for "land use intensity." Some forms of power production, like coal, have a much more dramatic impact on the lands they are sited on than others.

Table 2: Land requirements for various sources of electric power generation

Plant Type	Plant Size (MW)	Land Area (acres)	Unusable Land Size
Coal/Biomass or Gasification with Steam Turbine	500 – 1,000	640	640
Nuclear Steam	500 – 1,000	640	640
Natural Gas Combined-Cycle	200 – 500	320	320
Geothermal Steam	200 – 500	320	320
Concentrating Solar	500	3,000	Varies by Technology
	1,000	~6,000	Varies by Technology
Wind	500	23,000	640 acres
	1,000	46,000	1,280 acres
Photovoltaic	1,000	12,160	Varies by Technology

requirements would be even higher, a staggering 46,000 acres per 1,000 megawatts. Table 2 provides estimates of the land requirements for various types of technologies.

If the applications BLM has received for solar projects in Arizona are any indication, however, the land situation is much, much worse. The twenty-eight CSP applications that have not significantly begun the environmental impact statement (EIS) process have requested 425,873 acres of public land. These companies have estimated an aggregate generating capacity of 18,575 megawatts. Assuming the applicants have not woefully overestimated the amount of land they require, Arizona's new CSP systems would require approximately 22,927 acres for every 1000 megawatts of power produced — nearly four times the amount suggested by Sandia Labs. Tucson Electric Power's rule of thumb for PV requires eight acres of land per megawatt of power produced, putting the PV figure for 1000 megawatts at 8,000 acres. Either way, the landmass footprints necessary for utility-scale solar power are staggering.

Still, as one solar company pointed out in a June 2010 report, it is now possible to build solar fields without concrete foundations and extremely limited grading and leveling of land, allowing for vegetation in solar fields to co-exist with mirrors. In the future, such

---

See also Sara C. Bronin, *Curbing Energy Sprawl with Microgrids*, 43 Conn. L. Rev. 547 (2010) (arguing that small-scale distributed generation of energy through microgrids of neighbors may be able to cut back on the anticipated 67 million acres of land needed to develop energy projects over the next twenty years).

technological advances may be needed in order to overcome the large land requirements for solar energy production.

* * *

IV. THE PERMITTING PROCESS

In the Energy Policy Act of 2005,138 Congress instructed the Department of Interior and Department of Energy to collaborate in order to place at least 10,000 megawatts of nonhydroelectric renewable energy on federal land. The act has set off a frantic land-grab as solar and wind energy companies have rushed to obtain permits for projects in Arizona, California, Colorado, Nevada, New Mexico and Utah. In Arizona alone, BLM has received thirty-two solar energy applications that would encompass approximately 466,565 acres of public land. Nothing in the act, however, has changed the arduous permitting process that companies must navigate in order to break ground on public land. Currently, 585 megawatts of utility-scale solar power are operational—all of which are on private land. At the end of the 2009 fiscal year, oil, gas, and geothermal companies had received 31,133 leases for 27,800,932 acres of BLM-managed land—with 1,927 new leases issued in 2009—while solar had received zero permits. Though fourteen utility-scale solar projects were within striking distance of receiving BLM permits in 2010, only eight had been permitted as of November 2010.

The permitting process is both time and cost intensive; one commentator has noted that preparing a single Environmental Impact Statement (EIS) can cost millions of dollars and take up to twelve years. Additionally, a coalition of government agencies, including the National Park Service and the Fish and Wildlife Service, and environmental organizations, including the National Resources Defense Council, have urged that solar plants be located on disturbed lands, or lands that have already had significant use and where prior activities have ceased. Abandoned mines, developed oil and gas fields, fallowed agricultural lands, brownfields,148 former landfills, and inactive gravel pits illustrate the kinds of lands that would be desirable to use for solar projects. In response to the concerns of environmental groups worried about land impact and businesses fretting over the cost and length of the permitting process, BLM has:

- Removed from consideration sensitive lands, such as wilderness areas and other lands with high conservation values;

- Identified twenty-four Solar Energy Study Areas, where it seems most sensible to consider locating solar power plants;

- Embarked on a solar Programmatic Environmental Impact Statement (PEIS), aimed at addressing broad issues of policy in connection with all applications for solar plants on federal lands. It is then hoped that, in the future, the PEIS will enable developers to undergo a less time consuming permit process as they will already have a model to work from.

Consistent with these actions, BLM's Restoration Design Energy Project is attempting to identify disturbed or previously developed sites in Arizona. In concept, this is a great idea. In execution, the reality is somewhat different. After two years of trying to identify such lands, BLM has come up with fifty-nine potential "wastelands" totaling 156,366 acres. This sounds impressive, but in fact, only 25,360 acres of land on these proposed sites are managed by the BLM—just a tiny fraction of the 466,565 acres of land associated with the thirty-two pending solar power plant applications in Arizona.

Despite setbacks, BLM has received high praise from the business community *and* the environmental community. Yet, the jury is out as to how successful the PEIS will be in reducing the time between application and construction. Various factors have delayed the release date of the draft PEIS until late 2010, which in turn will push back the release of the final PEIS until 2011 or 2012. The PEIS will identify in advance particular areas that are likely candidates for solar projects, but that general conclusion is not going to satisfy the obligation of BLM to do an individual EIS with all of the attendant consultations with the Fish and Wildlife Service and other requirements under the National Environmental Policy Act (NEPA). A cynic might suggest that what the PEIS will have accomplished is to say: "Here is some land where maybe we will let you build." Moreover, even after applicants successfully survive the EIS process, they will then need to secure approval from state public utility commissions, something that can easily add another year to a project start-date.

This process does not allow for the kind of swift and definitive decisionmaking that the business community needs in a world where the time-value of money is critical and where many solar companies are thinly capitalized. In 2009, one California solar outfit, Ausra, for instance, abandoned plans for its Carrizo Energy Solar Farm as the permitting process continued to stall. Even more recently, Tessera Solar North America backed out of a planned partnership with the city of Phoenix to build a 250-megawatt power plant on a city-owned landfill. Peter Wilt, Tessera's senior director of development, explained that Arizona's utility companies have shown greater interest in smaller projects more likely to receive fast-track status for permits. "We're not getting a whole lot of traction on the market," Wilt said.160 Smaller companies have faced similar problems including Boulder, Colorado's Simple Solar, which filed for Chapter 11 bankruptcy in May 2010 and New Jersey-based EPV Solar, which filled out Chapter 11 paperwork in February 2010.

In early 2010, in an attempt to deal with these delays, Secretary of the Interior Ken Salazar announced plans for BLM to –fast-track certain solar projects. In April 2010, BLM released its draft EIS for the Sonoran Solar Energy Project, in Maricopa County, Arizona, a CSP trough project that would use 4,000 acres and generate 375 megawatts of power. This draft EIS could serve as a guide for future EISs and is thus a matter of considerable importance. As the preferred alternative in the draft EIS, BLM would permit the company to have a wet-cooled solar thermal project. BLM considered a dry-cooled system, but rejected it, in part, because the water is available and an analysis of the water needs of the project, between 2,300 and 3,000 acre-feet per year, would not result in a substantial drop in the water table or adverse impacts on adjacent groundwater wells. This may seem controversial or even absurd, given that the project is in the desert west of Phoenix, but with the particular hydrogeology of the site near the Gila River, there is substantial groundwater available. Thus, it would be premature to read into this draft EIS the assumption that BLM will be as sanguine when it comes to wet-cooled projects in other areas that do not have the same access to substantial quantities of groundwater.

The draft EIS also rejected as an alternative a utility-scale photovoltaic system, in part because no PV system on this scale has ever been constructed anywhere in the world. Here, BLM laid emphasis on the problem of PV not being dispatchable (i.e. able to be stored). The draft EIS also rejected alternative solar technologies, including Stirling engines and power towers because, according to BLM, they are development-stage options. Despite this (or possibly due to rapid advancements in technology), APS included a proposal in its 2010 Renewable Energy Standard and Tariff (REST) Implementation Plan to include Stirling technology within the Arizona Corporation Commission's (ACC) approved definitions of renewable technologies available for tax incentives. The ACC approved the plan indicating, hopefully, that BLM may soon change its tune concerning these technologies.

A final (though major) problem with the permitting process is the issue of transmission line right-of-ways. The nation's transmission grid is woefully outdated for the energy needs of the 21st Century. What the solar industry needs, though, is not long-term resolution of the transmission grid problem, but upgrading of certain smaller-length segments that will allow particular projects to come on-line promptly. But it is a thorny problem for BLM to figure out how to allocate right-of-way permits. In addition, the permitting, construction, and maintenance of transmission lines creates additional cost burdens that will likely be passed to consumers. The California Public Utilities Commission has estimated, for instance, that seven new major transmission lines will need to be built, at a cost of $12 billion, for the state to meet its 2020 RPS goal. The likelihood that such enormous costs will not affect utility rates seems, at best, far-fetched.

V. ENVIRONMENTAL AND POLITICAL OBJECTIONS

The environmental community, for years, has invested its political capital, as well as enormous sums of money, in trying to obtain climate change legislation and incentives for renewable energy. Every environmental organization supports the idea of utility-scale solar projects. But the consensus breaks down when specific sites are proposed for solar plants. The idea of solar plants seems to be more appealing than the reality. BrightSource Energy, for instance, found its Ivanpah CSP project being resisted by the very environmental groups that had previously proclaimed their support for renewable power facilities. In this process, some national environmental organizations are at loggerheads with local chapters.

The National Park Service is also concerned with the visual blight that will be created by incredibly tall solar towers; BrightSource Energy's towers, for instance, could range anywhere from 400 to 800 feet in height. The scale of several solar projects, as big as six square miles, is also a problem. The Park Service is also worried about the cumulative impact of multiple projects on the value and resources of the parks and monuments under its jurisdiction.

The environmental community has reacted with equal alarm to proposals for large numbers of wet-cooled CSP plants in the Southwestern deserts. Even modest amounts of groundwater pumping could dry up rare and critical seeps and springs, thus threatening endangered species. Environmental groups have criticized virtually every proposal for solar power plants due to their impact on federal land, which—in addition to concerns over scarce water—will be graded flat and sterilized in many cases.

To gauge how difficult it is to site a solar project on federal land, considering BrightSource Energy's Ivanpah project is useful. The company thought it had found the perfect site: it is adjacent to Interstate 15, across the highway from a natural gas power plant, next to a thirty-six-hole golf course, and five miles from a major casino and an outlet mall. The land itself has been used for decades for grazing and off-road vehicles, and a dozen eight-to twelve-foot wide trails criss-cross the site. A transmission corridor containing two high-voltage network lines bisects the site. The site does not contain any Desert Wildlife Management Areas (DWMA), Areas of Critical Environmental Concern (ACEC), Wildlife Habitat Management Areas (WHMA), or any other designated critical habitat.

The Ivanpah site has no endangered species, but a survey documented seventeen desert tortoises—a threatened species.[185] The BLM has classified the area as Category 3 ("least important") habitat for the desert tortoise. The site averages fewer than four tortoises per square mile. "Typical" habitat contains from ten to twenty tortoises and high-quality habitat has 250 tortoises per square mile. In the Ivanpah Valley, more than 630,000 acres are already designated as critical habitat for the tortoise.

In the EIS process, the Center for Biological Diversity (CBD), the local chapter of the Sierra Club, and Defenders of Wildlife (as well as other groups) intervened to express concerns about the Ivanpah proposal. In response, BrightSource Energy reduced the site's footprint by twelve percent in order to omit an area that the environmental organizations considered valuable tortoise habitat. This action also reduced the site's generating capacity from 440 megawatts to 392 megawatts. That loss of forty-eight megawatts represents more than one-quarter of all the PV installed in California in 2009. In July 2010, the California Energy Commission (CEC) staff report proposed a mitigation plan that BrightSource has endorsed that will relocate the tortoises, monitor them, and fence off the relocation area from predators. The CEC plan will require the company to spend more than $20 million on this relocation effort. The BLM's Supplemental Draft EIS endorsed this downsized project, but CBD still considered the project unacceptable.

The final decision on the Ivanpah project came from the BLM in October 2010. SCE and PG&E have signed purchase power agreements (PPA) to take the electricity generated at Ivanpah. In October 2010, BrightSource broke ground on the project just before the expiration of ARRA payments (in lieu of tax credits) for construction. Five other BLM projects in California—most notably Solar Millennium's Blythe Solar Power Project, a parabolic trough project with 1,000 megawatts of rated capacity, and Tessera Solar's Imperial Valley Project, a Stirling dish project with 709 megawatts of rated capacity—received final BLM approval in October and November of 2010 as well. These approvals (along with the approval of NextLight's Silver State North project and Amargosa's Farm Road Solar Project in Nevada) represent the first utility-scale solar projects that have ever been approved on public lands. These projects, in the aggregate, will have a rated capacity of approximately 3,500 megawatts of power upon completion and seemingly represent a fundamental shift in the BLM's commitment to approving renewable energy projects on public land—a change that should be applauded. Important to note, though, is that all eight of these projects had completed their Final EISs by September 2010. Of the other six projects "fast-tracked" by the BLM, only one has thus far completed its Final EISs: the Silver State South project in Nevada with a rated capacity of 267 megawatts. While it seems likely that this project will receive approval before year's end, thereby being eligible for Stimulus money, the other five fast-track projects may be in trouble. And Arizona, despite hoping to be a national leader in solar power, does not yet have any of its thirty-two projects proposed on public lands at the Final EIS stage, including the fast-tracked Sonoran Solar Project. After losing out on the $10 billion earmarked in the Stimulus for renewable energy projects, it will be interesting to see how many of these projects continue with their plans to move forward.

Economic and permitting concerns aside, the issue of transmission lines also creates interstate conflicts and resistance from the environmental community. For example, the Audubon Society is concerned about a proposed SunZia Southwest Transmission Project designed to carry power over two 500-kilovolt (kv) lines from central New Mexico to Phoenix, Arizona and eventually to Southern California. The proposed route would be through the lower San Pedro River Valley, an area designated "an Important Bird Area of Local Significance." The project is enormous in scale. It would involve constructing as many as 300 sixteen-story towers that would run the length of the valley with an access easement up to 1,000 feet wide, and access roads to each of the 300 towers. To put this in perspective, this is nearly ten times as many sixteen-story structures that currently exist in Arizona. The planners of SunZia have requested a one-mile-wide corridor from BLM for future expansion. Given the scale of this project, it is easy to understand the Audubon

Society's concern for an area that is home to more than 400 bird species, and is one of the most important north-south migratory bird flyways in North America.

Still, to the engineers and managers of solar power companies like BrightSource, who ardently believe they are changing the world by producing carbon-free electricity, it is naturally frustrating to have the environmental community oppose their specific sites. As *Newsweek* recently reported, the classic acronym for resistance to older power producing technologies such as coal and nuclear, NIMBY (Not In My Backyard), has been replaced among frustrated renewable energy developers with a newer one: BANANA (Build Absolutely Nothing Anywhere Near Anyone). Speaking at Yale University in 2008, Governor Arnold Schwarzenegger expressed his concern over this mentality: "They say that we want renewable energy, but we don't want you to put it anywhere. I mean, if we cannot put solar power plants in the Mojave Desert, I don't know where the hell we can put them."

The Governor's comments were in response to environmental organizations' complaints about proposed solar projects in the Mojave Desert. These groups range from relatively obscure ones, such as the Center for Community Action and Environmental Justice, to big-hitters such as the Sierra Club's California/Nevada Desert Committee. Terry Frewin, the committee's chairman, has criticized the Sierra Club's national leadership for its tacit support of large-scale solar projects, recently admonishing that "[r]emote solar arrays destroy all native resources on site, and have indirect and irreversible impacts on surrounding wildernesses ..." In response, Carl Zichella, then-western renewable projects director for the Sierra Club, said "We don't take a back seat to anyone in caring for the desert." The Club, however, did not withdraw its support for the project. Thus, on the national level, the Sierra Club's support for solar projects remains unchanged.

At the most basic level, *all* undisturbed land is habitat for some species. But not all habitat is equally valuable for the protection of critical species. Unfortunately, objective criteria do not exist for determining the size or locations of tracts of public land that should be sacrificed for solar projects.

In December 2009, the issue of the Mojave Desert was again catapulted to national significance as Senator Dianne Feinstein (D-CA) introduced the California Desert Protection Act of 2010 (S.2921). Although still in Committee, if passed the bill would essentially carve out another 1.7 million acres of public land for protection. [S. 2921 did not pass in 2010, but Senator Feinstein has vowed to raise the issue again—Ed.] Not surprisingly, based on previous reactions to large-scale solar projects, thirteen environmental groups (from the Death Valley Conservancy to The Wilderness Society) and the cities of Barstow, Desert Hot Springs, Hesperia, Indio, Palm Springs, San Bernardino and Yucaipa immediately expressed their support for the legislation. But it is also worth noting that some solar companies, like Abengoa Solar, and major utility companies, have expressed support for the bill as well. Edison International, the parent company of Southern California Edison, which provides power to 13 million Californians, recently expressed support for S.2921 and sent its Executive Vice President for Power Operations, Pedro Pizarro, to testify before the Senate Committee on Natural Energy and Resources. Pizarro stated that "when projects impact federally protected species or their habitat, the process for permitting renewable energy development on public lands is significantly slower than projects proposed on private lands, taking years instead of months. The bill addresses this inequity by allowing projects on public lands to mitigate environmental impacts by providing funding to help purchase or rehabilitate additional BLM lands."

Addressing these concerns, Senator Feinstein recently noted:

[T]he federal renewable energy permitting system [is] broken. Until recently, the BLM process has operated on a first-come, first-serve basis. And it didn't distinguish between a viable project and a speculative one. In fact, over the past five years, more than 100 applications have been submitted to build utility-scale renewable energy projects on public lands—and not a single project has received a permit. Under this status quo, no one wins.

In the proposed bill, Feinstein has called for streamlining the BLM permitting process and for requiring the Forest Service and the Department of Defense to research the possibility of locating solar projects on lands under their control. Whether these additions will successfully combat the "NIMBY/BANANA" effect is hard to predict but, at the moment, the proposal is generating substantial support, even from BLM. Regardless of what happens, though, something must change for the United States to become serious about developing utility-scale solar projects.

Notes and Questions

1. The excerpt outlines a theme we have seen throughout this book—the difficulty of balancing environmental concerns with the development of renewable energy projects—both large and small scale. Melissa A. Nigro created Figure 8.10 to summarize land and water use associated with most current forms of energy production. The water use includes the water used to mine and process the fuels for production (excluding biomass energy production, which does not include the water used to grow the biomass). Energy from solar PV, wind, and air-cooled Rankine steam cycle production techniques are the most efficient with respect to water use. The figure also shows that water use associated with the Rankine steam cycle energy production techniques (coal, natural gas, nuclear, geothermal and biomass) is highly dependent on the type of cooling process used. Therefore, advances in technology for cooling systems will be extremely important in the near future. Current research shows that the use of a hybrid air-water cooling system could reduce water use by up to 85 percent with only a 3 percent decrease in energy efficiency.

2. Based on the Figure 8.10 chart, Nigro concludes that air-cooled nuclear energy is the most efficient production technique if you attempt to combine the criteria on the chart—both water and land use. She also notes, however, that this energy production technique is extremely inefficient in producing electricity due to the difficulty of using air to cool the high temperatures that result from the nuclear fission process. In fact, only one air-cooled nuclear energy plant exists in the world.

3. BrightSource's Ivanpah project in California has encountered significant opposition. The BLM halted construction in the spring of 2011 because of greater than anticipated impacts on the desert tortoise. Construction restarted after BrightSoure worked with BLM on mitigation measures. In another section of their article, Glennon and Reeves note that Ivanpah's land use footprint is just 5.6 square miles. Because California would need at least 270 square miles of renewable CSP projects to meet its RPS standards, California will need to build approximately the equivalent of forty-eight Ivanpahs.

4. One of the solutions proposed by Glennon and Reeves is that there should be a heavy presumption toward requiring only dry cooling CSP plants on public lands unless the operator could find reclaimed water to use for cooling. Do you think this is a good solution? What do you think about the hybrid systems that Nigro mentions, which use water to cool only on the hottest days resulting in less water consumption than a wet-cooled system but avoiding some of the generation inefficiencies of the dry cooling process?

Figure 8.10: Summary of water and land use for energy production

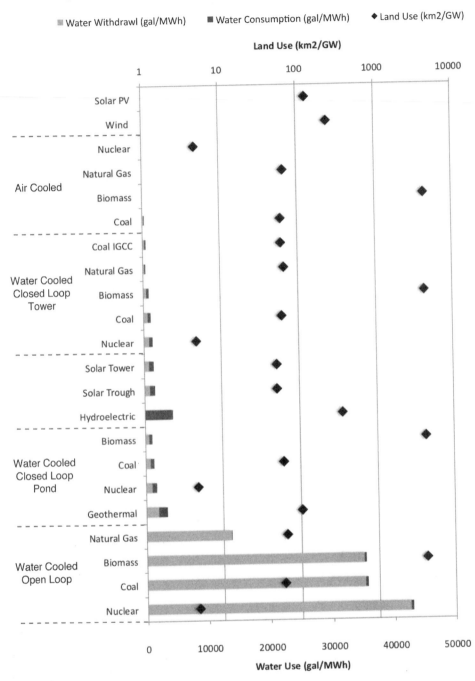

Graph compilation and summary by Melissa A. Nigro. Sources for data:U.S. Dep't of Energy, Energy Demand on Water Resources: Report to Congress on the Interdependency on Energy and Water (2006); R.I. McDonald et al., Energy Sprawl or Energy Efficiency: Climate Policy Impacts on Natural Habitat for the United States of America (2009), PLoS ONE, 4(8): e6802. doi:10.1371/journal.pone.0006802; U.S. Dep't of Energy, Concentrating Solar Power Commercial Application Study: Reducing Water Consumption of Concentrating Solar Power Electricity Generation: Report to Congress (2009)

5. Another proposed solution would be to adopt the Center for Biological Diversity's (CBD) position that we should put more PV on rooftops in urban environments instead of disturbing public lands for energy development. However, this solution has many hurdles as discussed in Chapter 2 above. Discuss each of the following: solar access rights, costs, and utility preferences and incentives to retain control of costs and production through a large centralized facility. Can you think of others?

6. Another solution proposed by CBD is conservation. How does the discussion in Chapter 7 above inform your response to whether conservation will be enough?

7. Glennon and Reeves's primary suggestion is to encourage developers of large-scale solar facilities to purchase land and water rights from already disturbed, private agricultural land for siting wet-cooled solar plants instead of developing them on public lands. What do you think about this solution from an economic standpoint considering that agriculture is a relatively low value and high-water-consumption activity in comparison to CSP electricity generation? Do you also think this process might be easier, faster, and cheaper than developing large-scale solar on public lands? Can you see any other advantages or disadvantages? Even if you site a project on private land, up to 90 percent of transmission will have to cross federal lands in the western United States. Does this change your previous assessment?

8. The final solution proposed by Glennon and Reeves is that renewable energy be developed on tribal lands. This will be the topic of our last chapter.

Chapter 9

Native Nations and Energy Justice

The U.S. Government has formally recognized over 500 tribal communities within its boundaries. Throughout U.S. history, these tribes have played a key role in energy development, providing significant coal, uranium, and hydrocarbon resources. They also are poised to play a key role in the future as some estimate the almost one hundred million acres of tribal lands in the United States could produce more than four times our electricity needs. "If you're going to reach the energy independence goals of the president [Obama], if you're going to reach those climate change goals, you can't do it without effective development of renewable energy in Indian Country." Keith Harper, *quoted in* Phil Taylor, *Public Lands: Tribes See Brightening of Once-Bleak Energy Development Prospects*, LAND LETTER, July 16, 2009.

This chapter first addresses key issues related to the development of renewable and alternative energy on lands directly under tribal control. Next, it explains some of the mechanisms native nations can use to impact renewable energy development on federal lands outside of reservations. Finally, as tribes have suffered from exploitation of their resources in the past and have received only "pennies on every dollar" that mining firms and electric utilities have made from the development of tribal resources, this chapter also addresses environmental and energy justice issues.

A. Alternative Energy Development on Tribal Lands

This section includes two excerpts providing an overview of some of the key issues related to the development of renewable and other alternative energy on tribal lands, including some context, some specific examples, and some impediments.

Elizabeth Ann Kronk, *Alternative Energy Development in Indian Country: Lighting the Way for the Seventh Generation*
46 IDAHO L. REV. 449, 450-70 (2010)

Native communities are facing major economic and cultural impacts directly related to climate change. As climate change forces many migratory species to leave their traditional ranges, Native communities, who may only have rights to hunt or fish in certain defined areas, may find it difficult, if not impossible, to survive in their traditional manner. Additionally, Native communities that rely on tourism may face the negative economic effects of a decline in tourism, as the changing environment decreases the desirability of tourism enterprises. Native communities may also face increased health effects related to climate change, including individual members' mental health issues that result from the loss of homes and cultural resources.

Alternative energy development within Indian country offers Native communities that have been uniquely harmed by the effects of climate change an opportunity to proactively address the effects of climate change in their communities. This is because alternative energies largely do not contribute to greenhouse gases, which are the primary contributors to climate change. Native communities now deserve the opportunity to participate in the development of alternative energy projects to help offset the harmful effects of climate change. In addition to the increasing threat from climate change to Native communities, there are other recent developments suggesting that now may be an ideal time to invest in alternative energy development within Indian country.

Studies increasingly show that Indian country may be uniquely positioned to develop alternative energy. Indian country in general constitutes a significant portion of land in the United States. "The [U.S. Department of the Interior Office of Indian Energy and Development—Ed.] (OIEED) estimates that an additional 15 million acres of undeveloped traditional energy mineral resources and over 22 million acres of undeveloped renewable energy resources exist on individual Indian and tribal lands." In recognition of the increasing interest in alternative energy development within Indian country, some Native Nations have developed critical financial infrastructure to support and foster alternative energy development.

The idea is ... a form of climate security. Indian tribes stand in a unique nexus between renewable energy resources and transmission of electricity in key areas of the West. Indian tribes would also be natural leaders for hosting and developing these key areas to promote climate security and energy security. This development would be a call to service that Indian tribes are absolutely ready to answer—and uniquely ready to do so.

* * *

II. INCREASING INTEREST IN ALTERNATIVE ENERGY DEVELOPMENT IN INDIAN COUNTRY

There are a myriad of reasons why Native Nations and outside investors are increasingly interested in alternative energy development within Indian country. As suggested above, many Native Nations may be acting because of a feeling that there is a moral imperative to act in order to address the negative effects of climate change on Native communities. Aside from that, there are many additional incentives that are causing Native Nations to act. There is also substantial interest from third-party investors in alternative energy opportunities within Indian country. Finally, the federal government has offered several initiatives, which if fully funded and supported, could have the combined effect of making alternative energy development in Indian country a very attractive proposal. Each of these factors contributing to the increasing interest in alternative energy development in Indian country will be discussed below.

A. Native Communities Are Uniquely Impacted by Climate Change, Creating Increasing Interest within Tribal Communities to Explore Alternative Energy Projects

Because of the unique character of Indian country, Native communities are more likely to be impacted by climate change. First, because Native communities are often tied to specific areas of land, such as reservations, it is impossible for Natives to leave these areas, whether to escape the effects of climate change or to follow migratory species moving to new ranges or for some other reason, without abandoning their land. For example, many of the inhabitants of the Inuit village of Shishmaref, Alaska, now find themselves homeless because their homes have fallen into the sea as a result of the eroding coastline related to warming in the Arctic. Furthermore, as species shift their ranges to follow their preferred climates, those shifts may threaten the culture of Native communities because Natives may no longer have access to these species. Alaskan Natives may be particularly hard hit by shifts in the species' ranges, such as caribou, because the Alaska Native Claims Settlement Act of 1971 extinguished Alaskan Natives' claims to aboriginal title and to hunting and fishing rights. Alaskan Natives will not be able to argue for their rights to hunt beyond existing boundaries, as they have lost claims to aboriginal territories and rights.

Furthermore, Native people are uniquely impacted by climate change because many Native cultures and traditions are tied to the environment. For many Native communities, land possesses certain spiritual and cultural aspects that are crucial to the community. Therefore, as the land changes many communities may be faced with devastating impacts on their culture and traditions due to the effect of climate change on their spiritual connections to their lands.

Given that Indian country is uniquely impacted by climate change, Native communities are actively pursuing initiatives to address climate change—both because it is imperative that Native perspectives be represented in larger discussions on climate change and because many Native communities have the natural resources and knowledge to become involved in climate change initiatives. Many states and regional organizations are moving forward with their own initiatives to address climate change. Native communities need to be involved in these initial discussions about climate change initiatives to ensure their perspectives are included and their interests are protected. This is especially relevant because the federal government will likely look to regional and state climate change programs when the time comes for the United States to adopt a nationwide climate change strategy.

Additionally, many Native communities possess the requisite natural resources and infrastructure to become involved in climate change initiatives and alternative energy development. For example, some Native communities possess large areas of land that may be used to develop carbon sequestration projects....

Moreover, the development of alternative energy projects in Indian country offers the potential for economic development. Many Native communities have sought to develop mechanisms for increasing and diversifying economic development. A recent Harvard Project on American Indian Economic Development study concluded that there are three keys to successful economic development in Indian country: sovereignty, culture, and tribal institutions. With regard to sovereignty, "[w]here [Native Nations] make their own decisions about what approaches to take and what resources to develop, they consistently out-perform outside decision-makers." With regard to culture, "[Native] culture is a resource that shores up the strength of government and has concrete impacts [on] ... bottom line results...." Finally, with regard to Native institutions, Native communities that have successfully engaged in economic development have done so by "settl[ing] disputes fairly, ... separate[ing] the functions of elected representation and business management, and ... implement[ing] tribal policies that advance tribal strategic goals." The Harvard Project on American Indian Economic Development study therefore concluded that economic development projects undertaken in Indian country that take into consideration sovereignty, culture, and tribal institutions are likely to be successful.

Potential opportunities for alternative energy projects in Indian country reflect well the first two of these three keys to successful development in Indian country. First, opportunities designed to address climate change through the development of alternative energy projects allow Native Nations the chance to affirmatively decide to participate in either domestic or international programs, thereby potentially increasing the perception of the Nation's sovereignty within the dominant society. By participating in the development of these programs, Native Nations would certainly be acting like sovereigns. Second, alternative energy opportunities can potentially protect Native cultural resources, as many of the proposed projects have limited impact on the environment. As a result, "Indian tribes are ready for 'nation building at home' by investing, developing, facilitating, and participating in building the infrastructure required to support green energy."

B. Alternative Energy Development Projects Present Incentives for Native Nations

Not only do alternative energy projects have a strong likelihood of success, as explained above, but there are many incentives for alternative energy development in Indian country. Moreover, as discussed, many tribal communities are uniquely situated to take advantage of alternative energy development opportunities because these communities may be flush with renewable energies. Alternative energy development in Indian country offers a new hope for some tribal communities, because "[w]ith Tribal communities economically hamstrung by inadequate infrastructure, no tax base, and population growth outpacing infrastructure growth, energy and infrastructure development ... will not just provide new revenue streams but also attract capital investments in manufacturing ... [and] new sustainable employment." The Honorable Steve Herrera, a Southern Ute Indian Tribal Council Member, also indicated that Native Nations may be interested in taking greater control over alternative energy development, because "[o]ne persistent theme reflected in the last thirty years of our tribe's history is the notion that ultimately we are the best protectors of our own resources and the best stewards of our own destiny."

Even if a Native Nation should determine that a large-scale commercial development of alternative energy resources is not plausible because of the obstacles discussed later in this article, it may be advantageous for the Native Nation to consider alternative energy development to support the energy needs of its own people, such as the Blackfeet nation did. Small scale wind projects, for example, can be relatively affordable for most Native Nations. While not presenting the same financial incentives of the large-scale alternative energy projects, these smaller alternative energy projects still offer Native Nations an opportunity to provide free or affordable energy for their citizens, and also to provide some employment opportunities. Additionally, because of their potentially smaller impact on the environment, small-scale alternative energy projects may avoid the extensive environmental impact assessment requirements the National Environmental Policy Act imposes on larger projects.

C. Interest in Alternative Energy Development from a Third Party Investor Perspective

Native Nations may also welcome alternative energy development within Indian country because of potential third-party investors' increasing interest. There are several reasons why third-party investors may be interested in developing alternative energy projects within Indian country. First, they may prefer to work with tribal governments because of the ease of access to most tribal governing bodies. "In other words, tribal councils make the rules, which benefits all parties when it comes to developing wind projects and avoiding the complexity and delay often attendant to the bureaucracy of federal guidelines and procedure." Moreover, the less intrusive the project, the less likelihood that other entities outside of the tribal council and the third-party investor would become involved in the project. Second, the improved stability of tribal judiciaries can improve investor confidence in a fair resolution of matters should conflicts associated with the project arise.

D. Federal Support for Alternative Energy Development in Indian Country

In addition to increasing interest from outside investors, federal initiatives have spurred interest in alternative energy development in Indian country. As a result of the incentives offered through the Energy Policy Act of 1992, several Native Nations developed alternative energy projects on their lands.... Additionally, the Department of Energy (DOE) has tools available to help encourage the development of alternative energy in Indian country, but many of these tools, such as the Office of Indian Energy Policy and Programs, have yet to be fully and effectively funded. Similarly, there is statutory language designed to encourage federal agencies to purchase renewable energy, and especially renewable energy developed within Indian country. However, to date, few federal agencies have purchased alternative energy produced in Indian country.

As suggested previously, there is significant political interest in developing alternative energy resources. The recently passed American Recovery and Reinvestment Act (ARRA) may be used to help foster alternative energy development in Indian country....

There are therefore tools in place that, although perhaps not yet fully funded or implemented at the federal level, should ultimately help spur alternative energy development in Indian country.

III. NATIVE NATIONS CAN DO IT! PRESENT-DAY ALTERNATIVE ENERGY DEVELOPMENT IN INDIAN COUNTRY

Recognizing the importance of combating climate change and taking advantage of many of the incentives discussed above, some Native communities are already participating in alternative energy projects. By reducing the dependence on fossil fuels through "clean" energy generation, renewable energy projects help combat climate change. Common renewable energy sources include: wind, hydroelectric, solar, landfill gas, geothermal, and biomass. Many of these renewable energy sources have a limited impact on the environment but the potential for substantial benefits for Native Nations. In addition to the wide array of renewable energy projects available to Native Nations, Native Nations have many choices as to how to finance alternative energy projects.

* * *

B. Alternative Energy Projects: The Navajo Nation, Campo Band of Kumeyaay Indians, and Blackfeet Wind Power Project

1. Navajo Nation

The Navajo Nation covers 27,000 square miles and is the largest reservation in the country. Located in the Southwest, the Nation also has substantial access to solar resources. Despite the alternative energy potential, approximately "18,000 out of 48,000 homes on the reservation are without electricity." As a result, the Nation has taken steps to utilize its alternative energy potential and supply electricity to all of its residents. The cost of expanding the transmission grid to give the Nation's residents access to traditional electricity was prohibitively expensive. Instead, with help from the DOE, the Nation installed 72 individual solar energy systems during the early 1990s. In 2000, the Nation purchased 200 more solar photovoltaic systems for use by individual residents of the reservation.

Recently, the Navajo Nation also began to develop its wind resources as a corollary to its solar resources. As a result, the Nation has incorporated wind turbines within many of its solar stations. After all, even in the Southwest, "[t]he sun doesn't shine all the time."

2. Campo Band of Kumeyaay Indians

Another model for Native Nations interested in developing alternative energy resources is the Campo Band of Kumeyaay Indians' project near San Diego. There, the Campo Band successfully developed the largest wind power project in Indian country. More so than perhaps other Native Nations, the Campo Band is uniquely situated to reap the benefits of wind energy development because the

> Campo [Indian Reservation] has ... potential for over 300 megawatts of wind generation capacity. ... The reservation has a substation that provides ready access to a 69 kilovolt (kV) utility line. The utility line was already scheduled for an update in its carrying capacity from 269 to 418 amps and Superior Renewable was able to negotiate an accelerated upgrade to handle the wind project load. Interstate 8 runs next to the project site, making access to the site simply a matter of grading three miles of existing dirt road 50 percent wider to handle the width of the shipping trucks.

* * *

The experience of the Campo Band is likely similar to the experience many Native Nations have encountered or will encounter while developing alternative energy resources. Native Nations are not taxable entities under federal law. As a result, a portion of an alternative energy project owned by a Native Nation will not receive the same tax benefits as a taxable entity would. Thus, there is little incentive from a tax perspective for a Native Nation to become involved in alternative energy development. Given this reality, Native Nations such as the Campo Band who opt to develop their alternative energy resources do so by leasing their land to third-party investors, who in turn reap the tax benefits. This arrangement allows both the Native Nation and its third-party investor to benefit financially—the Native Nation receives the lease payments and the third-party investor receives the associated tax benefits.

This arrangement is less than ideal for the Campo Band, however, as the Band does not receive as much income as it would if it were a partial owner of the project. Additionally, "the county receives more revenue from taxing the tribe's lessee business partners than the tribe receives from lease payments, and the county provides virtually no governmental services within the tribe's reservation." However, because of the tax obstacles to effective development of alternative energy as discussed below, the Campo Band opted to lease its land rather than develop its own project.

Because of equipment problems, the Band's project produced somewhat less electricity than expected. On a positive note, however, bird and bat deaths were kept to a minimum. The Band's project is also estimated to save approximately 110,000 tons per year in greenhouse gas emissions. Accordingly, alternative energy projects, such as the Band's wind project, allow Native Nations to counterbalance some of the negative effects of climate change.

* * *

IV. OBSTACLES TO ALTERNATIVE ENERGY DEVELOPMENT IN INDIAN COUNTRY

Despite the fact that some Native Nations have successfully developed alternative energy projects, obstacles to efficient development of these projects remain. Indeed, while

> [t]here is tremendous potential for renewable energy development in Indian Country ... actual projects have been slow to materialize. This is due to a variety

of obstacles ranging from overly complex and burdensome lease approval processes to difficult transmission access and ill-fitting financial incentives.

A. Lack of Necessary Infrastructure in Indian Country

One of the most significant obstacles to alternative energy development is the lack of infrastructure, notably transmission lines, to move energy developed within Indian country, which tends to be located in more rural portions of the country, to areas of higher population density where there is a corresponding higher demand for energy. As previously discussed, many Native Nations have only been able to develop relatively small alternative energy projects because of their lack of access to effective transmission infrastructure. However, "key transmission corridors currently run through Indian reservations—or could do so in the future—and many of these tribes are anxious to develop their critical infrastructure and participate in the new green economy." There may, therefore, be an opportunity for Native Nations to encourage the development of transmission infrastructure and capitalize on existing transmission facilities within Indian country.

B. Burdensome Lease and Siting Review Process

The development of alternative energy resources in Indian country is also hampered by the long process currently in place to lease and site facilities. "Indian lands lease review and approval processes can easily take as many as two to three years longer than the comparable processes for projects outside of reservations, even in comparison with projects on Federal lands...." As a result of these delays,

> [i]nvestors, developers and Tribes who seek to invest capital on renewable energy projects are finding that the lack of clarity with respect to trust and Indian land lease reviews and permitting, and the often severe delays and extraordinary and unpredictable length of time involved in such federal reviews and the federal issuance of permits, serve as a great disincentive to capital deployment.

C. Lack of Adequate Financial Incentives

Another substantial barrier to effective development of the alternative energy resources available within Indian country is the existing financial incentives, or lack thereof. "Nontaxable entities such as electric cooperatives, Indian tribes, municipal utilities, and their counterparts are deeply frustrated with ... financial incentives for using renewable energy, because the stringent rules regarding the use of these incentives do not easily allow these entities to participate in the financing or ownership of such projects." Furthermore,

> [o]ne major conundrum for many Indian tribes is that, although many now have capital they wish to invest in renewable energy projects, the current tax regime provides a disincentive for them to do so, because, in order to use tax credits most efficiently, tribes must usually bring on a tax-paying investor and owner ... for their costs to be competitive with those of other nontribal projects.

Thus, because they are tax-exempt entities, Native Nations cannot utilize tax incentives to offset liabilities under the existing scheme, which in many cases forces them to seek third party investors who can benefit from the tax incentives. And, the ability to effectively utilize these tax credits is essential to developing an economically viable renewable energy project. This ability is

> [s]o essential in fact that it is causing most Tribes, looking to develop and invest into these projects, to bring on tax partners who can utilize these credits. But this is a Catch-22 of sorts for the tribes—they need the partner to take advantage of the tax credit, but for an extraordinary long period of time the Tribal

governments [are], in essence, losing significant control over their own critical infrastructure.

Additionally, although investors may be interested in developing alternative energies within Indian country, the recent economic downturn may negatively impact their ability to do so.

It has been suggested that a solution to this obstacle to alternative energy development already exists. The Energy Policy Act of 2005 authorized the DOE Indian Energy Loan Guarantee Program. However, the Indian Energy Loan Guarantee Program has never been fully funded. Fully funding this program would go a long way toward creating the necessary incentives to adequately promote alternative energy development in Indian country.

Similarly, another provision of the Energy Policy Act of 2005, the Indian Tribal Energy Development and Self-Determination Act of 2005 (the "Act"), initially appeared to be a solution to some of the obstacles Native Nations face in alternative energy development, although that has not turned out to be the case. The Act also seemed likely to spur alternative energy development within Indian country because some of its provisions should have resulted in increased alternative energy development within Indian country. For example, the Act created the Office of Indian Energy Policy and Programs, with a Director who is required to "promote Indian tribal energy development, efficiency, and use." Likewise, the Act created the Tribal Energy Resource Agreement (TERA) mechanism, which appeared likely to streamline the approval process for alternative energy development in Indian country. If a Native Nation were to enter into a TERA with the Interior Secretary, any lease, business agreement, or rights-of-way involving energy development or transmission within that Native Nation's territory would not be subject to review by the Secretary. Many were hopeful that TERAs would help promote alternative energy development in Indian country by removing the cumbersome layer of bureaucratic review by the Department of Interior that had previously slowed down the process of energy development. However, despite the initial interest in TERAs, Native Nations have failed to take advantage of the ability to enter into a TERA with the Interior Secretary. This is likely a result of the provision of the Act indicating that once a TERA becomes effective, the United States "shall not be liable to any party (including any Indian tribe) for any negotiated term of, or any loss resulting from the negotiated terms of, a lease, business agreement, or right-of-way executed pursuant to and in accordance with a [TERA] approved by the Secretary...." One tribal councilman suggested that the implementation of the TERA provisions also undermined the congressional intent associated with the Indian Tribal Energy Development and Self-Determination Act of 2005. Although it is difficult to make generalizations about separate Native Nations, it may be possible to infer from tribes' failure to utilize the TERA option under the Act that Native Nations are loath to assume all liability for alternative energy projects within Indian country.

As the above discussion demonstrates, with the notable exception of the Campo Band, extensive large scale commercial alternative energy development in Indian country has not come to fruition. This type of development is likely hampered by the obstacles unique to Indian country, specifically that many Native communities have found it difficult to obtain adequate access to transmission lines. Moreover, review by the Department of Interior can significantly lengthen the time related to siting and leasing for a project. Finally, although there has been some federal support for alternative energy development in Indian country, adequate financial and tax incentives are still lacking.

———

Robert Glennon & Andrew M. Reeves, *Solar Energy's Cloudy Future*

1 Ariz. J. Envtl L. & Pol'y 92, 130-34 (2010)

Native American lands present another interesting possibility for siting solar projects. Keith Harper, a member of the Obama-Biden transition team, has stated that "Obama's top energy priorities ... will be difficult to accomplish without closer partnerships with the country's 562 federally recognized tribal communities." Also recognizing this reality, Congressman Raul Grijalva (D-AZ), in a 2007 hearing before the House Committee on Ways and Means, estimated the solar power potential of tribal lands to be about 4.5 times the annual electricity needs of the United States. Although his estimate seems quite optimistic in light of the large footprint of solar projects, his sentiment is on point and highlights the fact that tribal lands are a potentially untapped resource for solar projects. From the perspective of solar land requirements, it is worth noting that nearly thirty-five percent of the State of Arizona consists of tribal lands. Some of these reservations are located near the thirty-two projects proposed on BLM-managed land that investors have already expressed an interest in. Moreover, tribal lands may present far fewer hurdles to overcome in successfully implementing solar projects than BLM lands.

First, as part of the Energy Policy Act of 2005, federal agencies were granted authority to institute preferential purchase agreements for any "energy product" or "energy byproduct" produced by business entities that are majority-owned by an Indian tribe. The Act was "intended to provide support to tribal governments in the development of energy resources on Indian lands, ... to provide incentives for partnership with tribes that want to develop their resources[,]" and to "authorize individual Indians and tribal governments to enter into energy development leases or business agreements *without Federal review....*"

Thus, under certain scenarios, it is possible that solar projects on tribal land could be implemented and acted upon without the need for the costly and time-intensive NEPA review that has hindered so many solar proposals.

Second, unlike the NIMBY phenomenon witnessed in many communities where solar projects have been proposed, a number of tribes have already expressed interest in developing solar projects....

* * *

Third, siting solar projects on tribal land will not magically alleviate the energy-water nexus issues previously discussed—especially if these projects employ wet-cooled CSP rather than PV technology. Nevertheless, tribes may enjoy an advantage in this respect as well. In 1908, the United States Supreme Court decided a pivotal case in the history of tribal lands, *Winters v. United States*. The case involved the 1888 establishment of the Fort Belknap Reservation in Montana and addressed whether the Gros Ventre and Assiniboine tribes had relinquished their water rights to the land when they relinquished control of it (purportedly to shift from a nomadic to agrarian way of life) to the federal government. In oft-cited language, Justice McKenna, writing for the majority of the Court, determined that they did not, stating:

> [I]t would be extreme to believe that ... Congress destroyed the reservation and took from the Indians the consideration of their grant, leaving them a barren waste—took from them the means of continuing their old habits, yet did not leave them the power to change to new ones.

Since then, tribal water rights have often been referred to as "Winters rights." And, although tribes have often come upon a daunting chasm separating their legal rights to water (the "Winters rights") and the actual water itself ("wet water" is, sadly, actually employed to point out this distinction), courts have recently begun following through on the promise that the *Winters* decision presented over a century ago. In the seemingly endless battle over Colorado River water, for instance, the U.S. Supreme Court has remained steadfast in determining that approximately 950,000 acre-feet of the 7.5 million acre-feet of mainstream Colorado water allotted to Arizona, Colorado, and Nevada should go to the Chemehuevi, Cocopah, Fort Yuma, Fort Mojave, and Colorado River Indian Reservations. Nominally for "irrigable" use, there is some indication in the decision that tribes could, alternatively, utilize this water for energy production—a far more profitable endeavor.

Finally, although few large-scale solar projects have broken ground on tribal lands, that may soon change. In February 2009, one of the country's least populous tribes, the Augustine Band of Cahuilla Indians in California, began operating a 15,000 panel PV system on its land that is expected to produce up to 1.1 megawatts of power annually. As Michael Lombardi, Augustine Casino gaming commissioner, noted "[w]e've thrown a pebble in the pond that I'm sure will ripple across Indian Country." With the ARRA recently allotting $54.8 million to tribes for "energy efficiency improvements in Indian Country," we can only hope the ripple spreads far.

Still, as is seemingly true with all solar projects, there are problems that need to be addressed. For one, like a lot of BLM managed land, a number of reservations are remote, a situation that raises the problem of constructing new transmission lines. Additionally, there are major incentive issues that have kept private backers hesitant about throwing in with tribes rather than BLM. Because tribes, pursuant to the IRS Tax Code of 1986, are tax-exempt entities, they are ineligible for the 2.1 cent per kilowatt-hour tax benefit (for the first ten years of a facility's operation) guaranteed by the ARRA290 that has lured a number of private companies into the solar sector. For so-called "casino-rich" tribes able to build their own solar facilities, this does not present a problem because their tribal revenue is tax-exempt and, as a result, tax incentives are a moot point. For other tribes who would like to partner with private firms, the problem arises because companies pairing with tribes only receive 50 percent of the credit, rather than the full 100 percent they would receive by investing on state land. Rep. Grijalva has noted that "[t]his situation puts tribes at a tremendous disadvantage when trying to attract renewable energy projects to their lands" and has introduced a bill, the Fair Allocation of Internal Revenue Credit for Renewable Electricity Distribution by Indian Tribes Act, to combat the problem. Whether the Grijalva bill passes or not, the message concerning solar projects on tribal lands is clear: given the right incentive structure, these projects could be a successful component in moving toward a more sustainable future.

Notes and Questions

1. What are some of the reasons listed in each excerpt for why Indian country may be ideal for the development of alternative energy? Which of these are unique to tribal lands?

2. What are some of the special advantages tribal lands may have in the area of energy-water issues—especially for projects employing wet-cooled CSP technologies?

3. What are some of the obstacles to the development of alternative energy on tribal lands?

4. In the past, burdensome federal leasing and site-review requirements sometimes extended the approval process for projects on Indian lands two or three years beyond

comparable projects outside of reservations. In an effort to remedy this situation, Congress passed the Indian Tribal Energy Development and Self-Determination Act (ITEDSA) as part of the Energy Policy Act of 2005. The ITEDSA granted tribes authority over energy development on their lands without federal supervision and created the Tribal Energy Resource Agreement (TERA) as a mechanism to streamline the approval process. As discussed in the Russell excerpt in chapter 8.C, the Department of Interior must apply a NEPA analysis when considering a TERA, but once TERA approval is granted, no further NEPA review or DOI approvals are required for specific energy agreements under that TERA. As a result, it would seem that TERAs provide significant advantages to developers in terms of reducing the timing for approvals. However, as of mid-2011, not a single tribe had entered into a TERA. Why do you think this might be?

5. One of the biggest hurdles to establishing renewable energy on tribal lands is the lack of adequate financial incentives. Explain some of the problems with the current structure and possible solutions.

6. Some tribes, or some of their members, may welcome industrial-scale development of renewable resources on their lands both to improve their often-dire economic circumstances and to help address the U.S.'s energy-use and climate-change conundrum. However, others may view this as imputing western values on indigenous peoples who may place a higher priority on the survival of their communities and cultures. How would you propose to reconcile these two views? The next section raises some of the fundamental problems with the "consultation" and "accommodation" approach currently applied.

B. Tribal Input on Development Outside of Reservation Lands

This section addresses some tools Native Nations may use to impact development on federal lands outside of reservations. The first excerpt provides an overview, and the second is a specific example of how one of those tools, the National Historic Preservation Act, was recently employed to stop a solar project in the Mojave Desert.

1. Mechanisms for Protecting Native Values on Public Lands

The following excerpt explains the relationship between Native tribes and the federal government in areas of co-management. It also explains some of the legal arguments Native Nations can assert to protect their values in the use of federal lands.

Martin Nie, *The Use of Co-Management and Protected Land-Use Designations to Protect Tribal Cultural Resources and Reserved Treaty Rights on Federal Lands*
48 Nat. Resources J. 585, 585-602, 620-24 (2008)

INTRODUCTION

Several Native Nations in the United States have cultural resources and reserved treaty rights on federal lands. In many cases, these values and rights are threatened by resource development and recreational activities permitted by a federal land agency. A typical approach to such conflicts is for a tribe to legally challenge an agency's decision or to seek some type of accommodation by the agency through planning and other decision making processes....

* * *

II. CO-MANAGEMENT

A. Co-Management and Federal Indian Law

Options in tribal co-management cannot be understood without first recognizing some foundational principles of Indian law. These principles also explain why tribal co-management differs from other types of collaborative management for federal lands.

First, tribal governments are sovereign and have inherent powers of self- government. For this reason, there is a unique government-to-government relationship between federally-recognized tribes and the federal government. Several laws, regulations, executive orders, and internal agency management directives make clear how this relationship affects federal land management....

Also relevant to co-management is the trust relationship between tribes and the federal government. Though sovereign, Indian tribes are not foreign nations, but rather distinct political communities "that may, more correctly, perhaps be denominated domestic, dependent nations," whose "relation to the United States resembles that of a ward to his guardian." A less paternalistic way of thinking about this relationship is by thinking in terms of property; that the federal government has a duty to prevent harm to another sovereign's property. The federal government, in other words, has a responsibility to protect the rights, assets, and property of Indian tribes and citizens. Some courts, moreover, have used the trust doctrine as a way to force the federal government to protect tribal lands, resources, and off-reservation (property) rights. Klamath Tribes v. United States (1996) provides one relevant example where a tribe successfully stopped planned timber sales by the USFS to protect deer herds reserved by treaty. The Oregon District Court ruled that the federal government had a "substantive duty to protect 'to the fullest extent possible' the Tribes' treaty rights, and the resources on which those rights depend." This trust duty, enforced in this case and others, provides the context in which tribal co-management is taking place.

Another example of how the trust responsibility can foster intergovernmental cooperation is the Joint Secretarial Order on "American Indian Tribal Rights, Federal-Tribal Trust Responsibilities, and the Endangered Species Act." The Order was negotiated between tribal representatives and the federal government to harmonize "the federal trust responsibility to tribes, tribal sovereignty, and statutory missions of the Departments, and that strives to ensure that Indian tribes do not bear a disproportionate burden for the conservation of listed species, so as to avoid or minimize the potential for conflict and confrontation." Several principles are stated in the Order encouraging "cooperative assistance," "consultation," "the sharing of information," and the "creation of government-to-government partnerships to promote healthy ecosystems." Among other applicable provisions, the Order also calls for federal-tribal intergovernmental agreements: ...

Some commentators believe that an effective way to harmonize the trust responsibility with species conservation is through the use of such cooperative agreements, including co-management.

* * *

... [R]eserved treaty rights are central to [Native claims on off-reservation federal lands—Ed.]. Treaties are legally binding agreements between two or more sovereign governments. Three hundred and eighty-nine treaties precede the creation of the USFS. Sixty treaties contained provisions that reserved rights on what was then public domain land. The extent of off-reservation use rights reserved by a tribe depends on specific treaty language, but many treaties reserved various rights on ceded lands, and such lands are now managed by different federal land agencies. On national forest lands, for example, off-reservation treaty rights include hunting and fishing rights, gathering rights, water rights, grazing rights, and subsistence rights. It is critical to understand that the term "reserved rights" means just that; the federal government did not give such rights to the tribes, but rather the tribes reserved such rights as sovereigns. This is partly why such reserved rights constitute property, and why the governmental taking of this property requires financial compensation. When interpreting treaties, Courts use accepted canons of construction that are liberally construed in favor of tribes. Treaties are to be interpreted as the Indians who agreed to them understood them, and any ambiguities in the treaty are to be resolved in favor of the tribes. Congress has the plenary power, however, to abrogate treaty rights, though it must do so explicitly and with clear evidence for the Courts to recognize such change.

Also relevant to the forthcoming discussion is the U.S. Constitution's Establishment Clause and its relationship to cultural resources management. The Clause states that "Congress shall make no law respecting an establishment of religion, or prohibiting the free exercise thereof." It is within these parameters that the courts have decided a number of sacred lands disputes by applying different tests. For purposes here, the two most important are *Lyng v. Northwest Indian Cemetery Protective Association* (1988) and *Bear Lodge Multiple Use Association v. Babbitt* (1998).

In *Lyng*, the USFS planned to allow major timber harvesting activities in the high country held sacred by three California Indian tribes, and to construct 200 miles of logging roads in areas adjacent to the sacred Chimney Rock area. One section of road linking the towns of Gasquet and Orleans (known as the "G-O" road) would dissect the high country's sacred places. Indian plaintiffs argued that completion of this road and its attendant noise and environmental damage would violate the free exercise clause by degrading sacred lands and eroding the religious significance of this area. But the Supreme Court ruled in favor of the USFS, finding no free exercise violation because the government was not coercing Indians into religious beliefs. Similar free exercise-based arguments have basically been abandoned by Indian plaintiffs following this controversial decision. Property and ownership is also central to *Lyng*. The Supreme Court explained that federal ownership (of national forests and other federal lands) could be dispositive and shield the government against Indian free exercise claims. Writing for the majority, Justice O'Connor summarized that "[w]hatever rights the Indians may have to the use of the area, those rights do not divest the Government of its right to use what is, after all, its land."

The issue of accommodation was also addressed by the Court in *Lyng*: "nothing in our opinion should be read to encourage governmental insensitivity to the religious needs of any citizen" [and] "[t]he Government's rights to the use of its own land need not and should not discourage it from accommodating religious practices like those engaged in by the Indian respondents." But when it comes to accommodation, the *Bear Lodge* decision is most instructive. That case concerns NPS management of Devil's Tower National Monument in Wyoming (known to some Plains Indians as Bear Lodge). Bear Lodge is considered sacred by several Indian tribes and is also a very popular recreational climbing spot. Following tribal complaints, and a formal planning process, the NPS initially banned commercial rock climbing during the month of June, when most tribal ceremonies take place. The NPS then changed this ban to a voluntary closure upon a successful Establishment Clause challenge brought by the Bear Lodge Multiple-use Association and rock climbers. The Wyoming District Court and the Tenth Circuit upheld the voluntary closure and ruled that it was a legitimate accommodation of religious beliefs. The voluntary climbing ban, according to the district court, was "a policy that has been carefully crafted to balance the competing needs of individuals using Devil's Tower National Monument while, at the same time, obeying the edicts of the Constitution" and thus "constitutes a legitimate exercise of the Secretary of the Interior's discretion in managing the Monument."

Congress also has provided additional laws and resolutions that have been considered by the courts. The American Indian Religious Freedom Act of 1978 (AIRFA) makes the protection of American Indian religious freedom federal policy. Though symbolically important, this policy statement is mostly hollow and largely unenforceable. More substantive in nature is the Religious Freedom Restoration Act of 1993 (RFRA). It provides that "Government may substantially burden a person's exercise of religion only if it demonstrates that application of the burden to the person-(1) is in furtherance of a compelling governmental interest; and (2) is the least restrictive means of furthering that compelling governmental interest." Note that the RFRA goes beyond the Constitution's use of the word prohibiting the free exercise of religion to include the broader verb burden, thus providing more religious protection.

RFRA was central in a recent case involving the USFS in northern Arizona. The agency approved plans by a ski area to use recycled sewage effluent to make artificial snow on the San Francisco Peaks in the Coconino National Forest. The Peaks are sacred to the Navajo, Hopi, and several other Indian tribes, and are eligible for inclusion in the National Register of Historic Places as a TCP (as discussed below). In *Navajo Nation v. United States Forest Service* (2006), plaintiffs challenged this decision using RFRA and other laws. On appeal, the Ninth Circuit reversed the Arizona District Court, finding the agency's approval of the upgrade in violation of RFRA and the National Environmental Policy Act (NEPA). Among other findings, the circuit court concluded that the agency's authorization to use sewage effluent to make snow and expand the ski resort would impose a "substantial burden" on plaintiffs' exercise of religion and was not a "compelling governmental interest." Navajo Nation was petitioned for rehearing en banc. But at the time of this writing, it represents a significant shift from *Lyng*.

[In an en banc decision, the 9th Circuit overturned the original opinion and affirmed the district court's holding that the plan to use sewage effluent for snowmaking did not violate the RFRA. *Navajo Nation v. U.S. Forest Serv.*, 535 F.3d 1058 (9th Cir. 2008) (en banc), *cert. denied*, 129 S. Ct. 2763 (2009)—Ed.]

A few lessons can be drawn from these important cases. While *Lyng* basically put an end to First Amendment arguments as a way to protect sacred places, in some situations the RFRA might be successfully used as a way to protect them on federal lands. Courts, as made clear in *Bear Lodge* and subsequent cases, have found acceptable agency accommodations of religious practices. When such accommodations are voluntary in nature, and do not cause actual injury to other citizens, they generally withstand Establishment Clause challenges. This has left the protection of sacred places largely to the discretion of federal land managers-and this helps explain the interest in more predictable and permanent types of protection, as discussed in the following sections. Numerous laws, administrative regulations, internal directives, and an Executive Order instruct agencies about how to consult with tribes, manage cultural resources, and possibly make accommodations to safeguard sacred places. A few studies have exhaustively documented these sources of authority for federal land agencies, including the USFS, so there is no need to repeat them here. But the upshot is that, like the NPS in the *Bear Lodge* case, federal land agencies often have a great deal of discretion when making sacred land decisions, and can legally justify such choices if they are carefully crafted and within the constitutional parameters outlined above.

One quick example illustrates how the USFS can respond given such discretion. It concerns oil and gas leasing on the Rocky Mountain Front, managed by the Lewis and

Clark National Forest. Using a careful and thorough social assessment, among other tools, USFS supervisor Gloria Flora made the decision not to lease part of the Front for development. She based her decision on environmental laws and a "value of place" articulated by the Blackfeet Tribe and public comments made during the NEPA process. Said Flora, "The Forest has tried to recognize these social and emotional values and they have figured prominently in my decision not to lease the Rocky Mountain Division." The Rocky Mountain Oil and Gas Association litigated the decision, arguing that "value of place" was not a valid management criterion and that Flora's decision was based on land use for Indian religious practices and was therefore in violation of the Establishment Clause. The district court disagreed, and upon appeal the Ninth Circuit ruled that the no-lease decision had a secular purpose and did not advance or endorse religious beliefs nor foster excessive entanglement with religion. Moreover, said the court, "the government may, consistent with the Establishment Clause, accommodate religious practices in its decision-making processes."

This sort of accommodation is but one strategy that could be used to protect sacred lands in the future. Several scholars, advocates, and other interests promote others. Some emphasize the success and potential of using existing laws, policies, and agency decision making processes; viewing them as more flexible, site-specific, legitimate, and a less risky way to protect sacred sites than by using the highly uncertain and precedent-establishing judicial system. Others, however, remain skeptical of agency processes that essentially treat Indians as yet another stakeholder that must be consulted; some believe that "tribal rights to sacred sites are being collapsed into a series of procedural requirements" that do not go far enough. Legislative approaches have also been proposed, with debate centered on how prescriptive the law should be given constitutional constraints, and whether it should contain an enforceable cause of action, among other items. These approaches represent just a few potential options. This article explores the strengths and limitations of ... statutory and administrative land designations as ways to protect reserved treaty rights and sacred places on federal land.

* * *

III. Protected Land Policy Options

* * *

A. The National Historic Preservation Act

I will start by reviewing an administrative designation that has often been used as a way to consider, and sometimes protect, sacred places and cultural resources on federal land. The National Historic Preservation Act (NHPA, 1966) is the basic charter and method of historic preservation in the United States. Agencies implement the Act by determining whether a "federal undertaking" will "diminish the integrity of the property's location, setting, feeling, or association." The Act authorizes the Secretary of Interior "to expand and maintain a National Register of Historic Places composed of districts, sites, buildings, structures, and objects significant in American history, architecture, archeology, engineering, and culture."

Procedural protections are provided to properties that are listed on the National Register, or are determined eligible for listing. State Historic Preservation Officers and federal agencies nominate properties for inclusion on the National Register, though individuals and other entities may request nominations. The NHPA requires agencies to ensure that their historic properties are preserved to maintain their historic, archaeological, architectural, and cultural values. "Properties of traditional religious and cultural importance" to Indian tribes are types of properties eligible for listing. Though not defined by statute, the term traditional cultural properties (TCP) is used to describe a type of property that is eligible for listing because of its traditional cultural significance.

Two types of protection are provided by the Act, one substantive and the other more procedural in nature. Properties designated as National Historic Landmarks receive greater substantive protection. Before approving actions that would affect a landmark, Section 110 of the NHPA requires that the responsible federal agency "shall, to the maximum extent possible, undertake such planning and actions as may be necessary to minimize harm to such landmark."

The Bighorn Medicine Wheel in Wyoming's Bighorn National Forest provides an oft-used example of how this designation can help protect a sacred site on federal lands and influence agency decisions.... Upon a very critical reception of the proposal, the USFS began the NHPA consultation process. This process resulted in a long-term Historic Preservation Plan (HPP) that required consultation between the USFS and other parties for any project proposed within a 18,000-20,000 acre "area of consultation" surrounding the Medicine Wheel. The USFS approved the HPP by amending its existing forest plan in 1996.

This decision was also controversial because it had the potential of limiting timber harvesting activities in the Bighorn National Forest, even though the HPP does not prohibit logging in the area of consultation. A commercial timber company litigated the decision on constitutional and procedural grounds, arguing among other things that the HPP was a significant change to the forest plan that required full NEPA/NFMA (National Forest Management Act) compliance. But the district and circuit courts found in favor of the USFS, partly because the area of consultation comprises only 1.6 percent of the Bighorn National Forest, and was thus a non-significant change to the forest plan that did not require the full NEPA/NFMA process to be used by the agency.

Section 106 of the NHPA, on the other hand, provides procedural protection in that it requires effects on properties to be considered by agencies. This is basically a required consultation process whereby agencies consult "with any Indian tribe or Native Hawaiian organization that attaches religious and cultural significance" to an historic property that would be affected by a proposed federal undertaking. The Section 106 process also requires that agencies assess the effects of their undertakings on any eligible properties found, determine whether the effect will be adverse, and to avoid, minimize, or mitigate the adverse effects.

Though the courts have characterized Section 106 as a "stop, look, and listen" provision requiring agencies to consider the effects of their programs, this provision, along with others, is not to be taken lightly. In one case, for example, the court held that the USFS did not make a reasonable effort to identify traditional cultural properties or engage in a meaningful consultation process. And in another important decision, the court found that the USFS did not satisfy the NHPA's mitigation requirement when it proposed to map and photograph culturally significant land that was proposed to be exchanged with Weyerhaeuser timber corporation.

In other places, however, historic designation seems to have mattered little to agencies or the courts. Take, for example, in the *Lyng* case the USFS proposed road building and timber sales in the sacred high country managed by the Six Rivers National Forest (discussed above). At the time this proposal was made, the area was already part of the Helkau Historic District and determined eligible for listing on the National Register. Yet this did not dissuade the USFS from its plans to construct the road and allow timber harvesting.

A more recent example is provided by the Navajo Nation/San Francisco Peaks case in which the USFS permitted the expansion of the Snowbowl ski area and the use of sewage effluent to make snow on land held sacred by multiple tribes. The Peaks are eligible for inclusion on the National Register as a TCP. Because of this, the USFS began its required consultation process whereby consulting parties must consider feasible and prudent alternatives to the undertaking that could avoid, mitigate, or minimize adverse effects on a National Register for eligible property. For the Snowbowl project, a "finding of adverse effect" was made by the USFS. Its attempt to avoid, minimize, or mitigate this effect included allowing access for traditional cultural practitioners and free use of the ski lifts in the summer. Though the Ninth Circuit Court found in favor of Indian plaintiffs on other grounds, both it and the district court found the USFS in full compliance with the NHPA because it attempted to consult with affected tribes and adequately described ways to mitigate adverse effects.

There are several different perspectives on how effective the NHPA has been in protecting sacred sites on federal land. Much of the divergence stems from the considerable discretion afforded to agencies in determining eligibility, and how agencies manage the cultural properties that have been administratively designated as such. On one hand, the NHPA designation requires consultation, and this process is important. According to Dean Suagee, director of the First Nations Environmental Law Program at the University of Vermont, "[b]ecause many tribes attach religious and cultural importance to places that are not within the boundaries of their reservations, many tribes regard this as a very important right, even though it is just a procedural right. In essence, it is the right to have a seat at the table, a chance to persuade the responsible federal official to do the right thing." On the other hand, there are lots of cases in which such persuasion did not work, and tribal government representatives and other commentators often voice frustration at how little NHPA designation seems to matter at times.

Notes and Questions

1. What is the basis for Native claims on off-reservation federal lands? How do courts construe the concept of reserved rights and any ambiguities in treaties?

2. What is the relationship of federal government and tribal governments?

3. Some of the mechanisms employed to assert tribal rights have been the Establishment Clause of the U.S. Constitution, the American Indian Religious Freedom Act of 1978, and the Religious Freedom Restoration Act of 1993. How has each of these worked for protecting lands sacred to Native Americans?

4. The last mechanism set out in the excerpt above is the National Historic Preservation Act of 1966 (NHPA). What is the goal of the NHPA and what substantive and procedural protections does it provide?

5. Consider the discussion of water use for CSP from the previous chapter and the recommendation for using sewage effluent if wet-cooling is required in areas of water scarcity. How do you think such a proposal might correlate to the use of sewage effluent for snowmaking in the *Navajo Nation/San Francisco Peaks* case discussed in the Nie excerpt above?

6. Native peoples' relationship to the lands is fundamentally different from that of European-origin cultures. The land is not a resource to be dominated, exploited, and profited from, but a "place" that defines their identity and existence. This essential philosophical distinction on the relationship to the land was the heart of the indigenous arguments in the *Navajo Nation/San Francisco Peaks* and *Bear Lodge* cases. Does approaching the issues in this light help reconcile any of the conflicts?

2. Quechan Tribe v. U.S. Dep't of Interior

In a Herculean effort to move renewable energy development forward on public lands, as mandated by the 2005 EPAct and in time to receive Stimulus funds, the Department of Interior approved some of the first federal-land solar projects in the fall of 2010. Although the BLM process may have addressed the needs of the majority of interest groups, the following excerpt reflects how it may have failed to meet its obligations to Native Nations. This excerpt also illustrates a Tribe's employment of the National Historic Preservation Act to play a role in the energy-development process.

Quechan Tribe of Fort Yuma Indian Reservation v. U.S. Dep't of Interior

755 F. Supp. 2d 1104 (S.D. Cal. 2010)

Burns, J.

On October 29, 2010, Plaintiff (the "Tribe") filed its complaint, alleging Defendants' decision to approve a solar energy project violated various provisions of federal law. On November 12, the Tribe filed a motion for preliminary injunction, asking the Court to issue an order to preserve the status quo by enjoining proceeding with the project, pending the outcome of this litigation. After the motion was filed, Imperial Valley Solar LLC intervened as a Defendant.

* * *

Background

The Quechan Tribe is a federally-recognized Indian tribe whose reservation is located mostly in Imperial County, California and partly in Arizona. A large solar energy project is planned on 6500 acres of federally-owned land known as the California Desert Conservation Area ("CDCA"). The Department of the Interior, as directed by Congress, developed a binding management plan for this area.

The project is being managed by a company called Tessera Solar, LLC. Tessera plans to install about 30,000 individual "suncatcher" solar collectors, expected to generate 709 megawatts when completed. The suncatchers will be about 40 feet high and 38 feet wide, and attached to pedestals about 18 feet high. Support buildings, roads, a pipeline, and a power line to support and service the network of collectors are also planned. Most of the project will be built on public lands. Tessera submitted an application to the state of California to develop the Imperial Valley Solar project. The project is planned in phases.

After communications among BLM, various agencies, the Tribe, and other Indian tribes, a series of agreements, decisions, and other documents was published. The final EIS was issued some time in July, 2010. At the same time, a Proposed Resource Management Plan– Amendment, amending the Department of the interior's CDCA was also published. On September 14 and 15, certain federal and state officials, including BLM's field manager, executed a programmatic agreement (the "Programmatic Agreement") for management of the project. The Tribe objected to this. On October 4, 2010, Director of the Bureau of Land Management Robert Abbey signed the Imperial Valley Record of Decision ("ROD") approving the project, and the next day Secretary of the Interior Ken Salazar signed the ROD. The ROD notice was published on October 13, 2010.The area where the project would be located has a history of extensive use by Native American groups. The parties agree 459 cultural resources have been identified within the project area. These include over 300 locations of prehistoric use or settlement, and ancient trails that traverse the site. The tribes in this area cremated their dead and buried the remains, so the area also appears to contain archaeological sites and human remains. The draft environmental impact statement ("EIS") prepared by the BLM indicated the project "may wholly or partially destroy all archaeological sites on the surface of the project area."

The Tribe believes the project would destroy hundreds of their ancient cultural sites including burial sites, religious sites, ancient trails, and probably buried artifacts. Secondarily, it argues the project would endanger the habitat of the flat-tailed horned lizard, which is under consideration for listing under the Endangered Species Act and which is culturally important to the Tribe. The Tribe maintains Defendants were required to comply with the National Environmental Policy Act (NEPA), the National Historical Preservation Act (NHPA), and the Federal Land Policy and Management Act of 1976 (FLPMA) by making certain analyses and taking certain factors into account deciding to go ahead with the project. The Tribe now seeks judicial intervention under the Administrative Procedures Act (APA).

* * *

NHPA Consultation Requirements

The NHPA's purpose is to preserve historic resources, and early consultation with tribes is encouraged "to ensure that all types of historic properties and all public interests in such properties are given due consideration...." *Te–Moak Tribe v. U.S. Dept. of Interior*, 608 F.3d 592, 609 (9th Cir.2010) (quoting 16 U.S.C. § 470a(d)(1)(A)). The consultation process is governed by 36 C.F.R. § 800.2(c)(2), one of Section 106's implementing regulations. Under this regulation, "[c]onsultation should commence early in the planning process, in order to identify and discuss relevant preservation issues...." § 800.2(c)(2)(ii)(A). The Ninth Circuit has emphasized that the timing of required review processes can affect the outcome and is to be discouraged. (Citations omitted.) The consultation requirement is not an empty formality; rather, it "must recognize the government-to-government relationship between the Federal Government and Indian tribes" and is to be "conducted in a manner sensitive to the concerns and needs of the Indian tribe." § 800.2(c)(2)(ii)(C). A tribe may, if it wishes, designate representatives for the consultation. *Id.*

The Section 106 process is described in 36 C.F.R. §§ 800.2–800.6. After preliminary identification of the project and consulting parties, Section 106 requires identifying historic properties within a project's affected area, evaluating the project's potential effects on those properties, and resolving any adverse effects. The Tribe insists this consultation must be completed at least for Phase 1 of the project, before construction begins.

Throughout this process, the regulations require the agency to consult extensively with Indian tribes that fall within the definition of "consulting party," including here the Quechan Tribe. Section 800.4 alone requires at least seven issues about which the Tribe, as a consulting party, is entitled to be consulted before the project was approved. Under § 800.4(a)(3), BLM is required to consult with the Tribe identify issues relating to the project's potential effects on historic properties. Under § 800.4(a)(4), BLM is required to gather information from the Tribe to assist in identifying properties which may be of religious and cultural significance to it. Under § 800.4(b), BLM is required to consult with the Tribe to take steps necessary to identify historic properties within the area of potential effects. Under § 800.4(b)(1), BLM's official is required to take into account any confidentiality concerns raised by tribes during the identification process. Under § 800.4(c)(1), BLM must consult with the Tribe to apply National Register criteria to properties within the identified area, if they have not yet been evaluated for eligibility for listing in the National

Register of Historic Places. Under § 800.4(c)(2), if the Tribe doesn't agree with the BLM's determination regarding National Register eligibility, it is entitled to ask for a determination. And under § 800.4(d)(1) and (2), if BLM determines no historic properties will be affected, it must give the Tribe a report and invite the Tribe to provide its views. Sections 800.5 and 800.6 require further consultation and review to resolve adverse effects and to deal with failure to resolve adverse effects.

Furthermore, under § 800.2, consulting parties that are Indian tribes are entitled to *special consideration* in the course of an agency's fulfillment of its consultation obligations. This is spelled out in extensive detail in § 800.2(c). Among other things, that section sets forth the following requirements:

(A) The agency official shall ensure that consultation in the section 106 process provides the Indian tribe ... a reasonable opportunity to identify its concerns about historic properties, advise on the identification and evaluation of historic properties, including those of traditional religious and cultural importance, articulate its views on the undertaking's effects on such properties, and participate in the resolution of adverse effects.... Consultation should commence early in the planning process, in order to identify and discuss relevant preservation issues and resolve concerns about the confidentiality of information on historic properties.

(B) The Federal Government has a unique legal relationship with Indian tribes set forth in the Constitution of the United States, treaties, statutes, and court decisions. Consultation with Indian tribes should be conducted in a sensitive manner respectful of tribal sovereignty....

(C) Consultation with an Indian tribe must recognize the government-to-government relationship between the Federal Government and Indian tribes. The agency official shall consult with representatives designated or identified by the tribal government.... Consultation with Indian tribes ... should be conducted in a manner sensitive to the concerns and needs of the Indian tribe....

(D) When Indian tribes ... attach religious and cultural significance to historic properties off tribal lands, section 101(d)(6)(B) of the act requires Federal agencies to consult with such Indian tribes ... in the section 106 process. Federal agencies should be aware that frequently historic properties of religious and cultural significance are located on ancestral, aboriginal, or ceded lands of Indian tribes ... and should consider that when complying with the procedures in this part.

36 C.F.R. § 800.2(c)(2)(ii)(A)-(D) (emphasis added). The Tribe points out the significance of the "confidentiality" provisions, citing *Pueblo of Sandia v. United States,* 50 F.3d 856, 861–62 (10th Cir.1995) (noting that pueblo's reticence to share information about cultural and religious sites with outsiders was to be expected, and that federal government knew tribes would typically not answer general requests for information).

The Ninth Circuit has emphasized that federal agencies owe a fiduciary duty to all Indian tribes, and that at a minimum this means agencies must comply with general regulations and statutes. (Citations omitted.)(mentioning the "unique legal relationship" between federal government and Indian tribes). Violation of this fiduciary duty to comply with NHPA and NEPA requirements during the process of reviewing and approving projects vitiates the validity of that approval and may require that it be set aside. *Id.*

Defendants, citing 36 C.F.R. § 800.14(b)(1)(ii), argue that "the execution of a Programmatic Agreement completes the Section 106 process" (Opp'n to Mot. for Prelim.

Inj., 22:11–17) and is an acceptable way to resolve adverse effects from complex projects "[w]hen effects on historic properties cannot be fully determined prior to approval of an undertaking." (*Id.* at 9:10–11.) But this is true only if "executing" means "carrying out;" merely entering into a programmatic agreement does not satisfy Section 106's consultation requirements. 36 C.F.R. § 800.14(b)(2)(iii) ("Compliance with the procedures established by an approved programmatic agreement satisfies the agency's section 106 responsibilities for all individual undertakings of the program covered by the agreement....") The Tribe asks that consultation be completed at least for phase 1 before the project begins. That Defendants are resisting this suggests they are probably not prepared to do so.

The programmatic agreement must be negotiated in accordance with § 800.14(b), which itself requires an extensive consultation process. § 800.14(f). The Tribe has also argued a programmatic agreement is not authorized for this type of project.

Defendants are correct that under § 800.4(b)(2), identification of historic properties can be deferred if "specifically provided for" in a programmatic agreement negotiated pursuant to § 800.14(b). But this deferral is not indefinite, and entering into an appropriately-negotiated programmatic agreement does not relieve the BLM of all responsibility. The second half of § 800.4(b)(2) contemplates consultation on historic properties as it becomes feasible:

> The process should establish the likely presence of historic properties within the area of potential effects for each alternative or inaccessible area through background research, consultation and an appropriate level of field investigation, taking into account the number of alternatives under consideration, the magnitude of the undertaking and its likely effects, and the views of ... any other consulting parties. As specific aspects or locations of an alternative are refined or access is gained, the agency official shall proceed with the identification and evaluation of historic properties in accordance with paragraphs (b)(1) and (c) of this section.

In short, entering into an appropriately-negotiated programmatic agreement can result in deferral of the consulting process, but it would only allow a temporary delay in consultation, until it is feasible to identify and consult with the Tribe about the historic properties. *Compare Te–Moak,* 608 F.3d at 610 (explaining that assessment of impact on environmental resources could be deferred where drilling locations in mineral exploration project could not reasonably be determined at the time of approval, but where plan required assessment as drilling locations became known).

* * *

Analysis of Documentary Evidence

Preliminarily, several points bear noting. First, the sheer volume of documents is not meaningful. The number of letters, reports, meetings, etc. and the size of the various documents doesn't in itself show the NHPA-required consultation occurred.

Second, the BLM's communications are replete with recitals of law (including Section 106), professions of good intent, and solicitations to consult with the Tribe. But mere *pro forma* recitals do not, by themselves, show BLM actually complied with the law. As discussed below, documentation that might support a finding that true government-to-government consultation occurred is painfully thin.

At oral argument, the Tribe described the meetings as cursory information sessions and the reports and other communications as inadequate. Its briefing also argues that Defendants have confused "contact" with required "consultation." Defendants In response, Defendants argue that the Tribe "has been invited to government-to-government consultations since 2008" "BLM began informing the Tribe of proposed renewable energy projects within the California Desert District as early as 2007," and "[s]ince that time BLM has regularly updated the Tribe on the status of the [Imperial Valley Solar] project." (Opp'n, 5:26–6:3.)

The Tribe's first document contact with BLM was the tribal historical preservation officer's letter of February 19, 2008. That letter put BLM on notice that the historical and cultural sites within the project area would be considered important to the Tribe. It also asked BLM to provide a survey of the area and to meet with the Tribe's government, which would have constituted government-to-government consultation. BLM could not have provided the survey at that time, and apparently also didn't comply with the meeting request, because the historic preservation officer re-sent the letter the next month. In fact, the documentary evidence doesn't show BLM ever met with the Tribe's government until October 16, 2010, well after the project was approved. All available evidence tends to show BLM repeatedly said it would be glad to meet with the Tribe, but never did so.

Although BLM invited the Tribe to attend public informational meetings about the project, the invitations do not appear to meet the requirements set forth in 36 C.F.R. § 800.2(c)(2)(ii). This is particularly true because the Tribe first requested a more private, closed meeting between BLM and its tribal council. In later communications, the Tribe continued to request that BLM meet with its tribal council on the Tribe's reservation. In addition, the Tribe repeatedly complained that the properties hadn't been identified, and asked for a map showing where the identified sites were, requests that apparently went unanswered at least as late as June, 2010. The Tribe's letter of August 4, 2010 apparently acknowledges receipt of maps, but asks for an extension of the deadline so it could review them before responding.

The documentary evidence also confirms the Tribe's contention that the number of identified sites continued to fluctuate. Compare, *e.g.*, PI 008155 (BLM letter dated June 24, 2010 setting number of cultural sites in the project area at 446) and PI 00993 (Final EIS, stating Class III inventory identified 459 cultural sites). And Defendants have admitted the evaluation of sites eligible for inclusion in the National Register hasn't *yet* been completed.

BLM's invitation to "consult," then, amounted to little more than a general request for the Tribe to gather its own information about all sites within the area and disclose it at public meetings. Because of the lack of information, it was impossible for the Tribe to have been consulted meaningful as required in applicable regulations. The documentary evidence also discloses almost no "government-to-government" consultation. While public informational meetings, consultations with individual tribal members, meetings with government staff or contracted investigators, and written updates are obviously a helpful and necessary part of the process, they don't amount to the type of "government-to-government" consultation contemplated by the regulations. This is particularly true because the Tribe's government's requests for information and meetings were frequently rebuffed or responses were extremely delayed as BLM-imposed deadlines loomed or passed.

No letters from the BLM ever initiate government-to-government contact between the Tribe and the United States or its designated representatives, the BLM field managers.... Rather, the Tribe was invited to attend public informational meetings or to consult with two members of her staff, an archaeologist and a person identified only as a "point of contact." The BLM in fact rebuffed the Tribe's August 4 request that the BLM meet with the tribal council on its reservation, proposing instead that the tribal council call BLM staff.

The Tribe also repeatedly protested it was not being given enough time or information to consider the Programmatic Agreement, a matter it was also entitled to be consulted about. The letters sent to the Tribe's president make clear BLM had determined a programmatic agreement would be used and would be entered into no later than September, 2010. The Tribe's letter of February 4, 2010 suggests the Tribe had discovered on its own that BLM was already drafting the Programmatic Agreement. Furthermore, BLM insisted that consulting parties send their suggestions in writing. The Tribe's requests to consult about the Programmatic Agreement were obviously not granted.

Defendants have emphasized the size, complexity, and expense of this project, as well as the time limits, and the facts are sympathetic. Tessera hoped to qualify for stimulus funds under the American Recovery and Reinvestment Act of 2009 by beginning construction no later than the end of this year, which is about two weeks away. To that end, BLM apparently imposed deadlines of its own choosing. Section 106's consulting requirements can be onerous, and would have been particularly so here. Because of the large number of consulting parties (including several tribes), the logistics and expense of consulting would have been incredibly difficult. None of this analysis is meant to suggest federal agencies must acquiesce to every tribal request.

That said, government agencies are not free to glide over requirements imposed by Congressionally-approved statutes and duly adopted regulations. The required consultation must at least meet the standards set forth in 36 C.F.R. § 800.2(c)(2)(ii), and should begin early. The Tribe was entitled to be provided with adequate information and time, consistent with its status as a government that is entitled to be consulted. The Tribe's consulting rights should have been respected. It is clear that did not happen here.

The Court therefore determines the Tribe is likely to prevail at least on its claim that it was not adequately consulted as required under NHPA before the project was approved. Because the project was approved "without observance of procedure required by law," the Tribe is entitled to have the BLM's actions set aside under 5 U.S.C. § 706(2)(D).

Merits Analysis of Other Claims

The evidence shows, and the parties do not dispute, that the planned project is extensive. The size and number of suncatchers, not to mention roads, buildings, and other supporting infrastructure, ensures this will be a massive project. The undisputed evidence also shows the 459 historic properties extend from one end of the area to the other, so some type of impact on the properties is likely. In fact, phase 1 of the plan acknowledges that one such property *will* be adversely impacted; because of the property's size, power lines cannot span it, and one power pole must be installed on the property.

The Court therefore holds the FLPMA claim at least raises "serious questions" for purposes of injunctive relief.

The substance of the NEPA claim is less clear. Extensive environmental review has been conducted, so the chance that this project will harm the flat-tailed horned lizard appears to be reduced. At the same time, the Tribe was entitled to be consulted under NEPA as under NHPA, and its claims in this respect also raise "serious questions."

* * *

Public Interest

The final step in the *Winter* analysis requires the Court to consider whether a preliminary injunction is in the public interest. 129 S.Ct. at 374. Obviously there are many competing interests here. The interests the Tribe urges the Court to consider involve historic and cultural preservation, in this case of hundreds of prehistoric sites and other sites whose significance has yet to be completely evaluated. The Tribe itself is a sovereign, and both it and its members have an interest in protecting their cultural patrimony. The culture and history of the Tribe and its members are also part of the culture and history of the United States more generally.

The value of a renewal energy project of this magnitude to the public is also great. It provides the public with a significant amount of power while reducing pollution and dependence on fossil fuels. As Defendants point out, it is a goal of the federal government and the state of California to promote the development of such projects. Current federal policy as embodied in ARRA also favors the undertaking of projects of this time, as a way of creating jobs and stimulating the economy.

That being said, the Court looks to the statutes enacted by Congress rather than to its own analysis of desirable priorities in the first instance. *See, e.g., Marshall v. Barlow's, Inc.,* 436 U.S. 307, 331, 98 S.Ct. 1816, 56 L.Ed.2d 305 (1978) (refusing to question Congress' weighing of interests when enacting statute); *Salazar v. Buono,* ——— U.S. ———, 130 S.Ct. 1803, 1828, 176 L.Ed.2d 634 (2010) (Scalia, J., concurring in the judgment) ("Federal courts have no warrant to revisit [Congress' decision about what is in the public interest]—and to risk replacing the people's judgment with their own....."). Here, in enacting NHPA Congress has adjudged the preservation of historic properties and the rights of Indian tribes to consultation to be in the public interest. Congress could have, but didn't, include exemptions for renewable energy projects such as this one. And, as pointed out, Congress could determine this particular project is in the public interest and sweep aside ARRA deadlines as well as requirements under NHPA, NEPA, and FLPMA to get it built. But because Congress didn't do that, and instead made the determination that preservation of historical properties takes priority here, the Court must adopt the same view.

[The court granted the Tribe's motion for preliminary injunctive relief—Ed.]

Notes and Questions

1. This case involved approval of a concentrating solar power (CSP) plant on non-reservation federal lands. What is the basis for Native input into the permit approval process in this situation?

2. The previous section described the NHPA process. Were substantive or procedural protections the basis for the challenge in this lawsuit?

3. Upon what seven issues was the tribe entitled to be consulted before a project was approved? What is the basis for these consulting party requirements—case law, statute, or regulation? In addition to these seven issues, Indian tribes are entitled to what additional special considerations in the course of a federal agency's consultation obligations? What is the legal authority for these additional special requirements?

4. Did the BLM here meet the consultation requirements? Why or why not? Does inviting Tribes to attend public informational meetings about a project satisfy the regulatory requirements?

5. Do the logistics and expenses of consulting with numerous parties, which the court recognizes "would have been incredibly difficult," allow the BLM to relax any of the requirements, especially as the court acknowledged, "Section 106's consulting requirements can be onerous, and would have been particularly so here"?

6. In the final section of the order, the court addresses public interest. What would this court require to find that renewable energy development had priority over preservation

of historical properties? How does this court's holding compare to the court's holding concerning public interest in the *Western Watersheds Project* case set out in the previous chapter?

C. Energy Justice

In the first section of this chapter, Kronk mentioned a moral imperative to address the negative effects of climate change on Native communities. However, there is another moral imperative about developing renewables on tribal lands—and that is based on a legacy of shame:

> The toxic legacy left by fossil fuel and uranium development on tribal lands remains today and will persist for generations, even without additional development. Mines and electrical generation facilities have had devastating health and cultural impacts in Indian country at all stages of the energy cycle-cancer from radioactive mining waste to respiratory illness caused by coal-fired power plant and oil refinery air emissions on and near Native lands. Native communities have been targeted in all proposals for long-term nuclear waste storage.

Honor the Earth et al., *Energy Justice in Native America: A Policy Paper for Consideration by the Obama Administration and the 111th Congress* 2 (2009), http://treatycouncil.org/ PDF/EJ_in_NA_Policy_Paper_locked.pdf [hereinafter *Tribal Council Policy Paper*]

This Tribal Council Policy Paper further notes that despite the history of exploitation on tribal lands, tribes themselves have received little benefit in return for this development. Many Native peoples live in abject poverty. Unemployment and poverty rates for Native Americans are twice the national average. About a third of reservation homes are trailers without adequate weatherization, and more than 11 percent do not have complete plumbing. Most significantly, in the energy justice context, while many energy minerals—such as coal, uranium, and hydrocarbons—have been extracted from Indian lands, "about 14% of reservation households are without electricity, 10 times the national rate." *Id.* at 4.

Furthermore, Glennon and Reeves note:

> Previous energy-related projects on tribal lands ... have often been seen as disastrous. As reporter Phil Taylor has observed: "tribes are consistently shortchanged in the deals, earning pennies on every dollar that goes to the mining firms and electric utilities whose operations are fully dependent upon the reservations.... 90 percent of what tribes pay for their energy leaves the reservation. Still, a number of tribal leaders believe that, with the right training and support, tribally-owned solar projects could "change the energy paradigm in Native communities from one of exploitation to one of equity...."

Robert Glennon & Andrew M. Reeves, *Solar Energy's Cloudy Future*, 1 Arizona J. ENVTL L. & POL'Y 92, 131(2010).

This section includes three excerpts that trace the evolution of the environmental justice movement into climate justice and energy justice strains.

Randall S. Abate, *Public Nuisance Suits for the Climate Justice Movement: The Right Thing and the Right Time*
85 Wash. L. Rev. 197, 207-10 (May, 2010)

Climate justice has both domestic and international law underpinnings. First, the evolution of environmental justice in the United States helped lay a foundation for the climate justice field by recognizing an area outside of the traditional boundaries of environmental law for which the law should provide a remedy—namely, the disproportionate impacts of environmental regulation on minority and low-income communities. This theory encountered some obstacles when tested in the federal courts under Equal Protection Clause analysis and it ultimately failed to secure remedies for such disproportionate impacts through the court system. The litigation was not in vain, however, as it raised awareness of the need for a response to this inequity and prompted subsequent proactive measures at the federal and state levels to mitigate or avoid such disproportionate impacts in the future.

Beyond the realm of environmental justice in the United States, a parallel development under international law evolved concerning the growing recognition of the intersection between environmental law and human rights. The rise of the notion of sustainable development has helped fuel this awareness, and scenarios involving unsustainable growth that caused disproportionate impacts on indigenous populations (such as deforestation and development in the Amazon) have drawn international attention. More recently, climate change has created the potential for cultural genocide or may at least require the relocation of these peoples.

At the international level, the movement to recognize a human right to a healthy environment has enjoyed decades of support, and has increased significantly with the increase in awareness regarding climate change impacts. More specifically, the importance of the right to a healthy environment in developing nations has attracted attention, especially in developing nations that are either particularly vulnerable environmentally to climate change, or that lack the infrastructure to respond adequately to such threats.

The latest climate change science confirms the importance of an institutionalized climate justice framework as part of the post-Kyoto regime. The United Nations Environment Programme (UNEP) released a report in September 2009 entitled Climate Change Science Compendium 2009. This UNEP report underscores the need for immediate action to avoid the catastrophic climate change impacts that are projected by 2100, as well as the dangerous "tipping points" that could be reached within a few decades that would have tragic implications for the world's major ecosystems, such as the Sahara and the Amazon. The report notes that it still may be possible to avoid many of these catastrophic impacts, but only if there is "effective, efficient, and equitable" action to reduce greenhouse gas emissions and states take proactive measures to assist vulnerable countries adapt to the projected impacts.

Responding to the needs of vulnerable communities is not a "one size fits all" proposition. For example, the impacts of climate change on indigenous peoples raise difficult legal and ethical issues. Professor Rebecca Tsosie has suggested that the standard adaptation strategy of relocating a vulnerable population out of harm's way could be culturally genocidal for many groups of indigenous people when viewed in the climate justice context. As an alternative, she argues for recognition of an indigenous right to environmental self-determination, which would allow indigenous peoples to maintain their cultural and political status upon their traditional lands. In the context of climate change policy, such a right would impose affirmative requirements on nation-states to develop a plan to avoid catastrophic harm to indigenous peoples. Tsosie further recognizes that tort-based theories of compensation for the harms of climate change have only limited capacity to address the concerns of indigenous peoples. Ultimately, public nuisance claims in Kivalina-like scenarios are an important step, but only the beginning.

[In *American Power Company, Inc. v. Connecticut*, 131 S. Ct. 2527 (2011), the U.S. Supreme Court held that the Clean Air Act displaced any federal common-law right to sue to curtail greenhouse gas emissions. The question of whether the parties still had a remedy under state nuisance laws was left open for consideration on remand. The excerpt is included in this chapter for its general discussion of the concept of climate justice.—Ed.]

Alice Kaswan, *Greening the Grid and Climate Justice*
39 ENVTL. L. 1143, 1145-46, 1151-53 (2009)

... Climate justice is relevant not only to the issue of whether (and at what rate) to green the grid, it is highly relevant to the development of alternative energy policy itself. Climate policy presents a "democratic moment"—a time to consider our basic infrastructure and its ideal design. Green jobs advocate Van Jones states that "[t]oday the 'clean-tech' revolution and the transformation of our aging energy infrastructure are poised to become the next great engines for American innovation, productivity and job growth, and social equity gains." He argues that "we have the chance to build this new energy economy in ways that reflect our deepest values of inclusion, diversity, and equal opportunity for everyone." A comprehensive approach that integrates the environmental and economic ramifications of the new energy infrastructure can most effectively maximize the benefits and minimize the risks of the transition ahead.

Some of the opportunities created by alternative energy, like increasing U.S. energy security and stimulating green technology development, have been widely discussed. Less attention has been given to "climate justice"—to integrating environmental and economic justice into comprehensive energy planning. This Essay argues that as strategies to green the grid are developed, policymakers should integrate goals like reducing co-pollutants, ameliorating impacts on low-income consumers, and creating economic opportunities.

* * *

B. Potential Economic Benefits and Risks

1. Existing Disparities

With change comes opportunity. While climate change policies will undoubtedly impose certain economic costs, they could also create significant economic opportunities for disadvantaged communities. A comprehensive approach to climate policy would integrate and seek to maximize the economic opportunities presented by GHG reduction strategies in general, and greening the grid in particular.

The United States features stark contrasts in the distribution of wealth. Nor is that distribution random: Racial minorities are significantly more likely to be impoverished than whites. For example, the 2000 census identified 24.9 percent of African Americans in poverty, compared with 8.1 percent of the white population. Native Americans and Latinos have similarly high poverty rates. Inner cities and tribal lands suffer from high unemployment, demoralization, and all of the accompanying social problems. The impact is experienced not only by the poor but also by society as a whole. Poverty and its ills require government spending on unemployment, health care, housing, and the criminal justice system.

2. Energy Transformation's Economic Opportunities

a. Alternative Energy Development in Disadvantaged Areas

Native American groups in the windy northern plains are recognizing the opportunities that could flow from investing in wind energy on impoverished reservations. A critical question is whether the tribes themselves will be able to capture the benefits of that investment or whether, instead, private companies develop and profit from the resource.

Native American advocates view the wind potential in Indian Country as an opportunity for indigenous economic development and control. The environmental imperative of shifting to alternative energy is not their only focus. As Winona LaDuke has stated, "Alternative energy represents an amazing social and political reconstruction opportunity." If tribes control the renewable energy development on their lands, it could generate not only local and national energy supplies, but increased tribal revenue, employment, and control over their well-being.

Tribes generally do not, however, have sufficient capital to exploit and develop the existing alternative energy potential on their land. Explicit financing mechanisms are necessary to enable tribes to build capital-intensive alternative energy projects. In addition, tribes are likely to need job training to develop the skills to build and run energy projects. While some efforts have been made to support tribal control over renewable energy development, a more comprehensive and better-funded approach is necessary to realize its potential.

Lakshman Guruswamy, *Energy Justice and Sustainable Development*

21 Colo. J. of Int'l Envtl. L. & Pol'y 231, 233-38 (Spring 2010)

Energy Justice ("EJ") conjugates justice with energy. Justice is the first virtue of social institutions; energy is a fundamental need and the driving determinant of human progress. Energy justice seeks to apply basic principles of justice as fairness to the injustice evident among people devoid of life sustainable energy, hereinafter called the energy oppressed poor ("EOP"). EJ is an integral and inseparable dimension of the universally accepted foundational principle, or grundnorm, of international law and policy: Sustainable Development ("SD").

The original formulators of the concept, the World Commission on Sustainable Development, also known as the Brundtland Commission, pointed to the abject poverty of the developing world, and articulated a distributional principle which they called sustainable development. They reasoned that SD would meet the basic needs of the world's poor by providing economic and social development without which environmental protection could not be achieved. This distributional principle of SD is now re-affirmed and expressed in the most widely accepted energy and environmental treaties and declarations. EJ, however, has been egregiously ignored in international discourse and negotiations about energy and the environment....

The facts about energy justice are distressing. A disturbingly large swath of humanity is caught in a time warp. Between 2 and 2.5 billion people, amounting to nearly a third of the world, rely upon biomass-generated fire as their principal source of energy. These fires are made by burning animal dung, waste, crop residues, rotted wood, other forms of "bad" biomass, and raw coal. Unlike the rest of the world, the other third live without access to energy generated lighting, space heating, cooking, and mechanical power. They suffer from grinding poverty, lamentable diseases, lack of safe drinking water and sanitation, non-access to education, and barely experience economic and social development. Moreover, the biomass-generated fire they rely upon is an inadequate source of energy. It does not provide the kind of exogenous energy required for sustainable human development. Fire can be used for cooking and heating but fails to supply the majority of other basic energy needs. Fire does not power water pumps, grinding mills, vehicles, or agricultural equipment. Further, it does not provide clean lighting, water filtration, or more generally help create the goods and services required for food, clothing, and shelter.

In responding to this challenge, the nations of the world and the United Nations ("UN"), arrived at an obvious, rational, and integrated application of SD. In 2000, they agreed on the Millennium Development Goals ("MDGs") and Millennium Development Project ("MDP"). The objectives of the MDGs and MDP are to halve global poverty and hunger, increase access to safe water and sanitation, provide primary education, and improve gender equality. They further seek to reduce child and maternal mortality by sixty-six percent, and reverse the growth of malaria, HIV/ AIDS, and other major diseases. The target year for achieving these goals is 2015. Two aspects of the MDGs are worthy of special notice. First, they require access to energy, and second, they are a prerequisite for dealing with global warming.

The MDGs cannot be satisfied without access to energy. First, the goal of reducing poverty depends on the availability of energy because even the most rudimentary forms of income-generating activities, like agriculture and small businesses, need energy to power machines for milling or grinding, for transportation to market goods and services, for telecommunications, and for education. Second, the goal of reducing hunger requires that more food be grown and distributed. Most forms of irrigation require energy to power water pumps, as well as for machines that harvest crops. Processing food requires energy, as does transportation and distribution. Third, water treatment plants that provide safe drinking water require energy, and hospitals need energy for refrigeration of vital medications and vaccinations. Finally, in order to provide primary education, schools require energy for lighting and heating, and students need lighting at home to do their homework. It seems almost obvious that the MDGs, as an instrument of SD, should concentrate on the developmental objectives of the EOP.

The environmental and global warming implications of the MDG are equally clear. It empowers and enables healthier, more educated peoples, including women, to adapt to and mitigate global warming. There is no doubt that healthier, more educated peoples, are better able to combat global warming than an ill educated population dying from illness, disease, hunger and malnutrition. The MDGs should be used to further SD by fulfilling the developmental objectives of the EOP as a necessary first step in meeting their environmental and global warming challenges.

Particularly during the last five to ten years, however, the international agenda has been dominated by fervent and dedicated global warming crusaders and blinkered decision-makers from the developed world, who appear anaesthetized to the plight of the EOP. Consequently, the bulk of development assistance has been funneled toward reducing carbon dioxide and other greenhouse gas ("GHG") emissions at the expense of the MDGs. For example, Secretary Clinton recently confirmed that the U.S. Agency for International Development's ("USAID") key focus on development assistance for over a decade has been on environmental programs that have reduced growth in GHG emissions. Given that the EOP hardly emit any GHGs, left unsaid is the stark fact that those USAID resources are not available for the MDGs. The obvious result is that international resources for achieving the MDGs are drying up. A recent report of the UN Development Program ("UNDP") diplomatically emphasized this point. The report points out that economic growth, eradication of poverty, and the MDGs remain the highest priorities of developing countries, but that the focus of world leaders on reducing GHG emissions may constrain those priorities and efforts.

Climate change negotiations have ignored the EOP. In the most recent chapter of climate change negotiations under the UN Framework Convention on Climate Change ("UNFCCC") at Copenhagen in December 2009, the world's decision-makers, while paying lip service to SD, demonstrated once again that they remain impervious to the EOP and their lament of disease, public health problems, lack of safe drinking water, non-access to education, sickness, death, and economic deprivation that is not attributable to carbon dioxide. Consistent with their preoccupation with GHG reductions, world leaders continued to ignore the energy-based problems afflicting one-third of the world's population, which are caused by the absence of modern sustainable energy. The Copenhagen Accord stated in passing that "Developing countries, especially those with low emitting economies should be provided incentives to continue to develop on a low emission pathway." However, this provision was left without reference to any funds to help fulfill such an objective. Instead, the only reference to funding made available to developing countries was for mitigation, adaptation, technological development and transfer, and capacity-building. Once again, the primacy of global warming was emphasized and funded while the plight of the non carbon dioxide generating EOP—and the countries they inhabit—were almost totally ignored. The amaurosis afflicting climate change negotiators is perplexing for a number of reasons.

Indoor pollution is the clearest example of an energy problem that extracts a horrendous toll of death and sickness, especially among women and children. It blights the EOP who rely on fire as their sole source of energy for cooking, illumination, and heating. Using an open fire, or a traditional stove fueled by biomass, results in inefficient combustion that releases dangerous quantities of carbon monoxide, particulate matter, and other pollutants into the air. These indoor pollutants result in the premature death every year of 2 million women and children from pneumonia, chronic obstructive pulmonary diseases, lung cancer, and asthma. They also cause chronic respiratory ailments and debilitating sickness for many more millions.

With regard to indoor pollution, recent scientific investigations published in well established and respected peer-reviewed journals conclude that black carbon or black soot emitted by the burning of biomass makes the second strongest contribution to current global warming after carbon dioxide emissions. According to these studies, the particulates in black carbon absorb reflected solar radiation, as well as direct solar radiation, thus warming the atmosphere more severely than other greenhouse gases like methane, halocarbons, and tropospheric ozone. Moreover, black carbon can travel potentially thousands of miles on air currents, and eventually settle out of the air, onto land, water, and ice. Black carbon may lower the albedo, or reflectivity, of polar ice that covers vast stretches of the Arctic and Antarctica. The presence of overlying black carbon may result in ice retaining more heat, leading to increased melting and eventually a warmer Earth.

These scientific facts offer compelling evidence that the EOP unmistakably and objectively fall within the economic, social, and environmental dimensions of SD. Providing cook stoves for example, could save millions of people from premature death and sickness, and free them to embark upon income generating economic activities. Moreover, the environmental co-benefits are incontestable. Apart from establishing a healthier population that can fight global warming, reducing black soot or black carbon by using cook stoves, will positively and directly reduce global warming.

Furthermore, reducing black carbon will cost only a fraction of the price of carbon dioxide mitigation. Unlike carbon dioxide, which remains in the atmosphere from 50 to 200 years and is very costly to mitigate, black carbon is short-lived and significantly cheaper to remove. Even if all carbon dioxide emissions were miraculously stopped today, the effects of existing carbon dioxide will continue for a century. Conversely, black carbon dissipates and disappears within a week. Thus, the beneficial effects of the removal of black carbon will be felt within a short time frame.

Notes and Questions

1. Define Environmental Justice, Climate Justice, and Energy Justice. How are they similar? How are they distinguishable?

2. What are the foundations for these theories? What legal avenues might be used to apply them?

3. What are some of the specific climate or energy justice examples from these excerpts? Does the concept of sustainable "development" still connote exploitation and consumerism? Some indigenous energy development is aimed at providing basic resources for their peoples to provide economic sovereignty to decolonize at the grass roots level. Do you believe that the less-consumptive focus on harmony and balance with the earth that some indigenous peoples embrace may provide pathways for solving the problems of climate change and contamination of the earth?

4. Most of this book has focused on legal issues related to renewable energy in the United States—which is more than any one book can cover comprehensively. However, these final excerpts illustrate that what renewable energy solutions we develop for the United States will have significant impacts on energy for the entire world. With market economics often driving our choices, are there other ways these compelling moral and social issues can be integrated into the global energy equation?

5. In Chapter 1.A.1, I note that the International Energy Agency's World Energy Outlook 2010 juxtaposed the cost of consumption subsidies to fossil fuels in 2009 with the cost of support given to renewable energy in 2009, showing fossil fuel consumption subsidies were approximately 600 percent of the renewable energy subsidies. The WEO 2010 also included a third number in this juxtaposition—the cost of ending global energy poverty by 2030. At $36 billion, this figure is approximately one-tenth of the fossil fuel consumption subsidies. The report notes, "Adding under two percent to electricity tariffs in the OECD would raise enough money to bring electricity to the entire global population within twenty years...." WEO 2010, *supra*, at 3. What do you think about this proposal?

Index